P9-CBI-293

LARGE-SCALE DYNAMIC SYSTEMS

NORTH-HOLLAND SERIES IN

SYSTEM SCIENCE AND ENGINEERING

Andrew P. Sage, *Editor*

Wismer and Chattergy	Introduction to Nonlinear Optimization: A Problem Solving Approach
Sutherland	Societal Systems: Methodology, Modeling, and Management
Šiljak	Large-Scale Dynamic Systems: Stability and Structure

LARGE-SCALE DYNAMIC SYSTEMS

Stability and Structure

Dragoslav D. Šiljak

University of Santa Clara, Santa Clara, California

Elsevier North-Holland, Inc.
52 Vanderbilt Avenue, New York, New York 10017

Distributors outside the United States and Canada:

Thomond Books
(A Division of Elsevier/North-Holland
Scientific Publishers, Ltd)
P.O. Box 85
Limerick, Ireland

Library of Congress Cataloging in Publication Data
Siljak, Dragoslav D.
Large-scale dynamic systems: stability and structure
(North-Holland series in system science and engineering)
Bibliography: p.
Includes index.
1. System analysis. 2. Stability. I. Title.
QA402.S49 003 77-23932
ISBN 0-444-00246-4

Manufactured in the United States of America

To Dragana

Turning and turning in the widening gyre
The falcon cannot hear the falconer;
Things fall apart; the centre cannot hold.

W.B. Yeats
The Second Coming

CONTENTS

PREFACE

One of the foremost challenges to system theory brought forth by present-day technological, environmental, and societal processes is to overcome the increasing size and complexity of the relevent mathematical models. Since the amount of computational effort required to analyze a dynamic process usually grows much faster than the size of the corresponding system, the problems arising in large complex systems may become either impossible or uneconomical to solve even with modern computing machines. For these reasons, it has been recognized recently that for purposes of stability analysis, control, optimization, and so forth, it may be beneficial to decompose a large system into a number of interconnected subsystems. Presumably these subsystems can be considered independently, so that solutions to the corresponding subproblems can be combined with interconnection constraints to come up with a solution to the original problem of the overall large system. In this way, advantage can be taken of the special structural features of a given system to devise feasible and efficient "piece-by-piece" algorithms for solving large problems which were previously intractable or impractical to tackle by "one shot" methods and techniques.

Besides the purely computational aspect of large-scale dynamic systems, there is the equally important problem of determining the effects of complexity on system behavior and the role it plays in the systems with large interconnection structures. Fundamental to evolving of processes in physical, biological, and societal systems alike is stability, and a question often arises: Does an increase of complexity lead to an improvement of system stability, or is it the other way around? Since this question is asked under a wide variety of conditions in such diverse fields as ecology and power systems, economics and spacecraft engineering, no unequivocal general answer is possible. A good deal of evidence can be assembled, however, to show that complexity can enhance stability provided a system is subject to relatively mild changes in its interconnection structure and its environment. When a system is exposed to sudden structural perturbations in its composition or surroundings, the adverse effects of complexity become sharply pronounced, and increasing complexity tends to beget diminished reliability of the otherwise stable system.

The purpose of this book is to initiate a systematic inquiry into the intricate relationship between complexity, stability, and reliability of large-scale dynamic systems. We hope to provide a number of convincing arguments that although increased complexity may promote stability, it is only when complexity is limited that a large system can remain stable despite structural perturbations whereby groups of subsystems are disconnected from each other and again connected together in various ways during operation. Although this fact is more or less intuitively acceptable, it will take mathematical machinery from the theory of differential equations and inequalities (such as the comparison principle of E. Kamke and T. Ważewski, as well as the concept of vector Liapunov functions introduced by R. Bellman and V. M. Matrosov and developed by V. Lakshmikantham, F. N. Bailey, S. Leela, J. P. LaSalle, G. S. Ladde, A. N. Michel, Lj. T. Grujić, M. Araki, S. Weissenberger, and many others) to derive the conditions for stability under structural perturbations and prove this fact rigorously. It should be noted at this point that only when we introduce the notion of dynamic reliability of large-scale systems via the notion of stability under structural perturbations are we able to determine rationally a limit to complexity in systems and partially resolve the problem of complexity vs. stability.

"Largeness" is a subjective notion, and so is "large-scale systems". We take a pragmatic viewpoint and consider a system as large-scale whenever it is necessary to partition it into a number of interconnected subsystems either for conceptual or computational reasons. When partitioning is used to achieve conceptual simplifications and interpretations, it is quite common to strive for physical identity of the subsystems. In cases where partitioning is applied for purposes of reducing numerical complexities, the subsystems may have no physical interpretations at all. Both aspects of partitioning will be applied in this book to various models in economics, ecology, and engineering.

The plan of the book is the following:

The central concept of stability under structural perturbations (that is, connective stability), which permeates all of this book, is introduced immediately in Chapter 1 with a minimum of mathematical sophistication.

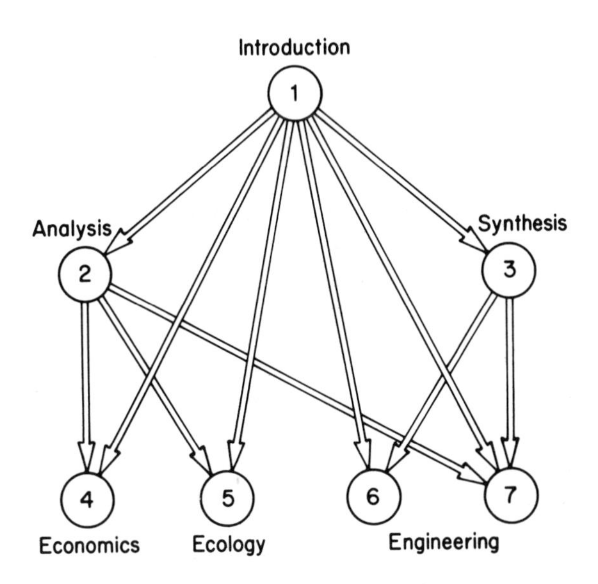

Book Plan.

By associating directed graphs with dynamic systems in an essential way, we will be able to define precisely the relation between system structure and stability in the sense of Liapunov. This opens up the possibility of using the powerful Liapunov's direct method to derive suitable conditions for connective stability. By decomposing a large-scale system into interconnected subsystems and aggregating the stability properties of subsystems by appropriate Liapunov functions, we will form an aggregate model of the system which involves a vector Liapunov function. Stability of each subsystem and stability of the aggregate model imply connective stability of the overall system.

The material of Chapter 1 is basically all that is needed in applications of the connective-stability concept to models in ecology, economics, and engineering, which are given in the later parts of the book. However, an advanced presentation of the concept is outlined in Chapter 2. In this chapter, we shall strive for mathematical rigor and use the modern methods of vector Liapunov functions developed by V. Lakshmikantham and his collaborators, to establish firmly the notion and content of connective stability of dynamic systems. Although Chapter 2 can be skipped entirely in the first reading of the book (as indicated by the Book Plan above), it should help the reader as a basis for possible extensions and applications not attempted in this book.

In his work on automata theory, John von Neumann recognized reliability as one of the central problems in complex systems. He proposed methods to overcome unreliability of the components in building complex computing machines, not by making the components more reliable, but by organizing them in such a way that the reliability of the whole computer is greater than the reliability of its parts. In another study of complexity in systems, H. A. Simon argued on rather intuitive grounds that complexity often takes the form of hierarchy and that the evolution of complex systems is highly reliable if it is carried out as a hierarchic process whereby systems are formed by interconnecting stable simple parts (subsystems). The objective of Chapter 3 is to show rigorously that by using a hierarchic feedback scheme, it is possible to synthesize complex systems which are reliably stable despite unreliable interconnections among the subsystems. The proposed control scheme is a decentralized one, because each subsystem is stabilized by a local feedback in such a way that the stability of each subsystem overrides any adverse effect of their mutual interactions, and stability of the overall system is implied by stability of its parts. Obviously, this feedback scheme is a rather primitive realization of von Neumann's reliability concept and Simon's intuitive recipe, but it is believed that the proposed decentralized approach is a step in the right direction, and that results obtained here indicate a considerable potential of the approach in building reliable control of large-scale dynamic systems.

The second part of the book, consisting of Chapters 4–7, illustrates the application of the concept of connective stability and decomposition-aggregation method to a variety of dynamic processes in economics, ecology and engineering. In Chapter 4 we start the application of the concept to the oldest studied decentralized system—the market. By relying on the results of numerous economists, notably P. A. Samuelson, K. J. Arrow, L. Hurwitz, J. Quirk, F. H. Hahn, O. Lange, P. K. Newman, and many others, we will formulate, analyze, and resolve the stability of competitive equilibrium under structural perturbations in the Hicks-Metzler algebraic setting. Nonstationary models are introduced for multiple markets of commodities or services in order to consider a reduction of the commodity space, changing of a commodity from a substitute to a complement to another commodity over a time interval, deterministic as well as stochastic shifts in demand schedules, etc. By using the modern mathematical machinery of vector Liapunov functions, we will analyze large market models by the Hicks-Leontief concept of composite commodities. The central result is that stable market models are inherently robust and can tolerate a wide range of nonlinear nonstationary phenomena.

Nowhere has the problem of complexity vs. stability been so explicitly formulated and systematically analyzed as in the ecological context of multispecies communities. Although some early considerations of complexity in the dynamic context of interconnected systems were given by W. R. Ashby in his *Design for a Brain*, it is R. H. MacArthur who formulated the problem precisely in terms of interacting populations. Besides MacArthur, there were a large number of people who investigated the effects of complexity in a variety of specific ecomodels. The most comprehensive and original treatment of the subject was given by R. M. May in his book *Stability and Complexity of Model Ecosystems*. By using May's exposition as a basis, we will develop in Chapter 5 an appropriate mathematical framework for considering community stability under both stochastic and structural perturbations. Although our results do not contradict May's conclusion that competition and mutualism between species is less conducive to overall web stability than is a predator-prey relationship, they point to the fact that mutualism coupled with density dependence enhances stability when structural changes occur. In fact, we shall show that if all the species are density-dependent and the size of interactions, which reflects the complexity of the community, is properly bounded, then the community can endure a wide range of disturbances such as random fluctuations in its environment, predator switching, saturation of predator attack capacities, etc. We remove the requirement of density dependence of species by considering hierarchial communities composed of interconnected blocks of species—a feature observed in many ecosystems. Again, connective stability is promoted by limited complexity. This is as far as we go with our analysis; the problem of complexity vs. stability remains an open one, and we hope that future efforts can exploit the proposed mathematical framework and produce new results.

As seen from the book plan, both Chapter 6 and Chapter 7, which present applications to engineering problems, are concerned with control systems. In Chapter 6, we shall use the decomposition-aggregation method to design control systems for two NASA spacecraft, the Large Space Telescope and the Skylab. A common feature in the two designs is the fact that the equations of motion are decomposed so that individual motions (roll, pitch, and yaw for the LST, and spin and wobble for the Skylab) are chosen as subsystems. Then, a decentralized feedback control is used to stabilize each subsystem separately, and an aggregate model involving the vector Liapunov function is used to determine the overall stability. An interesting conceptual aspect of the Skylab control system design is the fact that the decomposition-aggregation method is suitable for maximization of a salient system parameter which couples the spin and wobble subsystems.

In Chapter 7, we consider the two distinct problems in power systems: transient stability and automatic generation control. A common feature in the two applications is the fact that the notion of subsystem is broadened to allow for "overlapping" of subsystem spaces. That is, certain states of the system appear in more than one subsystem. In the case of transient stability of a multimachine system, a pairwise decomposition is applied, so that a reference machine appears in all subsystems. When automatic generation control is designed by the decentralized scheme, the area subsystems overlap the tie-lines. The overlapping of subsystems results in considerable flexibility in the way a large dynamic system can be decomposed into subsystems for purposes of stability analysis and stabilization by decentralized feedback control. A somewhat surprising result in multimachine stability analysis is that by partitioning the system and using a vector Liapunov function, we get a stability-region estimate expressed explicitly in terms of machine parameters.

Finally, in the Appendix, we collect all the properties of various classes of matrices that are used throughout the book. They are $\mathfrak{M}$-matrices, Metzler and Hicks matrices, and matrices with a dominant diagonal. These matrices are essential in providing appropriate models of the dynamic processes, on one hand, and in establishing suitable stability criteria, on the other.

I would like to express gratitude to my old friend and colleague, Stein Weissenberger, who by numerous discussions and comments at every stage helped me to sharpen the ideas contained in this book.

I am greatly indebted to the National Aeronautics and Space Administration, which has supported my research on large-scale systems. The research has been monitored with a great deal of encouragement and motivation by Brian Doolin at the Ames Research Center and by Sherm Seltzer at the Marshall Space Flight Center.

I am also grateful to my doctoral students Misha Vukčević and Ljubo Jocić for their comments, suggestions, and a devoted collaboration on many research problems and results presented in this book.

My sincere thanks are due to Mary Schlotterbeck for typing hundreds of pages of equations and unflinchingly making revision after revision of the manuscript.

Dragoslav D. Šiljak
Saratoga, California

1

INTRODUCTION

Stability, Complexity, Reliability

The conceptual ingredients in the analysis of large-scale systems which is presented in this book are borrowed from or applied to such diverse fields as the theory of differential equations and ecology, control engineering and mathematical economics, digraphs and spacecraft. To motivate the reader for an exposition of such varied subjects in the chaotic but exciting new area of large-scale dynamic systems, we introduce in this chapter the key topics of this book, using the minimum of mathematical sophistication. The introduction should justify by intuitive arguments our solution of the complexity-vs.-stability problem in large dynamic systems via system reliability. Using simple language, we shall explain our use of the terms structure, competition, cooperation, decentralization, aggregation, etc., and point out their interrelationship in the context of complex dynamic systems. Despite its elementary character, the material of this chapter provides an adequate background for understanding various applications of the theory to models in economics, ecology, and engineering, which are presented in the later parts of the book. Before we enter into technical considerations, let us outline in simple terms the reliability aspect of dynamic systems in the context of stability and structure, which permeates most of our analysis.

The major issue throughout the following development is the intricate interplay between complexity and stability in dynamic systems. There are numerous mathematical models in engineering, ecology, and social sciences which represent a dynamic interaction among a number of agents, elements, or subsystems in general. These subsystems influence each other through competition or cooperation, or both. If subsystems are in competition, their interaction is stable if they are stable when isolated and if the magnitude of their interactions is properly limited. If some subsystems are not self-sufficient or cease to be so in the course of operation, cooperation can produce overall stability at the price of increasing the complexity of the system by raising the degree of beneficial interactions among the subsystems. Now, large-scale systems, which are composed of a number of interconnected subsystems, quite commonly (either by design or by fault) do not stay together for all time, but are subject to structural perturbations whereby groups of subsystems are disconnected from and again connected to each other during operation. Since structural perturbations can destroy beneficial interactions among the subsystems, they can cause a breakdown of the systems with predominant cooperation effects. Therefore, such systems are vulnerable under structural perturbations and it is intuitively clear that

A dynamic system composed of interconnected subsystems is reliable if all subsystems are self-sufficient and their interdependence is properly limited.

This simple reliability principle of dynamic systems resolves the complexity-vs.-stability problem by requiring the systems to have a competitive structure where each subsystem is stable when isolated and the magnitude of the interactions does not exceed a certain limiting value. Besides stability, this recipe will be used in the context of optimality, controllability, reachability, and other qualitative properties of dynamic systems, and it will serve as the major tool for synthesizing reliable large-scale systems by decentralized control and estimation.

1.1. COMPETITION

The linear differential equation

$$\dot{x} = Ax + b \tag{1.1}$$

is often used to describe interactions of a number of agents in as diverse fields as economics, and arms race, model ecosystems and engineering. In Equation (1.1), $x = x(t)$ is a column n-vector

$$x = \begin{bmatrix} x_1 \\ x_2 \\ \vdots \\ x_n \end{bmatrix}, \tag{1.2}$$

which is the *state* of the system (1.1). The state $x(t)$ may be the price vector on a commodity market, the population vector in a multispecies community, or the armament vector in an arms race.

In (1.1), the matrix A is a constant $n \times n$ matrix and b is a constant n-vector,

$$A = \begin{bmatrix} a_{11} & a_{12} & \cdots & a_{1n} \\ a_{21} & a_{22} & \cdots & a_{2n} \\ \cdots & \cdots & \cdots & \cdots \\ a_{n1} & a_{n2} & \cdots & a_{nn} \end{bmatrix}, \qquad b = \begin{bmatrix} b_1 \\ b_2 \\ \vdots \\ b_n \end{bmatrix}. \tag{1.3}$$

The matrix $A = (a_{ij})$ describes by its elements a_{ij} how the jth agent acts on the ith agent, and the vector b represents how the external world affects the interaction process. If the system (1.1) is used to describe a multispecies community, A is the community matrix which shows how the species interact among themselves, and b can represent the food supply of the community.

A situation that is fairly common in the mentioned scientific disciplines is that of a matrix A such that

$$a_{ij} \begin{cases} < 0, & i = j, \\ \geqslant 0, & i \neq j, \end{cases} \tag{1.4}$$

that is, the matrix A has negative diagonal elements ($a_{ii} < 0$) and nonnegative off-diagonal elements ($a_{ij} \geqslant 0$, $i \neq j$). A matrix A with the sign pattern as in (1.4) is called in economics a *Metzler matrix* (Arrow, 1966), because it was Metzler (1945) who gave it an essential significance in his work on multiple markets. The Metzler matrices have quite interesting properties which we are going to use throughout the foregoing development, and which are summarized in the Appendix at the end of the book.

A system (1.1) with a Metzler matrix describes a competitive interaction among the agents represented by the states x_i. In economics, (1.1) is used to model multiple markets of commodities (or services) which are substitutes for each other and "compete for consumers". For example, sugar and honey are substitutes, and their interaction can be described by a second-order system of the form

$$\begin{aligned} \dot{x}_1 &= a_{11}x_1 + a_{12}x_2 + b_1, \\ \dot{x}_2 &= a_{21}x_1 + a_{22}x_2 + b_2, \end{aligned} \tag{1.5}$$

where x_1 and x_2 are the prices of sugar and honey, and $x = \{x_1, x_2\}$ is the price vector. The substitution effect is reflected by the positivity of the off-diagonal elements a_{12}, a_{21} as in (1.4). Then, if the price of sugar x_1 goes down while the price of honey x_2 is constant, the positivity of a_{21} will cause the rate $\dot{x}_2$ to decrease, which in turn will cause the price x_2 to decrease, too. A similar argument can be used to show that the positivity of a_{12}, a_{21} means that an increase of x_1 implies an increase of x_2 and vice versa.

The market model (1.5) is based upon the law of supply and demand (Arrow and Hahn, 1971). The right sides of (1.5) represent the differences between demand and supply (excess demand) for both commodities. When supply equals demand for each commodity, excess demands—and thus, price rates—are equal to zero, so that the prices are at their equilibrium values. At the equilibrium prices, the market forces are "in balance", and exchange of commodities takes place. This equilibrium situation is called a *competitive equilibrium* (Arrow and Hahn, 1971) and will be described in more detail in Chapter 4.

The negativity of the diagonal elements a_{11}, a_{22} in (1.5) reflects the law of supply and demand for each individual commodity. A rise in price of one commodity above the equilibrium price causes a negative excess demand (excess supply) of that commodity, and a fall of the price below the equilibrium price causes a positive excess demand for that commodity, provided the price of the other commodity is adjusted at its equilibrium value. Under such conditions, a displacement of prices from the equilibrium tends to be self-corrective, and prices will eventually approach equilibrium values. Whether the price adjustment process would do the same when both self-corrective and substitution effects are considered simultaneously is a basic problem of stability of competitive equilibrium. This problem is considered in Chapter 4, where competitive equilibrium is given a much broader meaning than this sketchy consideration can suggest.

We should note, however, that dynamic interactions represented by Metzler matrices, which may or may not have explicit competitive interpretations, will be of basic interest throughout this book. For example, Equation (1.5) can be used to describe a chemical reaction involving two reactants whose concentrations are x_1 and x_2; then b_1 and b_2 are constant input rates of the two substances from external sources (Rosen, 1970). Again, a_{11} and a_{22} are assumed negative to reflect the consumption of each substance in the reaction. The positivity of a_{12} and a_{21} means that the reactants are activators for each other, so that an increase in the instantaneous value of x_1 causes the rate of production of x_2 to increase and vice versa. Therefore, the chemical reaction is characterized by the system (1.5) with a Metzler-type matrix.

It is of interest to note that in population dynamics (May, 1973), where the components x_1 and x_2 in (1.5) stand for the numbers of the two species which coexist in the same habitat, competition is specified by negativity of the off-diagonal elements a_{12} and a_{21}. In this context, a Metzler matrix would describe a symbiotic interaction among the two species (see Chapter 5).

In the following analysis, competition is represented by the system (1.1) with a Metzler matrix A. The term "competition" is borrowed from economic studies, but its meaning will vary depending on a particular application under study. On occasion, the term will be reduced to a simple synonym for the Metzlerian sign pattern of the system matrix, and no other interpretations will be implied or attempted.

1.2. STRUCTURE

Let us for a moment ignore the influence of the "outside world" on the competitive model (1.1) represented by the vector b, and consider a "free" system

$$\dot{x} = Ax. \tag{1.6}$$

While the elements a_{ij} of the system matrix A specify quantitatively the relationship among pairs of agents represented by the states (x_i, x_j), the matrix A itself displays the structure of the competitive system. To study structural properties of competitive systems, it is convenient to describe the system structure by directed graphs and networks and exploit the corresponding mathematical theory about the abstract notion "structure" (Harary, Norman, and Cartwright, 1965).

Let us consider a pair of interacting agents with activity described by equations

$$\begin{aligned} \dot{x}_1 &= a_{11}x_1 + a_{12}x_2, \\ \dot{x}_2 &= a_{21}x_1 + a_{22}x_2, \end{aligned} \tag{1.7}$$

which is the system (1.5) with $b_1 = b_2 = 0$. The interactions among the agents represented by the states x_1 and x_2 can be described by a *graph* shown in Figure 1.1, which consists of two points connected by *directed*

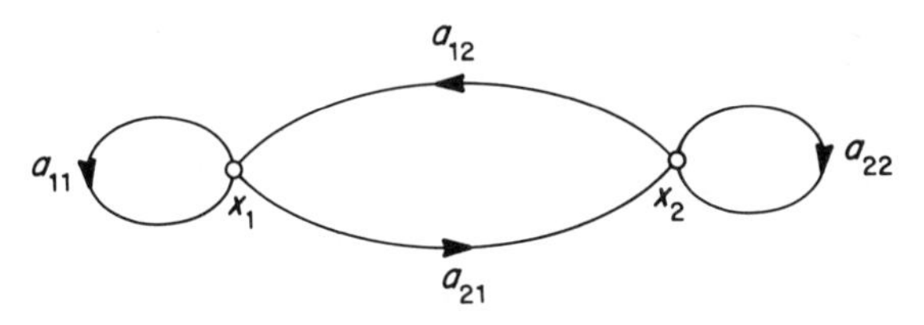

FIGURE 1.1. Digraph.

lines or simply *lines*. The points of the graph are labeled x_1 and x_2, while the directed lines are labeled a_{11}, a_{12}, a_{21}, a_{22}. The direction of each line is indicated by an arrowhead. If again (1.7) describes a community of two species that coexist in the same habitat, the graph of Figure 1.1 can be used to describe the interconnection structure. The points x_1 and x_2 stand for the populations of the two species.

To concentrate strictly on the structure of the system (1.7), we ignore the values a_{11}, a_{12}, a_{21}, a_{22} and consider the lines to be simply either present or absent in the graph. Then we do not use the labeled network of Figure 1.1 to represent the system structure, but merely the *directed graph*, or briefly *digraph*. If, in the system (1.7), $a_{21} = 0$, the matrix A takes the form

$$A = \begin{bmatrix} a_{11} & a_{12} \\ 0 & a_{22} \end{bmatrix}. \tag{1.8}$$

The corresponding directed graph is shown in Figure 1.2(a), where there is no line going from x_1 to x_2, and there is no indication of the "strength" in interactions between x_1 and x_2. The 2×2 matrix $\bar{E} = (\bar{e}_{ij})$ that is associated with the digraph of Figure 1.2(a) is the *adjacency matrix*, or the *interconnection matrix*,

$$\bar{E} = \begin{bmatrix} 1 & 1 \\ 0 & 1 \end{bmatrix}. \tag{1.9}$$

It is a binary matrix with the property that the entry $\bar{e}_{ij} = 1$ if x_j "acts" on x_i, and $\bar{e}_{ij} = 0$ when x_j "does not act" on x_i. It should be noted here that we deviate slightly from the theory of digraphs (Harary, Norman, and Cartwright, 1965) in allowing for the *loops* in Figure 1.2(a) and changing the ordering of the pair (x_1, x_2). These deviations are introduced to accommodate our special interest in applications, and can be easily incorporated in the digraph theory.

It is quite common in dynamic models of physical, social, and biological processes that agents are disconnected and again connected in various ways during the process. For example, a predator in a multispecies community stops preying on another species, causing a *line removal* in the corresponding digraph. This is a *structural perturbation* of the system and is described by another adjacency matrix where the corresponding unit element is changed to zero. In the case of the system (1.7) with A as in (1.8), when a_{22} becomes zero, the matrix $\bar{E}$ of (1.9) is changed to

$$E_1 = \begin{bmatrix} 1 & 1 \\ 0 & 0 \end{bmatrix}. \tag{1.10}$$

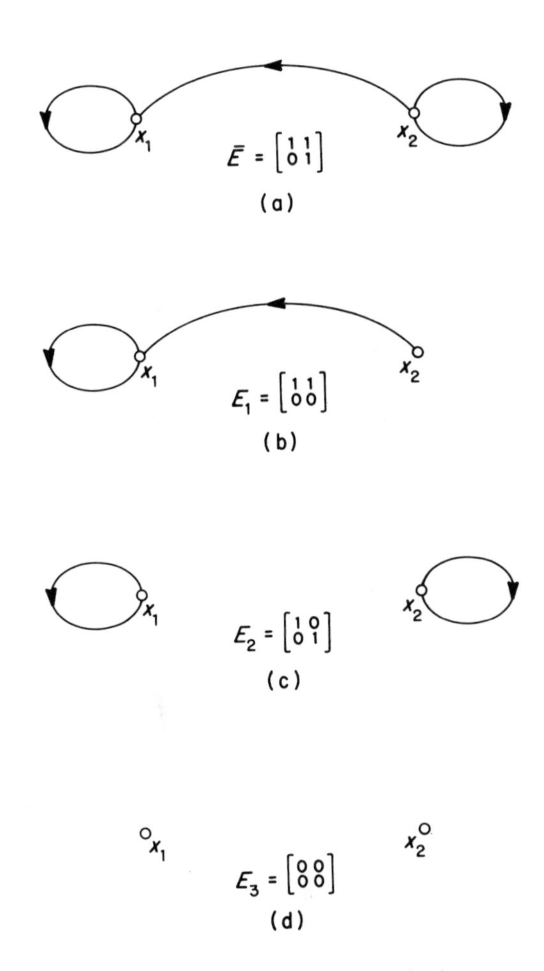

FIGURE 1.2. Structural perturbations.

The corresponding digraph is shown in Figure 1.2(b).

If the two agents are disconnected from each other, then the situation is represented by the digraph of Figure 1.2(c). This case can be also viewed as a consequence of a removal of one agent from the interaction process, thus having the other agent to represent the structurally perturbed system. This is equivalent to the removal of a point from the corresponding digraph, and it is called a *principal structural perturbation* to distinguish it from structural perturbations caused by line removals. Obviously, the *point removals* can be performed by removing all the lines connected to that point except the line which forms a loop at that point. This kind of structural perturbation is represented by the adjacency matrix

$$E_2 = \begin{bmatrix} 1 & 0 \\ 0 & 1 \end{bmatrix}, \tag{1.11}$$

and the corresponding digraph is shown in Figure 1.2(c).

The *total disconnection* is represented by the zero adjacency matrix

$$E_3 = \begin{bmatrix} 0 & 0 \\ 0 & 0 \end{bmatrix} \tag{1.12}$$

and the digraph shown in Figure 1.2(d).

The adjacency matrix $\bar{E} = (\bar{e}_{ij})$ of (1.9) represents the basic structure of the system; in the following developments, it is called the *fundamental interconnection matrix*. The adjacency matrices E_1, E_2, E_3 representing the structural perturbations of the system are obtained from the fundamental interconnection matrix $\bar{E}$ by replacing unit elements with zeros. We may drop the subscripts and use the symbol E to stand for any of the matrices E_1, E_2, E_3. Then the matrix $E = (e_{ij})$ is simply *an interconnection matrix* associated with $\bar{E}$.

The concept of structural perturbations can be widened to incorporate the "environment" of the system (1.7). Consider the system (1.5) where b_1 and b_2 are not zero. When the system

$$\begin{aligned} \dot{x}_1 &= a_{11}x_1 + a_{12}x_2 + b_1, \\ \dot{x}_2 &= a_{21}x_1 + a_{22}x_2 + b_2 \end{aligned} \tag{1.5}$$

is used to model two interacting species, b_1 and b_2 may either represent availability of food in the habitat, or reflect the consequences of pesticides applied to the community.

If for some reason the system (1.5) is such that

$$A = \begin{bmatrix} a_{11} & a_{12} \\ 0 & a_{22} \end{bmatrix}, \qquad b = \begin{bmatrix} 0 \\ b_2 \end{bmatrix}, \tag{1.13}$$

we can represent the structure of the model by the digraph in Figure 1.3(a). The *fundamental interconnection vector* $\bar{l} = \{\bar{l}_1, \bar{l}_2\}$ is

$$\bar{l} = \begin{bmatrix} 0 \\ 1 \end{bmatrix}, \tag{1.14}$$

and the structural perturbation corresponding to the removal of lines associated with $\bar{l}$ is the zero vector

$$l = \begin{bmatrix} 0 \\ 0 \end{bmatrix}. \tag{1.15}$$

Such a situation is shown in Figure 1.3(b).

A systematic treatement of the structure of dynamic systems in the context of structural perturbations and vulnerability is outlined in Section 3.1.

1.3. STABILITY

We are now interested in finding out how stability of the competitive systems is affected by structural perturbations. We again consider the system

$$\dot{x} = Ax, \tag{1.6}$$

and recall the well-known fact that for any initial time t_0 and any initial state $x_0 = x(t_0)$, there is one (and only one) solution $x(t; t_0, x_0)$ of Equation (1.6),

$$x(t; t_0, x_0) = e^{(t-t_0)A} x_0, \tag{1.16}$$

where $e^{(t-t_0)A}$ is the matrix exponential (e.g. Hahn, 1967).

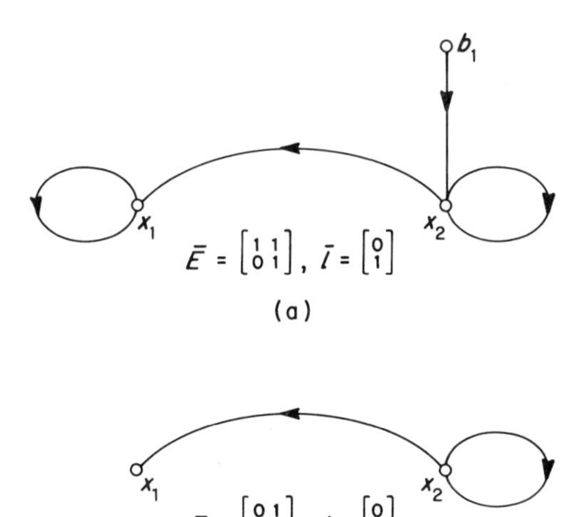

FIGURE 1.3. Structural perturbations.

Of special interest in the analysis of the linear system (1.6) is the equilibrium $x^* = 0$, which is the state of the system where all the rates $\dot{x}_i$ are zero and the system (1.6) is "in balance". Therefore, $x^* = 0$ is the constant solution of Equation (1.6) for which we have $x(t; t_0, 0) = 0$ for all $t \geqslant t_0$. That is, if the system starts at an equilibrium it stays there for all future time.

The most significant interpretation attached to the equilibrium in a wide variety of applications in engineering, economics, model ecosystems, social dynamics, etc., is *stability*. Stability is commonly understood as a situation where the system is in equilibrium, and if perturbed, returns in time to equilibrium. That is, the "distance" between the perturbed process and the equilibrium approaches zero as time goes on. Now, the distance between the perturbed process $x(t; t_0, x_0)$ and the equilibrium $x^* = 0$ of (1.6) can be taken as the length of the vector $x = x(t; t_0, x_0)$, that is, as the Euclidean norm

$$\|x\| = (x_1^2 + x_2^2 + \cdots + x_n^2)^{1/2}. \tag{1.17}$$

Stability, as defined verbally above, can now be mathematically formulated as a limiting process

$$\|x(t; t_0, x_0)\| \to 0 \quad \text{as} \quad t \to +\infty. \tag{1.18}$$

It is a well-known fact which can be verified easily from the explicit solution (1.16), that in the linear system (1.6), stability (if present) is always exponential. That is, we can be more specific about the convergence (1.18) and show that there exist always two positive numbers Π and π independent of initial conditions (t_0, x_0), such that

$$\|x(t; t_0, x_0)\| \leqslant \Pi \|x_0\| \exp[-\pi(t - t_0)], \qquad t \geqslant t_0, \tag{1.19}$$

for any initial conditions (t_0, x_0). Therefore, exponential stability of $x^* = 0$ is always a global property of the system (1.6). As we well know (e.g. Hahn, 1967; Šiljak, 1969), this is not necessarily true when the models are not linear, or even linear but time-varying.

One of the basic problems of stability analysis is to find (necessary and sufficient) conditions on the system parameters so that the convergence of solutions (1.18) takes place. The most powerful method of solving such a problem is the *Liapunov direct method* (Kalman and Bertram, 1960; LaSalle and Lefschetz, 1961; Šiljak, 1969). The method answers questions of stability without the explicit solution of the related differential equations, such as (1.16). The most attractive property of the method is its geometric insight into stability and other pertinent characteristics of system behavior.

The underlying idea of the Liapunov method is to find a positive scalar function $v(x)$ with the rate of change $\dot{v}(x)$ negative for every possible state x of the system except for a single equilibrium state x^* where $v(x)$ attains its minimum $v(x^*)$. Then the function $v(x)$ will continuously decrease along the solutions $x(t; t_0, x_0)$ of the system until it assumes its minimum $v(x^*)$ and the system reaches the equilibrium x^*. In general, however, the Liapunov function is not available, and there are no systematic methods for choosing an appropriate one. The name "direct method" is, therefore, somewhat misleading with respect to the content it represents, since it offers a powerful general approach rather than a systematic method for stability analysis.

To illustrate the application of the Liapunov method, let us find the stability conditions on the coefficients of the system

$$\begin{aligned} \dot{x}_1 &= a_{11}x_1 + a_{12}x_2, \\ \dot{x}_2 &= a_{21}x_1 + a_{22}x_2, \end{aligned} \tag{1.7}$$

when

$$a_{11} < 0, \qquad a_{22} < 0, \qquad a_{12} \geqslant 0, \qquad a_{22} \geqslant 0. \tag{1.20}$$

Therefore, the matrix A corresponding to (1.7) is a Metzler matrix, and (1.7) represents a competitive system.

A candidate for Liapunov's function $v(x)$ can be chosen as the Euclidean distance (1.17) squared,

$$v(x) = x_1^2 + x_2^2, \tag{1.21}$$

which is positive for all possible states except at the origin and, thus, is a positive definite function of x. A geometric representation of such a function is the concave cup shown in Figure 1.4(a). If only the x_1x_2-plane is considered, another interpretation is possible. In the x_1x_2-plane, the level curves $v(x) = \text{const}$ are represented by a family of concentric circles which surround the origin as shown in Figure 1.4(b). These circles are obtained by cutting the cup of Figure 1.4(a) by horizontal planes and then projecting the sections on the x_1x_2-plane.

The total time derivative $\dot{v}(x)$ of the function $v(x)$,

$$\dot{v}(x) = \frac{\partial v}{\partial x_1}\dot{x}_1 + \frac{\partial v}{\partial x_2}\dot{x}_2 = 2x_1\dot{x}_1 + 2x_2\dot{x}_2, \tag{1.22}$$

is calculated along the solutions of the system (1.7) when we replace $\dot{x}_1$, $\dot{x}_2$ in (1.22) by the right-hand side of (1.7). That is,

$$\dot{v}(x) = 2x_1(a_{11}x_1 + a_{12}x_2) + 2x_2(a_{21}x_1 + a_{22}x_2). \tag{1.23}$$

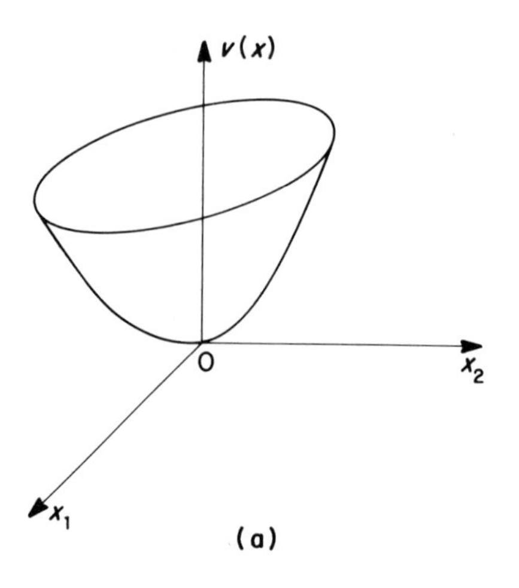

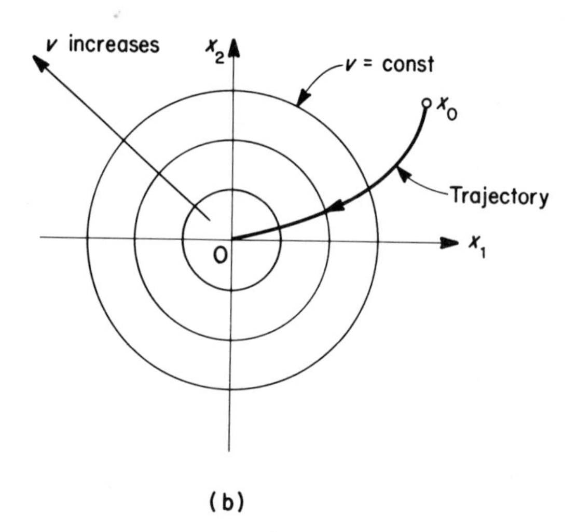

FIGURE 1.4. Liapunov function.

Taking into account that $a_{11}, a_{22} < 0$, we can rewrite (1.23) as

$$\dot{v}(x) = -2|a_{11}|x_1^2 + 2a_{12}x_1x_2 - 2|a_{22}|x_2^2 + 2a_{21}x_1x_2. \tag{1.24}$$

By using the simple fact that $x_1^2 - 2x_1x_2 + x_2^2 = (x_1 - x_2)^2 > 0$ everywhere except at $x_1 = x_2 = 0$, and its consequence

$$x_1^2 + x_2^2 > 2x_1x_2, \tag{1.25}$$

as well as the assumption $a_{12}, a_{21} \geqslant 0$, after simple manipulations, we get $\dot{v}(x)$ as

$$\dot{v}(x) \leqslant -[2|a_{11}| - (a_{12} + a_{21})]x_1^2 - [2|a_{22}| - (a_{12} + a_{21})]x_2^2, \quad (1.26)$$

which implies that $\dot{v}(x)$ is always negative provided the algebraic conditions

$$2|a_{11}| > a_{12} + a_{21}, \qquad 2|a_{22}| > a_{12} + a_{21} \quad (1.27)$$

are satisfied.

From (1.21) and (1.26), we conclude that $v(x) > 0$, $\dot{v}(x) < 0$ for all states x except at the equilibrium $x^* = 0$, which implies the convergence (1.18) of the solution $x(t; t_0, x_0)$ for any initial values (t_0, x_0).

We can now interpret the conditions (1.27) in terms of the system (1.7) representing a two-commodity market. The conditions (1.27) are obviously implied by

$$\begin{aligned} |a_{11}| > a_{12}, \qquad |a_{11}| > a_{21}, \\ |a_{22}| > a_{12}, \qquad |a_{22}| > a_{21}, \end{aligned} \quad (1.28)$$

which means that we have stability whenever the change in price of a commodity influences itself more than it influences and is influenced by the price of the other commodity. The conditions (1.28) qualify the system matrix A of (1.5) as a row and column *diagonal-dominant* matrix, and they are common stability conditions for Metzler matrices (Newman, 1959).

We should notice, however, that the conditions are only *sufficient* for stability, and (at least for linear systems) this is not satisfactory. That we can do better for linear systems is a well-known fact (Newman, 1959). We should not expect the situation to change necessarily when considering nonlinear time-varying systems. The failure of the Liapunov direct method to provide necessary conditions for such systems is its weakness, and reducing its conservativeness is a major goal in system analysis.

A geometric interpretation of stability concluded from the fact that $v(x) > 0$, $\dot{v}(x) < 0$ is given in Figure 1.4(b). Any state trajectory $x(t; t_0, x_0)$, starting at an initial state x_0 at time t_0, must penetrate the $v(x) = \text{const}$ circles from outside until it reaches the equilibrium x^* at the origin. The Liapunov function $v(x)$ can provide additional information about the rapidity of the convergence of $x(t; t_0, x_0)$ to $x^* = 0$. That is, we can use $v(x)$ and derive an inequality

$$\|x(t; t_0, x_0)\| \leqslant \Pi \|x_0\| \exp[-\pi(t - t_0)], \qquad t \geqslant t_0, \quad (1.19)$$

where Π and π are positive constants which are independent of the initial conditions (t_0, x_0). This inequality establishes the global exponential stability of the equilibrium $x^* = 0$. That is, all trajectories $x(t; t_0, x_0)$ of the system (1.7) approach the origin $x^* = 0$ faster than an exponential.

If the inequalities (1.28) are rewritten as

$$\begin{aligned} |a_{11}| - a_{12} &\geqslant \pi, \qquad & |a_{11}| - a_{21} &\geqslant \pi, \\ |a_{22}| - a_{12} &\geqslant \pi, \qquad & |a_{22}| - a_{21} &\geqslant \pi, \end{aligned} \tag{1.29}$$

where π is a sufficiently small (but finite) positive number, then the inequality (1.26) can be given as a *differential inequality*

$$\dot{v} \leqslant -2\pi v, \tag{1.30}$$

with $v_0 = v(t_0)$. To get a solution inequality (1.19) from (1.30), we show first that (1.30) implies

$$v(t) \leqslant v_0 \exp[-2\pi(t - t_0)], \qquad t \geqslant t_0. \tag{1.31}$$

Multiplying both sides of (1.30) by $e^{2\pi t}$ and transposing, we get

$$0 \geqslant [\dot{v}(t) + 2\pi v(t)]\exp(2\pi t) = \frac{d}{dt}\{v(t)\exp(2\pi t)\}, \qquad t \geqslant t_0. \tag{1.32}$$

Thus, the function $v(t)\exp(2\pi t)$ has a nonpositive derivative, and is nonincreasing for $t \geqslant t_0$. Consequently, $v(t)\exp(2\pi t) \leqslant v_0 \exp(2\pi t_0)$, and we get (1.31).

Now, we recall from (1.21) that $v(x) = x_1^2 + x_2^2 = \|x\|^2$, and (1.31) can be rewritten as

$$\|x(t)\|^2 \leqslant \|x_0\|^2 \exp[-2\pi(t - t_0)], \qquad t \geqslant t_0. \tag{1.33}$$

The solution inequality (1.19) follows directly from (1.33) with $\Pi = 1$. Therefore, the conditions (1.29) imply global exponential stability of the equilibrium $x^* = 0$ which is expressed by (1.19).

Before concluding this section we should note that the diagonal dominance conditions (1.29) can be used as a measure of the *complexity* of the system (1.20), since they indicate the level of interactions among the constituents of the system relative to the level of influence that each individual constituent has on itself. The number π in (1.29) can serve as an "aggregate" index of complexity of competitive systems. Then from (1.29) we conclude that if complexity is limited, so that self-regulatory effects exceed interaction effects for each constituent ($\pi > 0$), then the corresponding competitive system is stable. Other interpretations of the diagonal-dominance conditions with regard to complexity will be given in applications.

For an introduction of Liapunov's method see LaSalle and Lefschetz (1961), as well as, Kalman and Bertram (1960). Comprehensive presentations of the method are given in the books of Hahn (1967) and Lakshmikantham and Leela (1969). For applications of the method to analysis and design of feedback control systems, see Kalman and Bertram (1960), Hahn (1967), and Šiljak (1969).

1.4. COMPARISON PRINCIPLE

In this section, we will outline a simple version of the *comparison principle*, which is an effective tool for solving qualitative problems in the theory of differential equations (Birkoff and Rota, 1962). The principle will be used to resolve the problem of stability under structural perturbations and will be widely applied throughout this book.

With the differential inequality (1.30) we associate the differential equation

$$\dot{r} = -2\pi r, \tag{1.34}$$

with $r(t_0) = r_0$. If the function $v(t)$ satisfies the differential inequality (1.30) for $t \geqslant t_0$, and $r(t)$ is a solution of the differential equation (1.34) satisfying the initial condition $r_0 = v_0$, then

$$v(t) \leqslant r(t), \qquad t \geqslant t_0, \tag{1.35}$$

which is (1.31). That is, the solutions $r(t)$ of the equation (1.34) majorize the functions $v(t)$ that satisfy the inequality (1.30) whenever they start at the same place, that is, $r_0 = v_0$.

Now, it is easy to see how the comparison principle applies to stability. If the equilibrium $r^* = 0$ of the equation (1.34) is stable and all the solutions $r(t) \rightarrow 0$ as $t \rightarrow +\infty$, then $v(t)$ must do the same, since it is positive by definition (cannot go negative) and is lying under the function $r(t)$.

This result can readily be extended to nonlinear differential inequalities by using the well-known Lipschitz condition, and comparing functions that satisfy the inequality

$$\dot{v} \leqslant f(t, v) \tag{1.36}$$

with exact solutions of the equation

$$\dot{r} = f(t, r). \tag{1.37}$$

We assume that for some constant $L > 0$, the function $f(t, v)$ satisfies a Lipschitz condition

$$|f(t, v) - f(t, w)| \leqslant L|v - w| \tag{1.38}$$

for all point-pairs (t, v) and (t, w) having the same t-coordinate such that $t \geqslant t_0$. Then we can establish the following simple fact:

Any function $v(t)$ that satisfies the differential inequality (1.36) *for $t \geqslant t_0$ satisfies also the inequality*

$$v(t) \leqslant r(t), \qquad t \geqslant t_0 \tag{1.35}$$

where $r(t)$ is a solution of the differential equation (1.37) *satisfying the initial condition $r_0 = v_0$.*

Suppose that (1.35) is violated and that for some $t_2 \geqslant t_0$, $v(t_2) > r(t_2)$. Define t_1 to be the largest t in the interval $t_0 \leqslant t \leqslant t_2$ for which $v(t) \leqslant r(t)$, so that $v(t_1) = r(t_1)$. We now define the function

$$d(t) = v(t) - r(t), \tag{1.39}$$

which is nonnegative, that is,

$$d(t) \geqslant 0, \qquad t_1 \leqslant t \leqslant t_2, \tag{1.40}$$

and moreover,

$$\begin{aligned} \dot{d}(t) &= \dot{v}(t) - \dot{r}(t) \\ &\leqslant f[t, v(t)] - f[t, r(t)] \\ &\leqslant L[v(t) - r(t)] \\ &\leqslant Ld(t). \end{aligned} \tag{1.41}$$

The last inequality in (1.41) implies

$$d(t) \leqslant d(t_1)\exp[L(t - t_1)] = 0, \qquad t_1 \leqslant t \leqslant t_2, \tag{1.42}$$

in view of the inequality (1.31) with -2π replaced by L. Since $d(t)$ is nonnegative, from (1.40) we conclude that it vanishes identically. However, this is a contradiction to our assumption that $v(t_2) > r(t_2)$, and thus the inequality (1.35) is established.

Finally, let us formulate a simple form of the *comparison principle* (Birkoff and Rota, 1962) established above, which we will use in the next sections to study stability properties of dynamic systems under structural perturbations, that is, connective stability:

If a continuous function $v(t)$ satisfies the differential inequality

$$\dot{v} \leqslant f(t, v), \tag{1.36}$$

for $t \geqslant t_0$, $f(t, v)$ *satisfies a Lipschitz condition* (1.38) *for* $t \geqslant t_0$, *and* $r(t)$ *is a solution of the differential equation*

$$\dot{r} = f(t, r) \tag{1.37}$$

satisfying the initial condition

$$r_0 = v(t_0), \tag{1.43}$$

then

$$v(t) \leqslant r(t), \qquad t \geqslant t_0. \tag{1.35}$$

In qualitative theory of differential equations, the comparison principle was used in early 1930's by Kamke (1932) to establish certain properties of the unknown solutions of one differential equation from the known solutions of another. It should be noted here that the comparison principle has been considerably generalized by a number of authors, as reported by Lakshmikantham and Leela (1969) as well as by Walter (1970). These extended versions of the principle will be presented in Chapter 2, since they are needed to get more general results.

1.5. CONNECTIVE STABILITY

We are now in a position to introduce the concept of connective stability (Šiljak, 1972a). Furthermore, with the comparison principle in our hands, we will be able to get conditions under which the equilibrium state $x^* = (x_1^*, x_2^*)^T = 0$ of the second-order system

$$\begin{aligned} \dot{x}_1 &= a_{11} x_1 + a_{12} x_2, \\ \dot{x}_2 &= a_{21} x_1 + a_{22} x_2, \end{aligned} \tag{1.7}$$

is stable under structural perturbations, that is, it is connectively stable.

Let us start with the minor modification of the coefficients a_{ij} in (1.7). We assume that they have the form

$$a_{ij} = \begin{cases} -\alpha_i + e_{ii}\alpha_{ii}, & i = j, \\ e_{ij}\alpha_{ij}, & i \neq j, \end{cases} \tag{1.44}$$

where we split the diagonal elements into two parts α_i and $e_{ii}\alpha_{ii}$. Such a modification can be justified, for example, in the case where (1.7) describes an arms race involving two countries and x_1, x_2 are their armaments (Richardson, 1960). Then $\alpha_{ii} x_i$ ($\alpha_{ii} \geqslant 0$) represents a force in the ith country which is interested in the arms buildup regardless of the hostilities, and which thus opposes the "expense and fatigue" effects represented by the terms $-\alpha_i x_i$ ($\alpha_i > 0$). Since we are interested in stability, it is easy to show that we need the condition $\alpha_i > \alpha_{ii}$ for $i = 1, 2$, that is,

$$\alpha_1 > \alpha_{11} \geqslant 0, \qquad \alpha_2 > \alpha_{22} \geqslant 0. \tag{1.45}$$

For the off-diagonal elements $a_{ij} = e_{ij}\alpha_{ij}$, we assume that $\alpha_{ij} \geqslant 0$, that is,

$$\alpha_{12} \geqslant 0, \qquad \alpha_{21} \geqslant 0. \tag{1.46}$$

Due to the assumptions (1.45) and (1.46) on the coefficients a_{ij} in (1.7), they satisfy the conditions (1.4), and the corresponding 2×2 system matrix $A = (a_{ij})$ is a Metzler matrix. In other words, $x^* = 0$ is the equilibrium state of a competitive system represented by the equations (1.7).

With the system (1.7) defined by (1.44), we can associate the same digraph of Figure 1.1 if a_{ij}'s are replaced by α_{ij}'s. The fundamental interconnection matrix $\bar{E} = (\bar{e}_{ij})$ of (1.9) can be chosen to represent the structure of the system, and the corresponding structural perturbations and interconnection matrices $E = (e_{ij})$ are those of Figure 1.2. Now, we can simply formulate the notion of connective stability as follows:

The equilibrium $x^* = 0$ is connectively stable if it is stable for all interconnection matrices E.

To establish this kind of stability of $x^* = 0$ in (1.7), we need to prove stability for the fundamental interconnection matrix $\bar{E}$ only. Then, by the comparison principle, stability for any E derived from $\bar{E}$ (that is, connective stability) follows automatically. This is an important qualitative result, since we establish the stability of a class of dynamic systems corresponding to the set of interconnection matrices E by establishing it for a member of that class corresponding to the fundamental interconnection matrix $\bar{E}$.

Let us now denote by $\bar{A} = (\bar{a}_{ij})$ the 2×2 system matrix corresponding to (1.7) and $\bar{E} = (\bar{e}_{ij})$. The coefficients $\bar{a}_{ij}$ of $\bar{A}$ are defined as

$$\bar{a}_{ij} = \begin{cases} -\alpha_i + \bar{e}_{ii}\alpha_{ii}, & i = j, \\ \bar{e}_{ij}\alpha_{ij}, & i \neq j. \end{cases} \tag{1.47}$$

The system (1.7) is now

$$\begin{aligned} \dot{x}_1 &= \bar{a}_{11}x_1 + \bar{a}_{12}x_2, \\ \dot{x}_2 &= \bar{a}_{21}x_1 + \bar{a}_{22}x_2. \end{aligned} \tag{1.48}$$

We choose again

$$v(x) = x_1^2 + x_2^2 \tag{1.21}$$

as a candidate for the Liapunov function, and calculate the total time derivative $\dot{v}(x)$ along the solutions of (1.48) to get

$$\dot{v}(x) \leqslant -[2|\bar{a}_{11}| - (\bar{a}_{12} + \bar{a}_{21})]x_1^2 - [2|\bar{a}_{22}| - (\bar{a}_{12} + \bar{a}_{21})]x_2^2, \tag{1.49}$$

as in (1.26). The stability conditions are again those of (1.27), which we write in terms of the elements $\bar{e}_{ij}$ of the matrix $\bar{E}$ using (1.44),

$$\begin{aligned} 2|-\alpha_1 + \bar{e}_{11}\alpha_{11}| &> \bar{e}_{12}\alpha_{12} + \bar{e}_{21}\alpha_{21}, \\ 2|-\alpha_2 + \bar{e}_{22}\alpha_{22}| &> \bar{e}_{12}\alpha_{12} + \bar{e}_{21}\alpha_{21}. \end{aligned} \tag{1.50}$$

Now, it is obvious that if conditions (1.50) are true, they are also true for any interconnection matrix $E = (e_{ij})$ obtained from $\bar{E} = (\bar{e}_{ij})$ by replacing the unit elements of E with zeros. That is, to establish connective stability of the equilibrium $x^* = 0$ of the system (1.7) defined by (1.44), it is sufficient to test stability only for the fundamental interconnection matrix $\bar{E}$, by verifying the algebraic conditions (1.50).

It is of interest to see how the above result follows from the comparison principle, which will allow us to generalize the result for interconnection matrices $E = (e_{ij})$ the coefficients of which are not necessarily constant binary numbers, but can be functions of time t and state x: $e_{ij} = e_{ij}(t, x)$. We assume, however, that the functions $e_{ij}(t, x)$ are continuous and bounded as

$$0 \leqslant e_{ij}(t, x) \leqslant 1 \tag{1.51}$$

for all pairs (t, x). The fundamental interconnection matrix $\bar{E} = (\bar{e}_{ij})$ is still a constant (binary) matrix, and the matrices E are obtained from $\bar{E}$ by replacing the unit elements with the corresponding functions $e_{ij}(t, x)$.

To show that we can establish connective stability of the system (1.7) with nonlinear time-varying interconnection elements $e_{ij}(t, x)$ from the stability of the linear time-constant system (1.48), let us assume that the conditions (1.29) are satisfied by the coefficients $\bar{a}_{ij}$ defined by (1.47). That is,

$$\begin{aligned} |\bar{a}_{11}| - \bar{a}_{12} \geqslant \pi, \qquad |\bar{a}_{11}| - \bar{a}_{12} \geqslant \pi \\ |\bar{a}_{22}| - \bar{a}_{12} \geqslant \pi, \qquad |\bar{a}_{22}| - \bar{a}_{21} \geqslant \pi. \end{aligned} \tag{1.52}$$

Let us denote by $a_{ij}(t, x)$ the coefficients of (1.7) defined by (1.44), and note that the conditions (1.45), (1.46), and (1.51) imply

$$
\begin{aligned}
|a_{11}(t,x)| &\geqslant \bar{a}_{11}, \qquad & |a_{22}(t,x)| &\geqslant \bar{a}_{22} \\
a_{12}(t,x) &\leqslant \bar{a}_{12}, \qquad & a_{21}(t,x) &\leqslant \bar{a}_{21}
\end{aligned}
\tag{1.53}
$$

for all pairs (t,x). With (1.53) we have the inequality

$$\dot{v} \leqslant -2\pi v, \tag{1.30}$$

valid for all pairs (t,x) where $t \geqslant t_0$, whenever the conditions (1.52) are satisfied. Therefore, these conditions are sufficient for the connective stability of the system (1.7) where the interconnection parameters e_{ij} are nonlinear time-varying functions.

There are two conceptually important conclusions to be reached here. First, the connective stability property establishes *reliability* of the corresponding dynamic system, that is, it establishes the ability of the system to withstand sudden changes in its fundamental structural composition. To distinguish this kind of reliability from the common notion of reliability (Barlow, Fussell and Singpurwalla, 1975), we refer to it as *dynamic system reliability* or, simply, *dynamic reliability*. Second, stability of the system (1.7) holds also for the case when the interconnection elements e_{ij} are replaced by functions $e_{ij}(t,x)$ the actual shape of which is not specified save that they are continuous and bounded between zero and one. This fact implies a considerable degree of *robustness* of the system. That is, the system remains stable despite (unexpected) variations in, or incomplete information about, nonlinear characteristics, parameter settings, inaccurate measurements and computations, etc., which may take place during system operation.

Connective stability concept can be used in a suitable way to explore conditions for a *breakdown* of complex dynamic system due to failures of their components or subsystems. This use, however, should involve the system structure in an essential way (Šiljak, 1977c, 1977d), as discussed in Chapter 3. Furthermore, this approach opens up a real possibility to apply various computer algorithms developed in the fault-tree analysis (Barlow, Fussell and Singpurwalla, 1975) to study the breakdown phenomena in dynamic systems when it is interpreted as dynamic instability.

1.6. COOPERATION

Let us for a moment consider again the system

$$
\begin{aligned}
\dot{x}_1 &= a_{11}x_1 + a_{12}x_2, \\
\dot{x}_2 &= a_{21}x_1 + a_{22}x_2 .
\end{aligned}
\tag{1.7}
$$

We said that this system represents the competition of two agents if a_{12} and

a_{21} are nonnegative and the corresponding 2×2 system matrix $A = (a_{ij})$ is a Metzler matrix. It is easy to show that the necessary and sufficient condition for connective stability of the competitive equilibrium $x^* = 0$ in (1.7) are

$$a_{11} < 0, \qquad \begin{vmatrix} a_{11} & a_{12} \\ a_{21} & a_{22} \end{vmatrix} > 0. \tag{1.54}$$

In economics these conditions are known as Hicks conditions (Arrow and Hahn, 1971); they are stronger than those of (1.27) used to determine stability by the comparison principle. That is, the conditions (1.27) are not necessary for stability and therefore can be violated by stable systems.

From (1.54) we easily conclude that the necessary conditions for stability are

$$a_{11} < 0, \qquad a_{22} < 0. \tag{1.55}$$

This simply means that if the interconnection matrix $E = 0$ and the structurally perturbed system (1.7) becomes

$$\begin{aligned} \dot{x}_1 &= -\alpha_1 x_1, \\ \dot{x}_2 &= -\alpha_2 x_2, \end{aligned} \tag{1.56}$$

then the equilibrium $x^* = 0$ remains stable. That is, if the two states x_1 and x_2 are decoupled from each other, and the two equations (1.56) represent two distinct systems, the respective equilibria $x_1^* = 0$ and $x_2^* = 0$ are both stable. This fact is a simple consequence of connective stability, since if the system (1.7) is connectively stable, it is stable for all E and thus for $E = 0$. Taking this argument the other way around, we conclude that stability of the disjoint systems (1.56) is a necessary condition for connective stability.

Let us now consider the case when a diagonal element a_{11} or a_{22} is nonnegative. This surely means instability of the equilibrium $x^* = 0$ in (1.7), since the Hicks conditions (1.54) are violated. A way to restore stability in this case is to make the off-diagonal interaction coefficients a_{12} or a_{21} negative. To illustrate the situation, let us consider a version of the classic model for a one-predator, one-prey (or one-parasite, one-host) system with continuous growth, formulated by Lotka (1925) and Volterra (1926),

$$\begin{aligned} \dot{y}_1 &= \alpha y_1 - \gamma y_1 y_2, \\ \dot{y}_2 &= \delta y_1 y_2 - \beta y_2 - \theta y_2^2, \end{aligned} \tag{1.57}$$

where $y_1(t)$ and $y_2(t)$ are the populations of prey and predator, respectively, at time t. The parameter α represents the birth rate of the prey; β represents the death rate of the predator; γ, δ represent the interaction between the two species; and θ represents the death rate of the predator due to direct competition within the predator population. All parameters are assumed positive. Obviously, in the absence of the predator, the prey population would grow as a Malthusian exponential process at the rate determined by α. Similarly, if the prey population disappeared, the predator would become extinct. Now, the question arises: Could the predator-prey combination (1.57) be stable about constant positive populations y_1^*, y_2^* which are the components of the equilibrium population y^*? In other words, can the presence of a predator stabilize the prey population about the value y_1^*?

There are two equilibrium populations which are obtained from (1.57) by setting $\dot{y}_1 = \dot{y}_2 = 0$. That is, by solving the equations

$$\begin{aligned} y_1(\alpha - \gamma y_2) &= 0, \\ y_2(\delta y_1 - \beta - \theta y_2) &= 0, \end{aligned} \tag{1.58}$$

we get one equilibrium at the origin $y^* = 0$, and the other at $y^* = c$, where the constant vector c has the components

$$c_1 = \frac{1}{\delta}\left(\beta + \theta\frac{\alpha}{\gamma}\right), \qquad c_2 = \frac{\alpha}{\gamma}, \tag{1.59}$$

which are both positive. It is stability of this equilibrium $y^* = c$ that is of interest.

Since there are two equilibria of the model (1.57), none of them can be globally stable, and we proceed to study small deviations $x(t)$ about the equilibrium $y^* = c$ using linearization. Substituting the perturbed populations

$$y(t) = c + x(t) \tag{1.60}$$

into (1.57) and neglecting the nonlinear terms of $x(t)$, we get the linear model

$$\begin{aligned} \dot{x}_1 &= a_{11}x_1 + a_{12}x_2, \\ \dot{x}_2 &= a_{21}x_1 + a_{22}x_2, \end{aligned} \tag{1.7}$$

where

$$\begin{aligned} a_{11} &= 0, & a_{12} &= -(\alpha\theta + \beta\gamma)\delta^{-1}, \\ a_{21} &= \alpha\gamma^{-1}\delta, & a_{22} &= -\alpha\gamma^{-1}\theta. \end{aligned} \tag{1.61}$$

Due to the negativity of a_{12}, the system matrix $A = (a_{ij})$ corresponding to (1.61) is not a Metzler matrix, and the Hicks conditions (1.54) do not apply.

However, it is easy to show that the system (1.7) with (1.61) is stable if (and only if)

$$a_{22} < 0, \qquad a_{12}a_{21} < 0. \tag{1.62}$$

Due to the positivity of the parameters of the system (1.57), the conditions (1.62) are satisfied, and the nonzero equilibrium population $y^* = c$ corresponding to $x^* = 0$ in (1.7) is stable.

Therefore, we conclude that although the two species y_1 and y_2 could not have stable positive populations if separated, their "cooperation" has stable equilibrium populations. It is somewhat strange to refer to a predator-prey (parasite-host) system as a cooperation, but the fact that the two species cannot have stable equilibrium levels unless they interact justifies the use of the term. In economics (Samuelson, 1974), the term is more plausible on its face, since negativity of off-diagonal elements is related to complementary commodities. While coffee and tea are "substitutes" because we can drink one or the other, tea and lemon are "complements" because people like to drink tea with lemon. A rise in the price of tea will reduce the demand for lemon and thus reduce its price. In the linear market model, this fact is reflected in the negativity of the corresponding off-diagonal element.

Although we were able to demonstrate that in the simple system (1.7), negativity of the off-diagonal element was conducive to system stability, it is difficult to extend this result to systems with more than two agents. A discussion of the effect of the signs of the off-diagonal elements on the system stability has been given by May (1973) in the context of multispecies communities. No general result is available along these lines, and the sign effect on system stability remains an open question. We should note, however, that stabilization of a system by cooperation requires the coalition to stay together for all time, and such coalitions are not connectively stable. Thus, in the context of the overall system stability, we should make sure that under structural perturbations, such coalitions remain as a solid part of the system.

Before we conclude this section, we should point out that connective stability of coalitions is possible if the constituents are stable when isolated. To see this, we assume in (1.7) that the a_{ij}'s are defined as in (1.44), and that (1.45) holds, that is a_{11}, $a_{22} < 0$. Then, we assume further that the interaction coefficients are no longer constants but are nonlinear time-varying functions defined as

$$a_{12} = e_{12}\varphi_{12}(t, x_2), \qquad a_{21} = e_{21}\varphi_{21}(t, x_1), \tag{1.63}$$

where the functions φ_{12}, φ_{21} are continuous and bounded as

$$|\varphi_{12}(t, x_2)| \leqslant \alpha_{12}|x_2|, \qquad |\varphi_{21}(t, x_1)| \leqslant \alpha_{21}|x_1|. \tag{1.64}$$

In (1.64), α_{12} and α_{21} are again nonnegative numbers. Now, using once more the function

$$v(x) = x_1^2 + x_2^2 \tag{1.21}$$

as a candidate for the Liapunov function, and relying on the constraints (1.64), we can compute

$$\begin{aligned} \dot{v}(x) &= 2x_1(a_{11}x_1 + a_{12}x_2) + 2x_2(a_{21}x_1 + a_{22}x_2) \\ &= -2|a_{11}|x_1^2 + 2a_{12}x_1x_2 - 2|a_{22}|x_2^2 + 2a_{21}x_1x_2 \\ &\leqslant -2|a_{11}|x_1^2 + 2|a_{12}||x_1||x_2| - 2|a_{22}|x_2^2 + 2|a_{21}||x_1||x_2| \\ &\leqslant -[2|\bar{a}_{11}| - (|\bar{a}_{12}| + |\bar{a}_{21}|)]x_1^2 - [2|\bar{a}_{22}| - (|\bar{a}_{12}| + |\bar{a}_{21}|)]x_2^2 \\ &\leqslant -2\pi v(x), \end{aligned} \tag{1.65}$$

which implies that v, $-\dot{v} > 0$ everywhere (except at $x^* = 0$) and for all interconnection matrices E, provided

$$\begin{aligned} |\bar{a}_{11}| - |\bar{a}_{12}| &\geqslant \pi, & |\bar{a}_{11}| - |\bar{a}_{21}| &\geqslant \pi, \\ |\bar{a}_{22}| - |\bar{a}_{12}| &\geqslant \pi, & |\bar{a}_{22}| - |\bar{a}_{21}| &\geqslant \pi, \end{aligned} \tag{1.66}$$

where the $\bar{a}_{ij}$'s are defined for the fundamental interconnection matrix $\bar{E}$ as in (1.47). Therefore, the equilibrium $x^* = 0$ of the system (1.7) is connectively stable regardless of the sign of the off-diagonal interaction coefficients a_{12}, a_{21}, so long as the conditions (1.66) hold. Furthermore, such systems can tolerate considerable nonlinear and time-varying phenomena in the interactions among the constituents, and the actual shape of the interaction functions $\varphi_{12}(t, x_2)$, $\varphi_{21}(t, x_1)$ need not be specified so long as their absolute values are properly bounded. That is, they exhibit a considerable degree of robustness in their interconnection structure.

1.7. DECOMPOSITION

Systems involving a large number of variables are difficult to consider in one piece. Despite the high efficiency of modern computers, the formidable complexity of a large system can make the problem numerically intractable even with the most valuable "one shot" techniques. It may be simply either impossible (insufficient computer memory), or too costly (excessive computer time) to apply such techniques to a large system.

It has long been recognized that certain complex systems made of interacting elements can be decomposed into subsystems of lower dimensionality. Then the system is considered "piece by piece", and the separate solutions of the subsystems are combined together in some way to provide a solution for the overall system. This *decomposition principle* was used explicitly at least as early as 1950 by Kron (1963) in the analysis of electrical networks, and it was reported recently by Himelblau (1973) that a decomposition strategy was used as far back as 1843 by Gerling to solve systems of equations with predominant principle diagonal.

While the decomposition principle can bring about a great saving in solution time over solving the whole system in one piece, it is still highly dependent on the choice of a particular decomposition. The difficulty is in choosing easily solvable subproblems without doing too much violence to the system: the overall solution may be either inaccurate or impossible to get by putting together the solutions of the subproblems. Therefore, the decomposition of large systems is a difficult task, and needs to be considerably simplified by physical insight into the particular problem at hand. That is why the Kron tearing technique was so successful in network analysis where it originated, but was much less helpful in other scientific disciplines. The application of computers to the decomposition problem (Himelblau, 1973; Sage, 1977) has advanced the methods of tearing systems of equations considerably and made the decomposition principle an important tool in the analysis of complex systems.

Most decomposition methods have been developed for algebraic equations, and the question how to tear a dynamic large-scale system is still unresolved. There are, however, two basic situations to be recognized: *physical decomposition* and *mathematical decomposition*.

When the system represents a structure of distinct interconnected elements (subsystems) which have physical meaning, such as electric power networks, then tearing the interconnections in the course of analysis can not only bring about numerical simplifications, but also provide information about the important structural properties of the system. On the other hand, a reason for decomposition can be entirely numerical, in which case various transformations before and after the decomposition can remove all physical meaning of the variables involved, and only the final solution of the overall system is to be interpreted. Both aspects of the decomposition principle will be of interest to us. In the case of physical decomposition, we shall start with a decomposed system and then investigate its connective structural characteristics. We shall also present a method for decentralizing a dynamic system for the purpose of stabilization and optimization of large-scale systems by local control functions associated with each element of the system, which may have no physical meaning at all.

Let us now illustrate by simple examples the principle of decomposition. Suppose that a linear constant system $\mathbb{S}$ is given as

$$\dot{x} = Ax, \tag{1.6}$$

where again $x = (x_1, x_2, \ldots, x_n)^T$ is the state vector and $A = (a_{ij})$ is the constant $n \times n$ system matrix. We can partition the state vector x into two vector components: $x_1 = (x_{11}, x_{12}, \ldots, x_{1n_1})^T$, $x_2 = (x_{21}, x_{22}, \ldots, x_{2n_2})^T$, where we have renamed the components of the vector x in an obvious way. The dimensions of the two vectors x_1, x_2 are n_1, n_2, so that $n_1 + n_2 = n$. This partition yields the system (1.6) in the form

$$\begin{bmatrix} \dot{x}_1 \\ \dot{x}_2 \end{bmatrix} = \begin{bmatrix} A_{11} & A_{12} \\ A_{21} & A_{22} \end{bmatrix} \begin{bmatrix} x_1 \\ x_2 \end{bmatrix}, \tag{1.67}$$

which can be further rewritten as two vector equations

$$\begin{aligned} \dot{x}_1 &= A_{11} x_1 + A_{12} x_2, \\ \dot{x}_2 &= A_{21} x_1 + A_{22} x_2. \end{aligned} \tag{1.68}$$

In (1.68), the matrices A_{11}, A_{22} have the dimensions $n_1 \times n_1$, $n_2 \times n_2$, respectively, while the matrices A_{12}, A_{21} have the dimensions $n_1 \times n_2$, $n_2 \times n_1$, respectively.

Now, if we identify the state vectors x_1, x_2 with two subsystems $\mathbb{S}_1$ and $\mathbb{S}_2$, then

$$\begin{aligned} \dot{x}_1 &= A_{11} x_1, \\ \dot{x}_2 &= A_{22} x_2, \end{aligned} \tag{1.69}$$

describe the *decoupled subsystems*, and $A_{12} x_2$, $A_{21} x_1$ represent the interactions between the two subsystems. Therefore, (1.68) describes the two *interconnected subsystems.*

If we split the diagonal blocks in (1.67) into two parts, one corresponding to the two subsystems and the other related to "self-interactions", then the equations (1.67) become

$$\begin{bmatrix} \dot{x}_1 \\ \dot{x}_2 \end{bmatrix} \begin{bmatrix} A_1 + A_{11} & A_{12} \\ A_{21} & A_2 + A_{22} \end{bmatrix} \begin{bmatrix} x_1 \\ x_2 \end{bmatrix}, \tag{1.70}$$

and the two subsystems $\mathbb{S}_1$ and $\mathbb{S}_2$ are described by

$$\begin{aligned} \dot{x}_1 &= A_1 x_1, \\ \dot{x}_2 &= A_2 x_2, \end{aligned} \tag{1.71}$$

while the equations

$$\begin{aligned} \dot{x}_1 &= A_1 x_1 + A_{11} x_1 + A_{12} x_2, \\ \dot{x}_2 &= A_2 x_2 + A_{21} x_1 + A_{22} x_2 \end{aligned} \tag{1.72}$$

represent the two interconnected subsystems.

To display the structure of the system in (1.72), we associate with it the 2×2 fundamental interconnection matrix $\bar{E} = (\bar{e}_{ij})$, and rewrite the equations (1.72) as

$$\begin{aligned} \dot{x}_1 &= A_1 x_1 + \bar{e}_{11} A_{11} x_1 + \bar{e}_{12} A_{12} x_2, \\ \dot{x}_2 &= A_2 x_2 + \bar{e}_{21} A_{21} x_1 + \bar{e}_{22} A_{22} x_2. \end{aligned} \tag{1.73}$$

The "weighted" digraph of the system (1.73) is shown in Figure 1.5. Now the points x_1 and x_2 represent the two subsystems $\mathbb{S}_1$ and $\mathbb{S}_2$, and the lines $x_j x_i$ are labeled as interactions $\bar{e}_{ij} A_{ij} x_j$ among the subsystems. Thus, the elements $\bar{e}_{ij}$ of the fundamental interconnection matrix $\bar{E}$ are defined as

$$\bar{e}_{ij} = \begin{cases} 1, & \mathbb{S}_j \text{ can act on } \mathbb{S}_i \\ 0, & \mathbb{S}_j \text{ cannot act on } \mathbb{S}_i. \end{cases} \tag{1.74}$$

In other words, $\bar{e}_{ij} = 1$ if there is a line $x_j x_i$ from the subsystem $\mathbb{S}_j$ to the subsystem $\mathbb{S}_i$ and $A_{ij} x_j \not\equiv 0$, and $\bar{e}_{ij} = 0$ if there is no line $x_j x_i$ and $A_{ij} x_j \equiv 0$.

The interconnection matrices $E = (e_{ij})$ and the structural perturbations corresponding to the fundamental interconnection matrix $\bar{E} = (\bar{e}_{ij})$ are defined as in Section 1.2. A matrix E is formed by replacing the unit elements $\bar{e}_{ij}$ of $\bar{E}$ by the elements e_{ij} of E.

The decomposition principle is not limited to linear systems. However, our understanding of how to tear a nonlinear system is superficial at present and almost all efficient decomposition techniques (Himelblau, 1973) are developed for linear systems. Therefore, a tearing of nonlinear systems is almost entirely guided by the physical configuration of the system and the insight one has into the system structural characteristics.

To illustrate some aspects of the decomposition of physical systems, let us consider the simple mechanical device shown in Figure 1.6, which is a disc fixed to a rotating shaft. The device is regarded as a massless elastic shaft with a mass particle attached at the center. Friction is assumed to be internal to the shaft. If ρ and μ represent deflections of the mass particle in

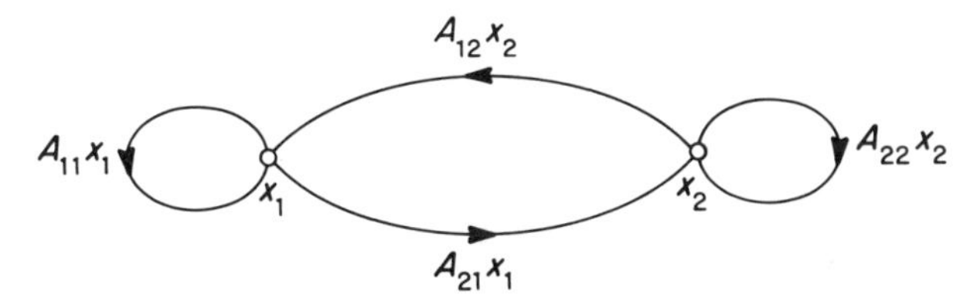

FIGURE 1.5. Weighted digraph.

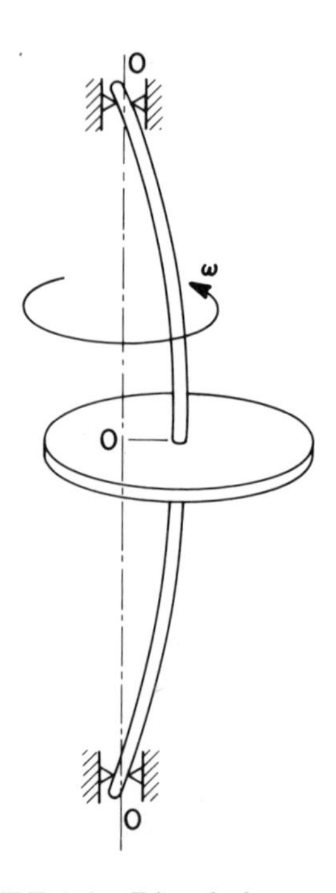

FIGURE 1.6. Disc-shaft system.

a coordinate system rotating at the angular velocity ω of the shaft, then the linearized equations of motion of the system $\mathbb{S}$ are

$$\begin{aligned} m\ddot{\rho} + f\dot{\rho} + (k - m\omega^2)\rho - 2m\omega\dot{\mu} &= 0, \\ m\ddot{\mu} + f\dot{\mu} + (k - m\omega^2)\mu + 2m\omega\dot{\rho} &= 0, \end{aligned} \tag{1.75}$$

where m is the mass, f is the damping coefficient, and k is the stiffness coefficient of the shaft (Ziegler, 1968).

The equations of motion (1.75) can be rewritten in the state form as

$$\begin{aligned} \dot{x}_1 &= x_2, \\ \dot{x}_2 &= (-\alpha + \omega^2)x_1 - \beta x_2 + 2\omega x_4, \\ \dot{x}_3 &= x_4, \\ \dot{x}_4 &= (-\alpha + \omega^2)x_3 - \beta x_4 - 2\omega x_2, \end{aligned} \tag{1.76}$$

where $x = (\rho, \dot{\rho}, \mu, \dot{\mu})^T$ is chosen as the state vector, and $\alpha = k/m$, $\beta = f/m$.

By identifying the subsystems $\mathbb{S}_1$ and $\mathbb{S}_2$ with the deflections ρ and μ, as

$$\begin{aligned} \dot{x}_{11} &= x_{12}, & \dot{x}_{21} &= x_{22}, \\ \dot{x}_{12} &= -\alpha x_{11} - \beta x_{12}, & \dot{x}_{22} &= -\alpha x_{21} - \beta x_{22}, \end{aligned} \tag{1.77}$$

we can decompose (1.76) into two interconnected subsystems

$$\dot{x}_1 = \begin{bmatrix} 0 & 1 \\ -\beta & -\alpha \end{bmatrix} x_1 + \begin{bmatrix} 0 & 0 \\ \omega^2 & 0 \end{bmatrix} x_1 + \begin{bmatrix} 0 & 0 \\ 0 & 2\omega \end{bmatrix} x_2,$$
$$\dot{x}_2 = \begin{bmatrix} 0 & 1 \\ -\beta & -\alpha \end{bmatrix} x_2 + \begin{bmatrix} 0 & 0 \\ 0 & -2\omega \end{bmatrix} x_1 + \begin{bmatrix} 0 & 0 \\ \omega^2 & 0 \end{bmatrix} x_2, \tag{1.78}$$

where the states of the subsystems are $x_1 = (x_{11}, x_{12})^T = (\rho, \dot{\rho})^T$ and $x_2 = (x_{21}, x_{22})^T = (\mu, \dot{\mu})^T$.

From the equations (1.78), we see that the angular velocity ω of the shaft appears only in the interactions between the two subsystems (1.77). Therefore, as shown in the following section, the chosen decomposition will allow us to study explicitly the effect on overall system stability of varying the velocity ω. It is obvious, however, that there is no connective stability in the true sense of the term, since the two subsystems cannot be physically disconnected from each other (although the two motions corresponding to the two deflections become decoupled as the angular velocity decreases to zero).

To illustrate the connectivity aspect of decompositions, let us consider a two-predator, two-prey community described by the equations

$$\begin{aligned} \dot{h}_1 &= \alpha_1 h_1 - \alpha_{11} h_1^2 - \gamma_1 h_1 p_1 - \alpha_{12} h_2, \\ \dot{p}_1 &= \delta_1 h_1 p_1 - \beta_1 p_1 - \theta_1 p_1^2, \\ \dot{h}_2 &= \alpha_2 h_2 - \alpha_{22} h_2^2 - \gamma_2 h_2 p_2 - \alpha_{21} h_1, \\ \dot{p}_2 &= \delta_2 h_2 p_2 - \beta_2 p_2 - \theta_2 p_2^2. \end{aligned} \tag{1.79}$$

The digraph of the multispecies community (1.79) is given in Figure 1.7(a). We may be interested in structural aspects of the community when there is a possibility of cutting the link between the two subcommunities (h_1, p_1) and (h_2, p_2). Then the two subsystems $\mathbb{S}_1$ and $\mathbb{S}_2$ given as

$$\begin{bmatrix} \dot{h}_1 \\ \dot{p}_1 \end{bmatrix} = \begin{bmatrix} \alpha_1 - \alpha_{11} h_1 & -\gamma_1 \\ \delta_1 p_1 & -\beta_1 - \theta_1 p_1 \end{bmatrix} \begin{bmatrix} h_1 \\ p_1 \end{bmatrix},$$
$$\begin{bmatrix} \dot{h}_2 \\ \dot{p}_2 \end{bmatrix} = \begin{bmatrix} \alpha_2 - \alpha_{22} h_2 & -\gamma_2 \\ \delta_2 p_2 & -\beta_2 - \theta_2 p_2 \end{bmatrix} \begin{bmatrix} h_2 \\ p_2 \end{bmatrix} \tag{1.80}$$

are encircled by dashed lines in Figure 1.7(a). Now, connective properties of the community structure involve the two subsystems $\mathbb{S}_1$, $\mathbb{S}_2$, which are themselves cooperative systems (coalitions), and may be regarded as two interacting agents. This produces the digraph in Figure 1.7(b), which corresponds to the fundamental interconnection matrix

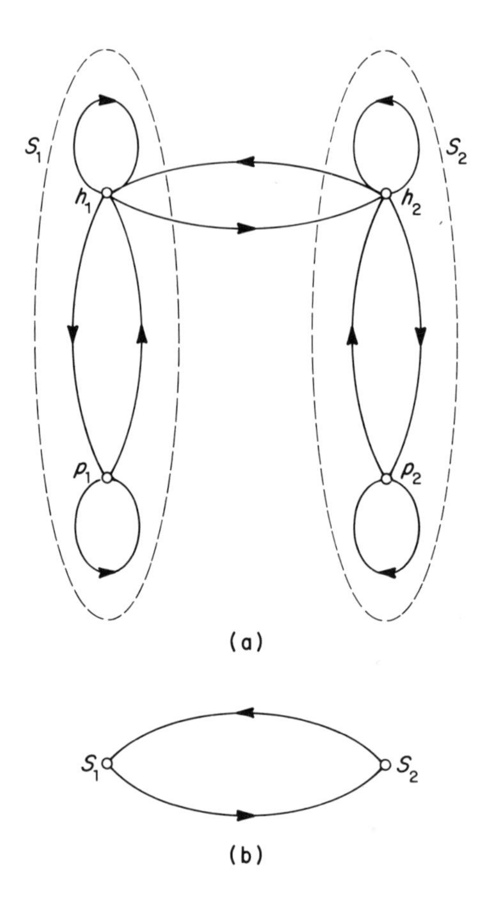

FIGURE 1.7. Community digraphs.

$$\bar{E} = \begin{bmatrix} 0 & 1 \\ 1 & 0 \end{bmatrix}, \tag{1.81}$$

and the interactions are

$$\begin{bmatrix} 0 & -\alpha_{12} \\ 0 & 0 \end{bmatrix} \begin{bmatrix} h_2 \\ p_2 \end{bmatrix}, \quad \begin{bmatrix} 0 & -\alpha_{21} \\ 0 & 0 \end{bmatrix} \begin{bmatrix} h_1 \\ p_1 \end{bmatrix}. \tag{1.82}$$

Since we included the self-interacting terms $-\alpha_{11} h_1^2$, $-\alpha_{22} h_2^2$, $-\theta_1 p_1^2$, $-\theta_2 p_2^2$ inside the subsystems, the digraph of Figure 1.7(b) has no loops, and the diagonal elements of $\bar{E}$ in (1.81) are zero.

Other possibilities of decomposing the system (1.79) are shown in Figure 1.8. The community can be decomposed into a predator and a prey subsystem as in Figure 1.8(a), or into a predator subsystem and two prey subsystems as shown in Figure 1.8(b).

When the decomposition principle is applied for numerical reasons only, it is recommended to perform decompositions so as to make the interactions among the subsystems as small as possible. That is because our methods are directed toward building competitive structures in which each subsystem views the others as competitors, and the interactions with the other subsystems are regarded as perturbations. Then, as indicated in Section 1.5 on connective stability, the smaller the interactions among the subsystems, the higher the degree of stability of the overall system. Methods for performing decompositions that do the least violence to the system and produce weakly coupled subsystems by identifying sparsity of the system were introduced by Steward (1962, 1965) and later developed by other authors (see Himelblau, 1973).

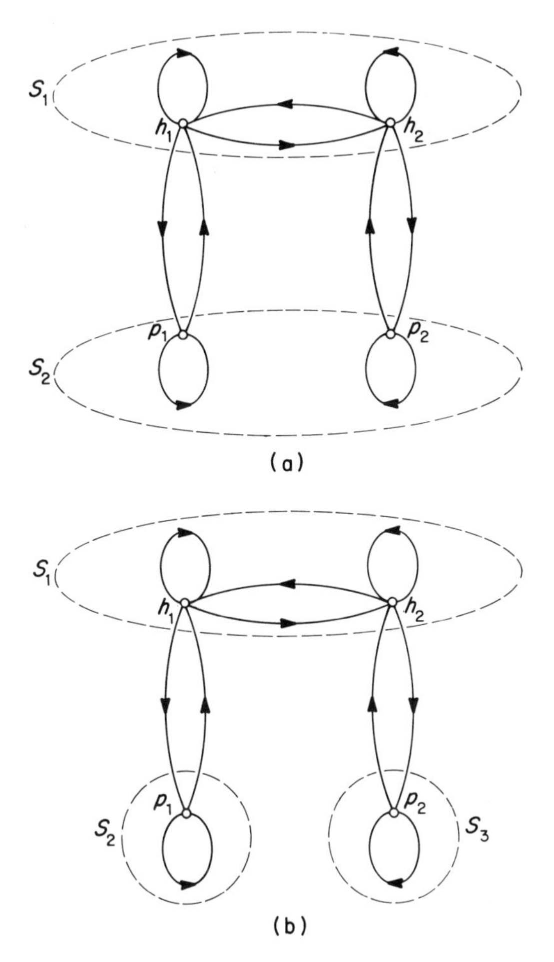

FIGURE 1.8. Decompositions.

1.8. AGGREGATION

In economic studies of large systems, it has been long recognized that aggregation is an effective process by which we can get both conceptual simplification in abstract analysis and numerical feasibility in actual computations (e.g. Theil, 1954; Green, 1964). An excellent example for the need of aggregation is provided by Simon and Ando (1961):

> Suppose that government planners are interested in the effects of a subsidy to a basic industry, say the steel industry, on the total effective demand in the economy. Strictly speaking, we must deal with individual producers and consumers, and trace through all interactions among the economic agents in the economy. This being an obviously impossible task, we would use such aggregated variables as the total output of the steel industry, aggregate consumption and aggregate investment.

A few comments are in order at this point. Even in the case when a direct analysis is possible, it may become bogged down in the welter of detail, requiring excessive computer storage and time to complete the investigation. Furthermore, treating a large system "in one piece" increases the likelihood of errors. However, we should be aware of the fact that aggregation often involves approximations and that we should provide conditions under which such approximations are justified. The aggregation process should be judged satisfactory to the extent that the conceptual and numerical gains outweigh the inaccuracy induced by the approximations.

For the reasoning behind the decomposition-aggregation procedure, we again quote Simon and Ando (1961):

> (1) We can somehow classify all the variables in the economy into a small number of groups;
>
> (2) we can study the interactions within the groups as though the interaction among groups did not exist;
>
> (3) we can define indices representing groups and study the interaction among these indices without regard to the interactions within each group.

Liapunov's second method is an ideal mechanism for accomplishing the above aggregation plan in the stability analysis of large-scale dynamic systems. Actually, the Liapunov method itself, as sketched in Section 1.3, can be viewed as an aggregation process. A stability property, involving several state variables, is entirely represented by a single variable—the Liapunov function. As mentioned above, however, this approach simplifies the stability problem, but sacrifices detailed information about the size of variations of each separate state variable.

To follow the above decomposition-aggregation plan in the connective-stability analysis of large systems, we need the concepts of vector differential inequalities and vector Liapunov functions. Although the possibility of using several Liapunov functions was discovered by Duhem as early as 1902. the concept of vector Liapunov functions was explicitly introduced only relatively recently by Matrosov (1962) and Bellman (1962); it has since been extended in various directions by many people, as outlined in the book by Lakshmikantham and Leela (1969). Roughly speaking, the concept associates with a dynamic system several scalar functions (say s) in such a way that each of them determines the desired stability properties in the system state space (of dimension $n \geqslant s$) wherever the others fail to do so. These scalar functions are considered as components of a vector Liapunov function, and a differential inequality is formed in terms of this function, using the original system of equations. Now, as in the case of one Liapunov function, the stability properties of an nth-order system are determined by considering only the s-vector differential inequality involving Liapunov functions. This may bring about a considerable reduction in the dimensionality of a stability problem.

It should be mentioned immediately that there is no general systematic procedure for choosing vector Liapunov functions, and that is at once the greatest advantage and the most serious drawback of the approach (as with the original Liapunov direct method). There are, however, several constructions for obtaining vector Liapunov functions, initiated by Bailey (1966) and developed later on by Michel (1974), Araki (1975), Weissenberger (1973), and many others, as surveyed by Šiljak (1972b) and Michel (1974) (see also Chapter 2). We will use here the construction proposed by Šiljak (1972a, b), which is concerned with stability of large-scale systems under structural perturbations, that is, connective stability.

We now outline the decomposition-aggregation method for connective-stability analysis, imitating the Simon-Ando plan, but using an entirely different technique—that of vector Liapunov functions. We consider a linear time-varying system $\mathbb{S}$ described by

$$\dot{x} = A(t)x, \tag{1.83}$$

where $x(t)$ is again the state n-vector and $A(t)$ is an $n \times n$ matrix whose elements are continuous functions of time t for the entire interval $t_0 \leqslant t < +\infty$. The system (1.83) is a time-varying version of (1.6).

To study connective stability of the equilibrium $x^* = 0$ of (1.83), we proceed as follows:

(1) We assume that the system (1.83) is decomposed into two interconnected subsystems $\mathbb{S}_1$ and $\mathbb{S}_2$,

$$\begin{aligned} \dot{x}_1 &= A_1 x_1 + e_{12}(t) A_{12} x_2, \\ \dot{x}_2 &= A_2 x_2 + e_{21}(t) A_{21} x_1, \end{aligned} \tag{1.84}$$

which can be obtained from (1.73) by assuming the matrix $\bar{E}$ to have the form

$$\bar{E} = \begin{bmatrix} 0 & 1 \\ 1 & 0 \end{bmatrix} \tag{1.85}$$

and replacing its unit elements with the elements $e_{12}(t)$, $e_{21}(t)$ of the matrix E. We also recall that $x = (x_1, x_2)^T$.

(2) Now, we consider the stability of each decoupled linear constant subsystem $\mathbb{S}_1$ and $\mathbb{S}_2$,

$$\begin{aligned} \dot{x}_1 &= A_1 x_1, \\ \dot{x}_2 &= A_2 x_2. \end{aligned} \tag{1.71}$$

Since we want to establish the connective stability of $\mathbb{S}$, we require that $\mathbb{S}_1$ and $\mathbb{S}_2$ be stable. Otherwise, for $E = 0$ the system $\mathbb{S}$ can be represented by (1.71), and instability of either $\mathbb{S}_1$ or $\mathbb{S}_2$ implies instability of $\mathbb{S}$.

To have stability of $\mathbb{S}_1$ and $\mathbb{S}_2$ in (1.71), we propose the scalar functions

$$v_1(x_1) = (x_1^T H_1 x_1)^{1/2}, \qquad v_2(x_2) = (x_2^T H_2 x_2)^{1/2} \tag{1.86}$$

as candidates for the Liapunov functions of $\mathbb{S}_1$ and $\mathbb{S}_2$, and require that for any choice of the positive definite matrices G_1, G_2 there exist positive definite matrices H_1, H_2 as solutions of the Liapunov matrix equations

$$A_1^T H_1 + H_1 A_1 = -G_1, \qquad A_2^T H_2 + H_2 A_2 = -G_2. \tag{1.87}$$

It is easy to show (e.g. Hahn, 1967) that the above requirement is no more and no less than what one needs for stability of both $\mathbb{S}_1$ and $\mathbb{S}_2$ (Theorem A.13). Therefore, we have the total time derivatives $\dot{v}_1(x_1)$, $\dot{v}_2(x_2)$ of the functions $v_1(x_1)$, $v_2(x_2)$ along the solutions of (1.71) as

$$\begin{aligned} \dot{v}_i &= (\text{grad } v_i)^T \dot{x}_i \\ &= (\text{grad } v_i)^T A_i x_i \\ &= (v_i^{-1} H_i x_i)^T A_i x_i \\ &= -\tfrac{1}{2} v_i^{-1/2} (x_i^T G_i x_i), \qquad i = 1, 2, \end{aligned} \tag{1.88}$$

where we have used

$$\text{grad } v_i = v_i^{-1} x_i^T H_i, \qquad i = 1, 2 \tag{1.89}$$

and the equations (1.87).

Finally, from Equations (1.86), (1.88), and (1.89), we produce the estimates for the Liapunov functions $v_1(x_1)$ and $v_2(x_2)$ as

$$\begin{aligned} \eta_{i1}\|x_i\| \leqslant v_i \leqslant \eta_{i2}\|x_i\|, \\ \dot{v}_i \leqslant -\eta_{i3}\|x_i\|, \qquad i = 1, 2, \\ \|\text{grad } v_i\| \leqslant \eta_{i4}, \end{aligned} \tag{1.90}$$

where the positive numbers η_{ij} are calculated as

$$\begin{aligned} \eta_{i1} &= \lambda_m^{1/2}(H_i), & \eta_{i2} &= \lambda_M^{1/2}(H_i), \\ \eta_{i3} &= \frac{1}{2}\frac{\lambda_m(G_i)}{\lambda_M^{1/2}(H_i)}, & \eta_{i4} &= \frac{\lambda_M(H_i)}{\lambda_m^{1/2}(H_i)}, \end{aligned} \qquad i = 1, 2. \tag{1.91}$$

Here λ_m and λ_M are the minimum and maximum eigenvalues of the indicated matrices.

(3) Now we use the Liapunov functions $v_1(x_1)$ and $v_2(x_2)$ as indices of stability for each subsystem $\mathbb{S}_1$ and $\mathbb{S}_2$, and show how one can study the stability of the overall system $\mathbb{S}$ by considering "the interaction among these indices without regard to the interactions within each group" (subsystem).

Let us take the total time derivative $\dot{v}_{i(1.84)}$ along the solutions of the interconnected subsystem $\mathbb{S}_i$ of (1.84),

$$\begin{aligned} \dot{v}_{i(1.84)} &= (\text{grad } v_i)^T \dot{x}_i \\ &= (\text{grad } v_i)^T [A_i x_i + e_{ij}(t) A_{ij} x_j] \\ &= (\text{grad } v_i)^T A_i x_i + (\text{grad } v_i)^T e_{ij}(t) A_{ij} x_j \\ &= \dot{v}_{i(1.71)} + (\text{grad } v_i)^T e_{ij}(t) A_{ij} x_j, \qquad i, j = 1, 2, \end{aligned} \tag{1.92}$$

where by $\dot{v}_{i(1.71)}$ we denote the total time derivative of the function v_i along the solutions of the decoupled subsystem $\mathbb{S}_i$ of (1.71), which is $\dot{v}_i$ given in (1.88) and (1.90).

By taking the norm on the right-hand side of (1.92) and using the estimate $-\eta_{i3}\|x_i\|$ for $\dot{v}_{i(1.71)}$ of (1.90), we get from (1.92),

$$\dot{v}_i \leqslant -\eta_{i3}\|x_i\| + e_{ij}(t)\|\text{grad } v_i\| \, \|A_{ij} x_j\| \qquad i, j = 1, 2. \tag{1.93}$$

Here we have dropped the index (1.84) associated with $\dot{v}_i$ in (1.92), and from now on we remember that $\dot{v}_i$ is taken along the solution of the interconnected subsystem $\mathbb{S}_i$ of (1.84).

If we use the constraint on the interactions as

$$\|A_{ij}x_j\| \leqslant \xi_{ij}\|x_j\|, \qquad i, j = 1, 2, \tag{1.94}$$

where $\xi_{ij} = \lambda_M^{1/2}(A_{ij}^T A_{ij})$, and from the estimates (1.90) express $\|x_i\| \geqslant \eta_{i2}^{-1} v_i$, then we can rewrite the inequalities (1.93) as

$$\begin{aligned} \dot{v}_1 &\leqslant -\eta_{12}^{-1}\eta_{13}v_1 + e_{12}(t)\xi_{12}\eta_{14}\eta_{21}^{-1}v_2, \\ \dot{v}_2 &\leqslant e_{21}(t)\xi_{21}\eta_{24}\eta_{12}^{-1}v_1 - \eta_{22}^{-1}\eta_{23}v_2. \end{aligned} \tag{1.95}$$

Now we define the vector Liapunov function as

$$v = (v_1, v_2)^T \tag{1.96}$$

and rewrite the scalar inequalities (1.95) as one vector inequality

$$\dot{v} \leqslant Wv, \tag{1.97}$$

which involves only the vector Liapunov function v and represents the aggregate model $\mathcal{A}$ for the overall system $\mathbb{S}$ defined in (1.83). The aggregate 2×2 matrix $W = (w_{ij})$ is defined as

$$w_{ij} = \begin{cases} -\eta_{i2}^{-1}\eta_{i3}, & i = j, \\ e_{ij}(t)\xi_{ij}\eta_{j1}^{-1}\eta_{i4}, & i \neq j, \end{cases} \qquad i,j = 1,2. \tag{1.98}$$

To complete our investigation, we should prove that connective stability of the overall system $\mathbb{S}$ follows from stability of the aggregate model $\mathcal{A}$ given in (1.97). To show this, we will first establish a comparison result for vector differential inequalities and majorize the functions $v(t)$ that satisfy the inequality (1.97) by solutions $r(t)$ of the equation

$$\dot{r} = \overline{W}r \tag{1.99}$$

when $v_0 = r_0$. Here $\overline{W} = (\overline{w}_{ij})$ is the aggregate matrix which corresponds to the fundamental interconnection matrix $\overline{E}$ defined in (1.85),

$$\overline{w}_{ij} = \begin{cases} -\eta_{i2}^{-1}\eta_{i3} & i = j, \\ \overline{e}_{ij}\xi_{ij}\eta_{j1}^{-1}\eta_{i4}, & i \neq j, \end{cases} \qquad i,j = 1,2. \tag{1.100}$$

If the matrix $\overline{W}$ is stable, and for all $E = E(t)$,

$$v(t) \leqslant r(t), \qquad t \geqslant t_0, \tag{1.101}$$

whenever $v_0 = r_0$, then from $v(t) \geqslant 0$, $t \geqslant t_0$, we conclude

$$\lim_{t \to +\infty} v(t) = 0 \qquad \text{for all } E(t), \tag{1.102}$$

and thus the connective stability of $\mathbb{S}$.

Let us first observe from (1.100) that

$$\bar{w}_{ij}(t) \geqslant 0, \qquad i \neq j, \tag{1.103}$$

and $\bar{W}$ is a Metzler matrix. Then we use a simple argument of Beckenbach and Bellman (1965) to show that the solution

$$r(t) = e^{(t-t_0)\bar{W}} r_0 \tag{1.104}$$

is nonnegative for all $t \geqslant t_0$ whenever $r_0 \geqslant 0$, if and only if (1.103) holds. To show the "if" part, we suppose without loss of generality that $r_0 > 0$, and let $r_1(t)$ be the component that is zero for the first time at time t_1. Then, from (1.99) at time $t = t_1$, we have

$$\dot{r}_1 = \bar{w}_{12} r_2 \geqslant 0, \tag{1.105}$$

which contradicts the fact that $r_1(t)$ must be decreasing as t tends to t_1. To establish the "only if" part, we notice that $\bar{w}_{12} < 0$ in (1.105) would force $r_1(t)$ to go negative for $t > t_1$. This argument also proves that a necessary and sufficient condition for the matrix exponential $e^{t\bar{W}}$ to be nonnegative for all $t \geqslant 0$ is $\bar{w}_{ij} \geqslant 0$, $i \neq j$.

To establish the comparison result (1.101), we notice that from the matrix inequality

$$E(t) \leqslant \bar{E}, \qquad t \geqslant t_0, \tag{1.106}$$

taken element by element, and from (1.98) and (1.100), it follows that

$$W(t) \leqslant \bar{W}, \qquad t \geqslant t_0. \tag{1.107}$$

From the nonnegativity of $v(t)$ and (1.107), it further follows that

$$Wv \leqslant \bar{W}v, \qquad t \geqslant t_0, \tag{1.108}$$

and we conclude that

$$\dot{v} \leqslant \bar{W}v \tag{1.109}$$

implies the inequality (1.97). That is, any function $v(t)$ which satisfies (1.97) also satisfies (1.109).

We show now a "vector version" of the comparison principle outlined in Section 1.5:

Let $v(t)$ be any function that satisfies the differential inequality

$$\dot{v} \leqslant \overline{W}v, \tag{1.109}$$

for $t \geqslant t_0$, and let $r(t)$ be the solution of the differential equation

$$\dot{r} = \overline{W}r, \tag{1.99}$$

for $r_0 = v_0$. Then a necessary and sufficient condition for the inequality

$$v(t) \leqslant r(t), \qquad t \geqslant t_0 \tag{1.101}$$

to hold is that

$$\overline{w}_{ij} \geqslant 0, \qquad i \neq j. \tag{1.103}$$

To establish this result, we use the simple argument given by Beckenbach and Bellman (1965). We recall that the solution of the inhomogeneous differential equation

$$\dot{v} = \overline{W}v + u(t) \tag{1.110}$$

is

$$v(t) = e^{(t-t_0)\overline{W}}v_0 + \int_{t_0}^{t} e^{(t-\tau)\overline{W}}u(\tau)\,d\tau, \tag{1.111}$$

where $v_0 = v(t_0)$. By taking $u(t) \leqslant 0$, $t \geqslant t_0$, we see that the nonnegativity of $e^{t\overline{W}}$, established above, proves the comparison principle and its consequence (1.101).

Finally, we conclude from the inequality (1.101) and the choice of nonnegative functions $v_1(x_1)$, $v_2(x_2)$, which satisfy the aggregate model $\mathcal{A}$ of (1.97), that stability of the constant matrix $\overline{W}$ is sufficient for the limit (1.102) to hold, and thus is sufficient for connective stability of the overall system $\mathbb{S}$. In other words, we can summarize the above results of the decomposition-aggregation analysis by vector Liapunov functions as follows:

Given:

(1) *stability of each decoupled subsystem $\mathbb{S}_1$ and $\mathbb{S}_2$ established by the estimates* (1.90) *obtained for the functions v_1 and v_2,*

(2) *the constraints* (1.94) *on the interactions* $A_{12}x_2$ *and* $A_{21}x_1$ *between the subsystems* $\mathbb{S}_1$ *and* $\mathbb{S}_2$, *and*

(3) *stability of the constant aggregate matrix* $\overline{W}$ *corresponding to the fundamental interconnection matrix* $\overline{E}$.

Then the system $\mathbb{S}$ *is stable for all interconnection matrices* $E(t)$, *that is, it is connectively stable.*

In the next chapter, we will prove that under the above conditions we have more than connective stability, and that $\mathbb{S}$ is exponentially stable under structural perturbations. We show now that this property of connective stability is implied by the diagonal dominance conditions (1.52),

$$|\overline{w}_{jj}| - \overline{w}_{ij} \geqslant \pi, \qquad i, j = 1, 2, \quad i \neq j, \tag{1.112}$$

which are only sufficient for stability of $\overline{W}$.

Let us consider the function

$$\nu(x) = v_1(x_1) + v_2(x_2) \tag{1.113}$$

as a single "second-level" Liapunov function for the overall system $\mathbb{S}$. Taking the total time derivative $\dot{\nu}(x)$ of the function $\nu(x)$ along the solutions of the equations (1.84), we get from (1.109)

$$\begin{aligned}
\dot{\nu} &= \dot{v}_1 + \dot{v}_2 \\
&\leqslant (w_{11}v_1 + w_{12}v_2) + (w_{21}v_1 + w_{22}v_2) \\
&\leqslant (\overline{w}_{11}v_1 + \overline{w}_{12}v_2) + (\overline{w}_{21}v_1 + \overline{w}_{22}v_2) \\
&\leqslant -(|\overline{w}_{11}| - \overline{w}_{21})v_1 - (|\overline{w}_{22}| - \overline{w}_{12})v_2 \\
&\leqslant -\pi(v_1 + v_2) \\
&\leqslant -\pi\nu \qquad \text{for all } E(t),
\end{aligned} \tag{1.114}$$

which is valid for all $t \geqslant t_0$. As in the case of the inequality (1.30), we obtain from (1.114)

$$\nu(t) \leqslant \nu_0 \exp[-\pi(t - t_0)], \qquad t \geqslant t_0, \tag{1.115}$$

where $\nu(t_0) = \nu_0 = v_{10} + v_{20}$. Then, using the estimates (1.90) for the functions $v_1(x_1)$ and $v_2(x_2)$, as well as the definition (1.113) of the function $\nu(x)$, we get from (1.115) the following inequalities:

$$\begin{aligned}
\nu &\geqslant \eta_{11}\|x_1\| + \eta_{21}\|x_2\|, \\
\nu_0 &\leqslant \eta_{12}\|x_{10}\| + \eta_{22}\|x_{20}\|,
\end{aligned} \tag{1.116}$$

where $x_{10} = x_1(t_0)$, $x_{20} = x_2(t_0)$. Using the simple relationship among the norms

$$\|x\| \leqslant \|x_1\| + \|x_2\| \leqslant 2^{1/2}\|x\|, \tag{1.117}$$

and denoting

$$\eta_{m1} = \min_i \eta_{i1}, \quad \eta_{M2} = \max_i \eta_{i2}, \qquad i = 1, 2, \tag{1.118}$$

we get from (1.116)

$$\nu \geqslant \eta_{m1}\|x\|, \qquad \nu_0 \leqslant 2^{1/2}\eta_{M2}\|x_0\|, \tag{1.119}$$

where $x_0 = x(t_0)$. Finally, by applying the inequalities (1.119) to the inequality (1.115), we get

$$\|x(t; t_0, x_0)\| \leqslant \Pi\|x_0\|\exp[-\pi(t - t_0)], \qquad t \geqslant t_0, \tag{1.120}$$

where $x(t; t_0, x_0)$ is the solution of the system $\mathbb{S}$ of (1.84) for the initial conditions (t_0, x_0). The inequality (1.120) is valid for all x_0 and all interconnection matrices $E(t)$ obtained from the fundamental interconnection matrix $\bar{E}$ specified in (1.85). Therefore, the system $\mathbb{S}$ is globally and exponentially connectively stable.

It is now trivial to extend the above arguments and establish the comparison principle and the vector Liapunov function for s ($s \leqslant n$) subsystems. Moreover, these concepts can be generalized a great deal further than the above development suggests. An advanced exposition of the concepts will be given in the following chapter, which is based upon the work of Lakshmikantham and Leela (1969) and the connectivity aspect of these concepts proposed by Šiljak (1972a, b).

Let us now apply the decomposition-aggregation method to determine the stability of the mechanical system shown in Figure 1.6. We start with the equations for the interconnected subsystems (1.78) and assume the parameter values $\alpha = \beta = 1$. Then, from (1.78), we have

$$\begin{aligned}\dot{x}_1 &= \begin{bmatrix} 0 & 1 \\ -1 & -1 \end{bmatrix} x_1 + \begin{bmatrix} 0 & 0 \\ \omega^2 & 0 \end{bmatrix} x_1 + \begin{bmatrix} 0 & 0 \\ 0 & 2\omega \end{bmatrix} x_2, \\ \dot{x}_2 &= \begin{bmatrix} 0 & 1 \\ -1 & -1 \end{bmatrix} x_2 + \begin{bmatrix} 0 & 0 \\ 0 & -2\omega \end{bmatrix} x_1 + \begin{bmatrix} 0 & 0 \\ \omega^2 & 0 \end{bmatrix} x_2.\end{aligned} \tag{1.121}$$

We are interested in estimating the region of stability for the interconnecting parameter ω.

Let us follow our plan above step by step:

(1) We choose the functions $v_1(x_1)$, $v_2(x_2)$ as in (1.86) and solve the Liapunov matrix equations (1.87) for A_i $(i = 1, 2)$ given in (1.121) and

$$G_i = \begin{bmatrix} 1 & 0 \\ 0 & 1 \end{bmatrix}, \qquad i = 1, 2, \tag{1.122}$$

to get

$$H_i = \begin{bmatrix} 1.5 & 0.5 \\ 0.5 & 1 \end{bmatrix}, \qquad i = 1, 2, \tag{1.123}$$

and the estimates (1.90) as

$$\eta_{i1} = 0.83, \quad \eta_{i2} = 1.35, \quad \eta_{i3} = 0.37, \quad \eta_{i4} = 2.18, \qquad i = 1, 2. \tag{1.124}$$

(2) The constraints (1.94) are calculated for $A_{ij} (i, j = 1, 2)$ of (1.121) to get

$$\xi_{11} = \xi_{22} = \omega^2, \qquad \xi_{12} = \xi_{21} = 2\omega. \tag{1.125}$$

(3) By using the numbers given in (1.124) and (1.125) and the definition for the element $\bar{w}_{ij}$ of the matrix $\bar{W}$,

$$\bar{w}_{ij} = \begin{cases} -\eta_{i2}^{-1}\eta_{i3} + \bar{e}_{ii}\xi_{ii}\eta_{i1}^{-1}\eta_{i4}, & i = j, \\ \bar{e}_{ij}\xi_{ij}\eta_{j1}^{-1}\eta_{i4}, & i \neq j, \end{cases} \qquad i, j = 1, 2, \tag{1.126}$$

—which is a simple modification of (1.100), since now

$$\bar{E} = \begin{bmatrix} 1 & 1 \\ 1 & 1 \end{bmatrix} \tag{1.127}$$

—we get the aggregate model $\mathscr{A}$ of (1.109) as

$$\dot{v} \leqslant \begin{bmatrix} -0.276 + 2.64\omega^2 & 5.28\omega \\ 5.28\omega & -0.276 + 2.64\omega^2 \end{bmatrix} v. \tag{1.128}$$

Instead of the diagonal-dominance conditions (1.112), we use the sharper Hicks conditions (1.55) to determine stability of the aggregate matrix $\bar{W}$ in (1.128). Thus,

$$-0.276 + 2.64\omega^2 < 0,$$

$$\begin{vmatrix} -0.276 + 2.64\omega^2 & 5.28\omega \\ 5.28\omega & -0.276 + 2.64\omega^2 \end{vmatrix} > 0, \tag{1.129}$$

and we have the stability region for the parameter ω,

$$0 \leqslant \omega \leqslant 0.05. \tag{1.130}$$

The above example illustrates the parametric aspect of the vector Liapunov function: The effect on stability of changing parameters in the interactions can be studied explicitly. This aspect was exploited in the stability analysis of the Space Orbiting Laboratory (Skylab), which is outlined in Chapter 6.

The connectivity property in the case of the above example can be established in terms of time-varying angular velocity. In (1.126), we can replace the binary elements $\bar{e}_{ij}$ of the constant matrix $\bar{E}$ by the elements $e_{ij}(t)$ of the interconnection matrices $E(t)$, and include time-varying $\omega(t)$ as $e_{11}(t) = e_{22}(t) = 400\omega^2(t)$, $e_{12}(t) = e_{21}(t) = 10\omega(t)$. Then we conclude that the overall system $\mathbb{S}$ is stable for the range (1.130) of the parameter $\omega(t)$, that is, for $0 \leqslant \omega(t) \leqslant 0.05$.

1.9. DECENTRALIZATION

When we introduced the system

$$\dot{x} = Ax + b \tag{1.1}$$

in Section 1.1, we said that the vector b may represent an external influence of the environment on the system state $x(t)$. Such an influence may be at our disposal, so that we can actually control the motion $x(t)$ of the system by choosing the vector b. To incorporate such possibilities, we consider another version of (1.1),

$$\dot{x} = Ax + Bu, \tag{1.131}$$

where $x(t)$ is the state n-vector of the system, $u(t)$ is the value at time t of the m-vector

$$u = (u_1, u_2, \ldots, u_m)^T, \tag{1.132}$$

and the constant matrices A and B have appropriate dimensions. The vector $u(t)$ is called the *input* to the system described by (1.131), and represents the forcing function which "forces" the system to move in a certain way. That is, given the initial conditions (t_0, x_0), $u(t)$ determines the motion $x(t; t_0, x_0)$ of the system. If we cannot participate in the choice of the forcing function $u(t)$, then it is usually termed a *perturbation function.* When, however, the choice of $u(t)$ is at our disposal, then it represents a *control function.* It is a central problem of control theory (see, for example, Kalman, Falb and Arbib, 1969) to select a suitable control function $u(t)$ which forces the system to behave according to preassigned characteristics such as stability, optimality, insensitivity, etc. A fundamental principle for solving this problem is to choose the control $u(t)$ as a function of the state $x(t)$, thus producing a state-variable feedback control.

How to choose a suitable feedback control function for $u(t)$ will be shown in the next section and in Chapter 3. In this section, however, we are interested in examining the structure of the system (1.131) with regard to the control vector u. Since our objective is to consider large systems and thus rely on the decomposition-aggregation approach, we are immediately interested in the effects on control structure of decomposing the system (1.131). For that purpose, let us assume that $\mathbb{S}$ described by (1.131) is specialized as

$$\dot{x} = \left[\begin{array}{c|c} 2 & -10 \\ \hline 0 & 9 \end{array}\right] x + \left[\begin{array}{cc} 2 & -1 \\ \hline -1 & 1 \end{array}\right] u, \tag{1.133}$$

where $x = (x_1, x_2)^T$ is the state of $\mathbb{S}$ and $u = (u_1, u_2)^T$ is its input. The system $\mathbb{S}$ can be decomposed into two interconnected subsystems as indicated in (1.133) by dashed lines, to get

$$\begin{aligned} \dot{x}_1 &= 2x_1 - 10x_2 + [2 \quad -1]u, \\ \dot{x}_2 &= 9x_2 + [-1 \quad 1]u, \end{aligned} \tag{1.134}$$

where x_1 and x_2 are the states of the subsystems $\mathbb{S}_1$ and $\mathbb{S}_2$, and u is the common input. With the interconnected subsystems (1.134) we can associate a digraph shown in Figure 1.9. The digraph has three points: two for the states of the subsystems x_1, x_2, and one for the common input u. In this case, when subsystems have a common input, we call the interconnected system (1.134) an *input-centralized system* (Šiljak and Vukčević, 1976c).

Let us now identify the matrices A and B in (1.133) as

$$A = \begin{bmatrix} 2 & -10 \\ 0 & 9 \end{bmatrix}, \qquad B = \begin{bmatrix} 2 & -1 \\ -1 & 1 \end{bmatrix}, \tag{1.135}$$

and let us apply the linear transformation

$$x = T\tilde{x}, \tag{1.136}$$

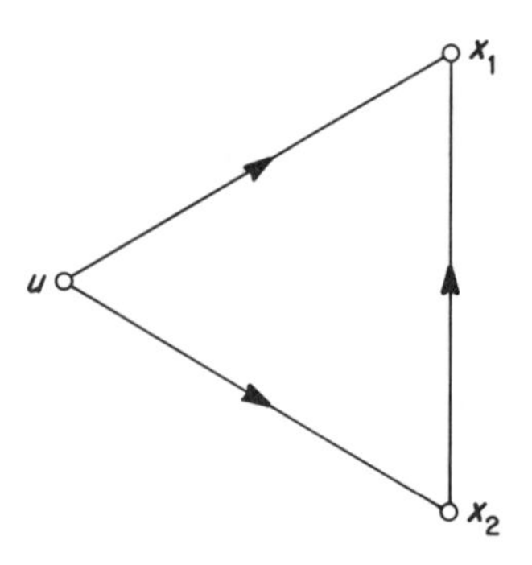

FIGURE 1.9. Input-centralized system.

where T is a nonsingular matrix

$$T = \begin{bmatrix} 1 & 1 \\ 1 & 2 \end{bmatrix}. \tag{1.137}$$

The transformed system $\tilde{\mathbb{S}}$,

$$\dot{\tilde{x}} = \tilde{A}\tilde{x} + \tilde{B}u, \tag{1.138}$$

defined by

$$\tilde{A} = T^{-1}AT, \qquad \tilde{B} = T^{-1}B, \tag{1.139}$$

is obtained as

$$\dot{\tilde{x}} = \left[\begin{array}{c|c} 5 & -3 \\ \hline -4 & 6 \end{array}\right]\tilde{x} + \left[\begin{array}{cc} 1 & 0 \\ \hline 0 & 1 \end{array}\right]u. \tag{1.140}$$

If we decompose the transformed system $\tilde{\mathbb{S}}$ and bear in mind that $u = (u_1, u_2)^T$, we can write the two interconnected subsystems $\tilde{\mathbb{S}}_1$ and $\tilde{\mathbb{S}}_2$ as

$$\begin{aligned} \dot{\tilde{x}}_1 &= 5\tilde{x}_1 - 3\tilde{x}_2 + u_1, \\ \dot{\tilde{x}}_2 &= -4\tilde{x}_1 + 6\tilde{x}_2 + u_2. \end{aligned} \tag{1.141}$$

We see that now each subsystem $\tilde{\mathbb{S}}_1$, $\tilde{\mathbb{S}}_2$ has its own distinct or *local input*. When in an interconnected system all subsystems have distinct (and only distinct) inputs, we say that it is an *input-decentralized system* (Šiljak and Vukčević, 1976c).

With the system (1.141) we can associate a digraph shown in Figure 1.10. We see that the digraph has again two points representing the states x_1, x_2

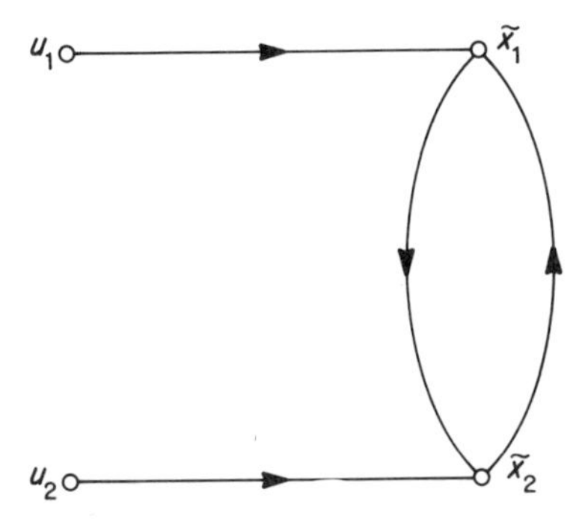

FIGURE 1.10. Input-decentralized system.

of the subsystems, which we may call the *state points*. We also have two points for the two distinct inputs u_1, u_2, which we call the *input points*. In both centralized and decentralized systems, input points are always *transmitters*, that is, there are no lines to the input points. If a digraph represents an input-decentralized system, the number of input points is equal to the number of the subsystems and every input point is a transmitter with one and only one line. That is, the input points of an input-decentralized system have *outdegree* equal to one and *indegree* equal to zero.

The equations (1.134) of the input-centralized system can be rewritten as

$$\begin{aligned} \dot{x}_1 &= 2x_1 - 10x_2 + 2u_1 - u_2, \\ \dot{x}_2 &= 9x_2 - u_1 + u_2, \end{aligned} \tag{1.142}$$

by which we indicate the fact that both components u_1, u_2 of u are inputs of each subsystem $\mathbb{S}_1$, $\mathbb{S}_2$. Then the digraph of Figure 1.9 is given in the form of Figure 1.11. This new form of the digraph does not justify in an obvious way the "centralized" structure of system (1.133), but offers a possibility of examining in detail how various components of the input u effect the states of the subsystems.

Given the digraph on Figure 1.9, we can associate with it a pair of interconnection matrices $(\bar{E}, \bar{L})$. The matrix $\bar{E}$ is the familiar fundamental interconnection matrix defined in (1.74) to describe the internal structure of a large-scale system. The matrix $\bar{L}$ is a more general version of the interconnection vector $\bar{l}$ defined at the end of Section 1.2, which is introduced to describe the external structure from the side of the input. The elements $\bar{l}_{ip}$ of the *fundamental input-connection matrix* $\bar{L}$ are defined as

$$\bar{l}_{ip} = \begin{cases} 1, & u_p \text{ can act on } \mathbb{S}_i, \\ 0, & u_p \text{ cannot act on } \mathbb{S}_i. \end{cases} \tag{1.143}$$

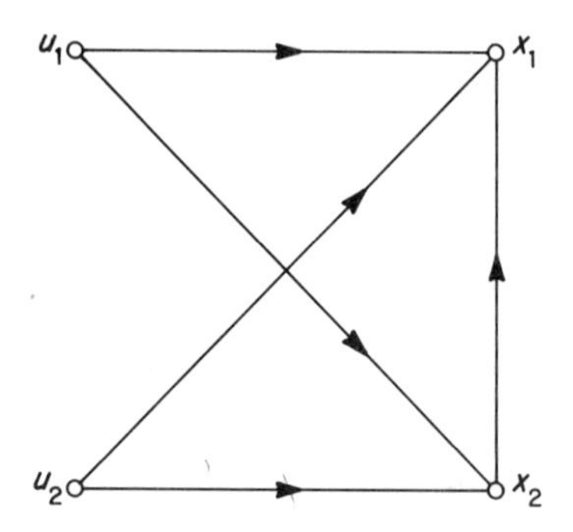

FIGURE 1.11. Input-centralized system.

In other words, $\bar{l}_{ip} = 1$ if there is a line $u_p x_i$ from the input point up to the state point x_i, and $\bar{l}_{ip} = 0$ if there is no such line. If the structural perturbations can take place at the inputs, we can use the input-interconnection matrix $L = (l_{ip})$, which is obtained from $\bar{L} = (\bar{l}_{ip})$ by replacing the unit elements $\bar{l}_{ip}$ with l_{ip} such that $0 \leqslant l_{ip} \leqslant 1$. If the perturbations are considered during the motion of the system, the matrix L is time-varying and $0 \leqslant l_{ip}(t) \leqslant 1$ for all t.

It is now easy to see that for the input-decentralized system (1.141), the matrix $\bar{L}$ is the identity matrix, as shown explicitly in (1.140). Generalization of the interconnection matrices to systems composed of more than two subsystems is straightforward. It is not at all clear, however, how one can choose the transformation matrix T of (1.137) to get a general system in an input-decentralized form. Such generalizations of the described concepts will be outlined in Chapter 3 on multilevel control and estimation.

It is a well-known fact that in general, not all the variables of a physical system can be observed or measured directly. The state of a system is an abstract notion, and only the outputs of the system are available externally. And yet we often need the states to construct the feedback control. Therefore, we resort to building state estimators. When a system is of high dimension, building a single estimator for the entire system may become uneconomical or even impossible, requiring excessive computer time and storage. In Chapter 3, we will show how a number of low-order estimators can be built to replace a single large estimator. For this task to be possible, we require the large system to be in an output-decentralized form (Šiljak and Vukčević, 1976c). That is, each subsystem has one and only one distinct output, which is the "dual" situation to that of an input-decentralized system.

To illustrate the output-decentralization scheme, let us consider a simple second-order system $\mathbb{S}$ described by the equations

$$\dot{x} = \begin{bmatrix} 5 & | & -4 \\ -- & | & -- \\ -3 & | & 6 \end{bmatrix} x + \begin{bmatrix} 1 & 0 \\ -- & -- \\ 0 & 1 \end{bmatrix} u, \tag{1.144}$$
$$y = \begin{bmatrix} 1 & 1 \\ -- & -- \\ 1 & 2 \end{bmatrix} x,$$

where $x(t)$ is the state of the system $\mathbb{S}$, $u(t)$ is its input, and $y(t)$ is its output. By means of the dashed lines in (1.144), we can decompose $\mathbb{S}$ into two interconnected subsystems $\mathbb{S}_1$ and $\mathbb{S}_2$ described by

$$\begin{aligned} \dot{x}_1 &= 5x_1 - 4x_2 + u_1, & \dot{x}_2 &= -3x_1 + 6x_2 + u_2, \\ y_1 &= [1 \quad 1]x, & y_2 &= [1 \quad 2]x, \end{aligned} \tag{1.145}$$

where x_1, x_2 are the states of the subsystems $\mathbb{S}_1$, $\mathbb{S}_2$, and u_1, u_2 are the distinct inputs to the two subsystems. The digraph for the interconnected system $\mathbb{S}$ is shown in Figure 1.12, which corresponds to the obvious modification of the equations (1.145),

$$\begin{aligned} \dot{x}_1 &= 5x_1 - 4x_2 + u_1, & \dot{x}_2 &= 3x_1 + 6x_2 + u_2, \\ y_1 &= x_1 + x_2, & y_2 &= x_1 + 2x_2. \end{aligned} \tag{1.146}$$

If we now use the transformation (1.36) with another matrix

$$T = \begin{bmatrix} 2 & -1 \\ -1 & 1 \end{bmatrix}, \tag{1.147}$$

we get the transformed system $\tilde{\mathbb{S}}$ as

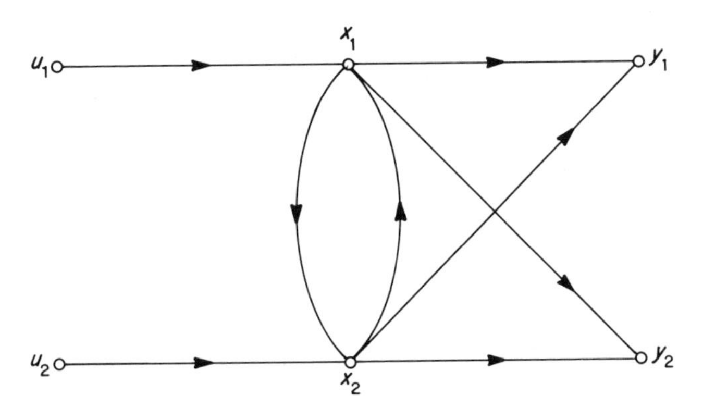

FIGURE 1.12 Output-centralized system.

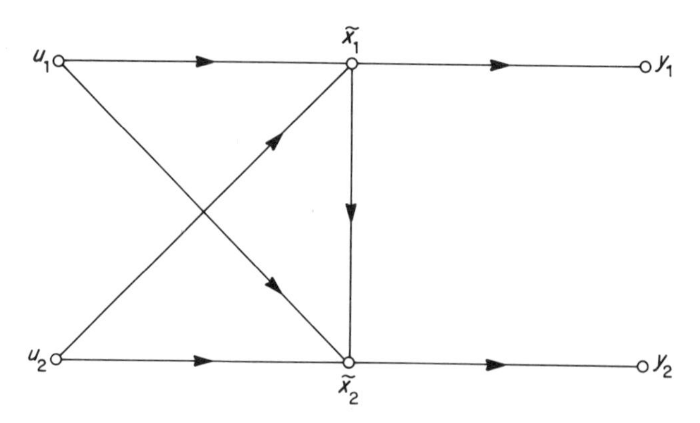

FIGURE 1.13. Output-decentralized system.

$$\dot{\tilde{x}} = \begin{bmatrix} 2 & 0 \\ -10 & 9 \end{bmatrix} \tilde{x} + \begin{bmatrix} 1 & 1 \\ 1 & 2 \end{bmatrix} u,$$
$$y = \begin{bmatrix} 1 & 0 \\ 0 & 1 \end{bmatrix} \tilde{x}. \tag{1.148}$$

If we decompose $\tilde{\mathfrak{S}}$ and remember that $u = (u_1, u_2)^T$, $y = (y_1, y_2)^T$, we can write the two interconnected subsystems $\tilde{\mathfrak{S}}_1$ and $\tilde{\mathfrak{S}}_2$ as

$$\begin{aligned} \dot{\tilde{x}}_1 &= 2\tilde{x}_1 + u_1 + u_2, & \dot{\tilde{x}}_2 &= -10\tilde{x}_1 + 9\tilde{x}_2 + u_1 + u_2, \\ y_1 &= x_1, & y_2 &= x_2. \end{aligned} \tag{1.149}$$

Each subsystem $\tilde{\mathfrak{S}}_1$, $\tilde{\mathfrak{S}}_2$ has now its own distinct or *local output*. Such an interconnected system $\tilde{\mathfrak{S}}$ we call an *output-decentralized system.* With such a system, we associate the digraph of Figure 1.13. The digraph again has two output points, as in Figure 1.12, and they are again *receivers*, as expected. That is, there are no lines from those points. However, in the digraph of Figure 1.13, each output point has indegree one and outdegree zero.

It is of interest to notice that the output-centralized system of Figure 1.12 is input-decentralized. After the output decentralization, however, the system became input-centralized. In general, it is highly unlikely that after a transformation, we obtain a system that is both input- and output-decentralized. However, many physical systems have that dual property to start with, as will be shown.

1.10. SYNTHESIS

"Roughly, by a complex system I mean one made up of a large number of parts that interact in a nonsimple way", is Simon's (1962) description of a complex system. He goes on further to show by intuitive arguments that the evolution of complex systems is highly reliable if it is carried out as a hierarchic process whereby complex systems are formed by interconnecting stable simple parts (subsystems). One of the ultimate goals of the present exposition is to show rigorously that a complex system, when synthesized of interconnected stable substems, indeed has highly reliable stability properties. In this section, however, we will not consider the general synthesis problem, but rather illustrate by simple examples the intuitive arguments of Simon. The general problem is postponed until Chapter 3, where a multilevel control and estimation scheme is outlined for the synthesis of large dynamically reliable systems.

Let us state again what we mean by a "dynamically reliable large-scale system". We propose to call a *connectively stable* system dynamically reliable, since it is stable under structural perturbations whereby subsystems are disconnected and again connected in various ways during the operation of the system. This is a reasonable definition, since large ecomodels, complex electric power networks, interconnected electronic systems, and many other real systems do not stay "in one piece" for all time, but rather are expected to disintegrate and then reconstitute themselves for various natural and technological reasons. Such changes in system structure may disrupt stability and cause a breakdown of the system. Therefore, to function properly under structural perturbations, complex systems should obviously be connectively stable. This property, however, requires that stable competitive structures be used, since then (and only then) any group of subsystems is a stable formation. That is none of the subsystem groups or subsystems themselves are dependent on other subsystem groups or subsystems for stability—they regard each other as competitors rather than cooperators. This is where a competitive structure has its greatest weakness and, at the same time, its greatest strength. If the subsystems are regarding each other as competitors, they are abandoning the possibility of utilizing their interactions with other subsystems, and thus causing an inferior performance of the overall system. On the other hand, if the beneficial aspects are used explicitly, and a cooperation is formed among the subsystems to improve the performance of the overall system, they may become so dependent on each other that if some of the subsystems quit, the entire system may collapse. Therefore, intuitively we expect the following statement to be true:

Competition *vs.* cooperation *means* (*among other things*) reliability *vs.* optimality.

Let us now use a simple example to show how this statement can be interpreted in the context of dynamic systems. We would like to synthesize the system $\mathbb{S}$ composed of two agents,

$$\begin{aligned} \dot{x}_1 &= 5x_1 - 3x_2 + u_1, \\ \dot{x}_2 &= -4x_1 + 6x_2 + u_2, \end{aligned} \tag{1.150}$$

by choosing the control functions u_1, u_2 so as to minimize the quadratic performance index

$$J = \int_0^\infty [(x_1^2 + x_2^2) + (u_1^2 + u_2^2)]\, dt. \tag{1.151}$$

That is, J is a measure of the cost involved in operation of the system (1.150), and we would like to come up with the "best system" with respect to that measure. The digraph of the system (1.150) is shown on Figure 1.14.

By using the standard regulator theory (e.g. Anderson and Moore, 1971), the optimal control

$$\begin{aligned} u_1^\oplus &= -10.49x_1 + 7.14x_2, \\ u_2^\oplus &= 7.14x_1 - 11.18x_2, \end{aligned} \tag{1.152}$$

is easily found, which minimizes the index J and produces the "best closed-loop system"

$$\begin{aligned} \dot{x}_1 &= -5.49x_1 + 4.14x_2, \\ \dot{x}_2 &= 3.14x_1 - 5.18x_2, \end{aligned} \tag{1.153}$$

the digraph of which is shown in Figure 1.15(a). It is a well-known fact that the optimal control (1.152) produces at the same time a stable system (1.153).

Now, let us take a different attitude towards the system (1.150) and consider it explicitly as an interconnected structure of two subsystems $\mathbb{S}_1$, $\mathbb{S}_2$ governed by the equations

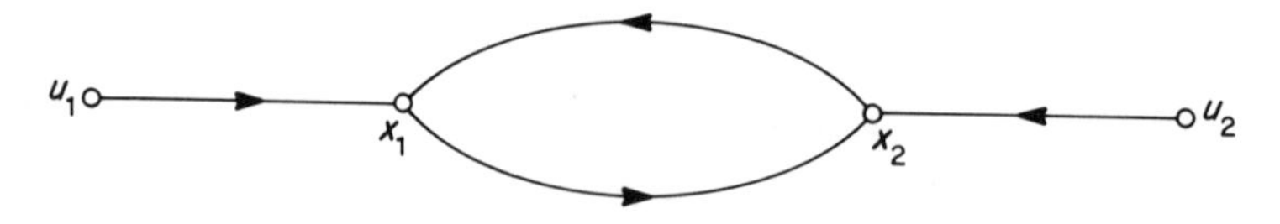

FIGURE 1.14. Open-loop system.

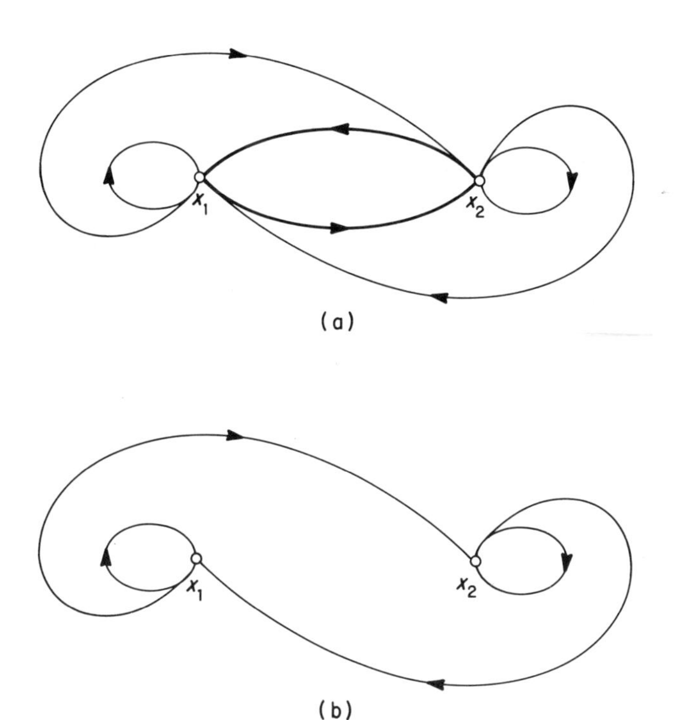

FIGURE 1.15. "One shot" optimization: (a) Optimal and stable. (b) Unstable.

$$
\begin{aligned}
\dot{x}_1 &= 5x_1 + u_1, \\
\dot{x}_2 &= 6x_2 + u_2,
\end{aligned}
\tag{1.154}
$$

where $-3x_2$ and $-4x_1$ are the interactions between $\mathbb{S}_1$ and $\mathbb{S}_2$. If during the operation of the overall system (1.150) disconnection of the subsystems (1.154) is an expected possibility, then the application of the computed optimal control is not appropriate. When the interconnections are removed, the use of the control (1.152) for the disjoint subsystems (1.154) results in a closed-loop system

$$
\begin{aligned}
\dot{x}_1 &= -5.49x_1 + 7.14x_2, \\
\dot{x}_2 &= 7.14x_1 - 5.18x_2,
\end{aligned}
\tag{1.155}
$$

which is unstable, and we have a system breakdown. The digraph representing this structurally perturbed system is shown in Figure 1.15(b).

Therefore, the standard "one shot" optimization will not provide a satisfactory solution for the synthesis of large-scale systems which are expected to undergo structural perturbations. In this case, we propose an alternative approach, which produces a *dynamically reliable overall system* that is expected to be stable under arbitrary structural perturbations. The price of achieving reliability is that the resulting system is suboptimal with respect to the specified performance index. The suboptimality index is a direct measure of the cost introduced by the trade-off between dynamic reliability and optimality of the overall system (Šiljak and Sundareshan, 1976a, b).

To incorporate the structural perturbations, we rewrite the original equations (1.150) for the interconnected subsystems $\mathcal{S}_1$ and $\mathcal{S}_2$ as

$$\begin{aligned} \dot{x}_1 &= 5x_1 - 3e_{12}x_2 + u_1, \\ \dot{x}_2 &= -4e_{21}x_1 + 6x_2 + u_2, \end{aligned} \tag{1.156}$$

where e_{12} and e_{21} are the elements of the interconnection matrix

$$E = \begin{bmatrix} 0 & e_{12} \\ e_{21} & 0 \end{bmatrix}, \tag{1.157}$$

which corresponds to the fundamental interconnection matrix

$$\bar{E} = \begin{bmatrix} 0 & 1 \\ 1 & 0 \end{bmatrix}. \tag{1.158}$$

With each of the decoupled subsystems $\mathcal{S}_1$ and $\mathcal{S}_2$ in (1.154) we associate a performance index:

$$J_1 = \int_0^{\infty} (x_1^2 + u_1^2)\, dt, \qquad J_2 = \int_0^{\infty} (x_2^2 + u_2^2)\, dt, \tag{1.159}$$

and we proceed to compute the local optimal control for each decoupled subsystem in (1.154) with respect to the performance indices in (1.159), using the standard linear regulator solution. This gives the local optimal controls as

$$u_1^0 = -10.10x_1, \qquad u_2^0 = -12.05x_2, \tag{1.160}$$

resulting in the optimal costs J_1^0 and J_2^0.

The controls u_1^0 and u_2^0, when substituted in the equations (1.156), produce the closed-loop subsystems

$$\begin{aligned} \dot{x}_1 &= -5.10x_1 - 3x_2, \\ \dot{x}_2 &= -4x_1 - 6.05x_2, \end{aligned} \tag{1.161}$$

which correspond to the fundamental interconnection matrix $\bar{E}$ in (1.158). The digraph of the system is shown in Figure 1.16(a). Since (1.161) satisfies the diagonal-dominance conditions (1.52), the system is connectively stable. Therefore, for the zero matrix

$$E = \begin{bmatrix} 0 & 0 \\ 0 & 0 \end{bmatrix}, \tag{1.162}$$

the two closed-loop decoupled subsystems $\mathbb{S}_1$ and $\mathbb{S}_2$, shown in Figure 1.16(b), remain stable.

The dynamic reliability of the system $\mathbb{S}$ in (1.161) is achieved at the price of deterioration in the system performance. With controls u_1^0 and u_2^0, the system (1.156) cannot attain the optimal value

$$J^0 = J_1^0 + J_2^0, \tag{1.163}$$

because of the interactions. The number

$$\varepsilon = \frac{\tilde{J} - J^0}{J^0} = 0.85 \tag{1.164}$$

can serve as a suboptimality index for the system under structural perturbations, where $\tilde{J}$ is the actual value of the performance index for the overall system (1.161) which corresponds to the matrix $\bar{E}$ in (1.158). Therefore, ε represents the price we have to pay, in terms of the performance index, to achieve connective stability and, thus, dynamic reliability of the overall system. How a more general type of optimization under structural perturbations can be developed, we will show in Chapter 3.

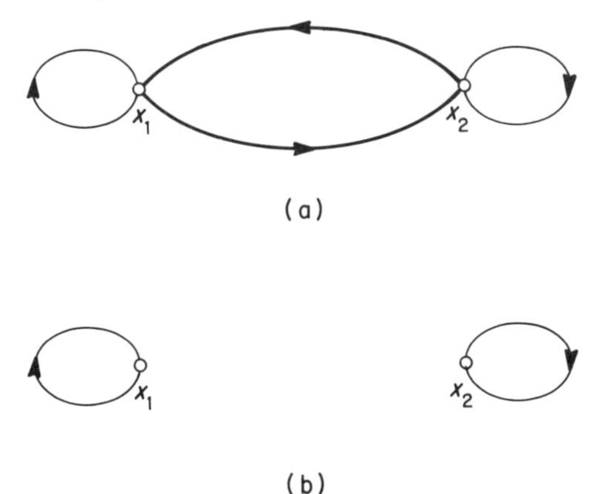

FIGURE 1.16. Decentralized optimization: (a) Suboptimal and stable. (b) Stable.

Let us now take note of an important fact which is implicit in the above example and is related to the stabilization problem of large-scale systems. We can disregard the optimization part of the example, and view the choice of controls $u_1^{\oplus}$, $u_2^{\oplus}$ in (1.152) as a stabilizing control which makes the unstable system (1.150) into a stable system (1.153). This "one shot" stabilization, however, falls short of making the system connectively stable, that is, dynamically reliable under structural perturbations. Alternatively, if we stabilize each decoupled subsystem $\mathbb{S}_1$ and $\mathbb{S}_2$ in (1.154) by local controls (1.160), and form a system with a stable competitive structure (1.161), then the overall system $\mathbb{S}$ is connectively stable and thus stable under arbitrary structural perturbations (including, of course, $E = 0$).

Going back to Section 1.5, and recalling the inequalities (1.53), it is not difficult to conclude that a locally stabilized system is ***robust***: It has a high tolerance to nonlinearities in the interactions among the two subsystems $\mathbb{S}_1$ and $\mathbb{S}_2$. That this is true in general was shown by Weissenberger (1974), and we are going to enlarge on that issue in Chapter 3. At present, however, for the sake of completeness, we conclude that the chosen control

$$u_1^0 = -10.10x_1, \qquad u_2^0 = -12.05x_2 \tag{1.160}$$

stabilizes the nonlinear system

$$\begin{aligned} \dot{x}_1 &= 5x_1 - e_{12}h_{12}(x_2) + u_1, \\ \dot{x}_2 &= 6x_2 - e_{21}h_{21}(x_1) + u_2, \end{aligned} \tag{1.165}$$

provided the nonlinear functions $h_{12}(x_2)$, $h_{21}(x_1)$ satisfy the "conical" constraints

$$|h_{12}(x_2)| \leqslant 3|x_2|, \qquad |h_{21}(x_1)| \leqslant 4|x_1|. \tag{1.166}$$

Again, we do not need to know the actual shape of the nonlinear interactions except that they are bounded as in (1.166). A graphical interpretation of the conditions (1.166) is shown in Figure 1.17 for two hypothetical functions $h_{12}(x_2)$ and $h_{21}(x_1)$.

Furthermore, the local controls (1.160) produce a closed-loop system

$$\begin{aligned} \dot{x}_1 &= -5.10x_1 - e_{12}h_{12}(x_2), \\ \dot{x}_2 &= -6.05x_2 - e_{21}h_{21}(x_1), \end{aligned} \tag{1.167}$$

which is connectively stable if $h_{12}(x_2)$, $h_{21}(x_1)$ satisfy the constaints (1.166).

The above example suggests that in the synthesis of robust systems, it is better to use local control associated with each subsystem. That this conclusion is true will be clear in Chapter 3 when we consider the general

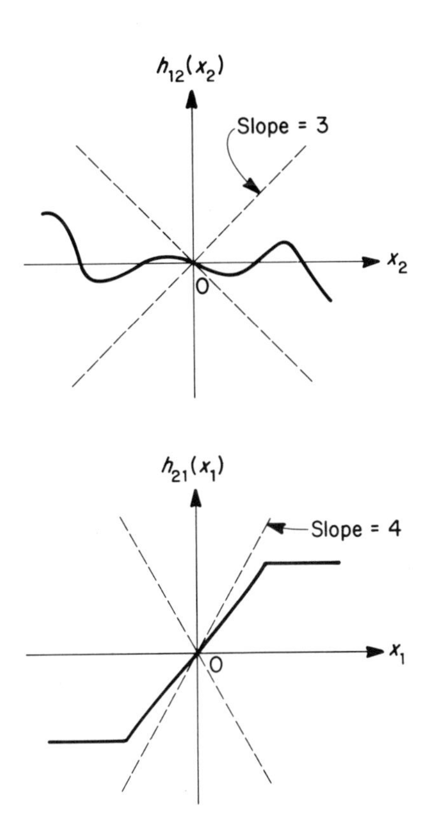

FIGURE 1.17. Nonlinear interactions.

problem of stabilization and optimization by local controls only. In this context, the input-decentralization scheme of the preceeding section becomes a highly useful design tool.

Let us now go a step further and propose *multilevel control* of large-scale systems, using the same example. We recall the fact that in competitive structures, each subsystem is stable if left alone, and views the interactions with other subsystems as undesirable perturbations. Therefore, it is logical to use a part of the control locally to stabilize each subsystem, and another part to neutralize the interactions.

We consider the control functions in (1.156) as consisting of two components:

$$u_1 = u_1^l + u_1^g, \qquad u_2 = u_2^l + u_2^g, \tag{1.168}$$

where u_1^l, u_2^l are *local controls* and u_1^g, u_2^g are *global controls*.

Let us now consider the system (1.165) when $E = \bar{E}$,

$$\begin{aligned} \dot{x}_1 &= 5x_1 - 3x_2 + u_1, \\ \dot{x}_2 &= -4x_1 + 6x_2 + u_2, \end{aligned} \tag{1.169}$$

and substitute the controls (1.168) into (1.169) to get

$$\begin{aligned} \dot{x}_1 &= 5x_1 - 3x_2 + u_1^l + u_1^g, \\ \dot{x}_2 &= -4x_1 + 6x_2 + u_2^l + u_2^g. \end{aligned} \tag{1.170}$$

By associating the local controls u_1^l, u_2^l with the subsystems $\mathbb{S}_1$ and $\mathbb{S}_2$, and the global controls u_1^g, u_2^g with the interactions, we obtain

$$\begin{aligned} \dot{x}_1 &= (5x_1 + u_1^l) + (-3x_2 + u_1^g), \\ \dot{x}_2 &= (6x_2 + u_2^l) + (-4x_1 + u_2^g). \end{aligned} \tag{1.171}$$

We ignore the interactions $-3x_2 + u_1^g$, $6x_2 + u_2^g$, and choose the local controls to stabilize (or optimize) the decoupled subsystem $\mathbb{S}_1$ and $\mathbb{S}_2$ given by (1.154), that is,

$$u_1^{l0} = u_1^0, \qquad u_2^{l0} = u_2^0, \tag{1.172}$$

where u_1^0 and u_2^0 are given in (1.160). This choice of the local control is optimal for the decoupled subsystems, and interconnections will degrade the optimality by an amount proportional to their size. Therefore, we choose the global control u_1^g, u_2^g to make the effective interactions $-3x_2 + u_1^g$, $6x_2 + u_2^g$ as small as possible. In this simple case, it is possible to choose

$$u_1^{g0} = 3x_2, \qquad u_2^{g0} = 4x_1 \tag{1.173}$$

and completely neutralize the interactions to get the decoupled system shown in Figure 1.16(b). When the subsystems are of order higher than one, the total disconnection should not be expected, as is discussed in Chapter 3.

The control functions in (1.169) are chosen as

$$u_1^0 = -10.10x_1 + 3x_2, \qquad u_2^0 = -12.05x_2 + 4x_1, \tag{1.174}$$

and the closed-loop system $\mathbb{S}$ is

$$\begin{aligned} \dot{x}_1 &= -5.10x_1, \\ \dot{x}_2 &= -6.05x_2. \end{aligned} \tag{1.175}$$

Since the subsystems $\mathbb{S}_1$, $\mathbb{S}_2$ in (1.175) are decoupled, the system $\mathbb{S}$ is optimal with the value J^0, and the suboptimality index ε is zero.

Several remarks are in order. To construct the global control, the states of other subsystems should be available locally at each subsystem. This is not always possible. However, the global control can be partially implemented by those states that are available. That is, if x_2 is not available locally at the input of $\mathcal{S}_1$, we choose the global control as

$$u_1^{g0} = 0, \qquad u_2^{g0} = 4x_1. \tag{1.176}$$

Then the control functions are

$$u_1^0 = -10.10x_1, \qquad u_2^0 = -12.05x_2 + 4x_1, \tag{1.177}$$

and the closed-loop system shown in Figure 1.18 becomes

$$\begin{aligned} \dot{x}_1 &= -5.10x_1 - 3x_2, \\ \dot{x}_2 &= -6.05x_2, \end{aligned} \tag{1.178}$$

which is connectively stable, but suboptimal with respect to the performance index $J = J_1 + J_2$ specified in (1.159), that is, the suboptimality index ε is no longer zero.

When the global control is applied, the connective stability should include both the interactions and the global control itself. That is, for the system (1.156), we choose the controls

$$u_1 = u_1^l + e_{12}u_1^g, \qquad u_2 = u_2^l + e_{21}u_2^g \tag{1.179}$$

so that the system (1.156) becomes

$$\begin{aligned} \dot{x}_1 &= (5x_1 + u_1^l) + e_{12}(-3x_2 + u_1^g), \\ \dot{x}_2 &= (6x_2 + u_2^l) + e_{21}(-4x_1 + u_2^g). \end{aligned} \tag{1.180}$$

Then, in case x_2 is not available at the subsystem $\mathcal{S}_1$, the choice

$$u_1^0 = -10.10x_1, \qquad u_2^0 = -12.05x_2 + e_{21}4x_1 \tag{1.181}$$

produces the closed-loop system

$$\begin{aligned} \dot{x}_1 &= -10.10x_1 - e_{12}3x_2, \\ \dot{x}_2 &= -6.05x_2, \end{aligned} \tag{1.182}$$

FIGURE 1.18. Partial neutralization.

which is connectively suboptimal.

It is also of interest to point out that in stabilization and optimization by multilevel control, we use the states of the subsystem. They may not be all available for measurement, and we need estimators to handle unmeasurable states. The outlined multilevel stabilization scheme can be used to construct low-order asymptotic estimators as shown in Chapter 3. The scheme is attractive in that it reduces significantly the dimensionality of the estimation problems.

REFERENCES

Anderson, B. D. O., and Moore, J. B. (1971), *Linear Optimal Control*, Prentice-Hall, Englewood Cliffs, New Jersey.

Araki, M. (1975), "Application of $\mathcal{M}$-Matrices to the Stability Problems of Composite Nonlinear Systems", *Journal of Mathematical Analysis and Applications*, 52, 309–321.

Arrow, K. J. (1966), "Price-Quantity Adjustments in Multiple Markets with Rising Demands", *Proceedings of a Symposium on Mathematical Methods in the Social Sciences*, K. J. Arrow, S. Karlin, and P. Suppes (eds.), Stanford University Press, Palo Alto, California, 3–15.

Arrow, K. J., and Hahn, F. H. (1971), *General Competitive Analysis*, Holden-Day, San Francisco.

Bailey, F. N. (1966), "The Application of Lyapunov's Second Method to Interconnected Systems", *SIAM Journal of Control*, 3, 443–462.

Barlow, R. E., Fussell, J. B., and Singpurwalla, N. D. (eds.) (1975), *Reliability and Fault Tree Analysis*, SIAM, Philadelphia, Pennsylvania.

Beckenbach, E. F., and Bellman, R. (1965), *Inequalities*, Springer, New York.

Bellman, R. (1962), "Vector Lyapunov Functions", *SIAM Journal of Control*, 1, 32–34.

Bertziss, A. T. (1975), *Data Structures: Theory and Practice*, Academic, New York.

Birkhoff, G., and Rota, G-C. (1962), *Ordinary Differential Equations*, Blaisdell, Waltham, Massachusetts.

Deo, N. (1974), *Graph Theory with Application to Engineering and Computer Science*, Prentice-Hall, Englewood Cliffs, New Jersey.

Duhem, M. P. (1902), "Sur les conditions nécessaires pour la stabilité de l'équilibre d'un systéme visqueus", *Comptes Rendus*, 939–941.

Fiedler, M., and Pták, V. (1962), "On Matrices with Nonpositive Off–Diagonal Elements and Positive Principal Minors", *Czechoslovakian Mathematical Journal*, 12, 382–400.

Green, H. A. J. (1964), *Aggregation in Economic Analysis*, Princeton University Press, Princeton, New Jersey.

Hahn, W. (1967), *Stability of Motion*, Springer, New York.

Harary, F., Norman, R. Z., and Cartwright, D. (1965), *Structural Models: An*

Introduction to the Theory of Directed Graphs, Wiley, New York.
Hicks, J. R. (1939), *Value and Capital*, Oxford University Press, Oxford, England.
Himelblau, D. M. (ed.) (1973), *Decomposition of Large-Scale Problems*, North-Holland, Amsterdam, Holland.
Kalman, R. E., and Bertram, J. E. (1960), "Control System analysis and Design via the 'Second Method' of Lyapunov", *Transactions of ASME*, 82; Part I: 371–393; Part II: 394–400.
Kalman, R. E., Falb, P. L., and Arbib, M. A. (1969), *Topics in Mathematical System Theory*, McGraw-Hill, New York.
Kamke, E. (1932), "Zur Theorie der Systeme gewöhnlicher Differentialgleichungen", II, *Acta Mathematica*, 58, 57–85.
Kron, G. (1963), *Diakoptics*, Macdonald, London.
Ladde, G. S. (1976a), "Cellular Systems–I. Stability of Chemical Systems", *Mathematical Biosciences*, 29, 309–330.
Ladde, G. S. (1976b), "Cellular Systems–II. Stability of Compartmental Systems", *Mathematical Biosciences*, 30, 1–21.
Lakshmikantham, V., and Leela, S. (1969), *Differential and Integral Inequalities*, Vols. I and II, Academic, New York.
LaSalle, J. P., Lefschetz, S. (1961), *Stability by Liapunov's Direct Method with Applications*, Academic, New York.
Lotka, A. J. (1925), *Elements of Physical Biology*, Williams and Wilkins, Baltimore, Maryland. (Reissued as *Elements of Mathematical Biology*, Dover, New York, 1956).
Matrosov, V. M. (1962), "On the Theory of Stability of Motion" (in Russian), *Prikladnaia Matematika i Mekhanika*, 26, 992–1002.
Matrosov, V. M. (1965), "Development of the Liapunov Functions in Stability Theory" (in Russian), *Transactions of the Second Symposium on Theoretical and Applied Mechanics*, Nauka, Moscow, pp. 112–125.
May, R. M. (1973), *Stability and Complexity in Model Ecosystems*, Princeton University Press, Princeton, New Jersey.
McKenzie, L. (1966), "Matrices with Dominant Diagonals and Economic Theory", *Proceedings of a Symposium on Mathematical Methods in the Social Sciences*, K. J. Arrow, S. Karlin, and P. Suppes (eds.), Stanford University Press, Palo Alto, California, 47–62.
Metzler, L. (1945), "Stability of Multiple Markets: The Hicks Conditions", *Econometrica*, 13, 277–292.
Michel, A. N. (1974), "Stability Analysis of Interconnected Systems", *SIAM Journal of Control*, 12, 554–579.
Newman, P. K. (1959), "Some Notes on Stability Conditions", *Review of Economic Studies*, 72, 1–9.
Quirk, J., and Saposnik, R. (1968), *Introduction to General Equilibrium Theory and Welfare Economics*, McGraw-Hill, New York.
Richardson, L. F. (1960), *Arms and Insecurity*, Boxwood, Pittsburgh, Pennsylvania.
Rosen, R. (1970), *Dynamical System Theory in Biology*, Vol. I, Wiley, New York.
Sage, A. P. (1977), *Methodology for Large-Scale Systems*, McGraw-Hill, New York.

Samuelson, P. A. (1947), *Foundations of Economic Analysis*, Harvard University Press, Cambridge, Massachusetts.

Samuelson, P. A. (1974), "Complementarity: An Essay on the 40th Anniversary of the Hicks-Allen Revolution in Demand Theory", *The Journal of Economic Literature*, 12, 1255–1289.

Šiljak, D. D. (1969), *Nonlinear Systems*, Wiley, New York.

Šiljak, D. D. (1972a), "Stability of Large-Scale Systems Under Structural Perturbations", *IEEE Transactions*, SMC-2, 657–663.

Šiljak, D. D. (1972b), "Stability of Large-Scale Systems", *Proceedings of the Fifth IFAC Congress*, Paris, pp. C-32:1–11.

Šiljak, D. D. (1973), "On Stability of Large-Scale Systems Under Structural Perturbations", *IEEE Transactions*, SMC-3, 415–417.

Šiljak, D. D. (1974), "Connective Stability of Complex Ecosystems", *Nature*, 249, 280.

Šiljak, D. D. (1975a), "When is a Complex Ecosystem Stable?", *Mathematical Biosciences*, 25, 25–50.

Šiljak, D. D. (1975b), "Stabilization of Large-Scale Systems: A Spinning Flexible Spacecraft", *Proceedings of the Sixth IFAC Congress*, Boston, Massachusetts, 35–1:1–10. (See also: *Automatica*, 12, 309–320).

Šiljak, D. D. (1975c), "Connective Stability of Competitive Equilibrium", *Automatica*, 11, 389–400.

Šiljak, D. D. (1975d), "On Total Stability of Competitive Equilibrium", *International Journal of Systems Science*, 6, 951–964.

Šiljak, D. D. (1976a), "Competitive Economic Systems: Stability, Decomposition, and Aggregation", *IEEE Transactions*, AC-21, 149–160.

Šiljak, D. D. (1976b), "Competitive Analysis of the Arms Race", *Annals of Economic and Social Measurement*, 5, 283–295.

Šiljak, D. D. (1977a), "On Stability of the Arms Race", *Proceedings of the NSF Conference on System Theory in International Relations Research*, Edited by J. V. Gillespie and D. A. Zinnes (eds.), Praeger, New York, 264–304.

Šiljak, D. D. (1977b), "On Reachability of Dynamic Systems", *International Journal of Control*, 8, 321–338.

Šiljak, D. D. (1977c), "On Pure Structure of Dynamic Systems", *Nonlinear Analysis, Theory, Methods, and Applications*, 1, 397–413.

Šiljak, D. D. (1977d), "Vulnerability of Dynamic Systems", *Proceedings of the IFAC Workshop on Control and Management of Integrated Industrial Complexes*, Toulouse, France, 133–144.

Šiljak, D. D. (1977e), "On Stochastic Stability of Competitive Equilibrium", *Annals of Economic and Social Measurement*, 6, 315–323.

Šiljak, D. D. (1977f), "On Decentralized Control of Large-Scale Systems", *Proceedings of the IUTAM Symposium on Dynamics of Multibody Systems*, Munchen, Germany.

Šiljak, D. D. (1978), "Structure and Stability of Model Ecosystems", *Mathematical Systems Ecology*, E. Halfon (ed.), Academic Press, New York (to appear).

Šiljak, D. D., and Sundareshan, M. K. (1976a), "A Multilevel Optimization of Large-Scale Dynamic Systems", *IEEE Transactions*, AC-21, 79–84.

Šiljak, D. D., and Sundareshan, M. K. (1976b), "Large-Scale Systems: Optimality vs. Reliability", *Proceedings of a Conference on Directions in Decentralized Control, Many-Person Optimization and Large-Scale Systems*, Y. C. Ho and S. K. Mitter (eds.), Plenum Press, New York, 411–425.

Šiljak, D. D., and Vukčević, M. B. (1976a), "Large-Scale Systems: Stability, Complexity, Reliability", *Journal of the Franklin Institute*, 301, 49–69.

Šiljak, D. D., and Vukčević, M. B. (1976b), "Multilevel Control of Large-Scale Systems: Decentralization, Stabilization, Estimation and Reliability", *Large-Scale Dynamical Systems*, R. Saeks (ed.), Point Lobos Press, Los Angeles, California, 33–57.

Šiljak, D. D., and Vukčević, M. B. (1976c), "Decentralization, Stabilization and Estimation of Large-Scale Linear Systems", *IEEE Transactions*, AC-21, 363–366.

Simon, H. A. (1962), "The Architecture of Complexity", *Proceedings of the American Philosophical Society*, 106, 467–482.

Simon, H. A., and Ando, A. (1961), "Aggregation of Variables in Dynamic Systems", *Econometrica*, 29, 111–138.

Steward, D. V. (1962), "On an Approach to Techniques for the Analysis of the Structure of Large Systems of Equations", *SIAM Review*, 4, 321–342.

Steward, D. V. (1965), "Partitioning and Tearing Systems of Equations", *SIAM Journal of Numerical Analysis*, 2, 345–365.

Theil, H. (1954), *Linear Aggregation of Economic Relations*, North-Holland, Amsterdam, Holland.

Volterra, V. (1926), "Variazioni e fluttuazioni del numero d'individui in specie animali conviventi", *Memorie della R. Academia Nazionale del Lincei*, 2, 31–113. (Translation in an appendix to R. N. Capman's *Animal Ecology*, McGraw-Hill, New York, 1931).

Walter, W. (1970), *Differential and Integral Inequalities*, Springer, New York.

Waževski, T. (1950), "Systemes des équations et des inéqualités differentielles ordinaires aux deuxièmes membres monotones et leurs applications", *Annales de la societe Polonaise de mathématique*, 23, 112–166.

Weissenberger, S. (1973), "Stability Regions of Large-Scale Systems", *Automatica*, 9, 653–663.

Weissenberger, S. (1974), "Tolerance of Decentrally Optimal Controllers to Nonlinearity and Coupling", *Proceedings of the Twelfth Annual Allerton Conference on Circuits and System Theory*, University of Illinois, Monticello, Illinois, 87–95.

Ziegler, H. (1968), *Principles of Structural Stability*, Blaisdell, Waltham, Massachusetts.

2

ANALYSIS

Connective Stability

The purpose of this chapter is to provide a mathematical foundation for the concept of the connective stability of dynamic systems. We shall strive for rigor in presentation and shall not discuss the applicability of the concept to the real world. This is postponed for the later chapters, where we consider structural properties of various models in social sciences, ecology, and engineering.

A dynamic system is regarded as a heterogeneous composition of interacting components (subsystems) with interactions which can vary widely in strength and shape. A group of subsystems can be temporarily disassociated from the rest of the subsystems, causing a change in structure of the overall system. In order to maintain proper functioning of the system, we should be able to answer questions of the following type: To what extent is the system performance affected by structural changes? Is it possible to secure stability of the system under preselected class of structural perturbations?

Since classical stability theory is limited to initial-condition, forcing-function, and parameter-perturbation analysis, we begin by defining the structural perturbations of dynamic systems in terms of directed graphs and interconnection matrices. According to the definition, a system is considered connectively stable if it is stable in the sense of Liapunov under all admissable structural perturbations. Such a formulation is mathematically expedient, since it opens up a real possibility for using Liapunov's stability theory in studying the structural properties of dynamic systems. It is also sufficiently general to allow the other kinds of stability—such as exponential, stochastic, practical, etc.—to be considered in connection with connectivity when that is dictated by applications.

After fixing the definition of connective stability in the context of Liapunov stability theory within which we work, we develop the decomposition-aggregation method for connective-stability analysis of dynamic systems, using the modern mathematical machinery of the comparison principle and vector Liapunov functions. We start with some general considerations of vector differential inequalities, in order to set the stage for the application of vector Liapunov functions, the components of which are assigned to individual subsystems. Each scalar Liapunov function aggregates the stability properties of the corresponding subsystem, and when these functions are combined to form a system of differential inequalities, we obtain an aggregate model of the original system. The interactions among the subsystems are presented explicitly, and stability of the aggregate model implies connective stability of the original system.

A very important aspect of the decomposition-aggregation method is in that it promises to solve stability problems "piece by piece", and thereby not only makes the use of computers more economical, but also reduces the likelihood of errors in the analysis. Furthermore, if the system is composed of subsystems that have a distinct physical meaning, the method yields significant structural information about the system behavior, which is not generally available in a straightforward ("one shot") stability investigation. It is important to point out, however, that if the connectivity aspect is not taken into account, then the method yields more conservative results than the standard "one shot" procedures. It is for this reason that we shall be able to establish in many cases not only sufficient, but also necessary and sufficient conditions for connective stability of nonlinear and time-varying dynamic systems.

The level of mathematical sophistication in this chapter is relatively high, and when compared to a quick and painless introduction of the preceding chapter, the derivations seem sometimes long and tedious. The development is gratifying, however, since it leads to practical results and interpretations which are, in the end, quite simple. Some (possibly collateral) reading of the following chapters would be helpful in appreciating the breadth of applications of the results obtained, and may motivate the reader for a detailed study of the exposition.

In the following development, we use the notation that is standard in the theory of differential equations and inequalities (see for example the books of Hale, 1969, and Lakshmikantham and Leela, 1969). By $\mathfrak{R}^n$ we denote the real n-dimensional Euclidean space of elements $x = \{x_1, x_2, \dots, x_n\}$. As usual, we shall use $\mathfrak{R}$ instead of $\mathfrak{R}^1$. The symbol $\mathfrak{T}$ stands for the time interval $(\tau, +\infty)$, where τ is a number or the symbol $-\infty$, and $\mathcal{T}_0$ denotes the semi-infinite time interval $[t_0, +\infty)$ with $t_0 \in \mathcal{T}$ a fixed number. By $f\colon \mathcal{R}^n \to \mathcal{R}^n$ we shall mean a (single-valued) mapping from $\mathcal{R}^n$ into $\mathcal{R}^n$, and by $f \in C(\mathcal{R}^n)$ we shall mean the class of continuous mappings (functions) $f(x)$ on the domain $\mathcal{R}^n$. When $f\colon \mathcal{T} \times \mathcal{R}^n \to \mathcal{R}^n$, then $f \in C^{(0,1)}(\mathcal{T} \times \mathcal{R}^n)$ means that $f(t, x)$ is continuous with respect to t and continuously differentiable with respect to x on the Cartesian product $\mathcal{T} \times \mathcal{R}^n$. Quite often, we abuse the notation slightly and write $x(t) \in \mathcal{R}^n$ to mean "$x \in \mathcal{R}^n$ for any fixed $t \in \mathcal{T}$.". By $\|x\|$ we denote any convenient norm of the vector $x \in \mathcal{R}^n$, and most frequently it is the Euclidean norm $\|x\| = (x^T x)^{1/2}$, where $x = (x_1, x_2, \ldots, x_n)^T$ is a column vector and T denotes the transpose. We use Dini derivatives

$$D^+ x(t) = \limsup_{h \to 0^+} \frac{1}{h}[x(t+h) - x(t)],$$

$$D_+ x(t) = \liminf_{h \to 0^+} \frac{1}{h}[x(t+h) - x(t)],$$

$$D^- x(t) = \limsup_{h \to 0^-} \frac{1}{h}[x(t+h) - x(t)],$$

$$D_- x(t) = \liminf_{h \to 0^-} \frac{1}{h}[x(t+h) - x(t)],$$

where $x \in C(\mathcal{T})$. Furthermore, if $x, y \in \mathcal{R}^n$, then $x \leqslant y (x < y)$ means $x_i \leqslant y_i (x_i < y_i)$ for all $i = 1, 2, \ldots, n$. Finally, with some obvious exceptions, Greek letters denote scalars, lower-case italic letters denote vectors, capital italic letters denote matrices, and capital script letters denote sets.

2.1. STRUCTURAL PERTURBATIONS

A large dynamic system composed of several subsystems can hardly be expected to stay "in one piece" over long periods of operation. Quite commonly, either by design or by fault, the subsystems are disconnected and reconnected during the functioning of the system. Such on-off participations of the subsystems represent changes in the system structure which may destroy stability and cause the system to collapse. To prevent the collapse, systems should be built to have the desirable stability properties invariant under structural perturbations, that is, to be connectively stable.

In order to formulate precisely what we mean by connective stability, we need a detailed description of the structural perturbations of dynamic systems via digraphs and interconnection matrices. A rather informal description of the perturbations has already been outlined in Section 1.2, and we intend here to provide a more refined formulation.

Let us consider a dynamic system $\mathbb{S}$ described by an ordinary differential equation

$$\dot{x} = f(t, x), \tag{2.1}$$

where $x(t) \in \mathcal{R}^n$ is the state of the system $\mathbb{S}$ and $f\colon \mathcal{T} \times \mathcal{R}^n \to \mathcal{R}^n$ is a well-defined function on $\mathcal{T} \times \mathcal{R}^n$.

We assume that the system $\mathbb{S}$ represents an interconnection structure of s dynamic subsystems $\mathbb{S}_i$ described by the equations

$$\dot{x}_i = g_i(t, x_i) + h_i(t, x), \qquad i = 1, 2, \ldots, s, \tag{2.2}$$

where $x_i(t) \in \mathcal{R}^{n_i}$, $n_i \leqslant n$, is the state of the ith subsystem $\mathbb{S}_i$, the function $g_i\colon \mathcal{T} \times \mathcal{R}^{n_i} \to \mathcal{R}^{n_i}$ represents the decoupled subsystems $\mathbb{S}_i$ as

$$\dot{x}_i = g_i(t, x_i), \qquad i = 1, 2, \ldots, s, \tag{2.3}$$

and $h_i\colon \mathcal{T} \times \mathcal{R}^n \to \mathcal{R}^{n_i}$ is the interconnection between the subsystem $\mathbb{S}_i$ and the overall system $\mathbb{S}$.

To describe the structure of $\mathbb{S}$ with regard to its constituents $\mathbb{S}_i$, we use the notion of the fundamental interconnection matrix (Šiljak, 1972a) specified by the following:

Definition 2.1. *The $s \times s$ fundamental interconnection matrix $\bar{E} = (\bar{e}_{ij})$ corresponding to the system $\mathbb{S}$ has one row and one column for each subsystem, and the elements $\bar{e}_{ij}$ are defined as*

$$\bar{e}_{ij} = \begin{cases} 1, & x_j \text{ occurs in } h_i(t, x), \\ 0, & x_j \text{ does not occur in } h_i(t, x). \end{cases} \tag{2.4}$$

Therefore, $\bar{E}$ is the standard occurrence (interconnection) matrix (e.g. Harary, Norman, and Cartwright, 1965), which has binary elements $\bar{e}_{ij}$: 1 if $\mathbb{S}_j$ can act on $\mathbb{S}_i$, and 0 if $\mathbb{S}_j$ cannot act on $\mathbb{S}_i$. In applications, this means that $\bar{e}_{ij} = 1$ if there is a possibility for a (physical) interconnection from $\mathbb{S}_j$ to $\mathbb{S}_i$, and $e_{ij} = 0$ otherwise.

If in a system $\mathbb{S}$ all possible interconnections are present and its structure is described by a matrix $\bar{E}$, then removal of the interconnection from $\mathbb{S}_j$ to $\mathbb{S}_i$ represents a structural perturbation of the system $\mathbb{S}$. The new structure is described by a matrix E which is obtained from $\bar{E}$ by replacing the unit element $\bar{e}_{ij}$ with the zero element e_{ij}. Other structural perturbations can be described by interconnection matrices E which are generated from $\bar{E}$ by replacing certain number of unit elements in $\bar{E}$ by zeros.

Now, we introduce the following:

Definition 2.2. *An $s \times s$ interconnection matrix $E = (e_{ij})$ is said to be generated by an $s \times s$ fundamental interconnection matrix $\bar{E} = (\bar{e}_{ij})$ if $\bar{e}_{ij} = 0$ implies $e_{ij} = 0$ for all $i, j = 1, 2, \ldots, s$.*

The fact that E is generated by a given matrix $\bar{E}$ will be denoted for convenience by $E \in \bar{E}$ (notice the abuse of notation).

Obviously, Definition 2.2 includes $E = \bar{E}$ as a generated matrix when none of the unit elements of $\bar{E}$ are replaced by zeros. It also includes the opposite extreme $E = 0$, where 0 is the $s \times s$ zero matrix, which corresponds to the case when all unit elements of $\bar{E}$ are replaced by zeros, and all subsystems $\mathbb{S}_i$ are totally decoupled from each other. That is, for $E = 0$ the system $\mathbb{S}$ is described by the equations (2.3).

Interconnection matrices E, as specified by Definition 2.2, are binary constant matrices. In the following development, we will remove both of those restrictions. We shall consider situations in which the elements of E are functions $e_{ij}: \mathfrak{T} \times \mathfrak{R}^n \to \mathfrak{R}$ which may have an arbitrary form provided $e_{ij}(t,x) \in [0,1]\ \forall (t,x) \in \mathfrak{T} \times \mathfrak{R}^n$. That is, the only restriction on the elements $e_{ij}(t,x)$ is that on sign and magnitude,

$$0 \leqslant e_{ij}(t,x) \leqslant 1 \qquad \forall (t,x) \in \mathfrak{T} \times \mathfrak{R}^n \tag{2.5}$$

(besides, of course, the smoothness conditions which guarantee that $\mathbb{S}$ can represent a dynamic system; e.g. Šiljak, 1969).

Structural perturbations can also be described by digraphs (directed or oriented graphs), as shown in the preceeding chapter. In this section, we attempt a more precise description of structural properties of dynamic systems using digraphs (for an exposition of digraph theory, see Harary, Norman and Cartwright, 1965; Berztiss, 1975).

Let us consider a nonempty set $X = \{x_1, x_2, \dots, x_s\}$ of s points (nodes, vertices) x_j, and a set R of ordered pairs (x_j, x_i) called lines (arcs, branches), which is a relation in X. Now, we have the following:

Definition 2.3. *A digraph $\mathfrak{D}$ is the ordered pair* (X, R).

Digraphs have graphical interpretations which are useful in providing structural information about the systems they are associated with. The points $x_j \in X$ are drawn on the plane, and the two points x_j, x_i are joined by a line with an arrowhead pointing from x_j to x_i provided the pair $(x_j, x_i) \in R$ as shown in Figures 1.1 and 1.2. As distinct from the digraphs of Figure 1.2, the digraph in Figure 1.1 has lines "weighted" by the elements a_{ij} of the system matrix $A = (a_{ij})$. This is done to indicate the strength of interactions among the constituents $\mathbb{S}_i$ of the system $\mathbb{S}$ represented by the points x_j. Such digraphs are called weighted digraphs.

It is now simple to relate digraphs to interconnection matrices. An $s \times s$ interconnection matrix $E = (e_{ij})$ is associated with a digraph $\mathfrak{D} = (X, R)$ if the elements of E are given as

$$e_{ij} = \begin{cases} 1, & (x_j, x_i) \in R, \\ 0, & (x_j, x_i) \notin R, \end{cases} \qquad \forall i,j = 1,2,\dots,s. \tag{2.6}$$

Structural perturbations of the system $\mathbb{S}$ described by a fundamental digraph $\bar{\mathfrak{D}}$ can be interpreted as the removal of a number of lines from $\bar{\mathfrak{D}}$. To make the statement more precise, let us recall (Harary, Norman, and Cartwright, 1965) that a *subgraph* of a digraph $\bar{\mathfrak{D}}$ is a digraph whose points and lines are points and lines of $\bar{\mathfrak{D}}$. Furthermore, a *spanning subgraph* of $\bar{\mathfrak{D}}$ is a subgraph with the same set of points as $\bar{\mathfrak{D}}$. Now, we can state a digraph analog to Definition 2.2 as the following:

Definition 2.4. *A digraph $\mathfrak{D} = (X, R)$ is said to be generated by the fundamental digraph $\bar{\mathfrak{D}} = (X, \bar{R})$ if $\mathfrak{D}$ is a spanning subgraph of $\bar{\mathfrak{D}}$ and $(x_j, x_i) \notin \bar{R}$ implies $(x_j, x_i) \notin R$ for all $i, j = 1, 2, \dots, s$.*

As in Definition 2.2, we regard the system $\mathbb{S}$ to be under structural perturbations if its structure is described by a digraph $\mathfrak{D}$ which is generated by the fundamental digraph $\bar{\mathfrak{D}}$. Obviously, there are as many different structures of $\mathbb{S}$ as there are spanning subgraphs of $\bar{\mathfrak{D}}$. This is illustrated by Figure 1.2 in the preceeding chapter.

There is a great deal of freedom in choosing the state vectors for the subsystems (2.2), that is, (2.3). If a large system is composed of physically distinct subsystems, and it is of interest to study the influence of the interconnection structure on the stability of the overall system, then it is natural to associate a "mathematical" subsystem (2.3) with each physical component. However, when for some reason it is not possible to make this direct association (e.g., either because the physical components do not have a desired property necessary for further analysis, or because it is computationally unattractive), then other decompositions may be tried. One may associate a mathematical subsystem with two or more physical units, as is done in the transient-stability analysis of multimachine power systems in Chapter 7. Similarly, when there are dynamic elements in the interactions among the subsystems, it is possible to include the dynamic elements in some or all of the mathematical subsystems. Such "overlapping" of subsystems takes place in the design of automatic generation control by a decentralized feedback, outlined in Chapter 7.

2.2. CONNECTIVE STABILITY

Once structural perturbations are defined precisely, it is not difficult to formulate the kind of stability which corresponds to such perturbations, that is, connective stability (Šiljak, 1972a). We simply extend the notion of stability in the sense of Liapunov to accommodate structural perturbations.

Our interest will be mainly in global asymptotic and exponential stability, but other kinds of stability could be similarly modified to incorporate the connectivity aspects of dynamic systems.

Let us consider a dynamic system $\mathbb{S}$ described by the differential equation

$$\dot{x} = f(t, x), \tag{2.1}$$

where $x(t) \in \mathcal{R}^n$ is the state of $\mathbb{S}$, and the function $f: \mathcal{T} \times \mathcal{R}^n \to \mathcal{R}^n$ is defined, bounded, and continuous on the domain $\mathcal{T} \times \mathcal{R}^n$, so that solutions $x(t; t_0, x_0)$ of Equation (2.1) exist for all initial conditions $(t_0, x_0) \in \mathcal{T} \times \mathcal{R}^n$ and $t \in \mathcal{T}_0$. Furthermore, we assume that

$$f(t, 0) = 0 \qquad \forall t \in \mathcal{T} \tag{2.7}$$

and $x^* = 0$ is the unique equilibrium state of $\mathbb{S}$.

To introduce the connective aspects of stability, let us decompose the state vector $x \in \mathcal{R}^n$ into s vector components

$$x_i = (x_{i1}, x_{i2}, \ldots, x_{in_i})^T, \qquad i = 1, 2, \ldots, s, \tag{2.8}$$

where $x_i \in \mathcal{R}^{n_i}$, $n_i \leqslant n$. Therefore, the vector components x_i can share among themselves any number of components of x. Most often, however, we have no "overlapping" among x_i's and

$$\mathcal{R}^n = \mathcal{R}^{n_1} \times \mathcal{R}^{n_2} \times \cdots \times \mathcal{R}^{n_s}, \tag{2.9}$$

so that

$$x = (x_1^T, x_2^T, \ldots, x_s^T)^T, \tag{2.10}$$

and

$$n = \sum_{i=1}^{s} n_i. \tag{2.11}$$

The scalar components $f_i: \mathcal{T} \times \mathcal{R}^n \to \mathcal{R}$ of the function $f(t, x)$ in (2.1), are further specialized as

$$f_i(t, x) \equiv \hat{f}_i(t, x_i, e_{i1} x_1, e_{i2} x_2, \ldots, e_{is} x_s), \tag{2.12}$$

where e_{ij} are elements of a given $s \times s$ interconnection matrix E.

We formulate the following:

Definition 2.5. *The equilibrium $x^* = 0$ of the system $\mathbb{S}$ is connectively stable if for each number $\varepsilon > 0$ there exists a number $\delta > 0$ such that*

$$\|x_0\| < \delta \tag{2.13}$$

implies

$$\|x(t; t_0, x_0)\| < \varepsilon \qquad \forall t \in \mathcal{T}_0 \tag{2.14}$$

for all $E \in \bar{E}$.

In the above definition, t_0 appears as a parameter, but can be ignored because δ is considered as a function of t_0 and ε, that is, $\delta = \delta(t_0, \varepsilon)$. This freedom in the choice of δ as a function of t_0 allows us to conclude (e.g. Hahn, 1967) that stability of x^* at $t_0 \in \mathcal{T}$ implies stability of x^* at any initial time $t_1 \in \mathcal{T}_0$. When for each $\varepsilon > 0$ there exists a $\delta(\varepsilon) > 0$ independent of t_0 such that (2.13) implies (2.14) for all $E \in \bar{E}$, then x^* is *uniformly connectively stable*. Furthermore, if x^* is not stable according to Definition 2.5, it is *connectively unstable*. That is, if there is a single interconnection matrix $E \in \bar{E}$ such that x^* is not stable in the sense of Liapunov, then it is not connectively stable (i.e., it is connectively unstable).

Connective stability as defined above is a local concept, since it is relevant only near the equilibrium state. If a δ "works" for a given value of ε, it works for any other value that is greater than the given one. Therefore, implicit in Definition 2.5 is the expression "... for each number $\varepsilon > 0$, no matter how small, there exists a number $\delta > 0$ such that ...".

In most applications, we wish not only for the motions to be bounded as in (2.14), but for them to converge back to equilibrium after a small disturbance. Since we require that convergence be invariant to structural perturbations, we need the following:

Definition 2.6. *The equilibrium* $x^* = 0$ *of the system* $\mathcal{S}$ *is asymptotically connectively stable if it is connectively stable and, in addition, there exists a number* $\mu > 0$ *such that* $\|x_0\| < \mu$ *implies*

$$\lim_{t \to +\infty} x(t; t_0, x_0) = 0 \tag{2.15}$$

for all $E \in \bar{E}$.

The limiting process of (2.15) can be interpreted in more detail as follows: There exists a number $\mu > 0$, and to every number $\varepsilon > 0$ there corresponds a number $\tau > 0$, such that

$$\|x_0\| < \mu \tag{2.16}$$

implies

$$\|x(t; t_0, x_0)\| < \varepsilon \qquad \forall t \geqslant t_0 + \tau \tag{2.17}$$

for all $E \in \bar{E}$. Generally, the number μ is a function of t_0, and the number τ is a function of t_0, x_0, and ε, that is, $\mu = \mu(t_0)$ and $\tau = \tau(t_0, x_0, \varepsilon)$. If the numbers δ, μ, and τ in Definition 2.6 do not depend on t_0, then the asymptotic connective stability of $x^* = 0$ is also *uniform.*

Asymptotic connective stability is also a local concept, since we do not know *a priori* how small we should select μ. In most applications, however, our interest will be focused on cases when the number μ can be chosen as a fixed but arbitrarily large positive number. Then the stability property of Definition 2.6 holds globally, and $x^* = 0$ is *asymptotically connectively stable in the large*. Whether this kind of stability is uniform or not depends again on whether or not the numbers δ, μ, and τ can be selected irrespective of t_0. Our emphasis on the uniform aspect of connective stability is motivated by the fact that the Liapunov direct method is suitable for determining this kind of stability.

Quite often, in practical situations, we are not satisfied by the fact that system motions merely converge to the equilibrium, and we ask questions about the speed of convergence to determine how useful the system is. A common measure of the convergence speed is the exponential function, and we propose the following:

Definition 2.7. *The equilibrium $x^* = 0$ of the system $\mathfrak{S}$ is exponentially connectively stable in the large if there exist two numbers $\Pi > 0$ and $\pi > 0$, which are independent of the initial conditions (t_0, x_0), such that*

$$\|x(t; t_0, x_0)\| \leqslant \Pi \|x_0\| \exp[-\pi(t - t_0)] \qquad \forall t \in \mathfrak{T}_0, \quad \forall (t_0, x_0) \in \mathfrak{T} \times \mathfrak{R}^n \tag{2.18}$$

for all $E \in \bar{E}$.

That is, $x^* = 0$ is exponentially stable if all motions $x(t; t_0, x_0)$ converge to x^* faster than an exponential. Moreover, the global exponential convergence takes place for all interconnection matrices E generated by $\bar{E}$.

It is common in applications for the mathematical model of a physical system not to be determined precisely. Characteristics of elements and their interconnections may only be specified by magnitude constraints, either because they cannot be measured precisely or because they can change during system operation. Consequently, stability would be a useless indicator of the system performance if it were not robust enough to tolerate a certain degree of inaccuracy and variations of system characteristics.

To formulate a kind of stability which includes tolerance to model inaccuracies and to variations in its characteristics, let us assume that the system $\mathfrak{S}$ described by Equation (2.1) can be decomposed into s interconnected subsystems $\mathfrak{S}_i$ described by the equations

$$\dot{x}_i = g_i(t, x_i) + h_i(t, x), \qquad i = 1, 2, \ldots, s, \tag{2.2}$$

where $x_i(t) \in \mathcal{R}^{n_i}$ is the state of the ith subsystem $\mathbb{S}_i$, and $x(t) \in \mathcal{R}^n$ is the state of the overall system $\mathbb{S}$. In (2.2) the functions $g_i\colon \mathcal{T} \times \mathcal{R}^{n_i} \to \mathcal{R}^{n_i}$ describe the "decoupled" subsystems $\mathbb{S}_i$:

$$\dot{x}_i = g_i(t, x_i), \qquad i = 1, 2, \dots, s, \tag{2.3}$$

and the functions $h_i\colon \mathcal{T} \times \mathcal{R}^n \to \mathcal{R}^{n_i}$ represent the interactions among the subsystems $\mathbb{S}_i$, which have the form

$$h_i(t, x) \equiv h_i(t, e_{i1}x_1, e_{i2}x_2, \dots, e_{is}x_s), \tag{2.19}$$

where again e_{ij} are elements of the $s \times s$ interconnection matrix E. Furthermore, we assume that for all $i = 1, 2, \dots, s$,

$$g_i(t, 0) = 0 \qquad \forall t \in \mathcal{T} \tag{2.20}$$

so that $x_i^* = 0$ is the equilibrium point of $\mathbb{S}_i$. We also have

$$h_i(t, 0) = 0 \qquad \forall t \in \mathcal{T}, \tag{2.21}$$

which agrees with the assumption (2.7) regarding the function $f(t, x)$. Therefore, the decoupled subsystems (2.3) are obtained from (2.2) when $E = 0$.

Let us add another specialization of the interconnection functions, which allows a refinement of Definition 2.6 with respect to tolerances of variations in interactions. We assume that interconnection functions $h_i(t, x)$ have the form

$$h_i(t, x) = \sum_{j=1}^{s} e_{ij} h_{ij}(t, x_j), \tag{2.22}$$

where each function $h_{ij}\colon \mathcal{T} \times \mathcal{R}^{n_j} \to \mathcal{R}^{n_i}$ satisfies the constraint

$$\|h_{ij}(t, x_j)\| \leqslant \xi_{ij}\|x_j\| \qquad \forall (t, x_j) \in \mathcal{T} \times \mathcal{R}^{n_j}, \quad i, j = 1, 2, \dots, s \tag{2.23}$$

and ξ_{ij} are nonnegative numbers. We define a class $\mathcal{K}_{(2.23)}$ of functions $h_{ij}(t, x_j)$ which satisfy the inequality (2.23), that is,

$$\mathcal{K}_{(2.23)} = \left\{ h_{ij}(t, x_j)\colon \|h_{ij}(t, x_j)\| \leqslant \xi_{ij}\|x_j\| \ \forall (t, x_j) \in \mathcal{T} \times \mathcal{R}^{n_j} \right\}. \tag{2.24}$$

We arrive at the following modification of Definition 2.6:

Definition 2.8. *The equilibrium state $x^* = 0$ of the system $\mathbb{S}$ is absolutely connectively stable if it is asymptotically stable in the large for all $h_{ij}(t, x_j) \in \mathcal{K}_{(2.23)}$ and all $E \in \bar{E}$.*

The absolute aspect of stability reflected by the nonlinearity constraint (2.23) goes back to the work of Lur'e and Postnikov (see Lur'e, 1951). It is an important aspect in that it introduced a notion of stability for a family of systems characterized by a "sector" constraint of the type (2.23). The basic idea is to verify the stability of a control system for all possible nonlinear functions inside the sector by a single test involving only the bounds of the sector. The absolute-stability concept received an essential improvment when Popov in the 1960's (see Popov, 1973) invented a frequency criterion for absolute stability. By imitating the spirit of the absolute-stability concept, but otherwise using entirely different techniques, we will derive algebraic conditions for absolute stability of the system $\mathbb{S}$ in the context of structural perturbations.

Several general comments regarding the above stability definitions are now in order. Let us first reconsider the system $\mathbb{S}$,

$$\dot{x} = f(t, x), \tag{2.1}$$

and the assumption

$$f(t, 0) = 0 \qquad \forall t \in \mathfrak{T}, \tag{2.7}$$

which assures that the equilibrium x^* of $\mathbb{S}$ is at the origin of the state space. Although the above definitions are stated for the case $x^* = 0$, they apply to the stability of any fixed solution $x^*(t; t_0, x_0)$. To see this, we consider the difference

$$y = x - x^* \tag{2.25}$$

and write

$$\dot{y} = \dot{x} - \dot{x}^* = f(t, x^* + y) - f(t, x^*). \tag{2.26}$$

By introducing the function

$$g(t, y) = f(t, x^* + y) - f(t, x^*), \tag{2.27}$$

we rewrite (2.26) as

$$\dot{y} = g(t, y), \tag{2.28}$$

for which

$$g(t, 0) = 0 \qquad \forall t \in \mathfrak{T}. \tag{2.29}$$

Therefore, stability of $x^*(t; t_0, x_0)$ in (2.1) is replaced by stability of the equilibrium $y^* = 0$ in (2.28), which is formulated in this section.

The consideration above obviously includes, as a special case, any equilibrium solutions $x^* \neq 0$ which are constant solutions of Equation (2.1) defined as

$$x^*(t; t_0, x^*) = x^* \qquad \forall t \in \mathfrak{T}_0 \tag{2.30}$$

for any $t_0 \in \mathfrak{T}$, or

$$f(t, x^*) = 0 \qquad \forall t \in \mathfrak{T}. \tag{2.31}$$

Then, from (2.27), we have $g(t, y) = f(t, x^* + y)$, and again the system (2.28) with $y^* = 0$ replaces the system (2.1) with $x^* \neq 0$.

Although, in principle, stability of any fixed solution can be reduced by a simple transformation to stability of an equilibrium at the origin, the transformed equation is usually more complicated to analyze than the original one. In the context of connective stability there is another aspect of the transformation that should be mentioned. The equilibrium state $x^* = 0$ was assumed invariant under structural perturbations. There are, however, classes of systems for which this assumption is not valid, and there is a different equilibrium each time another interconnection matrix is in effect. Such cases require special care when above definitions are used, and we will explicitly point out these cases in later considerations of competitive equilibrium in Chapter 4.

Since we plan to use the conventional definitions of stability, it is of interest to note that these definitions can be obtained from their "connective" versions by ignoring the connectivity aspect introduced by interconnection matrices—that is, by simply omitting "for all $E \in \bar{E}$".

Stability definitions formulated above will be used most often in the following development. In certain situations, however, we shall need certain further refinements with regard to stochastic fluctuations, finiteness of the stability regions, new definitions of boundedness, etc. On those occasions, we will state the definitions immediately before their use.

2.3. DIFFERENTIAL INEQUALITIES

To derive conditions for the connective stability of dynamic systems, we need the modern mathematical machinery of the comparison principle developed in the theory of differential equations and inequalities. In this section, we shall present the principle following the exposition of Lakshmikantham and Leela (1969), but the generality of the presentation will be somewhat reduced to conform with our needs in applications.

Let us consider a differential equation

$$\dot{r} = w(t, r), \tag{2.32}$$

where $r \in \mathfrak{R}^s$, and the function $w\colon \mathfrak{D} \to \mathfrak{R}^s$ is defined and continuous on an open set $\mathfrak{D}$ in $\mathfrak{R}^{s+1}$ with an element written as (t, r). A function $r(t)$ is a *solution* of Equation (2.32) on an interval $\mathcal{T} \subset \mathcal{R}$ if $r(t)$ is a continuously differentiable function of t on $\mathcal{T}$, $[t, r(t)] \in \mathcal{D}$ for $t \in \mathcal{T}$, and $r(t)$ satisfies Equation (2.32) on $\mathcal{T}$, that is, $\dot{r}(t) = w[t, r(t)]$ for all $t \in \mathcal{T}$.

Most often we are not interested in finding all solutions of Equation (2.32), but only those that pass through a given point. That is, t_0 and $r(t_0) = r_0$ are given such that $(t_0, r_0) \in \mathcal{D}$, and we want to solve the *initial-value problem* for Equation (2.32), which consists of finding an interval $\mathcal{T}$ containing t_0 and a solution $r(t)$ of (2.32) satisfying $r(t_0) = r_0$. The well-known Cauchy-Peano existence theorem (e.g. Hale, 1969) provides the conditions for a solution of the initial-value problem for (2.32): If $w(t, r)$ is continuous in $\mathcal{D}$ [that is, $w \in C(\mathcal{D})$], then for any $(t_0, r_0) \in \mathcal{D}$ there is at least one solution $r(t)$ of Equation (2.32) passing through (t_0, r_0). If, in addition, $w(t, r)$ is bounded and $\|w(t, r)\| \leqslant \mu$ for all $(t, r) \in \mathcal{T} \times \mathcal{R}^s$, where $\mu \geqslant 0$ and $\mathcal{T} = [0, +\infty)$, then for any $(t_0, r_0) \in \mathcal{T} \times \mathcal{R}^s$ there exists a solution $r(t)$ of Equation (2.32) for $t \in \mathcal{T}_0$, where $\mathcal{T}_0 = [t_0, +\infty)$.

In various comparison results related to the solutions of the differential equation (2.32), a fundamental role is played by the class $\mathcal{W}$ of quasimonotonous increasing functions $w(t, r)$, which was introduced by Müler (1926), and used in an essential way by Kamke (1932), Ważewski (1950), and Szarski (1965).

Definition 2.9. *A function $w(t, r)$ belongs to the class $\mathcal{W}$ if for any fixed $t \in \mathcal{R}$ and all $r', r'' \in \mathcal{R}^s$ such that $r'_j \leqslant r''_j$, $r'_i = r''_i$ ($j = 1, 2, \ldots, s$, $i \neq j$), the inequality*

$$w_i(t, r') \leqslant w_i(t, r'') \qquad \forall i = 1, 2, \ldots, s \tag{2.33}$$

is satisfied.

To appreciate this definition, we may note that if a function $w(t, r)$ is linear in r, that is,

$$w(t, r) \equiv W(t) r, \tag{2.34}$$

where $W = (w_{ij})$ is an $s \times s$ matrix, then the function $W(t)r$ belongs to the class $\mathcal{W}$ if (and only if) the elements $w_{ij}(t)$ of $W(t)$ are such that

$$w_{ij}(t) \geqslant 0 \qquad \forall i, j = 1, 2, \ldots, s, \quad i \neq j, \tag{2.35}$$

for all $t \in \mathcal{T}$. In other words, $W(t)$ is a time-varying Metzler matrix defined by Šiljak (1975a) as a time-dependent analog to the original constant Metzler matrix discussed in Section 1.2 of the preceding chapter.

It should also be noted that any scalar ($s = 1$) function $w(t, r)$ belongs to the class $\mathcal{W}$. This fact is obvious from Definition 2.9.

The class of functions $\mathcal{W}$ is all one needs to state a "strict" version of the comparison principle.

Theorem 2.1. *Let $w(t, r)$ be continuous on an open set $\mathfrak{D}$ in $\mathfrak{R}^{s+1}$, and let $w(t, r)$ belong to the class $\mathfrak{W}$. Let $r(t)$ be a solution of the differential equation*

$$\dot{r} = w(t, r), \tag{2.32}$$

on an interval $[t_0, t_0 + \tau)$ passing through $(t_0, r_0) \in \mathfrak{D}$. If a function $v(t)$ is continuous on $[t_0, t_0 + \tau)$, $(t, v) \in \mathfrak{D}$ for $t \in [t_0, t_0 + \tau)$, and if it satisfies the differential inequality

$$D_- v < w(t, v) \qquad \forall t \in (t_0, t_0 + \tau), \tag{2.36}$$

then

$$v(t_0) < r_0 \tag{2.37}$$

implies

$$v(t) < r(t) \qquad \forall t \in [t_0, t_0 + \tau). \tag{2.38}$$

Proof. Since $v \in C\{[t_0, t_0 + \tau)\}$, we have

$$v(t) < r(t) \qquad \forall t \in [t_0, t_0 + \varepsilon), \tag{2.39}$$

where $\varepsilon > 0$ is a sufficiently small number. Suppose now that the assertion (2.38) is not true on the entire interval $[t_0, t_0 + \tau)$. Then the set

$$\mathfrak{I} = \bigcup_{i=1}^{s} \{t \in [t_0, t_0 + \tau): v_i(t) \geqslant r_i(t)\} \tag{2.40}$$

is nonempty. Since $\mathfrak{I}$ is closed, we have at least one value of i such that

$$v_i(t_1) = r_i(t_1) \tag{2.41}$$

and

$$v_j(t_1) \leqslant r_j(t_1), \qquad j \neq i, \tag{2.42}$$

where $t_1 = \inf \mathfrak{I}$, and $t_1 > t_0$. Then since $w \in \mathfrak{W}$,

$$D_- v_i(t_1) < w_i[t, v(t_1)] \leqslant w_i[t, r(t_1)] = \dot{r}_i(t_1), \tag{2.43}$$

and (2.41) implies that

$$v_i(t) > r_i(t) \tag{2.44}$$

for some values of $t < t_1$ arbitrarily close to t_1. This contradicts the definition of t_1, and the proof of Theorem 2.1 is complete.

To be able to weaken the inequalities in Theorem 2.1, we need the notion of *maximal solutions* of differential equations provided by the following:

Definition 2.10. *If a solution $r_M(t)$ of the equation*

$$\dot{r} = w(t, r), \tag{2.32}$$

passing through (t_0, r_0) and existing on an interval $\mathfrak{T}$ containing t_0, has the property that every other solution $r(t)$ of Equation (2.32) *passing through (t_0, r_0) and existing on $\mathfrak{T}$ is such that*

$$r(t) \leqslant r_M(t) \qquad \forall t \in \mathfrak{T}, \tag{2.45}$$

then $r_M(t)$ is called a maximum solution of Equation (2.32) *on the interval $\mathfrak{T}$ passing through (t_0, r_0).*

The existence of maximal solutions is guaranteed by continuity and quasimonotonicity of the function $w(t, r)$ in (2.32). This is convenient because we need no additional conditions for relaxing the inequalities in the comparison principle besides those required by Theorem 2.1. We have the following:

Theorem 2.2. *Let $w(t, r)$ be continuous and bounded on the rectangle*

$$\mathfrak{R}_0 = \{(t, r) \in \mathfrak{R}^{s+1}\colon t_0 \leqslant t \leqslant t_0 + \tau, \|r - r_0\| \leqslant \rho\}, \tag{2.46}$$

and let $w(t, r)$ belong to the class of functions $\mathfrak{W}$. Then there exists a maximal solution $r_M(t)$ on an interval $[t_0, t_0 + \theta]$ passing through any point $(t_0, r_0) \in \mathfrak{R}_0$, where $\theta = \min\{\tau, \rho/(2\mu + \rho)\}$.

Proof. Let us consider the initial-value problem

$$\dot{r} = w(t, r) + \frac{u}{n}, \qquad r(t_0) = r_0 + \frac{u}{n}, \tag{2.47}$$

where $u \in \mathfrak{R}^s$ is a constant vector and n is a positive integer such that $0 < u_i < n\rho/2$. We notice that $w_u \in C(\mathfrak{R}_u)$, where $w_u(t, r) = w(t, r) + u/n$, $\mathfrak{R}_u = \{(t, r) \in \mathfrak{R}^{n+1}\colon t_0 \leqslant t \leqslant t_0 + \tau,\ \ \|r - (r_0 + u/n)\| \leqslant \rho/2\}$, and $\mathfrak{R}_u \subset \mathfrak{R}_0$. Also, $\|w_u(t, r)\| \leqslant \mu + \rho/2$ on $\mathfrak{R}_u$. Then by Cauchy and Peano's existence theorem (e.g. Hale, 1969) the initial-value problem (2.47) has a solution $r(t, u/n)$ on the interval $[t_0, t_0 + \theta]$. If $n < m$, then by Theorem 2.1 we get

$$r(t, u/m) < r(t, u/n) \qquad \forall t \in [t_0, t_0 + \theta]. \tag{2.48}$$

Since the sequence $\{r(t, u/n)\}$ of functions $r(t, u/n)$ is equicontinuous in t and uniformly bounded, it follows that

$$\lim_{n \to +\infty} r(t, u/n) = r_M(t) \tag{2.49}$$

uniformly on $[t_0, t_0 + \theta]$, and $r_M(t)$ is a solution of Equation (2.32) on the interval $[t_0, t_0 + \theta]$ passing through (t_0, r_0).

It remains to show that $r_M(t)$ is the maximal solution of (2.32), that is, we have to prove that (2.45) is satisfied. Let $r(t)$ be any solution of (2.32) existing on $[t_0, t_0 + \theta]$ and passing through (t_0, r_0). Then

$$\dot{r} < w(t, r) + \frac{u}{n}, \tag{2.50}$$

and by Theorem 2.1 we get

$$r(t) < r(t, u/n) \qquad \forall t \in [t_0, t_0 + \theta]. \tag{2.51}$$

Consequently,

$$r(t) \leqslant \lim_{n \to +\infty} r(t, u/n) = r_M(t) \qquad \forall t \in [t_0, t_0 + \theta]. \tag{2.52}$$

This proves Theorem 2.2.

Finally, we arrive at a form of the comparison principle which will be used in the subsequent developments.

Theorem 2.3. *Let $w(t, r)$ be continuous on an open set $\mathfrak{D}$ in $\mathfrak{R}^{s+1}$, and let $w(t, r)$ belong to the class $\mathfrak{W}$. Let $r_M(t)$ be a maximal solution of the equation*

$$\dot{r} = w(t, r) \tag{2.32}$$

on an interval $[t_0, t_0 + \tau)$ passing through $(t_0, r_0) \in \mathfrak{D}$. If a function $v(t)$ is continuous on $[t_0, t_0 + \tau)$ with $(t, v) \in \mathfrak{D}$ and $t \in [t_0, t_0 + \tau)$, and satisfies the inequality

$$D_{-} v \leqslant w(t, v) \qquad \forall t \in [t_0, t_0 + \tau), \tag{2.53}$$

then

$$v(t_0) \leqslant r_0 \tag{2.54}$$

implies

$$v(t) \leqslant r_M(t) \qquad \forall t [t_0, t_0 + \tau). \tag{2.55}$$

Proof. The proof of Theorem 2.3 follows the proof of a similar theorem given by Coppel (1965).

Let t_1 be the largest value of t such that

$$v(\theta) \leqslant r_M(\theta) \qquad \forall \theta \in [t_0, t], \tag{2.56}$$

and suppose that (2.55) is false, that is, $t_1 < t_0 + \tau$. We again consider the initial-value problem (2.47) at $t = t_1$ on an interval $[t_1, t_1 + \varepsilon]$. By the proof of Theorem 2.2, $r(t, u/n)$ converges to $r_M(t)$ on $[t_1, t_1 + \varepsilon]$ when $n \to +\infty$. Now using Theorem 2.1, we conclude that

$$v(t) < r(t, u/n) \qquad \forall t \in [t_1, t_1 + \varepsilon]. \tag{2.57}$$

If $n \to +\infty$, then

$$v(t) \leqslant r_M(t) \qquad \forall t \in [t_1, t_1 + \varepsilon]. \tag{2.58}$$

But this contradicts the definition of t_1, and Theorem 2.3 is established.

Several remarks concerning Theorem 2.3 are now in order. First, we should mention the possibility of reversing the inequalities of the theorem if the notion of the *minimal solution* is used. The definition of a minimal solution $r_m(t)$ can be obtained directly from Definition 2.10 if the inequality (2.45) is reversed to get

$$r(t) \geqslant r_m(t) \qquad \forall t \in \mathcal{T}. \tag{2.59}$$

If $v(t)$ is continuous on $\mathcal{T} = [t_0, t_0 + \tau)$, and satisfies the differential inequality

$$D^- v > w(t, v) \qquad \forall t \in (t_0, t_0 + \tau), \tag{2.60}$$

then

$$v(t_0) \geqslant r_0$$

implies

$$v(t) \geqslant r_m(t) \qquad \forall t \in [t_0, t_0 + \tau). \tag{2.61}$$

A proof of this statement can be obtained by simple adjustments of the proof of Theorem 2.3. While Theorem 2.3 is useful in establishing stability, the preceding statement can be used to obtain the instability conditions as in Section 2.7.

Another important fact about Theorem 2.3 is that it is valid when the differential inequality (2.53) is replaced by

$$Dv \leqslant w(t, v) \qquad \forall t \in [t_0, t_0 + \tau) - \mathfrak{N} \tag{2.62}$$

where D is any fixed Dini derivative and $\mathfrak{N}$ is an at most countable subset of $[t_0, t_0 + \tau)$ (Lakshmikantham and Leela, 1969). This fact applies to (2.60).

Let us now outline the *existence and uniqueness* of solutions for the initial-value problem, and establish the relevant conditions using Theorem 2.3. This will serve as an illustration of the application of the theorem and at the same time provide the results needed in future developments.

To establish the uniqueness of solutions for initial-value problems based on the equation

$$\dot{r} = w(t, r), \tag{2.32}$$

it is necessary to impose additional conditions on the function $w(t, r)$. A standard condition for this purpose is the *Lipschitz condition*. We say that a function $w(t, r)$ defined on a domain $\mathfrak{D}$ in $\mathfrak{R}^{s+1}$ is *locally Lipschitzian* in r if for any closed bounded set $\mathfrak{U}$ in $\mathfrak{D}$, there is a constant κ such that

$$\|w(t, r') - w(t, r'')\| \leqslant \kappa\|r' - r''\| \tag{2.63}$$

for all (t, r'), $(t, r'') \in \mathfrak{U}$. If $w(t, r)$ has continuous first partial derivatives with respect to r in $\mathfrak{D}$, then one can show (e.g. Hale, 1969) that $w(t, r)$ is locally Lipschitzian. If there exists a constant $\kappa > 0$ such that the inequality (2.63) is satisfied for all (t, r'), $(t, r'') \in \mathfrak{D}$, then the function $w(t, r)$ is said to be *Lipschitzian*, and κ is called a *Lipschitz constant*. Obviously, if a function $w(t, r)$ is Lipschitzian, it is also locally Lipschitzian, but the opposite is not true in general. The existence and uniqueness theorem under the weaker hypothesis that $w(t, r)$ is locally Lipschitzian in r, is known as Picard-Lindelöf Theorem (e.g. Hale, 1969). We prove the following weaker version of that theorem as

Theorem 2.4. *If $w(t, r)$ is continuous on a rectangle $\mathfrak{R}_0 = \{(t, r) \in \mathfrak{R}^{s+1}: t_0 \leqslant t \leqslant t_0 + \tau, \|r - r_0\| \leqslant \rho\}$ and it is Lipschitzian on $\mathfrak{R}_0$, then the differential equation*

$$\dot{r} = w(t, r) \tag{2.32}$$

has at most one solution $r(t)$ on the interval $[t_0, t_0 + \tau]$ passing through (t_0, r_0).

Proof. Suppose, contrary to the theorem, that there are two solutions $r'(t)$ and $r''(t)$ of the equation (2.32) on $[t_0, t_0 + \tau]$ passing through $(t_0, r_0) \in \mathfrak{R}_0$. Define

$$q(t) = \|r'(t) - r''(t)\|. \tag{2.64}$$

Then, using the Lipschitz condition (2.63), we get

$$\begin{aligned} D^+q &\leqslant \|\dot{r}' - \dot{r}''\| \\ &\leqslant \|w(t,r') - w(t,r'')\| \\ &\leqslant \kappa\|r' - r''\| \\ &\leqslant \kappa q \qquad \forall t \in [t_0, t_0 + \tau]. \end{aligned} \tag{2.65}$$

Since $q_M(t) = q_0 \exp[\kappa(t - t_0)]$ is the maximal solution of the equation

$$\dot{q} = \kappa q \tag{2.66}$$

on the interval $[t_0, t_0 + \tau]$ passing through (t_0, q_0), and $q_0 = 0$, we have

$$q(t) = 0 \qquad \forall t \in [t_0, t_0 + \tau], \tag{2.67}$$

which proves Theorem 2.4.

The above proof could be used to integrate the differential inequalities (1.30) and (1.114) involving Liapunov functions, and deduce the stability of the corresponding systems. That is, in fact, how Theorem 2.3 will be used in the context of vector Liapunov functions, to study the connective stability of dynamic systems.

2.4. VECTOR LIAPUNOV FUNCTIONS

The possibility of using several Liapunov functions instead of a single one in stability studies of dynamic systems was proposed at least as far back as 1902 by Duhem (1902). The concept of vector Liapunov functions, however, was not introduced until 1962, when Bellman (1962) and Matrosov (1962) independently proposed the concept as providing more flexible mathematical machinery than the original Liapunov direct method. Following the work of Cordeanu (1960, 1961) for single Liapunov functions, Matrosov presented the concept in the framework of differential inequalities and the comparison principle due to Kamke (1932) and Wažewski (1950), thus providing numerous authors with the tools needed to obtain new important results in stability theory, as surveyed by Lakshmikantham and Leela (1969). The concept is of fundamental significance in connective-stability studies of complex large-scale systems which are inherently exposed to structural perturbations.

Let us consider again a dynamic system $\mathbb{S}$ described by the differential equation

$$\dot{x} = f(t, x), \tag{2.1}$$

where $x(t) \in \mathfrak{R}^n$ is the state of $\mathbb{S}$, and $f\colon \mathfrak{T} \times \mathfrak{R}_\rho \to \mathfrak{R}^n$ is defined, continuous, and bounded on $\mathcal{T} \times \mathcal{R}_\rho$, in which $\mathcal{T} = [\theta, +\infty)$, θ is a number or the symbol $-\infty$, and $\mathcal{R}_\rho = \{x \in \mathcal{R}^n: \|x\| < \rho\}$. We consider solutions $x(t; t_0, x_0)$ of Equation (2.1) on the interval $\mathcal{T}_0 = [t_0, +\infty)$ for the initial conditions $(t_0, x_0) \in \mathcal{T} \times \mathcal{R}_\rho$.

Let us consider a function $v: \mathcal{T} \times \mathcal{R}_\rho \to \mathcal{R}^s_+$, $\mathcal{R}^s_+ = \{v \in \mathcal{R}^s: v \geqslant 0\}$, which is continuous on $\mathcal{T} \times \mathcal{R}_\rho$, that is, $v \in C(\mathcal{T} \times \mathcal{R}_\rho)$. We define the vector function

$$D^+v(t,x) = \limsup_{h \to 0^+} \frac{1}{h}\{v[t + h, x + hf(t,x)] - v(t,x)\} \tag{2.68}$$

for $(t,x) \in \mathcal{T} \times \mathcal{R}^n$. We shall often write $D^+v(t,x)_{(2.1)}$ to indicate that the definition of $D^+v(t,x)$ is with respect to the Equation (2.1).

We now establish a basic comparison result for vector Liapunov functions following Lakshmikantham and Leela (1969).

Lemma 2.1. *Let $v \in C(\mathcal{T} \times \mathcal{R}_\rho)$, and let $v(t,x)$ be locally Lipschitzian in x. Let the vector function $D^+v(t,x)$ defined in* (2.68) *satisfy the differential inequality*

$$D^+v(t,x) \leqslant w[t, v(t,x)] \qquad \forall (t,x) \in \mathcal{T} \times \mathcal{R}_\rho, \tag{2.69}$$

where $w \in C(\mathcal{T} \times \mathcal{R}^s_+)$, and the function $w(t,x)$ belongs to the class $\mathcal{W}$. Let $r_M(t; t_0, r_0)$ be the maximal solution of the differential equation

$$\dot{r} = w(t,r) \tag{2.32}$$

on the interval $\mathcal{T}_0$ passing through $(t_0, r_0) \in \mathcal{T} \times \mathcal{R}^s_+$. If $x(t; t_0, x_0)$ is any solution of Equation (2.1) *on an interval $[t_0, t_0 + \tau)$ passing through $(t_0, x_0) \in \mathcal{T} \times \mathcal{R}_\rho$, then*

$$v(t, x_0) \leqslant r_0 \tag{2.70}$$

implies

$$v[t, x(t; t_0, x_0)] \leqslant r_M(t; t_0, r_0) \qquad \forall t \in [t_0, t_0 + \tau). \tag{2.71}$$

Proof. Let us define the vector function $z(t) = v[t, x(t; t_0, x_0)]$. For sufficiently small $h > 0$, we have

$$\begin{aligned} z(t + h) - z(t) \leqslant\ & v[t + h, x(t + h)] - v\{t + h, x(t) + hf[t, x(t)]\} \\ & + v\{t + h, x(t) + hf[t, x(t)]\} - v[t, x(t)]. \end{aligned} \tag{2.72}$$

Since $v(t,x)$ is locally Lipschitzian in x, from (2.72), using (2.68), we get the inequality (2.69) as

$$D^+z(t) \leqslant w[t, z(t)]. \tag{2.73}$$

Since $z(t_0) \leqslant r_0$, by Theorem 2.3 we obtain

$$z(t) \leqslant r_M(t; t_0, r_0) \qquad \forall t \in [t_0, t_0 + \tau), \tag{2.74}$$

which is (2.71). Thus, the proof of Lemma 2.1 is complete.

It is simple to imitate the proof of Lemma 2.1 and establish an analogous result with the inequalities reversed. Under the conditions of the lemma,

$$D^+ v(t, x) \geqslant w[t, v(t, x)] \qquad \forall (t, x) \in \mathcal{T} \times \mathcal{R}_\rho \tag{2.75}$$

and

$$v(t_0, x_0) \geqslant r_0 \tag{2.76}$$

imply

$$v[t, x(t; t_0, x_0)] \geqslant r_m(t; t_0, r_0) \qquad \forall t \in [t_0, t_0 + \tau), \tag{2.77}$$

where $r_m(t; t_0, r_0)$ is the minimal solution of Equation (2.32) on the interval $[t_0, t_0 + \tau)$ passing through $(t_0, r_0) \in \mathcal{T} \times \mathcal{R}_+^s$.

To be able to apply Lemma 2.1 to stability studies of the equilibria in dynamic systems, we need to impose additional constraints on the vector function $v(t, x)$. For this purpose, we need the notion of *comparison functions* (Hahn, 1967).

Let us consider a function ϕ: $\mathcal{I} \to \mathcal{R}_+$ defined on an interval $\mathcal{I} = [0, \gamma]$. We state

Definition 2.11. *A function $\phi(\zeta)$ belongs to the class $\mathcal{K}$ if $\phi \in C(\mathcal{I})$, $\phi(0) = 0$, and $\phi(\zeta_1) < \phi(\zeta_2)$ for all $\zeta_1, \zeta_2 \in \mathcal{I}$ such that $\zeta_1 < \zeta_2$.*

The last condition in Definition 2.11 means that $\phi(\zeta)$ is a strictly increasing function.

The class $\mathcal{K}$ of comparison functions $\phi(\zeta)$ can be used immediately to redefine connective stability as formulated by Definition 2.5. It is easy to show that the equilibrium $x^* = 0$ of the system $\mathbb{S}$ is connectively stable if there exists a function $\phi \in \mathcal{K}$ such that

$$\|x(t; t_0, x_0)\| \leqslant \phi(\|x_0\|) \qquad \forall t \in \mathcal{T}_0 \tag{2.78}$$

for all $(t_0, x_0) \in \mathcal{T} \times \mathcal{R}_\rho$ and all $E \in \bar{E}$.

To be able to define asymptotic connective stability by comparison functions, we consider a function ψ: $\mathcal{R}_+ \to \mathcal{R}_+$ defined on $\mathcal{R}_+$, and state the following:

Definition 2.12. *A function $\psi(\zeta)$ belongs to the class $\mathcal{L}$ if $\psi \in C(\mathcal{R}_+)$, $\psi(\zeta_1) > \psi(\zeta_2)$ for all $\zeta_1, \zeta_2 \in \mathcal{R}_+$ such that $\zeta_1 < \zeta_2$, and $\psi(\zeta) \to 0$ as $\zeta \to +\infty$.*

A function $\psi(\zeta)$ as defined as above is strictly decreasing.

Now, it is easy to establish a "local analog" to Definition 2.6 using the comparison functions of class $\mathcal{K}$ and $\mathcal{L}$. That is, the equilibrium state $x^* = 0$ of $\mathcal{S}$ is asymptotically connectively stable if there exist a number ρ, a function $\phi \in \mathcal{K}$, and for each $x_0 \in \mathcal{R}_\rho$ a function $\psi \in \mathcal{L}$, such that

$$\|x(t; t_0, x_0)\| \leqslant \phi(\|x_0\|)\psi(t - t_0) \qquad \forall t \in \mathcal{T}_0 \tag{2.79}$$

for all $(t_0, x_0) \in \mathcal{T} \times \mathcal{R}_\rho$ and all $E \in \bar{E}$.

Obviously, the $\psi(t - t_0)$ in (2.79) can be used to estimate how rapidly the solutions $x(t; t_0, x_0)$ of Equation (2.1) approach the equilibrium $x^* = 0$. In applications, the most common function $\psi(t - t_0)$ is the exponential, and we can state another version of Definition 2.8: The equilibrium $x^* = 0$ of the system $\mathcal{S}$ is exponentially connectively stable if in Definition 2.12, functions $\phi(\zeta)$ and $\psi(\zeta)$ can be found such that

$$\phi(\|x_0\|) = \Pi\|x_0\|, \qquad \psi(t - t_0) = \exp[-\pi(t - t_0)], \tag{2.80}$$

where Π and π are positive numbers independent of the initial conditions $(t_0, x_0) \in \mathcal{T} \times \mathcal{R}_\rho$ and interconnection matrices $E \neq \bar{E}$.

In formulating conditions for various kinds of connective stability, we shall use the comparison functions extensively to place restrictions on vector Liapunov functions. Therefore, it is convenient to review a few of the most common restrictions in this context (Hahn, 1967).

The class of functions $\mathcal{K}$ can be used to define positive (negative) definiteness of a function $\nu: \mathcal{T} \times \mathcal{R}_\rho \to \mathcal{R}_+$ on the interval $\mathcal{T} = [\tau, +\infty)$. A function $\nu(t, x)$ such that $\nu(t, 0) \equiv 0$ is *positive definite* if there exists a function $\phi \in \mathcal{K}$ such that

$$\nu(t, x) \geqslant \phi(\|x\|) \qquad \forall (t, x) \in \mathcal{T} \times \mathcal{R}_\rho. \tag{2.81}$$

Similarly, a function $\nu(t, x)$ such that $\nu(t, 0) \equiv 0$ is *negative definite* if $-\nu(t, x)$ is positive definite or, equivalently, if

$$\nu(t, x) \leqslant -\phi(\|x\|) \qquad \forall (t, x) \in \mathcal{T} \times \mathcal{R}_\rho, \tag{2.82}$$

where $\phi \in \mathcal{K}$.

In asymptotic stability analysis the notion of a *decrescent* function $\nu(t, x)$ is used. A function $\nu(t, x)$ is said to be decrescent if there exists a function $\phi \in \mathcal{K}$ such that

$$\nu(t, x) \leqslant \phi(\|x\|) \qquad \forall (t, x) \in \mathcal{T} \times \mathcal{R}_\rho. \tag{2.83}$$

When stability properties are required to hold globally, that is, when $\mathcal{R}_\rho$ is the entire space $\mathcal{R}^n$, then we need the following:

Definition 2.13. *A function $\phi(\zeta)$ belongs to the class $\mathcal{K}_\infty$ if $\phi \in \mathcal{K}$ for all $\zeta \in \mathcal{R}_+$ and $\phi(\zeta) \to +\infty$ as $\zeta \to +\infty$.*

A function $\nu(t,x)$ defined on $\mathcal{T} \times \mathcal{R}^n_+$ is *radially unbounded* if there exists a function $\phi \in \mathcal{K}_\infty$ such that

$$\nu(t,x) \geqslant \phi(\|x\|) \qquad \forall (t,x) \in \mathcal{T} \times \mathcal{R}^n. \tag{2.84}$$

In order to state and prove our fundamental theorem on connective stability in the context of vector Liapunov functions, let us consider again the dynamic system $\mathcal{S}$ described by the differential equation

$$\dot{x} = f(t,x), \tag{2.1}$$

where $x \in \mathcal{R}^n$ is the state of $\mathcal{S}$, and the function f: $\mathcal{T} \times \mathcal{R}_\rho \to \mathcal{R}^n$ is such that $f \in C(\mathcal{T} \times \mathcal{R}_\rho)$ and

$$f_i(t,x) \equiv \hat{f}_i(t, x_i, e_{i1}x_1, e_{i2}x_2, \ldots, e_{is}x_s). \tag{2.12}$$

Here $x_i \in \mathcal{R}^{n_i}$, and e_{ij}: $\mathcal{T} \to [0,1]$ is such that $e_{ij} \in C(\mathcal{T})$ are elements of $s \times s$ interconnection matrices $E \in \bar{E}$. Furthermore,

$$f(t,0) = 0 \qquad \forall t \in \mathcal{T}, \tag{2.7}$$

and the system $\mathcal{S}$ has an equilibrium $x^* = 0$ for all interconnection matrices $E \in \bar{E}$. Now we prove the following:

Theorem 2.5. *Suppose there exists a function v: $\mathcal{T} \times \mathcal{R}_\rho \to \mathcal{R}^s_+$ with the following properties: $v \in C(\mathcal{T} \times \mathcal{R}_\rho)$, $v(t,x)$ is locally Lipschitzian in x, and $v(t,0) \equiv 0$; a function ν: $\mathcal{T} \times \mathcal{R}_\rho \to \mathcal{R}_+$ defined as $d^T v(t,x)$ for some constant positive vector $d \in \mathcal{R}^s_+$ satisfies the inequalities*

$$\phi_1(\|x\|) \leqslant \nu(t,x) \leqslant \phi_2(\|x\|) \qquad \forall (t,x) \in \mathcal{T} \times \mathcal{R}_\rho, \tag{2.85}$$

where $\phi_1, \phi_2 \in \mathcal{K}$; and the function

$$D^+ v(t,x) = \limsup_{h \to 0^+} \frac{1}{h}\{v[t+h, x+hf(t,x)] - v(t,x)\}, \tag{2.86}$$

defined with respect to Equation (2.1), *satisfies a differential inequality*

$$D^+ v(t,x) \leqslant w[t, v(t,x)] \qquad \forall (t,x) \in \mathcal{T} \times \mathcal{R}_\rho \tag{2.87}$$

for all $E \in \bar{E}$, where the function w: $\mathcal{T} \times \mathcal{R}^s_+ \to \mathcal{R}^s$ belongs to the class $\mathcal{W}$, and $w(t,0) \equiv 0$, so that the differential equation

$$\dot{r} = w(t,r) \tag{2.32}$$

has a solution $r^* = 0$.

Then asymptotic stability of the solution $r^* = 0$ of the differential equation (2.32) *implies asymptotic connective stability of the equilibrium $x^* = 0$ of the system $\mathfrak{S}$ described by the differential equation* (2.1), *and $v(t, x)$ is a vector Liapunov function for the system $\mathfrak{S}$.*

Proof. Let us first show that under the conditions of the theorem, stability of $r^* = 0$ implies connective stability of $x^* = 0$.

Since the inequality (2.87) holds for all $E \in \bar{E}$, stability of $r^* = 0$ is connective. Therefore, if $0 < \varepsilon < \rho$, $t_0 \in \mathfrak{T}$ are given, then for all $E \in \bar{E}$ and $\phi_1(\varepsilon)$, $t_0 \in \mathfrak{T}$, there exists a positive number $\Delta = \Delta(t_0, \varepsilon)$ such that

$$\sum_{i=1}^{s} d_i r_i(t; t_0, r_0) < \phi_1(\varepsilon) \qquad \forall t \in \mathfrak{T}_0 \tag{2.88}$$

provided

$$\sum_{i=1}^{s} d_i r_{i0} < \Delta. \tag{2.89}$$

We now choose $r_{i0} = v_i(t_0, x_0)$ for all $i = 1, 2, \ldots, s$. Since $v(t, x)$ is continuous and $v(t, 0) \equiv 0$, we can always find a positive number $\delta = \delta(t_0, \varepsilon)$ such that

$$\|x_0\| < \delta, \qquad \sum_{i=1}^{s} d_i v_i(t_0, x_0) < \Delta \tag{2.90}$$

hold simultaneously. From (2.90), we conclude that $x_0 \in \mathfrak{R}_\delta$ implies

$$x(t; t_0, x_0) \subset \mathfrak{R}_\varepsilon \qquad \forall t \in \mathfrak{T}_0, \quad \forall E \in \bar{E}, \tag{2.91}$$

and thus connective stability of $x^* = 0$. Suppose that this conclusion is not true, and that there exists $t_1 > t_0$ such that for some $(t_0, x_0) \in \mathfrak{T} \times \mathfrak{R}_\rho$ and some $E \in \bar{E}$, we have $x(t; t_0, x_0) \subset \mathfrak{R}_\rho$, $t \in [t_0, t_1)$, but

$$\|x(t_1; t_0, x_0)\| = \varepsilon. \tag{2.92}$$

Then we have

$$\phi_1(\varepsilon) \leqslant \sum_{i=1}^{s} d_i v_i[t_1, x(t_1; t_0, x_0)]. \tag{2.93}$$

For $t \in [t_0, t_1]$, we can apply Lemma 2.1 to get

$$v[t, x(t; t_0, x_0)] \leqslant r_M(t; t_0, r_0) \qquad \forall t \in [t_0, t_1], \tag{2.94}$$

where $r_M(t; t_0, r_0)$ is the maximal solution of (2.32), and conclude that for all $E \in \bar{E}$,

$$\sum_{i=1}^{s} d_i v_i[t_1, x(t_1; t_0, x_0)] \leqslant \sum_{i=1}^{s} d_i r_{Mi}(t_1; t_0, r_0) \qquad \forall t \in [t_0, t_1]. \quad (2.95)$$

From the choice $r_{i0} = v_i(t_0, x_0)$ and (2.90), $x_0 \in \mathfrak{R}_\delta$ guarantees that (2.89) is satisfied. Thus, combining the inequalities (2.88) and (2.95), we get

$$\sum_{i=1}^{s} d_i v_i[t_1, x(t_1; t_0, x_0)] \leqslant \sum_{i=1}^{s} d_i r_i(t_1; t_0, r_0) < \phi_1(\varepsilon), \qquad (2.96)$$

which contradicts (2.93).

Let us now show that convergence of the solutions of Equation (2.32) implies connective convergence of solutions of Equation (2.1).

Assume again that $0 < \varepsilon < \rho$, $t_0 \in \mathfrak{T}$ are given. Then asymptotic stability of $r^* = 0$ implies that for all $E \in \overline{E}$, $\phi_1(\varepsilon) > 0$, and $t_0 \in \mathfrak{T}$, there exist two positive numbers $\Lambda = \Lambda(t_0, \varepsilon)$ and $\tau = \tau(t_0, \varepsilon)$ such that

$$\sum_{i=1}^{s} d_i r_{i0} \leqslant \Lambda \qquad (2.97)$$

implies

$$\sum_{i=1}^{s} d_i r_i(t; t_0, r_0) \leqslant \phi_1(\varepsilon) \qquad \forall t \geqslant t_0 + \tau. \qquad (2.98)$$

Again, we choose $r_{i0} = v_i(t_0, x_0)$, and conclude that there exists a $\gamma = \gamma(t_0, \varepsilon)$ for which

$$\|x_0\| < \gamma, \qquad \sum_{i=1}^{s} d_i v_i(t, x_0) \leqslant \Lambda \qquad (2.99)$$

hold simultaneously. Let $\mu = \min\{\gamma, \Delta_\rho\}$, where $\Delta_\rho = \Delta(t_0, \rho)$. Therefore, (2.97) is satisfied whenever $x_0 \in \mathfrak{R}_\mu$. Furthermore, since $x^* = 0$ is connectively stable, (2.94) holds for all $E \in \overline{E}$ and all $t \in \mathfrak{T}_0$. We can now assert that for all $(t_0, x_0) \in \mathfrak{T} \times \mathfrak{R}_\mu$,

$$x(t; t_0, x_0) \subset \mathfrak{R}_\varepsilon \qquad \forall t \geqslant t_0 + \tau. \qquad (2.100)$$

Let us suppose that contradictory to (2.100), there exists a sequence $\{t_k\}$, $t_k \geqslant t_0 + \tau$, with $t_k \to +\infty$ as $k \to \infty$, such that for some $E \in \overline{E}$ there is a solution $x(t; t_0, x_0)$ with $(t_0, x_0) \in \mathfrak{T} \times \mathfrak{R}_\mu$ and with the property

$$\|x(t_k; t_0, x_0)\| = \varepsilon. \qquad (2.101)$$

This implies

$$\phi_1(\varepsilon) \leqslant \sum_{i=1}^{s} d_i v_i[t_k; x(t_k; t_0, x_0)] \leqslant \sum_{i=1}^{s} d_i r_i(t_k; t_0, r_0) < \phi_1(\varepsilon), \quad (2.102)$$

which is absurd. Therefore, $x^* = 0$ is asymptotically connectively stable, and the proof of Theorem 2.5 is completed.

It is not difficult to show that under the conditions of Theorem 2.5, asymptotic connective stability is also uniform. That is, the numbers δ, μ, and τ can be chosen independently of t_0.

In applications, we are quite often interested in establishing global connective stability. For this purpose we have to assume first that in the equation (2.1), the function $f(t,x)$ is defined, continuous, and bounded on $\mathcal{T} \times \mathcal{R}^n$, so that solutions $x(t; t_0, x_0)$ of (2.1) exist for all $(t_0, x_0) \in \mathcal{T} \times \mathcal{R}^n$ and $\tau \in \mathcal{T}_0$. From Theorem 2.5, we derive

Theorem 2.6. *Suppose that all the conditions in Theorem 2.5 are valid for* $\mathcal{R}_\rho = \mathcal{R}^n$ *and* $\phi_1 \in \mathcal{K}_\infty$. *Then asymptotic stability in the large of the solution* $r^* = 0$ *of the differential equation* (2.32) *implies asymptotic connective stability in the large of the equilibrium* $x^* = 0$ *of the system* $\mathcal{S}$ *described by the differential equation* (2.1).

Proof. Since $\phi(\zeta) \to +\infty$ as $\zeta \to +\infty$, we have

$$\lim_{\varepsilon \to +\infty} \delta(t_0, \varepsilon) = +\infty \qquad \forall t_0 \in \mathcal{T} \tag{2.103}$$

and $\gamma = +\infty$ for all $E \in \bar{E}$. This proves Theorem 2.6.

By imitating this section and using a number of results on vector Liapunov functions outlined by Lakshmikantham and Leela (1969), one can extend and generalize considerably the concept of connective stability presented here. Our interest in the next section, however, is to apply the concept to the stability analysis of large-scale dynamic systems.

2.5. LARGE-SCALE DYNAMIC SYSTEMS

We turn our attention now to dynamic systems that are composed of a number of interconnected subsystems. Mathematical descriptions of such systems are obtained by adding more structure to the equation

$$\dot{x} = f(t, x) \tag{2.1}$$

and writing it in the form

$$\dot{x}_i = g_i(t, x_i) + h_i(t, x), \qquad i = 1, 2, \ldots, s. \tag{2.2}$$

In this new description of the system $\mathcal{S}$, the functions $g_i(t, x_i)$ represent the isolated subsystems $\mathcal{S}_i$, whereas the functions $h_i(t, x)$ describe the interactions among them.

As demonstrated in Section 1.8, the concept of the vector Liapunov function enables us to determine the stability of a large-scale dynamic system from the stability of its subsystems and the nature of their interactions (on the subsystem level), and the stability of the corresponding aggregate model (on the overall system level). Therefore, to be able to use the concept formalized by Theorems 2.5 and 2.6, we need first to establish the stability of each subsystem when decoupled, and then construct the aggregate model involving the vector Liapunov function. This is the spirit of stability analysis introduced by Matrosov (1962) and Bellman (1962), and used ingeniously by Bailey (1966) to study the stability of composite systems. Our interest in this section is to develop a decomposition-aggregation method for connective-stability analysis of large-scale systems in the context of vector Liapunov functions, and study the stability of the system $\mathbb{S}$ under structural perturbations whereby subsystems $\mathbb{S}_i$ are disconnected and again connected in various ways during the operation of $\mathbb{S}$.

For the system $\mathbb{S}$ to be connectively stable, it is obvious that each subsystem $\mathbb{S}_i$ should be stable when isolated. Therefore, let us consider the ith subsystem $\mathbb{S}_i$, described by the equation

$$\dot{x}_i = g_i(t, x_i), \tag{2.3}$$

where $x_i(t) \in \mathcal{R}^{n_i}$ is the state of $\mathbb{S}_i$, and $g_i\colon \mathcal{T} \times \mathcal{R}^{n_i} \to \mathcal{R}^{n_i}$ is defined and continuous on the domain $\mathcal{T} \times \mathcal{R}_\rho$ with $\mathcal{T} = [\tau, +\infty)$ and $\mathcal{R}_{\rho i} = \{x_i \in \mathcal{R}^{n_i}\colon \|x_i\| < \rho_i\}$. We consider solutions $x_i(t; t_0, x_{i0})$ of (2.3) on $\mathcal{T}_0$ for $(t_0, x_{i0}) \in \mathcal{T} \times \mathcal{R}_{\rho i}$. We recall the requirement

$$g_i(t, 0) = 0 \qquad \forall t \in \mathcal{T}, \tag{2.20}$$

so that $\mathbb{S}_i$ has an equilibrium at $x_i^* = 0$.

Now we establish the classical result that goes back to Liapunov himself:

Theorem 2.7. *Suppose there exists a function $v_i\colon \mathcal{T} \times \mathcal{R}_{\rho i} \to \mathcal{R}_+$ with the following properties: $v_i \in C(\mathcal{T} \times \mathcal{R}_{\rho i})$, $v_i(t, x_i)$ is locally Lipschitzian in x_i, $v_i(t, 0) \equiv 0$, and*

$$\left.\begin{aligned} \phi_{1i}(\|x_i\|) &\leqslant v_i(t, x_i) \leqslant \phi_{2i}(\|x_i\|), \\ D^+ v_i(t, x_i)_{(2.3)} &\leqslant -\phi_{3i}(\|x_i\|) \end{aligned}\right\} \quad \forall (t, x_i) \in \mathcal{T} \times \mathcal{R}_{\rho i}, \tag{2.104}$$

where $\phi_{ki} \in \mathcal{K}$, $k = 1, 2, 3$. Then the equilibrium $x_i^* = 0$ of the isolated subsystem is asymptotically stable, and $v_i(t, x_i)$ is a Liapunov function for $\mathbb{S}_i$.

Proof. We immediately notice that the last inequality in (2.104) can be rewritten as

$$D^+ v_i(t, x_i) \leqslant -\phi_{3i}(\phi_{1i}^I[v_i(t, x_i)]) \equiv -\tilde{\phi}_{3i}[v_i(t, x_i)], \tag{2.105}$$

where $\phi_{1i}^I(\mu)$ is the inverse function of $\phi_{1i}(\zeta)$, that is, $\phi_{1i}^I[\phi_{1i}(\zeta)] \equiv \zeta$. It is easy to see that if $\phi_{1i}(\zeta) \in \mathcal{K}$ for $\zeta \in [0, \zeta_0)$, and $\phi_{1i}(\zeta_0) = \mu_0$, then $\phi_{1i}^I(\mu)$ is defined (at least) for $\mu \in [0, \mu_0]$ and $\phi_{1i}^I \in \mathcal{K}$. Since $\phi_{i1}, \phi_{i1}^I \in \mathcal{K}$, from (2.105) we conclude that $\tilde{\phi}_{3i} \in \mathcal{K}$. Therefore, the inequalities (2.104) can be rewritten as

$$\left.\begin{aligned} \phi_{1i}(\|x_i\|) \leqslant v_i(t, x_i) \leqslant \phi_{2i}(\|x_i\|), \\ D^+ v_i(t, x_i) \leqslant -\tilde{\phi}_{3i}[v_i(t, x_i)] \end{aligned}\right. \quad \forall (t, x_i) \in \mathcal{T} \times \mathcal{R}_{\rho i}. \tag{2.106}$$

Following the proof of Theorem 2.5 when $s = 1$, we consider the scalar differential equation

$$\dot{r} = -\tilde{\phi}_{3i}(r) \tag{2.107}$$

for $(t_0, r_0) \in \mathcal{T} \times \mathcal{R}_+$ and $t \in \mathcal{T}_0$. Solutions of (2.107) are readily seen to be given as

$$r(t; t_0, r_0) = \varphi^I[\varphi(r_0) - (t - t_0)], \tag{2.108}$$

on $\mathcal{T}_0$, where the function φ: $\mathcal{R}_+ \to \mathcal{R}_+$ is determined as

$$\varphi(r) - \varphi(r_0) = \int_{r_0}^{r} [\tilde{\phi}_{3i}(\zeta)]^{-1} d\zeta = -(t - t_0) \tag{2.109}$$

and φ^I is the inverse function of φ. Now, given any $\varepsilon > 0$ we can choose $\delta(\varepsilon) = \varepsilon$ to establish the stability of the solution $r^* = 0$ in (2.107). Furthermore, we can easily show that for any $\varepsilon < \rho$, the choice $\gamma = \rho$ yields $r(t; t_0, r_0) < \varepsilon$ for all $t \geqslant t_0 + \tau$, where $\tau(\varepsilon) = \psi(\rho) - \psi(\varepsilon)$. Consequently, $r^* = 0$ is asymptotically stable. By Theorem 2.5 for $s = 1$, it follows that $x_i^* = 0$ is also asymptotically stable, and Theorem 2.7 is proved.

From the proof of Theorem 2.7, we see that the inequalities (2.104) or, equivalently, (2.106) actually imply uniform asymptotic stability of $x_i^* = 0$.

The last argument in the proof of Theorem 2.7, which follows Equation (2.109), can be reinterpreted using the comparison functions, as shown by Hahn (1967). If the function $[\tilde{\phi}_{3i}(\zeta)]^{-1}$ is integrable, then from (2.108) and (2.109) it follows that $r^* = 0$ is reached in finite time, which is incompatible with the Lipschitz condition. If $[\tilde{\phi}_{3i}(\zeta)]^{-1}$ is not integrable on an interval containing zero, then $-\varphi(r)$ is monotone increasing and unbounded as r tends to $r^* = 0$. Therefore, $-\varphi^I(r)$ is a function of class $\mathcal{L}$, and we can write (2.108) as $r(t; t_0, r_0) \leqslant \psi_1(t - t_0)\psi_2[-\varphi(r_0)]$, where $\psi_1, \psi_2 \in \mathcal{L}$. By using suitable notation, this last inequality can be rewritten as $r(t; t_0, r_0) \leqslant \phi_1(r_0)\psi_1(t - t_0)$ with $\phi_1 \in \mathcal{K}$, which implies the asymptotic stability of $r^* = 0$. From $v_{i0} = r_0$, $v_i[t, x_i(t; t_0, x_{i0})] \leqslant r(t; t_0, r_0)$, and the inequalities

(2.104), we get $\|x_i(t; t_0, x_{i0})\| \leqslant \phi_{1i}^I[\phi_1(r_0)\psi_1(t - t_0)] \leqslant \phi_{1i}^I\{\phi_1[\phi_{2i}(\|x_{i0}\|)]\psi_1(t - t_0)\} \leqslant \phi(\|x_{i0}\|)\psi(t - t_0)$, which is the inequality (2.79). Thus, the asymptotic stability of $x_i^* = 0$ is established.

It is now simple to extend Theorem 2.7 and get a global result like that of Theorem 2.6. The following theorem is almost automatic:

Theorem 2.8. *Suppose that all the conditions of Theorem 2.7 are valid for $\mathcal{R}_{pi} = \mathcal{R}^{n_i}$ and $\phi_{1i} \in \mathcal{K}_\infty$. Then the equilibrium $x_i^* = 0$ of $\mathcal{S}_i$ is asymptotically stable in the large.*

Proof. As in the proof of Theorem 2.6 (when $s = 1$), we show that $\phi_{1i} \in \mathcal{K}_\infty$ implies $\gamma = +\infty$ and $x_i^* = 0$ is globally asymptotically stable. This proves Theorem 2.8.

The global result of Theorem 2.8 was obtained by Barbashin and Krassovskii (1952).

By a single scalar Liapunov function $v_i(t, x_i)$ satisfying the inequalities (2.104) or (2.106), we aggregated the stability properties of the ith subsystem $\mathcal{S}_i$. In order to construct an aggregate model for the overall system $\mathcal{S}$, we impose constraints on the interaction functions $h_i(t, x)$ among the interconnected subsystems $\mathcal{S}_i$, described by

$$\dot{x}_i = g_i(t, x_i) + h_i(t, x), \qquad i = 1, 2, \ldots, s, \tag{2.2}$$

where the interconnection functions h_i: $\mathcal{T} \times \mathcal{R}^n \to \mathcal{R}^{n_i}$ are defined as

$$h_i(t, x) \equiv h_i(t, e_{i1}x_1, e_{i2}x_2, \ldots, e_{is}x_s). \tag{2.19}$$

Here, again, e_{ij}: $\mathcal{T} \to [0, 1]$, are continuous functions on $\mathcal{T}$ and represent elements of the $s \times s$ interconnection matrix E. We also impose certain magnitude constraints on the interactions $h_i(t, x)$ by the following:

Definition 2.14. *A continuous function h_i: $\mathcal{T} \times \mathcal{R}^n \to \mathcal{R}^{n_i}$ belongs to the class $\mathcal{K}_{(2.111)}$ if there exists a continuous function $\tilde{h}_i$: $\mathcal{T} \times \mathcal{R}_+^s \to \mathcal{R}_+$ defined as*

$$\tilde{h}_i(t, z) \equiv \tilde{h}_i(t, \bar{e}_{i1}z_1, \bar{e}_{i2}z_2, \ldots, \bar{e}_{is}z_s) \tag{2.110}$$

that is nondecreasing in z_i, $i = 1, 2, \ldots, s$, for each $t \in \mathcal{T}$, and such that for $z_i = \|x_i\|$ the inequality

$$\|h_i(t, x)\| \leqslant \tilde{h}_i(t, \bar{e}_{i1}\|x_i\|, \bar{e}_{i2}\|x_2\|, \ldots, \bar{e}_{is}\|x_s\|) \qquad \forall (t, x) \in \mathcal{T} \times \mathcal{R}_p \tag{2.111}$$

holds for all $E \in \bar{E}$.

Now we can prove the following result:

Theorem 2.9. *Suppose there exists a function* $v\colon \mathcal{T}\times\mathcal{R}_\rho \to \mathcal{R}_+^s$ *such that* $v \in C(\mathcal{T}\times\mathcal{R}_\rho)$, $v(t,x)$ *is locally Lipschitzian in* x, $v(t,0)\equiv 0$, *and*

$$\phi_{1i}(\|x_i\|) \leqslant v_i(t,x_i) \leqslant \phi_{2i}(\|x_i\|) \qquad \forall i = 1,2,\ldots,s, \quad \forall (t,x)\in\mathcal{T}\times\mathcal{R}_\rho, \tag{2.112}$$

where $\phi_{1i}, \phi_{2i} \in \mathcal{K}$, *and for each decoupled subsystem* $\mathcal{S}_i$ *we have*

$$D^+v_i(t,x_i)_{(2.3)} \leqslant \bar{g}_i[t, v_i(t,x_i)] \qquad \forall i = 1,2,\ldots,s, \quad \forall (t,x)\in\mathcal{T}\times\mathcal{R}_\rho, \tag{2.113}$$

where the function $\bar{g}\colon \mathcal{T}\times\mathcal{R}_+^s \to \mathcal{R}^s$ *is such that* $\bar{g}\in C(\mathcal{T}\times\mathcal{R}_+^s)$ *and* $\bar{g}(t,0) \equiv 0$. *Further suppose that the interconnection functions* $h_i\colon \mathcal{T}\times\mathcal{R}_\rho \to \mathcal{R}^{n_i}$ *are such that* $h_i \in C(\mathcal{T}\times\mathcal{R}_\rho)$, $h_i(t,0)\equiv 0$, *and* $h_i \in \mathcal{K}_{(2.111)}$ *for all* $i = 1, 2, \ldots, s$.

Then asymptotic stability of the solution $r^* = 0$ *of the equation*

$$\dot{r} = \bar{w}(t,r), \tag{2.32}$$

with

$$\bar{w}_i(t,r) \equiv \bar{g}_i(t,r_i) + \kappa_i \bar{h}_i(t,r), \qquad i = 1,2,\ldots,s, \tag{2.114}$$

where $\bar{h}(t,r) \equiv \tilde{h}_i[t, \bar{e}_{i1}\phi_{11}^I(r_1), \bar{e}_{i2}\phi_{12}^I(r_2), \ldots, \bar{e}_{is}\phi_{1s}^I(r_s)]$, $\phi_{1i}^I(r_i)$ *is the inverse function of* $\phi_{1i}(r_i)$, *and* κ_i *is the Lipschitz constant corresponding to* $v_i(t,x_i)$, *implies asymptotic connective stability of the equilibrium* $x^* = 0$ *of the system* $\mathcal{S}$; *and the differential equation* (2.32) *represents an aggregate model* $\mathcal{A}$ *for the system* $\mathcal{S}$.

Proof. Let us calculate the function $D^+v_i(t,x_i)_{(2.2)}$ as

$$\begin{aligned} D^+v_i(t,x_i)_{(2.2)} &= \limsup_{h\to 0^+} \frac{1}{h}\{v_i(t+h, x_i + h[g_i(t,x_i) + h_i(t,x)]) - v_i(t,x_i)\} \\ &= \limsup_{h\to 0^+} \frac{1}{h}\{v_i[t+h, x_i + hg_i(t,x_i)] - v_i(t,x_i) \\ &\qquad + v_i(t+h, h[g_i(t,x_i) + h_i(t,x)]) \\ &\qquad - v_i[t+h, x_i + hg_i(t,x_i)]\} \\ &\leqslant D^+v_i(t,x_i)_{(2.3)} + \kappa_i\|h_i(t,x)\| \\ &\leqslant \bar{g}_i[t, v_i(t,x_i)] + \kappa_i\bar{h}_i[t, v_i(t,x_i)] \\ &\leqslant \bar{w}_i[t, v_i(t,x_i)] \qquad \forall i = 1,2,\ldots,s, \quad \forall (t,x)\in\mathcal{T}\times\mathcal{R}_\rho \end{aligned} \tag{2.115}$$

and get the vector differential inequality

$$D^+ v(t,x) \leqslant \overline{w}[t, v(t,x)] \qquad \forall (t,x) \in \mathcal{T} \times \mathcal{R}_\rho, \tag{2.87}$$

where $\overline{w} \in \mathcal{W}$, and which is valid for all $E \in \overline{E}$. By applying Theorem 2.5, we establish Theorem 2.9.

It is again relatively simple to extend Theorem 2.9 and get the global result as

Theorem 2.10. *Suppose all the conditions of Theorem 2.9 are valid for $\mathcal{R}_\rho = \mathcal{R}^n$ and $\phi_{1i} \in \mathcal{K}_\infty$. Then asymptotic stability in the large of the equilibrium $r^* = 0$ of the aggregate representation $\mathcal{Q}$ implies asymptotic connective stability in the large of the equilibrium $x^* = 0$ of the system $\mathbb{S}$.*

Proof. The proof follows from Theorems 2.6 and 2.9.

There are several ways in which the aggregate $\mathcal{Q}$ can be constructed for a given system $\mathbb{S}$. They depend on the kind of connective stability that is of interest, and on the sort of constraints that are given for the interconnections. Let us first form an aggregate $\mathcal{Q}$ and establish conditions under which asymptotic stability of each subsystem $\mathbb{S}_i$ and the aggregate $\mathcal{Q}$ imply asymptotic connective stability of the overall system $\mathbb{S}$. We consider global stability, but the corresponding local result can be readily obtained imitating the corresponding theorems proved so far in this section. This construction of an aggregate $\mathcal{Q}$ was proposed first by Grujić and Šiljak (1973b).

We first need the following:

Definition 2.15. *A continuous function $h_i\colon \mathcal{T} \times \mathcal{R}^n \to \mathcal{R}^{n_i}$ belongs to the class $\mathcal{K}_{(2.116)}$ if there exist bounded functions $\xi_{ij}\colon \mathcal{T} \times \mathcal{R}^n \to \mathcal{R}$ such that*

$$\|h_i(t,x)\| \leqslant \sum_{j=1}^{s} \overline{e}_{ij} \xi_{ij}(t,x) \phi_{3j}(\|x_j\|) \qquad \forall (t,x) \in \mathcal{T} \times \mathcal{R}^n, \tag{2.116}$$

where $\overline{e}_{ij}$ are elements of an $s \times s$ fundamental interconnection matrix $\overline{E}$ and $\phi_{3j} \in \mathcal{K}$ for all $j = 1, 2, \ldots, s$.

Furthermore, we define an $s \times s$ constant aggretate matrix $\overline{W} = (\overline{w}_{ij})$ as

$$\overline{w}_{ij} = -\delta_{ij} + \kappa_i \overline{e}_{ij} \alpha_{ij}, \tag{2.117}$$

where

$$\delta_{ij} = \begin{cases} 1, & i = j, \\ 0, & i \neq j, \end{cases} \tag{2.118}$$

is the Kronecker symbol, κ_i is a positive number, and the nonnegative number α_{ij} is defined as

$$\alpha_{ij} = \max\{0, \sup_{\mathcal{T}\times\mathcal{R}^n} \xi_{ij}(t,x)\}. \tag{2.119}$$

We also need the following definition advanced by McKenzie (see Newman, 1959):

Definition 2.16. *An $s \times s$ constant matrix $\overline{W} = (\overline{w}_{ij})$ is said to be quasidominant diagonal if there exist positive numbers d_j such that*

$$d_j|\overline{w}_{jj}| > \sum_{\substack{i=1 \\ i\neq j}}^{s} d_i|\overline{w}_{ij}| \qquad \forall j = 1, 2, \ldots, s. \tag{2.120}$$

Let us prove the following (Šiljak, 1975b):

Theorem 2.11. *Suppose there exists a function $v\colon \mathcal{T}\times\mathcal{R}^n \to \mathcal{R}^s_+$ such that $v \in C(\mathcal{T}\times\mathcal{R}^n)$, $v(t,x)$ is locally Lipschitzian in x, $v(t,0) \equiv 0$, and for each decoupled subsystem $\mathcal{S}_i$ we have*

$$\begin{aligned} \phi_{1i}(\|x_i\|) \leqslant v_i(t,x_i) &\leqslant \phi_{2i}(\|x_i\|), \\ D^+v_i(t,x_i)_{(2.3)} &\leqslant -\phi_{3i}(\|x_i\|) \end{aligned} \qquad \forall(t,x_i) \in \mathcal{T}\times\mathcal{R}^{n_i}, \tag{2.121}$$

where $\phi_{1i}, \phi_{2i} \in \mathcal{K}_\infty$, $\phi_{3i} \in \mathcal{K}$. Suppose also that the interconnection functions $h_i(t,x)$ belong to the class $\mathcal{K}_{(2.116)}$ for all $i = 1, 2, \ldots, s$. Then the quasidominant diagonal property (2.120) *of the $s \times s$ aggregate matrix $\overline{W} = (\overline{w}_{ij})$ defined by* (2.117) *implies asymptotic connective stability in the large of the equilibrium $x^* = 0$ of the system $\mathcal{S}$.*

Proof. Following the proof of Theorem 2.9, we obtain

$$\begin{aligned} D^+v_i(t,x_i)_{(2.2)} \leqslant D^+v_i(t,x_i)_{(2.3)} + \kappa_i\|h_i(t,x)\| \qquad \forall i = 1, 2, \ldots, s, \\ \forall(t,x) \in \mathcal{T}\times\mathcal{R}^n. \end{aligned} \tag{2.122}$$

Since $h_i \in \mathcal{K}_{(2.116)}$, we can rewrite (2.122) as the differential inequality

$$D^+v(t,x) \leqslant \overline{W}q[v(t,x)] \qquad \forall(t,x) \in \mathcal{T}\times\mathcal{R}^n, \tag{2.123}$$

where the matrix W is defined by (2.117), and which is valid for all $E \in \overline{E}$. In (2.123) the vector function $q\colon \mathcal{R}^s_+ \to \mathcal{R}^s_+$ is defined as $q(v) \equiv (\phi_{31}[\phi^I_{11}(v_1)], \phi_{32}[\phi^I_{12}(v_2)], \ldots, \phi_{3s}[\phi^I_{1s}(v_s)])^T$. Then by Theorem 2.10, $x^* = 0$ is asymptotically connectively stable in the large if $r^* = 0$ of the equation

$$\dot{r} = \overline{W}q(r) \tag{2.124}$$

is globally asymptotically stable. To show that this last statement is true whenever $\overline{W}$ is quasidominant diagonal, let us consider the function $\nu: \mathcal{R}_+^s \to \mathcal{R}_+$ defined as

$$\nu(r) = \sum_{i=1}^{s} d_i r_i, \tag{2.125}$$

where the d_i's are positive numbers. The function $\nu(r)$ is positive definite, decrescent, and radially unbounded because of the inequalities (2.121). That is, there exist two functions $\phi_1, \phi_2 \in \mathcal{K}_\infty$ such that

$$\phi_1(\|r\|) \leqslant \nu(r) \leqslant \phi_2(\|r\|) \qquad \forall r \in \mathcal{R}_+^s. \tag{2.126}$$

Now we take the derivative $\dot{\nu}(r)$ along the solutions $r(t) = r(t; t_0, r_0)$ of (2.124) to get

$$\begin{aligned} \dot{\nu}[r(t)]_{(2.124)} &= \sum_{i=1}^{s} d_i \sum_{j=1}^{s} w_{ij} \phi_{3j}(\phi_{1j}^I[r_j(t)]) \\ &= \sum_{j=1}^{s} d_j w_{jj} \phi_{3j}[\phi_{1j}^I(r_j)] + \sum_{j=1}^{s} \sum_{\substack{i=1 \\ i\neq j}}^{s} d_i w_{ij} \phi_{3j}(\phi_{1j}^I[r_j(t)]) \\ &\leqslant -\sum_{j=1}^{s} \left[d_j |\overline{w}_{jj}| - \sum_{\substack{i=1 \\ i\neq j}}^{s} d_i |\overline{w}_{ij}| \right] \phi_{3j}(\phi_{1j}^I[r_j(t)]) \qquad \forall r \in \mathcal{R}_+^s. \end{aligned} \tag{2.127}$$

If $\overline{W}$ is quasidominant diagonal, then from (2.120) we conclude that there exist positive numbers d_j and π such that

$$|\overline{w}_{jj}| - d_j^{-1} \sum_{\substack{i=1 \\ i\neq j}}^{s} d_i |\overline{w}_{ij}| \geqslant \pi \qquad \forall j = 1, 2, \ldots, s, \tag{2.128}$$

and (2.127) can be rewritten as

$$\dot{\nu}(r)_{(2.124)} \leqslant \phi_3(\|r\|) \qquad \forall r \in \mathcal{R}_+^s, \tag{2.129}$$

where the function $\phi_3 \in \mathcal{K}$ can be chosen as

$$\phi_3(\|r\|) = \pi \sum_{j=1}^{s} d_j \phi_{3j}[\phi_{1j}^I(r_j)]. \tag{2.130}$$

By Theorem 2.8, the inequalities (2.126) and (2.129) imply global asymptotic stability of $r^* = 0$, corresponding to Equation (2.124). Therefore, using Theorem 2.6, we infer asymptotic connective stability of the equilibrium $x^* = 0$ of the system $\mathbb{S}$, and thus complete the proof of Theorem 2.11.

A comment should be made about the matrix $\overline{W}$ defined by (2.117). The matrix $\overline{W}$ is a Metzler matrix, since

$$\overline{w}_{ij}\begin{cases} < 0, & i = j, \\ \geqslant 0, & i \neq j, \end{cases} \tag{2.131}$$

which is (1.4) of the preceding chapter. Metzler matrices are treated in detail in Appendix. Here, however, we should mention a few facts that are relevant to Theorem 2.11. Since $\overline{W}$ is a Metzler matrix, the quasidominant-diagonal conditions (2.120) are necessary and sufficient for stability of $\overline{W}$ (all eigenvalues of $\overline{W}$ have negative real parts). Furthermore, for Metzler matrices the quasidominant-diagonal conditions (2.120) are equivalent to Hicks conditions (Newman, 1959): All even principal minors of $\overline{W}$ are positive, and all odd principal minors of $\overline{W}$ are negative (Theorem A.2). As shown by Gantmacher (1960), for matrices with nonnegative off-diagonal elements ($\overline{w}_{ij} \geqslant 0, i \neq j$)—and thus for Metzler matrices—Hicks conditions are equivalent to the Sevastyanov-Kotelyanskii conditions

$$(-1)^k \begin{vmatrix} \overline{w}_{11} & \overline{w}_{12} & \cdots & \overline{w}_{1k} \\ \overline{w}_{21} & \overline{w}_{22} & \cdots & \overline{w}_{2k} \\ \cdots & \cdots & \cdots & \cdots \\ \overline{w}_{k1} & \overline{w}_{k2} & \cdots & \overline{w}_{kk} \end{vmatrix} > 0 \qquad \forall k = 1, 2, \ldots, s, \tag{2.132}$$

which involve only the leading principal minors of $\overline{W}$. The conditions (2.132) are more useful in the sense that only the coefficients $\overline{w}_{ij}$ of $\overline{W}$ are needed to test stability of $\overline{W}$.

It is easy to see that the following corollary to Theorem 2.11 is also true:

Corollary 2.1. *Under the conditions of Theorem 2.11, the quasidominant diagonal property of the aggregate matrix $\overline{W}$ implies absolute connective stability of the equilibrium $x^* = 0$ of the system $\mathbb{S}$.*

Proof. From the proof of Theorem 2.11, it is obvious that $x^* = 0$ is asymptotically connectively stable in the large for all $h_i(t, x) \in \mathcal{K}_{(2.116)}$ and all $i = 1, 2, \ldots, s$. By Definition 2.8, $x^* = 0$ is absolutely connectively stable, which proves Corollary 2.1.

To illustrate the application of Theorem 2.11, let us consider a linear constant system $\mathbb{S}$,

$$\dot{x} = Ax, \tag{2.133}$$

where $x(t) \in \mathcal{R}^n$ is the state of the system, and $A = (a_{ij})$ is an $n \times n$

constant matrix. The system is decomposed into s linear subsystems $\mathbb{S}_i$,

$$\dot{x}_i = A_i x_i + \sum_{\substack{j=i \\ j\neq i}}^{s} A_{ij} x_j, \qquad i = 1, 2, \ldots, s, \tag{2.134}$$

where $x_i(t) \in \mathcal{R}^{n_i}$ is the state of the subsystem $\mathbb{S}_i$, and A_i, A_{ij} are $n_i \times n_i$ and $n_i \times n_j$ submatrices of A.

We assume that each isolated subsystem $\mathbb{S}_i$,

$$\dot{x}_i = A_i x_i, \tag{2.135}$$

is stable, so that for each $n_i \times n_i$ positive definite symmetric matrix G_i there exists an $n_i \times n_i$ positive definite symmetric matrix H_i which is a solution of the Liapunov matrix equation

$$A_i^T H_i + H_i A_i = -G_i. \tag{2.136}$$

Then the function

$$V_i(x_i) = x_i^T H_i x_i \tag{2.137}$$

is a Liapunov function for the system $\mathbb{S}_i$ in (2.135), and satisfies the following estimates:

$$\begin{aligned} \lambda_m(H_i)\|x_i\|^2 \leqslant V_i(x_i) &\leqslant \lambda_M(H_i)\|x_i\|^2, \\ \dot{V}_i(x_i)_{(2.135)} &\leqslant -\lambda_m(G_i)\|x_i\|, \end{aligned} \tag{2.138}$$

where λ_m and λ_M are the minimum and maximum eigenvalues of the indicated matrices.

To derive conditions under which stability of each isolated subsystem $\mathbb{S}_i$ implies stability of the overall system $\mathbb{S}$, we consider the function $v_i(x_i) = V_i^{1/2}(x_i)$, which is

$$v_i(x_i) = (x_i^T H_i x_i)^{1/2}, \tag{2.139}$$

as a Liapunov function for $\mathbb{S}_i$. The function $v_i(x_i)$ is a Lipschitz function, and

$$|v_i(x_i) - v_i(y_i)| \leqslant \kappa_i \|x_i - y_i\|, \tag{2.140}$$

where $\kappa_i = \lambda_M(H_i)/\lambda_m^{1/2}(H_i)$, and $x_i \neq 0$.

We also note that

$$\|A_{ij} x_j\| \leqslant \xi_{ij} \|x_j\|, \tag{2.141}$$

where $\xi_{ij} = \lambda_M^{1/2}(A_{ij}^T A_{ij})$.

Now, following the proof of Theorem 2.11, we get

$$
\begin{aligned}
D^+ v_i(x_i)_{(2.134)} &\leqslant D^+ v_i(x_i)_{(2.135)} + \kappa_i \sum_{\substack{j=i \\ j\neq i}}^{s} \xi_{ij} \|x_j\| \\
&\leqslant \tfrac{1}{2} V_i^{-1/2}(x_i)\dot{V}_i(x_i)_{(2.135)} + \kappa_i \sum_{\substack{j=i \\ j\neq i}}^{s} \xi_{ij} \|x_j\| \\
&\leqslant -\phi_{3i}(\|x_i\|) + 2\kappa_i \sum_{\substack{j=i \\ j\neq i}}^{s} \lambda_M^{1/2}(H_j)\lambda_m^{-1}(G_j)\xi_{ij}\,\phi_{3j}(\|x_j\|),
\end{aligned}
\tag{2.142}
$$

where $\phi_{3i}(\|x_i\|) = \frac{1}{2}\lambda_M^{-1/2}(H_i)\lambda_m(G_i)\|x_i\|$. When we consider the Liapunov function $\nu(v) = d^T v$ of (2.125) for the overall system $\mathbb{S}$, we get $\dot{\nu}(v)_{(2.133)} \leqslant d^T W q(v)$, where the $s \times s$ aggregate matrix $W = (w_{ij})$ is defined by

$$
w_{ij} = \begin{cases} -1, & i = j, \\ 2\kappa_i \lambda_M^{1/2}(H_j)\lambda_m^{-1}(G_j)\xi_{ij}, & i \neq j. \end{cases} \tag{2.143}
$$

By applying Theorem 2.11 we conclude that the overall system $\mathbb{S}$ is stable if the aggregate matrix W satisfies the conditions (2.132), that is, if W is stable. If in (2.134) $A_{ij}x_j$ is replaced by $e_{ij}A_{ij}x_j$, stability of $\overline{W}$ implies connective stability of $\mathbb{S}$.

Since the normlike function $v_i(x_i)$ in (2.139) is differentiable almost everywhere, in applications of the function to concrete problems the ordinary derivative $\dot{v}_i(x_i)$ will be used. Simple detailed arguments can be made to justify this practice and remove the ambiguity in $\dot{v}_i(x_i)$ at the point $x_i = 0$.

In the context of Theorem 2.11, it is possible to trade nondifferentiability of the vector Liapunov function for a somewhat different class of constraints imposed on the interactions among the subsystems. This alternative was made available by Grujić and Šiljak (1973b).

Let us denote grad $v_i(t, x_i) = (\partial v_i/\partial x_{i1}, \partial v_i/\partial x_{i2}, \ldots, \partial v_i/\partial x_{is})^T$ and introduce the following:

Definition 2.17. *A continuous function $h_i\colon \mathcal{T} \times \mathcal{R}^n \to \mathcal{R}^{n_i}$ belongs to the class $\mathcal{K}_{(2.144)}$ if there exist a continuously differentiable function $v_i\colon \mathcal{T} \times \mathcal{R}^{n_i} \to \mathcal{R}_+$ and bounded functions $\xi_{ij}\colon \mathcal{T} \times \mathcal{R}^n \to \mathcal{R}$ such that*

$$
[\text{grad } v_i(t, x_i)]^T h_i(t, x) \leqslant \sum_{j=1}^{s} \bar{e}_{ij}\xi_{ij}(t, x)\phi_{3j}(\|x_j\|) \qquad \forall (t, x) \in \mathcal{T} \times \mathcal{R}^n, \tag{2.144}
$$

where $\bar{e}_{ij}$ are elements of an $s \times s$ fundamental interconnection matrix $\overline{E}$ and $\phi_{3j} \in \mathcal{K}$ for all $j = 1, 2, \ldots, s$.

A somewhat different aggregate $s \times s$ matrix $\overline{W}$ is associated with the class $\mathcal{K}_{(2.144)}$, which is defined by

$$\overline{w}_{ij} = -\delta_{ij} + \overline{e}_{ij}\alpha_{ij}, \tag{2.145}$$

where α_{ij} is again computed as in (2.119).

Now, we can prove the following:

Theorem 2.12. *Suppose there exists a function v: $\mathcal{T} \times \mathcal{R}^n \to \mathcal{R}_+^s$ such that $v \in C^1(\mathcal{T} \times \mathcal{R}^n)$, $v(t, 0) \equiv 0$, and for each decoupled subsystem $\mathcal{S}_i$ we have*

$$\begin{aligned} \phi_{1i}(\|x_i\|) \leqslant v_i(t, x_i) \leqslant \phi_{2i}(\|x_i\|), \\ \dot{v}_i(t, x_i)_{(2.3)} \leqslant -\phi_{3i}(\|x_i\|) \end{aligned} \qquad \forall(t, x_i) \in \mathcal{T} \times \mathcal{R}^{n_i}, \tag{2.121}$$

where ϕ_{1i}, $\phi_{2i} \in \mathcal{K}_\infty$, $\phi_{3i} \in \mathcal{K}$. Suppose also that the interconnection functions $h_i(t, x)$ belong to the class $\mathcal{K}_{(2.144)}$ for all $i = 1, 2, \ldots, s$. Then the quasidominant diagonal property (2.120) *of the $s \times s$ aggregate matrix $\overline{W} = (\overline{w}_{ij})$ defined by* (2.145) *implies asymptotic connective stability in the large of the equilibrium $x^* = 0$ of the system $\mathcal{S}$.*

Proof. By taking the total time derivative of the function $v_i(t, x_i)$ along the solutions of Equation (2.2) and using the inequalities (2.121), (2.144), and (2.145), we obtain

$$\begin{aligned} \dot{v}_i(t, x_i)_{(2.2)} &= \dot{v}_i(t, x_i)_{(2.3)} + [\text{grad } v_i(t, x_i)]^T h_i(t, x) \\ &\leqslant -\phi_{3i}(\|x_i\|) + \sum_{j=1}^{s} \overline{e}_{ij}\xi_{ij}(t, x)\phi_{3j}(\|x_j\|) \\ &\leqslant \sum_{j=1}^{s} \overline{w}_{ij}\phi_{3j}(\|x_j\|) \qquad \forall(t, x) \in \mathcal{T} \times \mathcal{R}^n \end{aligned} \tag{2.146}$$

for $i = 1, 2, \ldots, s$ and all $E \in \overline{E}$. From (2.146), we can again get the aggregate model $\mathcal{Q}$ as

$$\dot{r} = \overline{W}q(r), \tag{2.124}$$

but with a different aggregate matrix $\overline{W}$. This new matrix $\overline{W}$ is again a Metzler matrix, as is obvious from (2.145). Repeating the same argument as in Theorem 2.11, we show that the quasidominant-diagonal property of the matrix $\overline{W}$ is sufficient for global asymptotic stability of $r^* = 0$ in (2.124). This, in turn, implies global asymptotic connective stability of the equilibrium $x^* = 0$ of the system $\mathcal{S}$, and the proof of Theorem 2.12 is complete.

We now turn our attention to the exponential property of connective stability. In various applications, it is not sufficient to guarantee that the motions of a large system converge to the equilibrium under all admissible structural perturbations. It is equally important to estimate the speed of the convergence, since that is a natural quantitative measure of the transient processes in the system.

Let us consider again the system $\mathcal{S}$,

$$\dot{x} = f(t, x), \tag{2.1}$$

which is composed of s interconnected subsystems $\mathcal{S}_i$,

$$\dot{x}_i = g_i(t, x_i) + h_i(t, x), \qquad i = 1, 2, \ldots, s, \tag{2.2}$$

where

$$\dot{x}_i = g_i(t, x_i), \qquad i = 1, 2, \ldots, s, \tag{2.3}$$

describe the isolated subsystems $\mathcal{S}_i$, and

$$h_i(t, x) \equiv h_i(t, e_{i1} x_1, e_{i2} x_2, \ldots, e_{is} x_s) \tag{2.19}$$

are the interactions. According to Definition 2.7, the system $\mathcal{S}$ is exponentially connectively stable in the large if all solutions $x(t; t_0, x_0)$ of (2.1) are such that the inequality

$$\|x(t; t_0, x_0)\| \leqslant \Pi \|x_0\| \exp[-\pi(t - t_0)] \qquad \forall t \in \mathcal{T}_0 \tag{2.18}$$

is satisfied for all $E \in \bar{E}$.

Let us first establish exponential stability for each subsystem when isolated. We prove the following:

Theorem 2.13. *Suppose there exists a function* $v_i\colon \mathcal{T} \times \mathcal{R}^{n_i} \to \mathcal{R}_+$ *such that* $v_i \in C(\mathcal{T} \times \mathcal{R}^{n_i})$, $v_i(t, x_i)$ *is Lipschitzian in* x_i *with a constant* $\kappa_i > 0$, $v_i(t, 0) \equiv 0$, *and*

$$\begin{aligned} \eta_i \|x_i\| \leqslant v_i(t, x_i) \leqslant \kappa_i \|x_i\|, \\ D^+ v_i(t, x_i)_{(2.3)} \leqslant -\pi_i v_i(t, x_i) \end{aligned} \qquad \forall (t, x_i) \in \mathcal{T} \times \mathcal{R}^{n_i}, \tag{2.147}$$

where η_i *and* π_i *are positive numbers. Then the equilibrium* $x_i^* = 0$ *of the isolated subsystem* $\mathcal{S}_i$ *is exponentially stable in the large.*

Proof. By applying the comparison principle of Theorem 2.3 to the last inequality in (2.147), we get

$$v_i[t, x_i(t; t_0, x_{i0})] \leqslant v_i(t_0, x_{i0}) \exp[-\pi_i(t - t_0)] \qquad \forall (t, x_i) \in \mathcal{T} \times \mathcal{R}^{n_i}. \tag{2.148}$$

From the first two inequalities (2.147), we obtain

$$\begin{aligned} \|x_i(t; t_0, x_{i0})\| &\leqslant \eta_i^{-1} v_i[t, x_i(t; t_0, x_{i0})], \\ v_i(t_0, x_{i0}) &\leqslant \kappa_i \|x_{i0}\|. \end{aligned} \tag{2.149}$$

This, together with (2. 147), yields

$$\|x_i(t; t_0, x_{i0})\| \leqslant \Pi_i \|x_{i0}\| \exp[-\pi_i(t - t_0)] \qquad \forall t \in \mathcal{T}_0, \qquad (2.150)$$

where $\Pi_i = \eta_i^{-1}\kappa_i$. The inequality (2.150) is valid for all $(t_0, x_{i0}) \in \mathcal{T} \times \mathcal{R}^{n_i}$ and implies global exponential stability of $x^* = 0$. Theorem 2.13 is proved.

It is simple to establish a "local" version of Theorem 2.13, when $\mathcal{R}^{n_i}$ is replaced by $\mathcal{R}_{\rho i}$ as in Theorem 2.7 on asymptotic stability.

When the subsystem Liapunov function $v_i(t, x_i)$ is continuously differentiable, we can use the following modification of Theorem 2.13:

Theorem 2.14. *Suppose there exists a function* $v_i: \mathcal{T} \times \mathcal{R}^{n_i} \to \mathcal{R}_+$ *such that* $v_i \in C^1(\mathcal{T} \times \mathcal{R}^{n_i})$, $v_i(t, 0) \equiv 0$, *and*

$$\begin{aligned} \eta_{1i}\|x_i\| &\leqslant v_i(t, x_i) \leqslant \eta_{2i}\|x_i\|, \\ \dot{v}_i(t, x_i)_{(2.3)} &\leqslant -\eta_{3i}\|x_i\|, \qquad\qquad \forall (t, x_i) \in \mathcal{T} \times \mathcal{R}^{n_i}, \quad (2.151) \\ \|\text{grad}\, v_i(t, x_i)\| &\leqslant \eta_{4i} \end{aligned}$$

where $\eta_{1i}, \eta_{2i}, \eta_{3i}, \eta_{4i}$ *are positive numbers. Then the equilibrium* $x_i^* = 0$ *of* $\mathcal{S}_i$ *is exponentially stable in the large.*

Proof. Following the steps of the proof of Theorem 2.13, but using the inequalities (2.151), we obtain the inequality (2.150) with $\Pi_i = \eta_{1i}^{-1}\eta_{2i}$, $\pi_i = \eta_{2i}^{-1}\eta_{3i}$, which establishes Theorem 2.14.

Theorem 2.14 is a slight modification of a result obtained by Krassovskii (1959), which has been used effectively for the stability analysis of large-scale systems when some (or all) subsystems are linear or nonlinear of the Lur'e-Postnikov type (Šiljak, 1972a). The use of estimates (2.151) in constructing aggregate models was illustrated in Section 1.8 of the preceding chapter. A generalization of this construction is presented next.

Let us now consider the system $\mathcal{S}$ of (2.2) and assume that the interconnections among the subsystems $\mathcal{S}_i$ belong to a class of functions specified by

Definition 2.18. *A continuous function* $h_i: \mathcal{T} \times \mathcal{R}^n \to \mathcal{R}^{n_i}$ *belongs to the class* $\mathcal{H}_{(2.152)}$ *if there exist nonnegative numbers* ξ_{ij} *such that*

$$\|h_i(t, x)\| \leqslant \sum_{j=1}^{s} \bar{e}_{ij}\xi_{ij}\|x_j\| \qquad \forall (t, x) \in \mathcal{T} \times \mathcal{R}^n, \qquad (2.152)$$

where $\bar{e}_{ij}$ *are elements of an* $s \times s$ *fundamental interconnection matrix* $\bar{E}$.

The aggregate matrix $\overline{W} = (\bar{w}_{ij})$ is specified by

$$\bar{w}_{ij} = -\pi_i\delta_{ij} + \bar{e}_{ij}\xi_{ij}\kappa_i\eta_j^{-1}, \qquad (2.153)$$

and we prove the following:

Theorem 2.15. *Suppose that there exists a function* $v\colon \mathcal{T} \times \mathcal{R}^n \to \mathcal{R}_+^s$ *such that* $v \in C(\mathcal{T} \times \mathcal{R}^n)$, $v(t,x)$ *is Lipschitzian in* x, $v(t,0) \equiv 0$, *and for each decoupled subsystem* $\mathcal{S}_i$ *we have*

$$\begin{aligned} \eta_i \|x_i\| \leqslant v_i(t,x_i) &\leqslant \kappa_i \|x_i\|, \\ D^+ v_i(t,x_i)_{(2.3)} &\leqslant -\pi_i v_i(t,x_i) \end{aligned} \qquad \forall (t,x_i) \in \mathcal{T} \times \mathcal{R}^{n_i}, \quad (2.147)$$

where η_i *and* π_i *are positive numbers, and* κ_i *is a Lipschitz constant associated with* $v_i(t,x_i)$. *Suppose also that the interconnection functions* $h_i(t,x)$ *belong to the class* $\mathcal{K}_{(2.152)}$ *for all* $i = 1, 2, \ldots, s$. *Let the* $s \times s$ *aggregate matrix* $\overline{W} = (\overline{w}_{ij})$ *defined by* (2.153) *satisfy the inequalities*

$$(-1)^k \begin{vmatrix} \overline{w}_{11} & \overline{w}_{12} & \cdots & \overline{w}_{1k} \\ \overline{w}_{21} & \overline{w}_{22} & \cdots & \overline{w}_{2k} \\ \cdots & \cdots & \cdots & \cdots \\ \overline{w}_{k1} & \overline{w}_{k2} & \cdots & \overline{w}_{kk} \end{vmatrix} > 0 \qquad \forall k = 1, 2, \ldots, s. \quad (2.132)$$

Then the equilibrium $x^* = 0$ *of the system* $\mathcal{S}$ *is exponentially connectively stable in the large.*

Proof. Using the inequalities (2.147) and (2.152), we compute the function $D^+ v_i(t,x_i)_{(2.2)}$ with respect to the equations (2.2) as in (2.115), to get

$$\begin{aligned} D^+ v_i(t,x_i)_{(2.2)} &\leqslant D^+ v_i(t,x_i)_{(2.3)} + \kappa_i \|h_i(t,x)\| \\ &\leqslant -\pi_i v_i(t,x_i) + \kappa_i \sum_{j=1}^{s} \bar{e}_{ij} \xi_{ij} \eta_j^{-1} v_j(t,x_j) \qquad \forall i = 1, 2, \ldots, s, \\ &\forall (t,x) \in \mathcal{T} \times \mathcal{R}^n. \end{aligned} \quad (2.154)$$

The last inequality is valid for all $E \in \overline{E}$.

From (2.154), we get the aggregate representation $\mathcal{A}$ of the system $\mathcal{S}$ as a linear constant differential equation

$$\dot{r} = \overline{W} r, \quad (2.155)$$

where $\overline{W}$ is the corresponding aggregate matrix. Since the conditions (2.132) imply stability of $\overline{W}$ and thus global asymptotic stability of $r^* = 0$, by using Theorem 2.10 we have asymptotic connective stability in the large of the equilibrium $x^* = 0$ of $\mathcal{S}$. To show that the stability is also exponential, let us consider again the function

$$\nu(r) = \sum_{i=1}^{s} d_i r_i, \quad (2.125)$$

where the d_i's are positive numbers. As in (2.127), but using (2.154), we get

$$\dot{\nu}[r(t)] \leqslant -\sum_{j=1}^{s}\left[d_j|\overline{w}_{jj}| - \sum_{\substack{i=1\\ i\neq j}}^{s} d_i|\overline{w}_{ij}|\right]r_j(t) \qquad \forall(t,r)\in \mathcal{T}\times\mathcal{R}_+^s. \tag{2.156}$$

Since $\overline{W}$ is a Metzler matrix, the fact that it satisfies the conditions (2.132) is equivalent to saying that $\overline{W}$ is a quasidominant diagonal matrix which, in turn, is equivalent to the conditions

$$|\overline{w}_{jj}| - d_j^{-1}\sum_{\substack{i=1\\ i\neq j}}^{s} d_i|\overline{w}_{ij}| \geqslant \pi \qquad \forall j = 1, 2, \ldots, s, \tag{2.128}$$

where $\pi > 0$. Now, from (2.156) and (2.128), we have

$$\dot{\nu}[r(t)] \leqslant -\pi\nu[r(t)] \qquad \forall(t,r)\in\mathcal{T}\times\mathcal{R}_+^s \tag{2.157}$$

and the inequality

$$\nu[r(t)] \leqslant \nu(r_0)\exp[-\pi(t-t_0)] \qquad \forall t\in\mathcal{T}_0, \tag{2.158}$$

which is valid for all $(t_0, r_0)\in\mathcal{T}\times\mathcal{R}_+^s$. By using the well-known relationship between the Euclidean norm $\|r\| = (\sum_{i=1}^{s} r_i^2)^{1/2}$ and the absolute-value norm $|r| = \sum_{i=1}^{s}|r_i|$, which is

$$\|r\| \leqslant |r| \leqslant s^{1/2}\|r\|, \tag{2.159}$$

and applying the inequalities (2.147), after simple calculations we arrive at the inequality

$$\|x(t; t_0, x_0)\| \leqslant \Pi\|x_0\|\exp[-\pi(t-t_0)] \qquad \forall t\in\mathcal{T}_0, \tag{2.18}$$

where

$$\Pi = s^{1/2} d_m d_M \eta_m^{-1}\kappa_M, \tag{2.160}$$

and $d_M = \max_i d_i$, $d_m = \min_i d_i$, $\eta_m = \min_i \eta_i$, $\kappa_M = \max_i \kappa_i$. Since the inequality (2.18) is valid for all $(t_0, x_0)\in\mathcal{T}\times\mathcal{R}^n$ and all $E\in\overline{E}$, $x^* = 0$ is exponentially connectively stable in the large, and Theorem 2.15 is established.

From the above proof the following corollary is automatic:

Corollary 2.2. *Under the conditions of Theorem 2.15, the equilibrium $x^* = 0$ of $\mathcal{S}$ is absolutely and exponentially connectively stable.*

Now, we can show how the exponential property of stability of each decoupled subsystem $\mathcal{S}_i$ guaranteed by Theorem 2.14, and the class of interconnections $\mathcal{K}_{(2.152)}$ specified in Definition 2.18, can be used to deduce the exponential property of connective stability for the overall system $\mathcal{S}$. We define the $s \times s$ aggregate matrix $\overline{W} = (\overline{w}_{ij})$ as

$$\overline{w}_{ij} = -\eta_{2i}^{-1}\eta_{3i}\delta_{ij} + \overline{e}_{ij}\xi_{ij}\eta_{4i}\eta_{1j}^{-1}, \tag{2.161}$$

and prove the following:

Theorem 2.16. *Suppose there exists a function* $v: \mathcal{T} \times \mathcal{R}^n \to \mathcal{R}_+^s$ *such that* $v \in C^1(\mathcal{T} \times \mathcal{R}^n)$, $v(t, 0) \equiv 0$, *and for each decoupled subsystem* $\mathcal{S}_i$ *we have*

$$\begin{aligned} \eta_{1i}\|x_i\| &\leqslant v_i(t, x_i) \leqslant \eta_{2i}\|x_i\|, \\ \dot{v}(t, x_i)_{(2.3)} &\leqslant -\eta_{3i}\|x_i\|, \qquad \forall (t, x_i) \in \mathcal{T} \times \mathcal{R}^{n_i}, \\ \|\text{grad } v_i(t, x_i)\| &\leqslant \eta_{4i} \end{aligned} \tag{2.151}$$

where $\eta_{1i}, \eta_{2i}, \eta_{3i}, \eta_{4i}$ *are positive numbers. Suppose also that the interconnection functions* $h_i(t, x)$ *belong to the class* $\mathcal{H}_{(2.152)}$ *for all* $i = 1, 2, \ldots, s$. *Let the* $s \times s$ *aggregate matrix* $\overline{W} = (\overline{w}_{ij})$ *defined by* (2.161) *satisfy the inequalities* (2.132). *Then the equilibrium* $x^* = 0$ *of the system* $\mathcal{S}$ *is exponentially connectively stable in the large.*

Proof. Let us prove the above theorem using the positive definite, decrescent, and radially unbounded function

$$\nu(t, x) = \sum_{i=1}^{s} d_i v_i(t, x_i) \tag{2.162}$$

as a candidate for a Liapunov function of the overall system $\mathcal{S}$. In (2.162), the d_i's are all positive (yet unspecified) numbers. Then the total time derivative $\dot{\nu}(t, x)_{(2.2)}$ along the solutions of the equations (2.2) is obtained as

$$\begin{aligned} \dot{\nu}(t, x)_{(2.2)} &= \sum_{i=1}^{s} d_i\{\dot{v}_i(t, x_i)_{(2.3)} + [\text{grad } v_i(t, x_i)]^T h_i(t, x)\} \\ &\leqslant \sum_{i=1}^{s} d_i\{\dot{v}_i(t, x_i)_{(2.3)} + \|\text{grad } v_i(t, x_i)\| \, \|h_i(t, x)\|\} \\ &\leqslant \sum_{i=1}^{s} d_i\left\{-\eta_{3i}\|x_i\| + \sum_{j=1}^{s} \overline{e}_{ij}\xi_{ij}\eta_{4i}\|x_j\|\right\} \\ &\leqslant \sum_{i=1}^{s} d_i\left\{-\eta_{3i}\eta_{2i}^{-1} v_i(t, x_i) + \sum_{j=1}^{s} \overline{e}_{ij}\xi_{ij}\eta_{4i}\eta_{1j}^{-1} v_i(t, x_i)\right\} \\ &\leqslant -\sum_{j=1}^{s}\left[d_j|\overline{w}_{jj}| - \sum_{\substack{i=1 \\ i \neq j}}^{s} d_i|\overline{w}_{ij}|\right] v_i(t, x_i) \\ &\leqslant -\pi\nu(t, x) \qquad \forall (t, x) \in \mathcal{T} \times \mathcal{R}^n. \end{aligned} \tag{2.163}$$

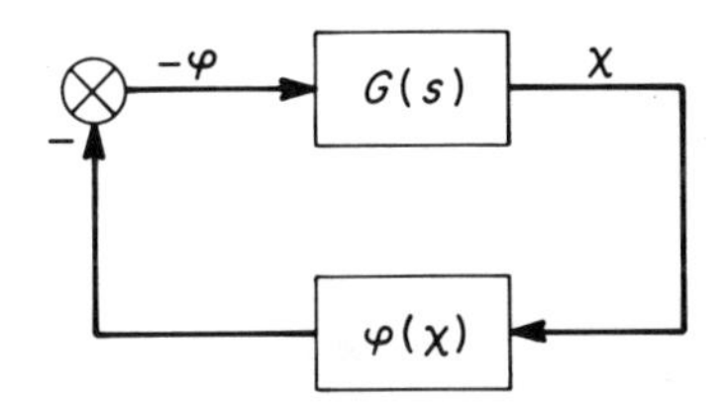

FIGURE 2.1. Lur'e–Postnikov system.

From the last inequality, which is valid for all $E \in \bar{E}$, we get

$$\nu[t, x(t)] \leqslant \nu(t_0, x_0)\exp[-\pi(t - t_0)] \qquad \forall t \in \mathcal{T}_0. \tag{2.164}$$

By applying the inequalities (2.159) relating the Euclidean and the absolute-value norm, we can manipulate (2.164) to get again the inequality (2.18) above, but now

$$\Pi = s^{1/2} d_M d_m^{-1} \eta_{2M} \eta_{1m}^{-1}, \tag{2.165}$$

where $d_M = \max_i d_i$, $d_m = \min_i d_i$, $\eta_{2M} = \max_i \eta_{2i}$, $\eta_{1m} = \min_i \eta_{1i}$. This proves Theorem 2.16.

It is apparent that the effectiveness of the decomposition-aggregation method for connective-stability analysis of large-scale systems, which is based upon the theorems presented in this section, depends crucially upon our ability to select appropriate Liapunov functions. Such selection is available in the case of linear systems and nonlinear systems of the Lur'e-Postnikov type. A connective stability analysis of linear systems was presented in Section 1.8 of the preceding chapter. In the rest of this section, we will consider the connective stability of interconnected Lur'e-Postnikov systems as proposed by Šiljak (1972a).

Let us first recall certain properties of subsystems which are of the Lur'e-Postnikov type (Šiljak, 1969), described by the equations

$$\dot{x}_i = A_i x_i + b_i \varphi_i(\chi_i), \qquad \chi_i = c_i^T x_i, \tag{2.166}$$

where $x_i(t) \in \mathcal{R}^{n_i}$ is the state of the subsystem $\mathcal{S}_i$, A_i is an $n_i \times n_i$ constant matrix, and b_i, $c_i \in \mathcal{R}^{n_i}$ are constant vectors. In presenting the properties of the subsystem $\mathcal{S}_i$ described by (2.166), it is convenient to drop the subscript i and refer to $\mathcal{S}_i$ as the system $\mathcal{S}$.

The block diagram of the system $\mathcal{S}$ described by (2.166) is shown in Figure 2.1. The *linear part* of $\mathcal{S}$ is represented by a scalar transfer function

$$G(s) = c^T(A - sI)^{-1}b, \tag{2.167}$$

which is a rational function in the complex variable $s = -\sigma + j\omega$. That is,

$$G(s) = \frac{r_m s^m + r_{m-1} s^{m-1} + \cdots + r_0}{s^n + q_{n-1} s^{n-1} + \cdots + q_0}, \tag{2.168}$$

where the coefficients r_k and q_k are all real and $n > m$. It is assumed that $G(s)$ is nondegenerate, which is equivalent to saying that the numerator and denominator polynomials have no common zeros. In terms of the Equations (2.166), it means that the pair (A, b) is completely controllable and the pair (A, c) is completely observable.

For the *nonlinear part* of the system $\mathbb{S}$, characterized by the nonlinear function $\varphi\colon \mathfrak{R} \to \mathfrak{R}$, we assume that $\varphi \in C(\mathfrak{R})$ and that $\varphi(\chi)$ belongs to the class of functions

$$\Phi = \{\varphi(\chi)\colon 0 \leqslant \chi\varphi(\chi) \leqslant \kappa\chi^2 \ \forall \chi \in \mathfrak{R};\ \varphi(0) = 0\}, \tag{2.169}$$

where κ is a constant such that $0 < \kappa < +\infty$. This simply means that the nonlinear function $\varphi(\chi)$ belongs to a sector in $\chi\varphi$ plane determined by the number κ, as shown in Figure 2.2. Furthermore, we assume that

$$\int_0^{c^T x} \varphi(\chi)\, d\chi \to +\infty \qquad \text{as} \quad \chi \to +\infty. \tag{2.170}$$

Now, we recall (Lur'e, 1951) that the system $\mathbb{S}$ is said to be *absolutely stable* if the equilibrium $x^* = 0$ of $\mathbb{S}$ is asymptotically stable in the large for

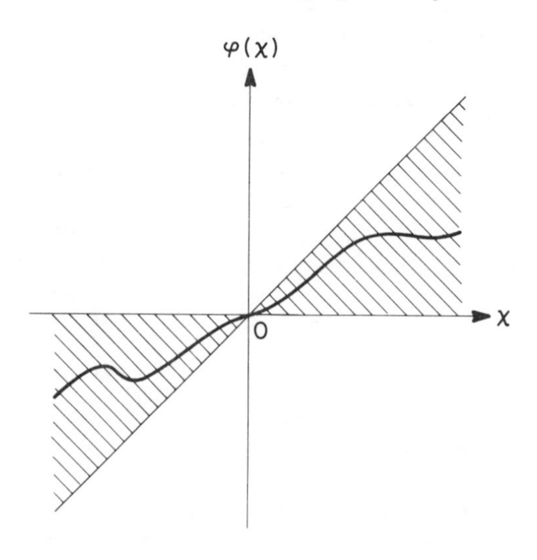

FIGURE 2.2. Nonlinear characteristic.

all $\varphi \in \Phi$. An effective criterion for this kind of stability was developed by Popov (1973). A system $\mathbb{S}$ is absolutely stable if the inequality

$$\kappa^{-1} + \mathrm{Re}[(1 + j\theta\omega)G(j\omega)] > 0 \qquad \forall\omega \in \mathcal{R}_+ \tag{2.171}$$

holds for a number $\theta \in \mathcal{R}$. The frequency condition (2.171) is appealing because it has a simple graphical interpretation (Popov, 1973) which leaves the choice of κ free and thus allows an evaluation of the largest possible class of functions Φ defined in (2.169). In that respect, the Popov frequency criterion is similar to that of Nyquist (1932).

It is also possible to formulate an algebraic criterion (Šiljak, 1969) for absolute stability, which is equivalent to Popov's condition. If we define a real, even polynomial

$$p(\omega) \equiv \sum_{k=0}^{n} p_{2k}\,\omega^{2k}, \qquad p_{2n} \neq 0, \tag{2.172}$$

as

$$p(\omega) = |\det(A - j\omega I)|^2\{\kappa^{-1} + \mathrm{Re}[(1 + j\theta\omega)c^T(A - j\omega I)^{-1}b]\} > 0 \qquad \forall\omega \in \mathcal{R}, \tag{2.173}$$

then it is obvious that the frequency condition (2.171) is equivalent to positivity of the polynomial $p(\omega)$, that is,

$$p(\omega) > 0 \qquad \forall\omega \in \mathcal{R}. \tag{2.174}$$

It is equally obvious that positivity of $p(\omega)$ is equivalent to the nonexistence of real zeros of $p(\omega)$ and $p_{2n} > 0$. Now, the existence of real zeros of $p(\omega)$ can be established by a modified Routh algorithm (Šiljak, 1969; Barnett and Šiljak, 1977). To see this, we note that zeros of $p(\omega)$ are symmetrically distributed with respect to both the imaginary and the real axis of the ω-plane. We consider the new polynomial

$$p(j\omega) \equiv \sum_{k=0}^{n} (-1)^k p_{2k}\,\omega^{2k} \tag{2.175}$$

and conclude that the previous symmetry is preserved, but real zeros of $p(\omega)$ (if there are any) become pure imaginary zeros of $p(j\omega)$. Consequently, if $p(j\omega)$ has n zeros with positive real parts, there are no positive real zeros of $p(\omega)$, and it is positive provided $p_{2n} > 0$. Therefore, we carry out the computation in the modified Routh table (Šiljak, 1969):

$$\begin{array}{c|ccccc} \omega^{2n} & (-1)^n p_{2n} & (-1)^{n-1} p_{2(n-1)} & \cdots & -p_2 & p_0 \\ \omega^{2n-1} & (-1)^n 2np_{2n} & (-1)^{n-1} 2(n-1) p_{2(n-1)} & \cdots & -2p_2 & \\ \vdots & \vdots & \vdots & & \vdots & \\ \omega^0 & p_0 & & & & \end{array} \tag{2.176}$$

where the first two rows are filled out by the coefficients of the two polynomials $p(j\omega)$ and $dp(j\omega)/d\omega$, and the other entries are computed as usual (e.g. Šiljak, 1969; Barnett and Šiljak, 1977).

Now, the system $\mathbb{S}$ is absolutely stable if the polynomial $p(\omega)$ produces the first column of the Routh table (2.176) so that

$$K[(-1)^n p_{2n}, (-1)^n 2np_{2n}, \ldots, p_0] = n, \qquad p_{2n} > 0, \tag{2.177}$$

where K denotes the number of sign changes in the sequence enclosed by the square brackets. It is clear that the algebraic condition (2.177) is equivalent to the Popov frequency condition (2.171).

For the stability analysis of large-scale systems composed of interconnected Lur'e-Postnikov systems, it is important to establish the existence of an appropriate Liapunov function, which can be used to aggregate the property of absolute stability of the system (2.166). It was shown by Šiljak and Sun (1972) that the conditions (2.171) or (2.177) can be used to establish the existence of a scalar Liapunov function of the Lur'e-Postnikov type,

$$V(x) = x^T Hx + \theta \int_0^{c^T x} \varphi(\chi)\, d\chi, \tag{2.178}$$

which satisfies the inequalities (2.151) and thus guarantees the exponential property of absolute stability for the system $\mathbb{S}$.

We compute the total time derivative $\dot{V}(x)$ of the function $V(x)$ along solutions of the equations (2.166) to get

$$-\dot{V}(x) = x^T Gx + (\gamma^{1/2}\varphi + l^T x)^2 + \Gamma, \tag{2.179}$$

where

$$\begin{aligned} -G &= A^T H + HA + ll^T, \\ -\gamma^{1/2} l &= Hb + \tfrac{1}{2}(\theta A^T + I)c, \\ -\gamma &= \theta c^T b - \kappa^{-1}, \\ -\Gamma &= (\kappa^{-1}\varphi - \chi)\varphi. \end{aligned} \tag{2.180}$$

By a result of Yakubovich (1962), we have that there exist a constant positive definite $n \times n$ matrix H and a constant n-vector l which satisfy the equations (2.180) if and only if the frequency condition (2.171) holds. Then, from (2.178) and (2.179), we conclude that (2.171) implies that for $\theta \geqslant 0$, the function $v(x)$ satisfies the inequalities

$$\begin{aligned} \lambda_m(H)\|x\|^2 \leqslant V(x) \leqslant \lambda_M(\tilde{H})\|x\|^2, \\ \dot{V}(x) \leqslant -\lambda_m(G)\|x\|^2, \qquad \forall x \in \mathcal{R}^n, \\ \|\text{grad } V(x)\| \leqslant 2\lambda_M(\tilde{H})\|x\| \end{aligned} \tag{2.181}$$

where λ_m and λ_M are the minimum and maximum eigenvalues of the indicated matrices and $\tilde{H} = H + \frac{1}{2}\kappa\theta cc^T$.

By substituting

$$v(x) = V^{1/2}(x) \tag{2.182}$$

and using the relations

$$\dot{v}(x) = \tfrac{1}{2}V^{-1/2}(x)\dot{V}(x), \qquad \text{grad } v(x) = \tfrac{1}{2}V^{-1/2}(x)\text{grad } V(x), \tag{2.183}$$

the inequalities (2.181) become

$$\begin{aligned} \eta_1\|x\| \leqslant v(x) \leqslant \eta_2\|x\|, \\ \dot{v}(x) \leqslant -\eta_3\|x\|, \qquad \forall x \in \mathcal{R}^n, \\ \|\text{grad } v(x)\| \leqslant \eta_4 \end{aligned} \tag{2.184}$$

which are those of (2.151), where

$$\begin{aligned} &\eta_1 = \lambda_m^{1/2}(H), \qquad \eta_2 = \lambda_M^{1/2}(\tilde{H}), \qquad \eta_3 = \tfrac{1}{2}\lambda_m(G)\lambda_M^{-1/2}(\tilde{H}), \\ &\eta_4 = \lambda_m^{-1/2}(H)\lambda_M(\tilde{H}). \end{aligned} \tag{2.185}$$

If $\theta < 0$, then the function $v(x)$ still satisfies the inequalities (2.184), but the roles of the matrices H and $\tilde{H}$ in the arguments of (2.185) are reversed.

Let us now consider a system $\mathbb{S}$ composed of s interconnected subsystems $\mathbb{S}_i$ of Lur'e-Postnikov type, described by the equations

$$\dot{x}_i = A_i x_i + b_i\varphi_i(\chi_i) + \sum_{j=1}^{s} e_{ij}h_{ij}, \qquad \chi_i = c_i^T x_i, \qquad i = 1, 2, \ldots, s. \tag{2.186}$$

For each decoupled subsystem (2.166) we assume that either the Popov condition (2.171) or the algebraic condition (2.177) is verified. Then, as shown by Šiljak and Sun (1972), we can solve the algebraic equations (2.180) and find a positive definite matrix H_i such that the function $v_i(x)$ defined in (2.178) is a Liapunov function for the decoupled subsystems $\mathbb{S}_i$. The matrices G_i, which were chosen in order to solve the equations (2.180), and the matrices H_i, $\tilde{H}_i$ are then used to compute the numbers in (2.185) necessasary for constructing an aggregate representation for the overall system $\mathbb{S}$. Furthermore, the interactions among the subsystems are assumed to belong to the class

$$\mathcal{H}_{(2.23)} = \left\{ h_{ij}(t, x_j) \colon \|h_{ij}(t, x_j)\| \leqslant \xi_{ij} \|x_j\| \ \forall (t, x_j) \in \mathcal{T} \times \mathcal{R}^{n_j} \right\}. \quad (2.24)$$

Now, by using Theorem 2.16, we conclude that if the $s \times s$ aggregate matrix $\overline{W} = (\overline{w}_{ij})$ defined by

$$\overline{w}_{ij} = -\tfrac{1}{2}\lambda_M^{-1}(\tilde{H}_i)\lambda_m(G_i)\delta_{ij} + \bar{e}_{ij}\xi_{ij}\lambda_m^{-1/2}(H_i)\lambda_M(\tilde{H}_i)\lambda_m^{-1/2}(H_j) \quad (2.187)$$

satisfies the inequalities (2.132), then the system $\mathbb{S}$ is absolutely and exponentially connectively stable. From Definition 2.8 and the definition of the class Φ in (2.169), this means the equilibrium state $x^* = 0$ of $\mathbb{S}$ is exponentially connectively stable in the large for all interconnections $h_{ij}(t, x_j) \in \mathcal{H}_{(2.23)}$ and all functions $\varphi_i(\chi_i) \in \Phi$.

It is interesting to note that the above system $\mathbb{S}$ is connectively stable even with respect to structural perturbations caused by the disconnection of nonlinear parts characterized by the functions $\varphi_i(\chi_i)$. This is clear from the definition of the class Φ in (2.169), which includes $\varphi_i(\chi_i) \equiv 0$.

Let us now be more specific, and consider a numerical example of the above type. A system $\mathbb{S}$ is composed of three interconnected second-order subsystems $\mathbb{S}_i$:

$$\dot{x}_i = \begin{bmatrix} 0 & 1 \\ -4 & -4 \end{bmatrix} x_i + \begin{bmatrix} 0 \\ -1 \end{bmatrix} \varphi_i(\chi_i) + \sum_{j=1}^{3} e_{ij} h_{ij}(\chi_j), \qquad i = 1, 2, 3. \quad (2.188)$$
$$\chi_i = [1 \quad 1] x_i$$

The subsystems are interconnected in a "ring" structure described by the 3×3 fundamental interconnection matrix

$$\overline{E} = \begin{bmatrix} 0 & 0 & 1 \\ 1 & 0 & 0 \\ 0 & 1 & 0 \end{bmatrix}, \quad (2.189)$$

and the digraph of Figure 2.3(a).

It was shown by Šiljak and Weissenberger (1970) that a free subsystem

$$\dot{x}_i = \begin{bmatrix} 0 & 1 \\ -4 & -4 \end{bmatrix} x_i + \begin{bmatrix} 0 \\ -1 \end{bmatrix} \varphi_i(\chi_i), \quad (2.190)$$
$$\chi_i = [1 \quad 1] x_i$$

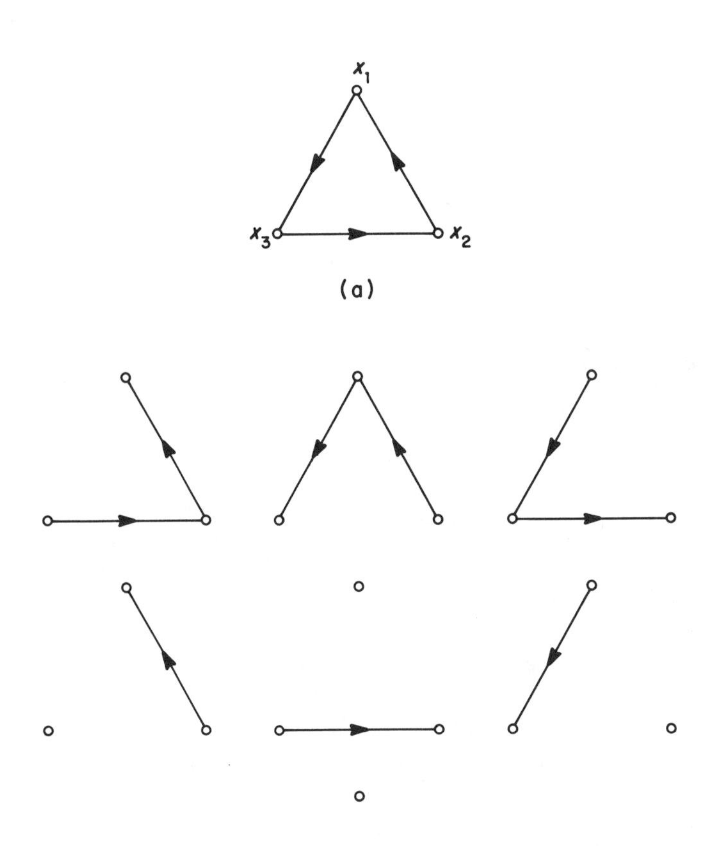

FIGURE 2.3. Structural perturbations.

is absolutely stable for all $\varphi_i \in \Phi$ where $\kappa_i = 41.6$, and that $V_i(x_i) = x_i^T H_i x_i$ is a Liapunov function for (2.188) with

$$H_i = \begin{bmatrix} 0.51 & 0.11 \\ 0.11 & 0.20 \end{bmatrix}, \qquad G_i = \begin{bmatrix} 0.35 & 0 \\ 0 & 0.35 \end{bmatrix}. \tag{2.191}$$

Since in this case $\theta_i = 0$, we have $\tilde{H} = H$, and

$$\lambda_m(H_i) = 0.27, \qquad \lambda_M(H_i) = 0.53, \qquad \lambda_m(G_i) = 0.35. \tag{2.192}$$

Now the function $v_i(x_i) = V_i^{1/2}(x_i) = (x_i^T H_i x_i)^{1/2}$ satisfies the inequalities (2.151) for

$$\eta_{1i} = 0.52, \quad \eta_{2i} = 0.73, \quad \eta_{3i} = 0.24, \quad \eta_{4i} = 3.23, \qquad i = 1, 2, 3. \tag{2.193}$$

Finally, from (2.187) we get the 3×3 aggregate matrix

$$\overline{W} = \begin{bmatrix} -0.33 & 0 & 1.92\xi_{13} \\ 1.92\xi_{21} & -0.33 & 0 \\ 0 & 1.92\xi_{32} & -0.33 \end{bmatrix}. \tag{2.194}$$

Applying the inequalities (2.132) to the matrix $\overline{W}$, we obtain the condition on interactions

$$\xi_{13}\xi_{21}\xi_{32} < 0.005, \tag{2.195}$$

which guarantees the absolute connective stability of the system $\mathbb{S}$. The corresponding structural perturbations of $\mathbb{S}$ are shown on Figure 2.3(b); they in turn correspond to binary interconnection matrices E obtained from $\overline{E}$ of (2.189) when the unit elements are replaced by zeros.

As a general comment concerning this secion, we can say that in a given system composed of a number of interconnected subsystems, the interactions among the subsystems can be viewed as perturbation terms in equations describing the subsystems. Then a large number of strong results from the theory of differential equations and inequalities (Lakshmikantham and Leela, 1969; Walter, 1970) become available for stability analysis of the overall system. Obviously, we have not exhausted all these results, but rather outlined those that we plan to use in the rest of this book.

Our use of vector Liapunov functions was made predominantly with regard to the connective stability of composite dynamic systems. Therefore, we did not exploit all the rich properties of such functions, nor did we point out possible analogous developments in other kinds of stability, such as input-output stability, stability on a finite time interval, practical stability, etc. After Bellman (1962) and Matrosov (1962) introduced the concept of vector Liapunov functions, Bailey (1966) proposed an efficient construction of such functions for composite systems. Bailey's results were significantly improved by Barbashin (1970), Thompson (1970), Michel (1970a, b, 1974), Weissenberger (1973), Grujić (1974, 1975b), LaSalle (1975), and many others, who considered a wide variety of stability properties of interconnected systems. A survey of these various results was given by Šiljak (1972b), and more recently up-to-date reviews were presented by Michel (1974) and by Athans, Sandell, and Varaiya (1975). Not included in these surveys are important input-output stability results obtained by Porter and

Michel (1974), Callier, Chan, and Desoer (1976), Willems (1976a), Vidyasagar (1977a), Sundareshan and Vidyasagar (1977), and Moylan and Hill (1978). It should be noted that most of the reported results on input-output stability of composite systems, can be rewritten one way or another in terms of connective stability as shown by Willems (1976a), and Sundareshan and Vidyasagar (1977). This adds yet another important aspect to the stability study of large-scale interconnected systems based upon the decomposition principle and techniques.

2.6. PARTIAL CONNECTIVE STABILITY

So far, we have required that the subsystems should be stable when isolated. This was a natural constraint, since we allowed the zero matrix to be an interconnection matrix, in which case all subsystems are decoupled from each other. As shown first by Grujić and Šiljak (1973b), unstable subsystems may be permitted to be parts of a large composite system provided the stabilizing negative feedback is present at all times. Now, if we carefully choose interconnection matrices which do not remove the stabilizing feedback paths, we can use the results of Grujić (1974) and permit unstable subsystems, at the price of achieving only *partial connective stability*.

Let us assume that a system $\mathcal{S}$ described by

$$\dot{x}_i = g_i(t, x_i) + h_i(t, x), \qquad i = 1, 2, \dots, s, \tag{2.2}$$

contains k stable subsystems $\mathcal{S}_i$ $(i = 1, 2, \dots, k)$, and $s - k$ unstable subsystems $\mathcal{S}_i$ $(i = k+1, k+2, \dots, s)$. We assume that the stability property of each isolated subsystem $\mathcal{S}_i$ is established by using a scalar function $v_i\colon \mathcal{T} \times \mathcal{R}^{n_i} \to \mathcal{R}_+$ such that $v_i \in C^1(\mathcal{T} \times \mathcal{R}^{n_i})$, $v_i(t, 0) \equiv 0$, and

$$\left.\begin{aligned} \phi_{1i}(\|x_i\|) &\leqslant v_i(t, x_i) \leqslant \phi_{2i}(\|x_i\|), \\ \mu_i \phi_{4i}(\|x_i\|) &\leqslant \dot{v}_i(t, x_i)_{(2.3)} \leqslant \mu_i \phi_{3i}(\|x_i\|) \end{aligned}\right\} \quad \forall (t, x_i) \in \mathcal{T} \times \mathcal{R}^{n_i}, \tag{2.196}$$

where $\phi_{1i} \in \mathcal{K}_\infty$ and $\phi_{2i}, \phi_{3i}, \phi_{4i} \in \mathcal{K}$. Here

$$\mu_i = \begin{cases} -1, & i = 1, 2, \dots, k \qquad \text{(stable } \mathcal{S}_i\text{)}, \\ +1, & i = k+1, k+2, \dots, s \qquad \text{(unstable } \mathcal{S}_i\text{)}. \end{cases} \tag{2.197}$$

That is, when $\mu_i = -1$, the inequalities (2.196) are those of (2.104) for $\mathcal{R}_{\rho i} = \mathcal{R}^{n_i}$, and by Theorem 2.8 the subsystem $\mathcal{S}_i$ is asymptotically stable in the large.

When $\mu_i = +1$, the inequalities (2.196) become

$$\begin{aligned}\phi_{1i}(\|x_i\|) &\leqslant v_i(t, x_i) \leqslant \phi_{2i}(\|x_i\|), \\ \phi_{4i}(\|x_i\|) &\leqslant \dot{v}_i(t, x_i)_{(2.3)} \leqslant \phi_{3i}(\|x_i\|)\end{aligned} \qquad \forall (t, x_i) \in \mathcal{T} \times \mathcal{R}^{n_i}. \tag{2.198}$$

From (2.198), we can get

$$\dot{v}_i(t, x_i)_{(2.3)} \geqslant \phi_{4i}(\phi_{1i}^I[v_i(t, x_i)]) \equiv \tilde{\phi}_{4i}[v_i(t, x_i)]. \tag{2.199}$$

The Lipschitzian nature of $v_i(t, x_i)$ and (2.199) yield

$$v_i(t, x_i) \geqslant v_i(t_0, x_{i0}) = r_0 > 0 \qquad \forall t \in \mathcal{T}_0 \tag{2.200}$$

for all $(t, x_i) \in \mathcal{T} \times \mathcal{R}^{n_i}$. From (2.199) and the comment following the proof of Theorem 2.3, we conclude that

$$v_i[t, x_i(t)] \geqslant r_m(t; t_0, r_0) \qquad \forall t \in \mathcal{T}_0, \tag{2.201}$$

where $r_m(t; t_0, r_0)$ is the minimal solution of

$$\dot{r} = \tilde{\phi}_{4i}(r), \tag{2.202}$$

with $r_0 > 0$ and $t_0 \in \mathcal{T}$. Similarly, as in the proof of Theorem 2.7, we can show that

$$r(t; t_0, r_0) \geqslant \phi_1(r_0)\varphi_1(t - t_0) \qquad \forall t \in \mathcal{T}_0, \tag{2.203}$$

where $\phi_1 \in \mathcal{K}$ and $\varphi_1(t - t_0)$ is a monotone increasing function. From (2.203), we can derive

$$\|x_i(t; t_0, x_{i0})\| \geqslant \phi(\|x_{i0}\|)\varphi(t - t_0) \qquad \forall t \in \mathcal{T}_0, \tag{2.204}$$

which is satisfied for all $(t_0, x_{i0}) \in \mathcal{T} \times \mathcal{R}^{n_i}$, with an appropriate choice of $\phi \in \mathcal{K}$ and a monotone increasing function $\varphi(t - t_0)$. The inequality (2.204) implies *global and complete instability* of $x_i^* = 0$ according to Hahn (1967).

In order to define partial connective stability, let us choose any binary $s \times s$ interconnection matrix $\hat{E} = (\hat{e}_{ij})$ generated from an $s \times s$ fundamental interconnection matrix $\bar{E} = (\bar{e}_{ij})$ by replacing a certain number of unit elements by zeros. We call the matrix $\hat{E}$ a *fixed interconnection matrix*. Now, we introduce the following:

Definition 2.19. *An $s \times s$ interconnection matrix $E = (e_{ij})$ is said to be generated by $s \times s$ fundamental and fixed interconnection matrices $\bar{E}$ and $\hat{E}$ if $\bar{e}_{ij} = 0$ implies $e_{ij} = 0$ and $\hat{e}_{ij} = 1$ implies $e_{ij} = 1$ for all $i, j = 1, 2, \ldots, s$.*

Obviously, Definition 2.2 follows from Definition 2.19 when $\hat{E} = 0$. Furthermore, classical stability follows from connective stability when $\hat{E} = \bar{E}$.

From Definition 2.19, it follows that interconnection matrices E are generated from $\bar{E}$ by replacing some unit elements of the matrix $\bar{E} - \hat{E}$ by zeros. This fact we denote by $E \in \bar{E} - \hat{E}$. Imitating Definition 2.5, but otherwise using $\bar{E} - \hat{E}$ instead of $\bar{E}$, we state the following:

Definition 2.20. *The equilibrium $x^* = 0$ of the system $\mathcal{S}$ is partially connectively stable for a pair $(\bar{E}, \hat{E})$ if it is stable in the sense of Liapunov for all $E \in \bar{E} - \hat{E}$.*

Let us now derive conditions for partial connective stability of the system $\mathcal{S}$ described by the equations (2.2) above. For this purpose, we assume that the interactions $h_i(t, x) \in \mathcal{K}_{(2.144)}$ such that

$$\xi_{ii}(t, x) < 0 \qquad \forall (t, x) \in \mathcal{T} \times \mathcal{R}^n, \quad \forall i = k+1, k+2, \ldots, s. \tag{2.205}$$

The $s \times s$ aggregate matrix $\bar{W} = (\bar{w}_{ij})$ is defined by

$$\bar{w}_{ij} = \mu_i \delta_{ij} + \bar{e}_{ij} \alpha_{ij}, \tag{2.206}$$

where now

$$\alpha_{ij} = \delta_{ij} \sup_{\mathcal{T} \times \mathcal{R}^n} \xi_{ij}(t, x) + (1 - \delta_{ij}) \max\{0, \sup_{\mathcal{T} \times \mathcal{R}^n} \xi_{ij}(t, x)\}, \tag{2.207}$$

and δ_{ij} is again the Kronecker symbol defined in (2.118).

Finally, we define the fixed $s \times s$ interconnection matrix $\hat{E} = (\hat{e}_{ij})$ as a diagonal binary matrix such that

$$\hat{e}_{ii} = \begin{cases} 0, & i = 1, 2, \ldots, k, \\ 1, & i = k+1, k+2, \ldots, s, \end{cases} \tag{2.208}$$

and $\hat{e}_{ij} = 0$ for all $i \neq j$.

Now, we can easily prove a corollary to Theorem 2.12 as the following:

Corollary 2.3. *Suppose there exists a function $v\colon \mathcal{T} \times \mathcal{R}^n \to \mathcal{R}^s_+$ such that $v \in C^1(\mathcal{T} \times \mathcal{R}^n)$, $v(t, 0) \equiv 0$, and $v_i(t, x_i)$ satisfies the inequalities* (2.196) *and* (2.144) *supplemented by* (2.205) *for each $i = 1, 2, \ldots, s$. Then the quasidominant diagonal property* (2.120) *of the $s \times s$ aggregate matrix $\bar{W} = (\bar{w}_{ij})$ defined by* (2.206) *implies that the equilibrium $x^* = 0$ of the system $\mathcal{S}$ is partially and asymptotically connectively stable in the large for a pair of interconnection matrices $(\bar{E}, \hat{E})$ defined by* (2.4) *and* (2.208).

Proof. By following the proof of Theorem 2.12, we get

$$\begin{aligned} \dot{v}_i(t, x_i)_{(2.2)} &= \dot{v}_i(t, x_i)_{(2.3)} + [\operatorname{grad} v_i(t, x_i)]^T h_i(t, x) \\ &\leqslant \mu_i \phi_{3i}(\|x_i\|) + \sum_{j=1}^{s} \bar{e}_{ij} \xi_{ij}(t, x) \phi_{3j}(\|x_j\|) \\ &\leqslant \sum_{j=1}^{s} \bar{w}_{ij} \phi_{3j}(\|x_j\|) \qquad \forall (t, x) \in \mathcal{T} \times \mathcal{R}^n \end{aligned} \tag{2.209}$$

for all $i = 1, 2, \ldots, s$ and all $E \in \bar{E}$. From (2.209), we get again the aggregate $\mathscr{A}$ as

$$\dot{r} = \bar{W}q(r), \tag{2.124}$$

but with the matrix $\bar{W}$ defined in (2.206). Now, since $\bar{W}$ is quasidominant diagonal, we have the inequalities

$$d_j|\bar{w}_{jj}| > \sum_{\substack{i=1\\i\neq j}}^{s} d_i|\bar{w}_{ij}| \qquad \forall j = 1, 2, \ldots, s, \tag{2.120}$$

which are obviously satisfied for all $E \in \bar{E} - \hat{E}$. This fact implies further that $x^* = 0$ of $\mathbb{S}$ is partially connectively stable in the large for the pair $(\bar{E}, \hat{E})$ and that stability is also asymptotic. The proof of Corollary 2.3 is completed.

To illustrate the application of Corollary 2.3, let us consider a system $\mathbb{S}$ composed of two interconnected subsystems $\mathbb{S}_1$ and $\mathbb{S}_2$, which is described by equations

$$\begin{aligned}\dot{x}_1 &= g_1(t, x_1) + h_1(x),\\ \dot{x}_2 &= g_2(t, x_2) + h_2(x),\end{aligned} \tag{2.210}$$

where $x = (x_1, x_2)^T$, $x_1 = (x_{11}, x_{12})^T$, $x_2 = (x_{21}, x_{22})^T$,

$$\begin{aligned}g_1(t, x_1) &= \begin{bmatrix} x_{12}\sin t - (2 + \cos t)x_{11}\|x_1\|^2 \\ -x_{11}\sin t - (2 + \cos t)x_{12}\|x_1\|^2 \end{bmatrix},\\ g_2(t, x_2) &= \begin{bmatrix} x_{22}\cos t + \tfrac{1}{3}(2 + \sin t)x_{21}\|x_2\|^2 \\ -x_{21}\cos t + \tfrac{1}{3}(2 + \sin t)x_{22}\|x_2\|^2 \end{bmatrix},\\ h_1(x) &= e_{12}\alpha\begin{bmatrix} x_{22}^3 \\ x_{21}^3 \end{bmatrix},\\ h_2(x) &= -e_{22}\beta\begin{bmatrix} x_{21}\|x_2\|^2\cosh x_{12} + x_{11}^3 \\ x_{22}\|x_2\|^2\cosh x_{11} + x_{12}^3 \end{bmatrix},\end{aligned} \tag{2.211}$$

and the digraph is shown in Figure 2.4 below.

The stability of the decoupled subsystems $\mathbb{S}_1$ and $\mathbb{S}_2$ was considered by Hahn (1967). Let us choose the following Liapunov functions for $\mathbb{S}_1$ and $\mathbb{S}_2$:

$$v_1(x_1) = \|x_1\|, \qquad v_2(x_2) = \|x_2\|. \tag{2.212}$$

By using the equations (2.210), we compute

$$\begin{aligned}\dot{v}_1(x_1) &= -(2 + \cos t)v_1^3(x_1), &\quad \text{grad } v_1(x_1) &= v_1^{-1}(x_1)x_1,\\ \dot{v}_2(x_2) &= \tfrac{1}{3}(2 + \sin t)v_2^3(x_2), &\quad \text{grad } v_2(x_2) &= v_2^{-1}(x_2)x_2.\end{aligned} \tag{2.213}$$

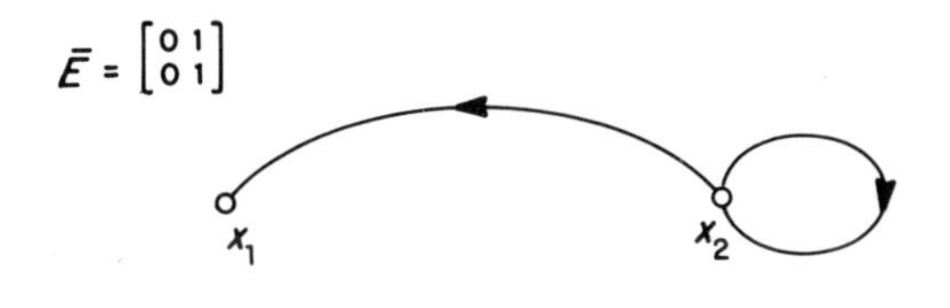

$\hat{E} = \begin{bmatrix} 0 & 0 \\ 0 & 1 \end{bmatrix}$

x_1 x_2

FIGURE 2.4. Partial connective stability.

From (2.212) and (2.213), we obtain

$$\begin{aligned} \phi_{11}(\|x_1\|) &= \phi_{21}(\|x_1\|) = \|x_1\|, & \phi_{31}(\|x_1\|) &= \|x_1\|^3 \\ & & \phi_{41}(\|x_1\|) &= 3\|x_1\|^3 \\ \phi_{12}(\|x_2\|) &= \phi_{22}(\|x_2\|) = \|x_2\|, & \phi_{32}(\|x_2\|) &= \|x_2\|^3, \\ & & \phi_{42}(\|x_2\|) &= \tfrac{1}{3}\|x_2\|^3, \end{aligned} \tag{2.214}$$

and $\mu_1 = -1$, $\mu_2 = +1$. Therefore, $\mathbb{S}_1$ is stable and $\mathbb{S}_2$ is unstable.

The inequalities (2.144) are in this case obtained as

$$\begin{aligned} [\text{grad } v_1(x_1)]^T h_1(x) &= e_{12}\alpha v_1^{-1}(x_1)(x_{11}x_{22}^3 + x_{12}x_{21}^3) \\ &\leqslant e_{12}\alpha v_1^{-1}(x_1)\|x_1\|(|x_{21}|^3 + |x_{22}|^3) \\ &\leqslant e_{12}\alpha\phi_{32}(\|x_2\|), \\ [\text{grad } v_2(x_2)]^T h_2(x) &= v_2^{-1}(x_2)[-e_{22}\beta\|x_2\|^2(x_{21}^2 \cosh x_{12} \\ &\quad + x_{22}^2 \cosh x_{11}) + x_{21}x_{11}^3 + x_{22}x_{12}^3] \\ &\leqslant v_2^{-1}(x_2)\|x_2\|[-e_{22}\beta\|x_2\|(x_{21}^2 + x_{22}^2) \\ &\quad + (|x_{11}|^3 + |x_{12}|^3)] \\ &\leqslant -e_{22}\beta\phi_{32}(\|x_2\|) + \phi_{31}(\|x_1\|), \end{aligned} \tag{2.215}$$

and we get

$$\begin{aligned} \dot{v}_1(x_1) &\leqslant -\phi_{31}(\|x_1\|) + e_{12}\alpha\phi_{32}(\|x_2\|), \\ \dot{v}_2(x_2) &\leqslant \phi_{31}(\|x_1\|) - (-1 + e_{22}\beta)\phi_{32}(\|x_2\|). \end{aligned} \tag{2.216}$$

From (2.216), the aggregate $\mathcal{Q}$ of the system $\mathbb{S}$ is obtained as

$$\dot{r} = \overline{W}q(r), \tag{2.124}$$

where

$$\overline{W} = \begin{bmatrix} -1 & \alpha \\ 1 & -(-1+\beta) \end{bmatrix}, \qquad q(r) = \begin{bmatrix} r_1^3 \\ r_2^3 \end{bmatrix}. \tag{2.217}$$

By applying the conditions (2.132), we conclude that the aggregate matrix W is stable if (and only if)

$$\alpha + 1 < \beta. \tag{2.218}$$

By Corollary 2.3, the system $\mathbb{S}$ is stable only under a structural perturbation described by the two matrices

$$\overline{E} = \begin{bmatrix} 0 & 1 \\ 0 & 1 \end{bmatrix}, \qquad \hat{E} = \begin{bmatrix} 0 & 0 \\ 0 & 1 \end{bmatrix}, \tag{2.219}$$

that is, $\mathbb{S}$ is partially connectively stable for the pair $(\overline{E}, \hat{E})$ defined by (2.219). To generate the interconnection matrices $E \in \overline{E} - \hat{E}$, one must satisfy the conditions

$$0 \leqslant e_{12} \leqslant 1, \qquad e_{22} = 1, \tag{2.220}$$

which means that only e_{12} can be varied between zero and one, while e_{22} must be fixed at one. The corresponding digraphs are shown in Figure 2.4.

We can extend the result of Corollary 2.3 so that the sign-negative interactions are represented by negative elements of the aggregate matrix, not only on the diagonal but off it as well. This extension, however, further restricts the allowable structural perturbations to principal perturbations only, which were introduced by Šiljak (1972a). In the remainder of this section, we will modify a result obtained by Grujić (1974) to include the principal structural perturbations, and get conditions for principal and asymptotic connective stability of dynamic systems.

Let us first define the principal structural perturbations by the use of interconnection matrices. We recall (Gantmacher, 1960) that for an $s \times s$ matrix $E = (e_{ij})$ the $q \times q$ principal submatrix is defined as

$$E_q = \begin{pmatrix} i_1\, i_2 \cdots i_q \\ i_1\, i_2 \cdots i_q \end{pmatrix} = \begin{bmatrix} e_{i_1 i_1} & e_{i_1 i_2} & \cdots & e_{i_1 i_q} \\ e_{i_2 i_1} & e_{i_2 i_2} & \cdots & e_{i_2 i_q} \\ \cdots & \cdots & \cdots & \cdots \\ e_{i_q i_1} & e_{i_q i_2} & \cdots & e_{i_q i_q} \end{bmatrix}, \tag{2.221}$$

where $1 \leqslant i_1 < i_2 < \cdots < i_q \leqslant s$, and $1 \leqslant q \leqslant s$. We state the following:

Definition 2.21. *An $s \times s$ interconnection matrix E is said to be a principal interconnection matrix E' if it can be transformed by only row and column transpositions into a quasidiagonal matrix*

$$\begin{bmatrix} E_{q_1} & & & 0 \\ & E_{q_2} & & \\ & & \ddots & \\ 0 & & & E_{q_l} \end{bmatrix} \tag{2.222}$$

with nonzero principal submatrices among $E_{q_1}, E_{q_2}, \ldots, E_{q_l}$ $(1 \leqslant l \leqslant s)$ equal to the corresponding principal submatrices of the $s \times s$ fundamental interconnection matrix $\bar{E}$.

We should note that in this case the zero $s \times s$ matrix is a principle interconnection matrix, since there are no nonzero principal submatrices to compare with the principal submatrices of the fundamental interconnection matrix. In terms of structural configurations, the system $\mathbb{S}$ is decomposed into (composed of) l independent groups of subsystems $\mathbb{S}_i$ under each principal structural perturbation. Equality of the corresponding principal submatrices of E' and $\bar{E}$ guarantees that the interconnections between the subsystems in each group remain the same as those represented by $\bar{E}$.

Now, we state the following:

Definition 2.22. *The equilibrium $x^* = 0$ of the system $\mathbb{S}$ is principally connectively stable if it is stable in the sense of Liapunov for all $E' \in \bar{E}$.*

Obviously, Definition 2.22 is broad enough to allow different kinds of Liapunov stability to be involved in a system analysis.

To be able to consider principal connective stability when unstable subsystems are present, it is necessary to combine Definition 2.22 with Definition 2.20 to get the following:

Definition 2.23. *The equilibrium $x^* = 0$ of the system $\mathbb{S}$ is partially and principally connectively stable for a pair $(\bar{E}, \hat{E})$ if it is stable in the sense of Liapunov for all $E' \in \bar{E} - \hat{E}$.*

The inclusion of the fixed interconnection matrix $\hat{E}$ is necessary in order to prevent the opening of a stabilizing feedback loop around unstable subsystems. Therefore, in the remainder of this section, the matrix $\hat{E}$ is defined by (2.208).

Let us now consider again the system $\mathbb{S}$ composed of s interconnected subsystems $\mathbb{S}_i$ described by the equations

$$\dot{x}_i = g_i(t, x_i) + h_i(t, x), \qquad i = 1, 2, \ldots, s. \tag{2.2}$$

In contrast with Corollary 2.3, we will now consider exponential stability of the system $\mathbb{S}$. Therefore, for the isolated subsystems $\mathbb{S}_i$ described by

$$\dot{x}_i = g_i(t, x_i), \qquad i = 1, 2, \ldots, s, \tag{2.3}$$

we assume that there exist continuously differentiable functions $V_i\colon \mathcal{T} \times \mathcal{R}^{n_i} \to \mathcal{R}_+$ which satisfy the inequalities

$$\begin{aligned} \eta_{1i}\|x_i\|^2 &\leqslant V_i(t, x_i) \leqslant \eta_{2i}\|x_i\|^2, \\ \mu_i\eta_{4i}\|x_i\|^2 &\leqslant \dot{V}_i(t, x_i)_{(2.3)} \leqslant \mu_i\eta_{3i}\|x_i\|^2 \end{aligned} \qquad \forall (t, x_i) \in \mathcal{T} \times \mathcal{R}^{n_i}, \tag{2.223}$$

where again

$$\mu_i = \begin{cases} -1, & i = 1, 2, \ldots, k \qquad \text{(stable } \mathbb{S}_i), \\ 1, & i = k+1, k+2, \ldots, s \qquad \text{(unstable } \mathbb{S}_i). \end{cases} \tag{2.197}$$

Let us specify a class $\mathcal{H}_{(2.224)}$ of interaction functions $h\colon \mathcal{T} \times \mathcal{R}^n \to \mathcal{R}^s$ defined by $h = (h_1^T, h_2^T, \ldots, h_s^T)^T$ by the following:

Definition 2.24. *A continuous function $h\colon \mathcal{T} \times \mathcal{R}^n \to \mathcal{R}^s$ belongs to the class $\mathcal{H}_{(2.224)}$ if there exist continuously differentiable functions $V_i\colon \mathcal{T} \times \mathcal{R}^{n_i} \to \mathcal{R}_+$ and numbers α_{ij} such that*

$$\sum_{i=1}^{s} [\operatorname{grad} V_i(t, x_i)]^T h_i(t, x) \leqslant \sum_{i,j=1}^{s} \bar{e}_{ij}\alpha_{ij}\|x_i\|\,\|x_j\| \qquad \forall (t, x) \in \mathcal{T} \times \mathcal{R}^n, \tag{2.224}$$

where $\bar{e}_{ij}$ are elements of an $s \times s$ fundamental interconnection matrix $\bar{E}$.

We also need to define the $s \times s$ aggregate matrix $\overline{W} = (\overline{w}_{ij})$ as

$$\overline{w}_{ij} = \mu_i\eta_{3i}\delta_{ij} + \bar{e}_{ij}\alpha_{ij}, \tag{2.225}$$

where again δ_{ij} is the Kronecker symbol defined in (2.118). We notice that sign-negative coupling among subsystems can be preserved in the majorization (2.224).

We prove the following:

Theorem 2.17. *Suppose there exists a function $V\colon \mathcal{T} \times \mathcal{R}^n \to \mathcal{R}_+^s$ such that $V \in C^1(\mathcal{T} \times \mathcal{R}^n)$, $v(t, 0) \equiv 0$, and for each decoupled subsystem $\mathbb{S}_i$ the inequalities (2.223) are satisfied. Suppose also that the interconnection function $h\colon \mathcal{T} \times \mathcal{R}^n \to \mathcal{R}^s$ belongs to the class $\mathcal{H}_{(2.224)}$. Then the negative definiteness property of the $s \times s$ aggregate matrix $\overline{W} = (\overline{w}_{ij})$ defined by (2.225) implies that the equilibrium $x^* = 0$ of the system $\mathbb{S}$ is partially and principally exponentially connectively stable in the large for a pair $(\overline{E}, \hat{E})$, where the fixed interconnection matrix $\hat{E}$ is defined by (2.208).*

Proof. Following Grujić (1974), we choose the positive definite, decrescent, and radially unbounded function

$$\nu(t,x) = \sum_{i=1}^{s} V_i(t,x_i) \tag{2.226}$$

as a candidate for Liapunov function of $\mathbb{S}$. By using (2.223), (2.224), and (2.225), we get from (2.226) the inequality

$$\dot{\nu}(t,x) \leqslant q^T(\|x\|)\overline{W}q(\|x\|) \qquad \forall(t,x) \in \mathcal{T}\times\mathcal{R}^n, \tag{2.227}$$

where $q(\|x\|) = (\|x_1\|, \|x_2\|, \ldots, \|x_s\|)^T$.

As usual, negative definiteness of $\overline{W}$ means that the quadratic form $q^T\overline{W}q$ is a negative definite function defined by (2.82). The inequality (2.227) can be further rewritten as

$$\dot{\nu}(t,x) \leqslant \lambda_M(\overline{Y})\|q(\|x\|)\|^2 \qquad \forall(t,x) \in \mathcal{T}\times\mathcal{R}^n, \tag{2.228}$$

where $\lambda_M(\overline{Y}) < 0$ is the maximum eigenvalue of the negative definite symmetric matrix $\overline{Y} = \frac{1}{2}(\overline{W} + \overline{W}^T)$. Since $\|q(\|x\|)\| = \|x\|$, (2.228) becomes

$$\dot{\nu}(t,x) \leqslant \lambda_M(\overline{Y})\|x\|^2 \qquad \forall(t,x) \in \mathcal{T}\times\mathcal{R}^n. \tag{2.229}$$

Using (2.223) again, we rewrite (2.228) as

$$\dot{\nu}(t,x) \leqslant \lambda_M(\overline{Y})\eta_{2M}^{-1}\nu(t,x) \qquad \forall(t,x) \in \mathcal{T}\times\mathcal{R}^n, \tag{2.230}$$

where $\eta_{2M} = \max_i \eta_{2i}$.

Now we recall the inclusion principle (Franklin, 1968) for negative (positive) definite matrices, which states that

$$\lambda_M(\overline{Y}) \geqslant \lambda_M(\overline{Y}'), \tag{2.231}$$

where $\overline{Y}'$ is any principal submatrix of the negative definite matrix $\overline{Y}$. Thus, the inequality (2.227), and therefore the inequality (2.230), are true for all $E' \in \overline{E} - \hat{E}$. This means principal connective stability.

To show that the stability of $\mathbb{S}$ is also exponential, we integrate (2.230) to get

$$\nu[t, x(t; t_0, x_0)] \leqslant \nu(t_0, x_0)\exp[-2\pi(t - t_0)], \tag{2.232}$$

where

$$\pi = -\tfrac{1}{2}\eta_{2M}^{-1}\lambda_M(\overline{Y}.) \tag{2.233}$$

From (2.232), we obtain

$$\|x(t; t_0, x_0)\| \leqslant \Pi \|x_0\| \exp[-\pi(t - t_0)] \qquad \forall t \in \mathcal{T}_0, \quad \forall (t_0, x_0) \in \mathcal{T} \times \mathcal{R}^n, \tag{2.18}$$

where

$$\Pi = \eta_{1m}^{-1/2} \eta_{2M}^{1/2}, \tag{2.234}$$

and $\eta_{1m} = \min_i \eta_{1i}$. This completes the proof of Theorem 2.17.

To illustrate the application of Theorem 2.17, let us consider a dynamic system $\mathcal{S}$ composed of two first-order subsystems described by the equations

$$\begin{aligned} \dot{x}_1 &= -\alpha_1 x_1 + e_{11} h_{11}(x) x_1 + e_{12} h_{12}(x) x_2, \\ \dot{x}_2 &= \alpha_2 x_2 + e_{21} h_{21}(x) x_1 + e_{22} h_{22}(x) x_2, \end{aligned} \tag{2.235}$$

where α_1, α_2 are positive numbers and the interconnection functions satisfy the constraints

$$\begin{aligned} h_{11}(x) x_1 &\leqslant \alpha_{11} x_1^2, & h_{12}(x) x_2 &\leqslant \alpha_{12} |x_1| |x_2|, \\ h_{21}(x) x_1 &\leqslant -\alpha_{21} |x_1| |x_2|, & h_{22}(x) x_2 &\leqslant -\alpha_{22} x_2^2, \end{aligned} \tag{2.236}$$

and α_{11}, α_{12}, α_{21}, α_{22} are all nonnegative numbers.

We choose

$$V_1(x_1) = \tfrac{1}{2} x_1^2, \qquad V_2(x_2) = \tfrac{1}{2} x_2^2 \tag{2.237}$$

and calculate

$$\begin{aligned} \text{grad } V_1(x_1) &= x_1, & \dot{V}_1(x_1) &= -\alpha_1 x_1^2, \\ \text{grad } V_2(x_2) &= x_2, & \dot{V}_2(x_2) &= \alpha_2 x_2^2 \end{aligned} \tag{2.238}$$

for the isolated subsystems $\mathcal{S}_1$ and $\mathcal{S}_2 (E = 0)$. Then, using (2.224) and (2.225), we compute the 2×2 aggregate matrix

$$\overline{W} = \begin{bmatrix} -\alpha_1 + \alpha_{11} & \alpha_{12} \\ -\alpha_{21} & \alpha_2 - \alpha_{22} \end{bmatrix}, \tag{2.239}$$

where the 2×2 fundamental and fixed interconnection matrices are

$$\overline{E} = \begin{bmatrix} 1 & 1 \\ 1 & 1 \end{bmatrix}, \qquad \hat{E} = \begin{bmatrix} 0 & 0 \\ 0 & 1 \end{bmatrix}. \tag{2.240}$$

Then $\overline{Y} = \frac{1}{2}(\overline{W} + \overline{W}^T)$ is computed as

$$\overline{Y} = \begin{bmatrix} -\alpha_1 + \alpha_{11} & \frac{1}{2}(\alpha_{12} - \alpha_{21}) \\ \frac{1}{2}(\alpha_{12} - \alpha_{21}) & \alpha_2 - \alpha_{22} \end{bmatrix}, \tag{2.241}$$

and the necessary and sufficient conditions for Y to be negative definite are obtained as

$$-\alpha_1 + \alpha_{11} < 0, \qquad (-\alpha_1 + \alpha_{11})(\alpha_2 - \alpha_{22}) - \tfrac{1}{4}(\alpha_{12} - \alpha_{21})^2 > 0, \tag{2.242}$$

which are equivalent to the condition $\lambda_M(\overline{Y}) < 0$. Thus, the conditions (2.242) imply that the system $\mathbb{S}$ described by (2.235) is partially and principally connectively stable for the pair $(\overline{E}, \hat{E})$ given in (2.241). By Theorem 2.17, the system is also exponentially stable.

The system digraph corresponding to the matrix $\overline{E}$ in (2.240) is shown in Figure 2.5(a). The partial and principal structural perturbations corresponding to the binary matrices E' that can be generated by the pair $(\overline{E}, \hat{E})$ of (2.240) are shown in Figure 2.5(b).

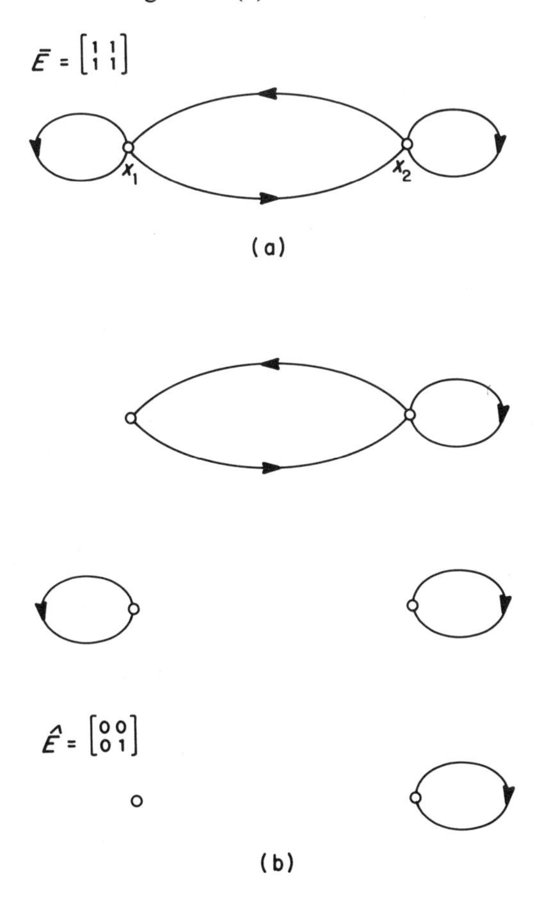

FIGURE 2.5. Principal and partial structural perturbations.

It is possible to modify the result of Grujić (1974) as proposed by Araki (1975) and remove the restriction "principal" from connective stability in Theorem 2.17. This modification, however, requires the matrix $\overline{W}$ of (2.25) to be a Metzler matrix defined by (2.131). This, as expected, implies that each subsystem is stable when isolated, that is, $\mu_i = -1$ for all $i = 1, 2, \dots, s$.

Following Araki (1975), we consider the function

$$\nu(t, x) = \sum_{i=1}^{s} d_i V_i(t, x), \tag{2.243}$$

where the d_i's are all positive, as a candidate for Liapunov's function for $\mathbb{S}$. By using (2.223), (2.224), and (2.225), we get from (2.243)

$$\dot{\nu}(t, x) \leqslant \tfrac{1}{2} q^T(\|x\|) \overline{U} q(\|x\|) \qquad \forall (t, x) \in \mathcal{T} \times \mathcal{R}^n, \tag{2.244}$$

where again $q(\|x\|) = (\|x_1\|, \|x_2\|, \dots, \|x_s\|)^T$. As distinct from (2.227), we have a matrix $\overline{U}$ defined as

$$\overline{W}^T D + D\overline{W} = \overline{U}, \tag{2.245}$$

where $D = \operatorname{diag}\{d_1, d_2, \dots, d_s\}$. As shown by Araki and Kondo (1974), $-\overline{W}$ is an $\mathfrak{M}$-matrix if and only if there is a positive definite matrix D such that the matrix $-\overline{U}$ is positive definite. We note that the Metzlerian property of W implies that $\overline{U}$ is also a Metzler matrix. Furthermore, $-\overline{W}$ being an $\mathfrak{M}$-matrix is equivalent to $\overline{W}$ being a negative quasidominant diagonal matrix (see Appendix), which is the matrix property used so far in connective stability analysis. Since $\overline{U}$ is a Metzler matrix, the positive definiteness of $-\overline{U}$ implies that it is also a Hicks matrix, that is, it satisfies the conditions (2.132). Therefore, we can have the same inequality (2.231) valid for the matrix U, which is valid not only for principal submatrices U' but for all matrices $E \in \overline{E}$ (see Appendix). If the inequality (2.244) is rewritten as that of (2.229), the inequality (2.231) for the matrix $\overline{U}$ implies exponential connective stability in the large of the system $\mathbb{S}$. Finally, if the norms are replaced by positive definite functions proposed by Araki (1975), we can broaden the obtained result to establish asymptotic connective stability in the large.

Let us illustrate the above obtained result using again the linear system $\mathbb{S}$,

$$\dot{x} = Ax, \tag{2.133}$$

which is represented as s interconnected subsystems $\mathbb{S}_i$,

$$\dot{x}_i = A_i x_i + \sum_{j=1}^{s} e_{ij} A_{ij} x_j, \qquad i = 1, 2, \dots, s. \tag{2.246}$$

we consider again

$$V_i(x_i) = x_i^T H_i x_i \tag{2.137}$$

as a Liapunov function for the ith isolated subsystem $\mathbb{S}_i$,

$$\dot{x}_i = A_i x_i. \tag{2.135}$$

We assume that $V_i(x_i)$ satisfies the following estimates:

$$\begin{aligned} \lambda_m(H_i)\|x_i\|^2 \leqslant V_i(x_i) &\leqslant \lambda_M(H_i)\|x_i\|^2, \\ \dot{V}_i(x_i)_{(2.135)} &\leqslant -\lambda_m(G_i)\|x_i\|, \\ \|\text{grad } V_i(x_i)\| &\leqslant 2\lambda_M(H_i)\|x_i\|. \end{aligned} \tag{2.247}$$

According to (2.224), (2.225) and (2.247), we have the $s \times s$ matrix $\overline{W} = (\overline{w}_{ij})$ specified as

$$\overline{w}_{ij} = -\lambda_m(G_i)\delta_{ij} + 2\bar{e}_{ij}\lambda_M(H_i)\xi_{ij}, \tag{2.248}$$

where ξ_{ij} is defined as in (2.141). The system $\mathbb{S}$ of (2.133) is connectively stable if the matrix $\overline{W}$ of (2.248) is a quasidominant diagonal matrix.

It is now of interest to compare the above result obtained using $V_i(x_i)$ of (2.137) with a result (Šiljak, 1972a) which is available when $v_i(x_i) = V_i^{1/2}(x_i)$, defined by

$$v_i(x_i) = (x_i^T H_i x_i)^{1/2}, \tag{2.249}$$

is used as in Section 2.5 in connection with the system $\mathbb{S}$ of (2.186). The function $v_i(x_i)$ has the estimates (2.184). When $b_i = 0$ for all $i = 1, 2, \ldots, s$, and $h_{ij}(t, x) \equiv A_{ij}x_j$, then the system $\mathbb{S}$ of (2.186) becomes that of (2.246). This implies that $\tilde{H}_i = H_i$ for all $i = 1, 2, \ldots, s$, and the corresponding aggregate matrix $\overline{W} = (\overline{w}_{ij})$ defined by (2.187) becomes $\hat{W} = (\hat{w}_{ij})$ specified by

$$\hat{w}_{ij} = -\lambda_M^{-1}(H_i)\lambda_m(G_i)\delta_{ij} + 2\bar{e}_{ij}\lambda_m^{-1/2}(H_i)\lambda_M(H_i)\lambda_m^{-1/2}(H_j)\xi_{ij}, \tag{2.250}$$

where each element of (2.187) is multiplied by two. Comparing (2.250) with (2.248), we conclude that $\overline{W} \leqslant \hat{W}$, where the inequality is taken element by element. This implies (see Appendix) that in the case of the linear system $\mathbb{S}$ of (2.246), $V_i(x_i)$ provides a less conservative stability criterion than $v_i(x_i)$. This conclusion also eliminates any question regarding the ambiguity of $\dot{v}_i(x_i)$ at the origin, such as was raised in Section 2.5. The two criteria, however, yield the same conservativeness when $v_i(x_i) = \|x_i\|$ is used for stabilization purposes in Chapter 3, since $H_i = I_i$, where I_i is the $n_i \times n_i$

identity matrix. Furthermore, it is not clear how to take advantage of the weaker restriction provided by $V_i(x_i)$ when stability regions are considered in Section 2.8. This is so because $\dot{v}_i(t, x_i)$ in (2.244) is expressed in terms of $\|x\|$ instead of $v_i(t, x_i)$, and the method of Weissenberger (1973) cannot be applied directly.

2.7. CONNECTIVE INSTABILITY

According to the connective-stability definitions, a system is connectively unstable if it is not stable for at least one interconnection matrix. This is a natural instability counterpart to stability under structural perturbations. One is tempted, however, to try to develop conditions under which a system is unstable for all interconnection matrices—this is *complete connective instability*. Our interest, therefore, is turned now to the following:

Definition 2.25. *The equilibrium $x^* = 0$ of the system $\mathcal{S}$ is completely connectively unstable if it is unstable in the sense of Liapunov for all interconnection matrices $E \in \bar{E}$.*

To derive conditions for complete connective instability of the system $\mathcal{S}$ given by equations

$$\dot{x}_i = g_i(t, x_i) + h_i(t, x), \qquad i = 1, 2, \ldots, s, \tag{2.2}$$

we assume that the interconnections $h_i(t, x)$ belong to the class $\mathcal{K}_{(2.251)}$ specified by the following:

Definition 2.26. *A continuous function h_i: $\mathcal{T} \times \mathcal{R}^n \rightarrow \mathcal{R}^{n_i}$ belongs to the class $\mathcal{K}_{(2.251)}$ if there exist a continuously differentiable function v_i: $\mathcal{T} \times \mathcal{R}^{n_i} \rightarrow \mathcal{R}_+$ and numbers α_{ij} such that*

$$[\text{grad } v_i(t, x_i)]^T h_i(t, x) \geqslant \sum_{j=1}^{s} \bar{e}_{ij} \alpha_{ij} \phi_{3j}(\|x_j\|) \qquad \forall (t, x) \in \mathcal{T} \times \mathcal{R}_\rho, \tag{2.251}$$

where $\bar{e}_{ij}$ are elements of an $s \times s$ fundamental interconnection matrix $\bar{E}$ and $\phi_{3j} \in \mathcal{K}$ for all $j = 1, 2, \ldots, s$.

We recall that $\mathcal{R}_\rho = \{x \in \mathcal{R}^n: \|x\| < \rho\}$ and $\mathcal{R}_{\rho i} = \{x_i \in \mathcal{R}^{n_i}: \|x_i\| < \rho\}$. We choose the $s \times s$ matrix $\bar{W} = (\bar{w}_{ij})$ as

$$\bar{w}_{ij} = \delta_{ij} + \bar{e}_{ij} \alpha_{ij} \tag{2.252}$$

with $\bar{w}_{ii}$'s all positive, and establish the following:

Theorem 2.18. *Suppose there exists a function v: $\mathcal{T} \times \mathcal{R}^n \rightarrow \mathcal{R}_+^s$ such that $v \in C^1(\mathcal{T} \times \mathcal{R}^n)$, $v(t, 0) \equiv 0$, and for each decoupled subsystem $\mathcal{S}_i$ we have*

$$\begin{aligned}&\phi_{1i}(\|x_i\|) \leqslant v_i(t,x_i) \leqslant \phi_{2i}(\|x_i\|),\\ &\dot{v}(t,x_i)_{(2.3)} \geqslant \phi_{3i}(\|x_i\|)\end{aligned} \qquad \forall(t,x_i) \in \mathcal{T}\times\mathcal{R}_{pi}, \tag{2.253}$$

where $\phi_{ki} \in \mathcal{K}(k = 1,2,3)$, *and the function* $v(t,x)$ *is such that* $h_i \in \mathcal{K}_{(2.251)}$ *for all* $i = 1, 2, \ldots, s$. *Then the quasidominant diagonal property* (2.128) *of the* $s \times s$ *aggregate matrix* $\overline{W} = (\overline{w}_{ij})$ *defined by* (2.252) *implies complete connective instability of the equilibrium* $x^* = 0$ *of the system* $\mathcal{S}$.

Proof. Following the results of Grujić and Šiljak (1973b), we choose again the function

$$\nu(t,x) = \sum_{i=1}^{s} d_i v_i(t,x_i) \tag{2.162}$$

as a candidate for Liapunov's function of the overall system $\mathcal{S}$. The derivative of $\nu(t,x)$ along the solutions of (2.2) is computed using (2.251), (2.252), and (2.253) as

$$\begin{aligned}\dot{\nu}(t,x)_{(2.2)} &= \sum_{i=1}^{s} d_i\{\dot{v}_i(t,x_i)_{(2.3)} + [\text{grad } v_i(t,x_i)]^T h_i(t,x)\}\\ &\geqslant \sum_{i=1}^{s} d_i\left\{\phi_{3i}(\|x_i\|) + \sum_{j=1}^{s} \bar{e}_{ij}\alpha_{ij}\phi_{3j}(\|x_i\|)\right\}\\ &\geqslant \sum_{j=1}^{s} d_j(1+\bar{e}_{jj}\alpha_{jj})\phi_{3j}(\|x_j\|) + \sum_{j=1}^{s}\phi_{3j}(\|x_j\|)\sum_{\substack{i=1\\ i\neq j}}^{s} d_i\bar{e}_{ij}\alpha_{ij}\\ &\geqslant \sum_{j=1}^{s} d_j\left[|\overline{w}_{jj}| - d_j^{-1}\sum_{\substack{i=1\\ i\neq j}}^{s} d_i|\overline{w}_{ij}|\right]\phi_{3j}(\|x_j\|)\\ &\geqslant \pi\sum_{j=1}^{s}\phi_{3j}(\|x_j\|) \qquad \forall(t,x)\in\mathcal{T}\times\mathcal{R}_p,\end{aligned} \tag{2.254}$$

which is valid for all $E \in \overline{E}$. Now, from (2.253) and (2.254), we conclude that for some $\phi_{\mathrm{I}}, \phi_{\mathrm{II}}, \phi_{\mathrm{III}} \in \mathcal{K}$ we have

$$\begin{aligned}&\phi_{\mathrm{I}}(\|x\|) \leqslant \nu(t,x) \leqslant \phi_{\mathrm{II}}(\|x\|),\\ &\dot{\nu}(t,x)_{(2.2)} \geqslant \phi_{\mathrm{III}}(\|x\|)\end{aligned} \qquad \forall(t,x) \in \mathcal{T}\times\mathcal{R}_p, \tag{2.255}$$

which is again valid for all $E \in \overline{E}$. Using the derivation following the conditions (2.198), we conclude from (2.255) that the equilibrium $x^* = 0$ is completely connectively unstable, and the proof of Theorem 2.18 is complete.

To illustrate the use of Theorem 2.18, let us consider a system $\mathcal{S}$ composed of two subsystems $\mathcal{S}_1$ and $\mathcal{S}_2$ given as

$$\begin{aligned} \dot{x}_1 &= g_1(x_1) + h_1(x), \\ \dot{x}_2 &= g_2(x_2) + h_2(x), \end{aligned} \tag{2.256}$$

where $x = (x_1^T, x_2^T)^T$, $x_1 = (x_{11}, x_{12})^T$, $x_2 = (x_{21}, x_{22})^T$,

$$g_1(x_1) = 2x_1, \qquad h_1(x) = \begin{bmatrix} x_{11}(x_{11}^2 + e_{12}\beta x_{21}) \\ x_{12}(x_{12}^2 + e_{12}\beta x_{22}) \end{bmatrix},$$

$$g_2(x_2) = x_2, \qquad h_2(x) = \begin{bmatrix} 2e_{21}\gamma \operatorname{sat}[\frac{1}{2}(x_{11} + e_{22}\delta x_{21})] \\ 2e_{21}\gamma \operatorname{sat}[\frac{1}{2}(x_{12} + e_{22}\delta x_{22})] \end{bmatrix}, \tag{2.257}$$

$$\operatorname{sat}\theta = \begin{cases} \theta, & |\theta| \leqslant 1 \\ 1, & |\theta| \geqslant 1 \end{cases}$$

β, γ, δ are positive numbers, and $\mathcal{R}_\rho = \{x \in \mathcal{R}^n\colon \|x\| < 1\}$. The structure of the system $\mathbb{S}$ is described by the digraph in Figure 2.6(a).

Let us choose

$$v_1(x_1) = \|x_1\|, \qquad v_2(x_2) = \|x_2\| \tag{2.258}$$

and calculate

$$\begin{aligned} \dot{v}_1(x_1) &= 2v_1(x_1), \qquad \operatorname{grad} v_1(x_1) = v_1^{-1}(x_1)x_1, \\ \dot{v}_2(x_2) &= v_2(x_2), \qquad \operatorname{grad} v_2(x_2) = v_2^{-1}(x_2)x_2. \end{aligned} \tag{2.259}$$

Now we calculate

$$\begin{aligned} [\operatorname{grad} v_1(x_1)]^T h_1(x) &= v_1^{-1}(x_1)[x_{11}^4 + x_{12}^4 + e_{12}\beta(x_{11}^2 x_{21} + x_{12}^2 x_{22}) \\ &\geqslant e_{12}\beta v_1^{-1}(x_1)(-\|x_1\|^2\|x_2\| - \|x_1\|^2\|x_2\|) \\ &\geqslant -2e_{12}\beta v_1^{-1}(x_1)\|x_1\|\,\|x_2\| \\ &\geqslant -2\bar{e}_{12}\beta v_2(x_2) \qquad \forall (t,x) \in \mathcal{T}\times\mathcal{R}_\rho, \\ [\operatorname{grad} v_2(x_2)]^T h_2(x) &= 2e_{21}\gamma v_2^{-1}(x_2)\{x_{21}\operatorname{sat}[\tfrac{1}{2}(x_{11} + e_{22}\delta x_{21})] \\ &\quad + x_{22}\operatorname{sat}[\tfrac{1}{2}(x_{11} + e_{22}\delta x_{22})] \\ &\geqslant -2e_{21}\gamma v_2^{-1}(x_2)[\tfrac{1}{2}\|x_2\|(\|x_1\| + e_{22}\delta\|x_2\|) \\ &\quad + \tfrac{1}{2}\|x_2\|(\|x_1\| + e_{22}\delta\|x_2\|) \\ &\geqslant -2e_{21}\gamma v_2^{-1}(x_2)\|x_2\|(\|x_1\| + e_{22}\delta\|x_2\|) \\ &\geqslant -2\bar{e}_{21}\gamma v_1(x_1) - 2\bar{e}_{21}\bar{e}_{22}\gamma\delta v_2(x_2) \qquad \forall (t,x) \in \mathcal{T}\times\mathcal{R}_\rho, \end{aligned} \tag{2.260}$$

which is valid for all $E \in \bar{E}$.

The inequalities (2.260) can be used to get the differential inequality

$$\dot{v} \geqslant \overline{W}v, \tag{2.261}$$

where $v = (v_1, v_2)^T$ is a vector Liapunov function and $\overline{W}$ is the 2×2 aggregate matrix

$$\overline{W} = \begin{bmatrix} 2 & -2\beta \\ -2\gamma & 1 - 2\gamma\delta \end{bmatrix}. \tag{2.262}$$

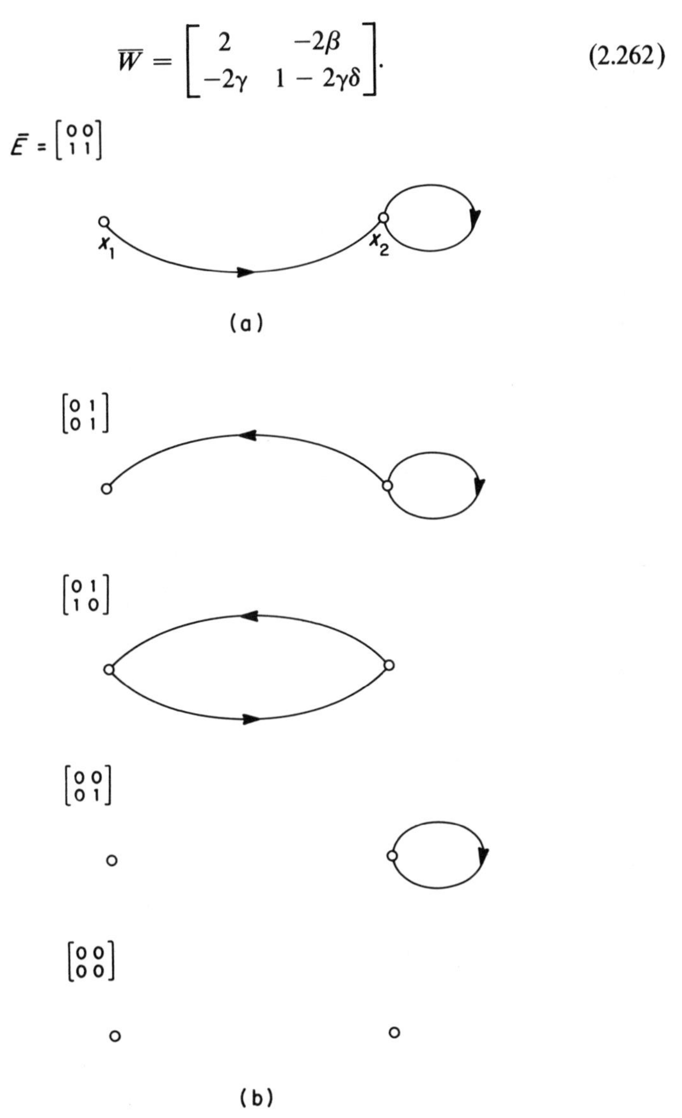

FIGURE 2.6. System structure and perturbations.

The matrix $\overline{W}$ has negative off-diagonal elements, and it is possible to show (Theorem A.2) that it has all eigenvalues with positive real parts if (and only if) $-\overline{W}$ satisfies the conditions (2.132), that is,

$$-1 + 2\gamma\delta < 0, \qquad 1 - 2\beta\gamma - 2\gamma\delta > 0. \tag{2.263}$$

Again, one can show (Theorem A.2) that the conditions (2.263) are equivalent to the quasidominant diagonal property (2.120) of the matrix $\overline{W}$. The conditions (2.263) are sufficient for complete connective instability of the equilibrium $x^* = 0$ of $\mathbb{S}$ with respect to the region $\mathfrak{R}_\rho = \{x \in \mathfrak{R}^4: \|x\| < 1\}$. The structural perturbations of the system structure shown in Figure 2.6(a) are listed in Figure 2.6(b). Again, we show only those perturbations which correspond to binary interconnection matrices.

Under the conditions (2.263), not only is $x^* = 0$ unstable for all $E \in \overline{E}$, but also every solution $x(t; t_0, x_0)$ of the equations (2.256) which starts in the region $\mathfrak{R}_\rho$ is increasing (leaving $x^* = 0$) as fast as an exponential. To see this, we note that in the case of the system $\mathbb{S}$, the last inequality (2.254) can be written as

$$\dot{\nu}(x) \geqslant \pi(2\|x_1\| + \|x_2\|) \qquad \forall x \in \mathfrak{R}_\rho, \tag{2.264}$$

where

$$\nu(x) = d_1 v_1(x_1) + d_2 v_2(x_2), \tag{2.265}$$

and d_1, d_2 are certain positive numbers whose existence is guaranteed by the inequalities (2.263). Now, from (2.264) and (2.265), we get the differential inequality

$$\dot{\nu}(x) \geqslant \tilde{\pi}\nu(x) \qquad \forall x \in \mathfrak{R}_\rho, \tag{2.266}$$

where $\tilde{\pi} = d_M^{-1}\pi$, $d_M = \max\{d_1, d_2\}$. The minimal solution of

$$\dot{r} = \tilde{\pi}r \tag{2.267}$$

is given as

$$r_m(t; t_0, r_0) = r_0 \exp[\tilde{\pi}(t - t_0)], \tag{2.268}$$

and we have from (2.266) and (2.268),

$$\nu(t; t_0, \nu_0) \geqslant \nu_0 \exp[\tilde{\pi}(t - t_0)] \qquad \forall t \in \mathfrak{T}_0. \tag{2.269}$$

Finally, using (2.258) and (2.266), we get from (2.269) the inequality

$$\|x(t; t_0, x_0)\| \geqslant \tilde{\Pi}\|x_0\| \exp[\tilde{\pi}(t - t_0)] \qquad \forall t \in \mathcal{T}_0, \tag{2.270}$$

where $\tilde{\Pi} = 2^{-1/2} d_m d_M^{-1}$, which implies exponential instability for all $(t_0, x_0) \in \mathcal{T} \times \mathcal{R}_\rho$ and all $E \in \bar{E}$. This result can be generalized whenever the comparison functions $\phi_{ki}(\|x_i\|)$ in (2.253) have the form $\eta_{ki}\|x_i\|$.

In establishing Theorem 2.18, we did not place any restriction on the sign of the numbers α_{ij} in the constraints (2.251). However, each subsystem $\mathcal{S}_i$ had to be unstable when isolated. It is of interest to show that in this context, there is a trade-off between stability properties of the subsystems and the nature of the constraints placed on the interactions. That is expressed by the following:

Theorem 2.19. *If the $s \times s$ aggregate matrix $\bar{W} = (\bar{w}_{ij})$ defined by* (2.252) *has elements specified by*

$$\bar{w}_{ij} \begin{cases} > 0, & i = j, \\ \geqslant 0, & i \neq j, \end{cases} \tag{2.271}$$

for some $i \in \{1, 2, \ldots, s\}$, then the equilibrium $x^* = 0$ of the system $\mathcal{S}$ is completely connectively unstable.

Proof. By following the proof of Theorem 2.12, but using the inequalities (2.251) and (2.271), we get

$$\begin{aligned} \dot{v}_i(t, x_i)_{(2.2)} &\geqslant \sum_{j=1}^{s} w_{ij}\phi_{3j}(\|x_j\|) \\ &\geqslant \phi_{3i}(\|x_i\|) \qquad \forall (t, x) \in \mathcal{T} \times \mathcal{R}_\rho, \end{aligned} \tag{2.272}$$

which implies instability of $x^* = 0$ for all $E \in \bar{E}$ and thus the assertion of Theorem 2.19.

While in Theorem 2.18 it was possible to show that all solutions near $x^* = 0$ diverge from $x^* = 0$ as time progresses, in Theorem 2.19 instability is established only with respect to the component x_i of the state vector. The reason this kind of "partial instability" takes place is that the interactions are all destabilizing because they are sign-positive. In Theorem 2.18, interactions could have arbitrary sign, but their total effect is dominated by the instability of the subsystems.

Further considerations of connective instability can be carried out for large-scale systems by applying various instability results outlined by Hahn (1967), Matrosov (1965), and Lakshmikantham and Leela (1969). Some new results along these lines were obtained by Vidyasagar (1977a,b).

2.8. REGIONS OF CONNECTIVE STABILITY

The connective-stability results obtained so far are valid either globally in the entire state space or locally in the neighborhood of the equilibrium state. Both types of results are unsatisfactory in a certain sense. The global results are unrealistic in that they cannot be realized in applications, on account of physical constraints. Furthermore, a large class of models (such as that of two-predator, two-prey model of Section 1.6), have multiple equilibria which exclude global asymptotic stability results, whether structural considerations are involved or not. On the other hand, local results are unsatisfactory in that it is not certain how far initial conditions can be allowed to vary without disrupting the stability properties established in the immediate vicinity of an equilibrium state. A compromise between these two extremes is provided by estimates of the actual stability regions using Liapunov's direct method. In this section, we shall use the results of Weissenberger (1973) which establish these estimates for large-scale systems using the vector Liapunov function in the context of the decomposition-aggregation method. We shall modify Weissenberger's results to include structural considerations and compute the regions of connective stability. We shall focus our attention on exponential stability, but the corresponding asymptotic connective properties can be easily formulated by simple technical adjustments.

Let us consider again the dynamic system $\mathbb{S}$ described by the equation

$$\dot{x} = f(t, x), \tag{2.1}$$

which can be represented by s interconnected subsystems $\mathbb{S}_i$,

$$\dot{x}_i = g_i(t, x_i) + h_i(t, x), \qquad i = 1, 2, \ldots, s, \tag{2.2}$$

so that $x \in \mathcal{R}$, $x_i \in \mathcal{R}^{n_i}$, and the state-space decomposition is given as

$$\mathcal{R}^n = \mathcal{R}^{n_1} \times \mathcal{R}^{n_2} \times \cdots \times \mathcal{R}^{n_s}, \tag{2.273}$$

which implies that the state x of $\mathbb{S}$ is expressed in terms of the states x_i of $\mathbb{S}_i$ simply as

$$x = (x_1^T, x_2^T, \ldots, x_s^T)^T, \tag{2.274}$$

and the order n of the system $\mathbb{S}$ is a direct summation of the orders n_i of the subsystems. That is,

$$n = \sum_{i=1}^{s} n_i. \tag{2.275}$$

We again assume that

$$f(t, 0) = 0 \qquad \forall t \in \mathcal{T} \tag{2.7}$$

and $x^* = 0$ is an equilibrium (not necessarily unique) of the system $\mathbb{S}$.

Now, we state

Definition 2.27. *The region of exponential connective stability for the equilibrium $x^* = 0$ of the system $\mathbb{S}$ is the set $\mathcal{X}$ of all points x_0 with the property*

$$\|x(t; t_0, x_0)\| \leqslant \Pi \|x_0\| \exp[-\pi(t - t_0)] \qquad \forall t \in \mathcal{T}_0 \tag{2.276}$$

for any $t_0 \in \mathcal{T}$ and all $E \in \bar{E}$, where Π, π are positive numbers independent of the initial values (t_0, x_0).

Our interest is in determining estimates $\tilde{\mathcal{X}} \subseteq \mathcal{X}$ of the region $\mathcal{X}$ by using Theorems 2.14 and 2.16 of Section 2.5. Therefore, let us assume again that the system $\mathbb{S}$ described by (2.1) is composed of s interconnected subsystems $\mathbb{S}_i$ represented by (2.2). We further assume that for each isolated subsystem $\mathbb{S}_i$ described by

$$\dot{x}_i = g_i(t, x_i), \qquad i = 1, 2, \ldots, s, \tag{2.3}$$

a function $v_i(x_i)$ is chosen such that $v_i \in C^1(\mathcal{R}_i^n)$, $v_i(0) \equiv 0$,

$$\begin{aligned} \eta_{1i}\|x_i\| \leqslant v_i(x_i) &\leqslant \eta_{2i}\|x_i\|, \\ \dot{v}_i(x_i)_{(2.3)} &\leqslant \eta_{3i}\|x_i\|, \qquad \forall x_i \in \tilde{\mathcal{X}}_i, \\ \|\text{grad } v_i(x_i)\| &\leqslant \eta_{4i} \end{aligned} \tag{2.277}$$

where

$$\tilde{\mathcal{X}}_i = \{x_i \in \mathcal{R}^{n_i} : v_i(x_i) < v_i^0\} \tag{2.278}$$

and v_i^0 is a positive number or the symbol $+\infty$. Furthermore, we require that each interaction $h_i(t, x)$ belong to the class $\mathcal{H}_{(2.152)}$ specified by the inequality

$$\|h_i(t, x)\| \leqslant \sum_{j=1}^{s} \bar{e}_{ij} \xi_{ij} \|x_j\| \qquad \forall (t, x) \in \mathcal{T} \times \tilde{\mathcal{X}}_0, \tag{2.279}$$

which have to be satisfied only in the region

$$\tilde{\mathfrak{X}}_0 = \tilde{\mathfrak{X}}_1 \times \tilde{\mathfrak{X}}_2 \times \cdots \times \tilde{\mathfrak{X}}_s, \tag{2.280}$$

which need not be the entire state space $\mathfrak{R}^n$, as was required by the original Definition 2.18 of the class $\mathfrak{K}_{(2.152)}$.

It is important to notice that $\tilde{\mathfrak{X}}_i$ is an estimate of the exponential stability region $\mathfrak{X}_i$ associated with the equilibrium $x_i^* = 0$ of the subsystem $\mathcal{S}_i$ of (2.3). That is, it has the property that $(t_0, x_{i0}) \in \mathfrak{T} \times \tilde{\mathfrak{X}}_i$ implies $x_i(t; t_0, x_{i0}) \subset \tilde{\mathfrak{X}}_i$ and

$$\|x_i(t; t_0, x_{i0})\| \leqslant \Pi_i \|x_{0i}\| \exp[-\pi_i(t - t_0)] \qquad \forall t \in \mathfrak{T}_0, \tag{2.150}$$

where Π_i, π_i are positive numbers independent of (t_0, x_{i0}). In the special case when $\tilde{\mathfrak{X}}_i = \mathfrak{R}^{n_i}$, the whole subsystem space $\mathfrak{R}^{n_i}$ is the region $\mathfrak{X}_i$ of exponential stability for the subsystem $\mathcal{S}_i$. There are quite a few effective procedures (e.g. Weissenberger 1966, 1973; Šiljak, 1969) for computing the estimates $\tilde{\mathfrak{X}}_i$ for wide classes of nonlinear systems $\mathcal{S}_i$.

We also note that when $E = 0$ and all subsystems $\mathcal{S}_i$ are decoupled from each other, the region $\tilde{\mathfrak{X}}_0$ of (2.280) is an estimate of the exponential-stability region $\mathfrak{X}$ for the overall system $\mathcal{S}$.

To produce an estimate $\tilde{\mathfrak{X}}$ of $\mathfrak{X}$ when $E = \bar{E}$, we use the inequalities (2.277) and (2.279) to form the aggregate $\mathcal{Q}$ as

$$\dot{r} = \overline{W} r, \tag{2.155}$$

where the $s \times s$ constant matrix $\overline{W} = (\overline{w}_{ij})$ is defined by

$$\overline{w}_{ij} = -\eta_{2i}^{-1} \eta_{3i} \delta_{ij} + \bar{e}_{ij} \xi_{ij} \eta_{4i} \eta_{1j}^{-1}. \tag{2.161}$$

We assume that $\overline{W}$ satisfies the inequalities (2.132) and is a stable matrix.

Now we consider a differentiable, positive definite scalar function $\nu(v)$, where $v(x)$ is a vector Liapunov function for the system $\mathcal{S}$. We also define the region

$$\tilde{\mathfrak{X}} = \{x \in \mathfrak{R}^n \colon \nu(v) < \gamma\}, \tag{2.281}$$

where

$$\gamma = \inf_{v \in \mathfrak{B}} \nu(v) \tag{2.282}$$

and

$$\mathfrak{B} = \bigcup_{i=1}^{s} \{v \in \mathfrak{R}_+^s \colon v_i = v_i^0\}. \tag{2.283}$$

If

$$\dot{\nu}(v)_{(2.155)} = (\text{grad } v)^T \overline{W} v \tag{2.284}$$

is negative definite on $\tilde{\mathfrak{X}}_0$, then $\nu[v(x)]$ is a "second level" Liapunov function for the overall system $\mathbb{S}$, and $\tilde{\mathfrak{X}}$ is an estimate of the exponential connective stability region $\mathfrak{X}$ for the system $\mathbb{S}$. That is, we have

Theorem 2.20. *Suppose there exists a function* $\nu\colon \mathfrak{R}_+^s \to \mathfrak{R}_+$ *such that* $\nu \in C^1(\mathfrak{R}_+^s)$, $\nu(0) \equiv 0$, *and*

$$\begin{aligned} \phi_{\mathrm{I}}(\|x\|) \leqslant \nu(x) \leqslant \phi_{\mathrm{II}}(\|x\|), \\ \dot{\nu}(x)_{(2.155)} \leqslant -\phi_{\mathrm{III}}(\|x\|) \end{aligned} \qquad \forall x \in \tilde{\mathfrak{X}}_0, \tag{2.285}$$

where $\tilde{\mathfrak{X}}_0$ *is defined in* (2.280). *Then* $x_0 \in \tilde{\mathfrak{X}}$ *implies* $x(t; t_0, x_0) \subset \tilde{\mathfrak{X}}$; *for any* $t_0 \in \mathfrak{T}$,

$$\|x(t; t_0, x_0)\| \leqslant \Pi \|x_0\| \exp[-\pi(t - t_0)] \qquad \forall t \in \mathfrak{T}_0 \tag{2.276}$$

for all $E \in \bar{E}$; *and the region* $\tilde{\mathfrak{X}}$ *defined by* (2.281) *is an estimate of the region* $\mathfrak{X}$ *of exponential connective stability corresponding to the equilibrium* $x^* = 0$ *of the system* $\mathbb{S}$.

Proof. By simple technical adjustments, the proof follows immediately from the proof of Theorem 2.5.

To apply Theorem 2.20 and compute effectively the estimate $\tilde{\mathfrak{X}}$ of connective-stability regions for large-scale systems, we can use the functions

$$\nu_1(v) = \sum_{i=1}^{s} d_i |v_i|, \tag{2.286}$$

$$\nu_2(v) = \max_{1 \leqslant i \leqslant s} \{d_i^{-1} |v_i|\}, \tag{2.287}$$

$$\nu_3(v) = v^T H v, \tag{2.288}$$

as proposed by Weissenberger (1973). In (2.286) and (2.287), the absolute-value signs are introduced in $\nu_1(v)$ and $\nu_2(v)$ to aid in the geometric visualization of these functions shown in Figure 2.7. In $\nu_1(v)$ and $\nu_2(v)$, the numbers d_i are positive for all $i = 1, 2, \ldots, s$, and in $\nu_3(v)$ the constant $s \times s$ matrix H is positive definite, that is, $v^T H v > 0$ for all $v \in \mathfrak{R}^s$, $v \neq 0$.

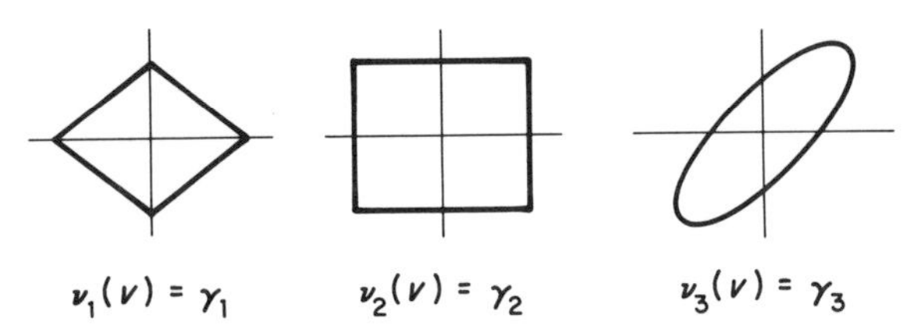

FIGURE 2.7. Liapunov functions.

Let us first consider the function $\nu_1(v)$ and form

$$\dot{\nu}_1(v)_{(2.155)} = d^T \overline{W} v, \tag{2.289}$$

using (2.155). Since the matrix $\overline{W}$ is a Metzler matrix and satisfies the conditions (2.132), a positive vector $d \in \mathcal{R}_+^s$ exists such that

$$d^T = -c^T \overline{W}^{-1} \tag{2.290}$$

for any positive vector $c \in \mathcal{R}_+^s$. This fact, which is proved in Appendix (Theorem A.2), ensures that $\nu_1(v)$ satisfies the inequalities (2.285). According to (2.281) the estimate $\tilde{\mathcal{X}}$ is determined by

$$\gamma_1 = \min_{1 \leqslant i \leqslant s} \{d_i v_i^0\} \tag{2.291}$$

where v_i^0 are determined on the subsystem level.

To consider the function $\nu_2(v)$, we first note that from (2.273) and (2.287), the region $\tilde{\mathcal{X}}$ (which is now assumed bounded) is given by

$$\begin{aligned} \tilde{\mathcal{X}} &= \{v \in \mathcal{R}_+^s : \nu(v) < \gamma\} \\ &= \{v \in \mathcal{R}_+^s : v < \hat{v} = \gamma d\}. \end{aligned} \tag{2.292}$$

We define the set of points $\mathcal{B}$ as

$$\mathcal{B} = \{v \in \mathcal{R}_+^s : \nu(v) = \gamma\}, \tag{2.293}$$

and from (2.292) get

$$\mathcal{B} = \bigcup_{i=1}^{s} \{v \in \mathcal{R}_+^s : v_i = \hat{v}_i;\ 0 \leqslant v_j \leqslant \hat{v}_j, j \neq i\}, \tag{2.294}$$

where $\hat{v}_i = \gamma d_i$. It follows readily that for $\nu_2(v) = \gamma_2$,

$$\dot{\nu}_2(v)_{(2.155)} \leqslant \max_{1 \leqslant i \leqslant s} \left\{ d_i^{-1} \sum_{j=1}^{s} \overline{w}_{ij} v_j \right\} \qquad \forall v \in \mathcal{B}. \tag{2.295}$$

Therefore,

$$\dot{\nu}_2(v)_{(2.155)} \leqslant 0 \qquad \forall v \in \tilde{\mathcal{X}} \tag{2.296}$$

whenever

$$\overline{W} d \leqslant 0. \tag{2.297}$$

Since $\overline{W}$ satisfies the conditions (2.132) and is a stable matrix, it is also a nonsingular matrix. Furthermore, $d > 0$ and the case $\overline{W} d = 0$ is excluded.

As in the case of the function $\nu_1(v)$, stability of $\overline{W}$ implies that for any positive $c \in \mathcal{R}_+^s$, there exists a positive $d \in \mathcal{R}_+^s$ as a solution of the equation

$$d = -\overline{W}^{-1}c. \tag{2.298}$$

Once a solution $d > 0$ of (2.298) is found, we compute γ_2 from (2.281) as

$$\gamma_2 = \min_{1 \leqslant i \leqslant s} \{d_i^{-1} v_i^0\}. \tag{2.299}$$

It is important to note that the region $\tilde{\mathfrak{X}}$ obtained by $\nu_2(v)$ is decoupled in the subsystem functions $v_i(x_i)$, which is a considerable simplification and makes it possible to achieve the exact estimate of $\tilde{\mathfrak{X}}_0$, that is, $\tilde{\mathfrak{X}} = \tilde{\mathfrak{X}}_0$. This case is illustrated in Figure 2.8, following Weissenberger (1973), and the relationship is shown between $\tilde{\mathfrak{X}}_1, \tilde{\mathfrak{X}}_2, \ldots, \tilde{\mathfrak{X}}_s, \tilde{\mathfrak{X}}_0$, and $\tilde{\mathfrak{X}}$ in both $\mathcal{R}^n$ and $\mathcal{R}_+^s$.

Finally, to calculate the region $\tilde{\mathfrak{X}}$ using the quadratic form $\nu_3(v)$, we choose a positive definite $s \times s$ matrix G and form

$$\dot{\nu}_3(v)_{(2.155)} = -v^T G v, \tag{2.300}$$

so that $\dot{\nu}_3(v)_{(2.155)}$ is a negative definite function. We recall (e.g. Hahn, 1967)

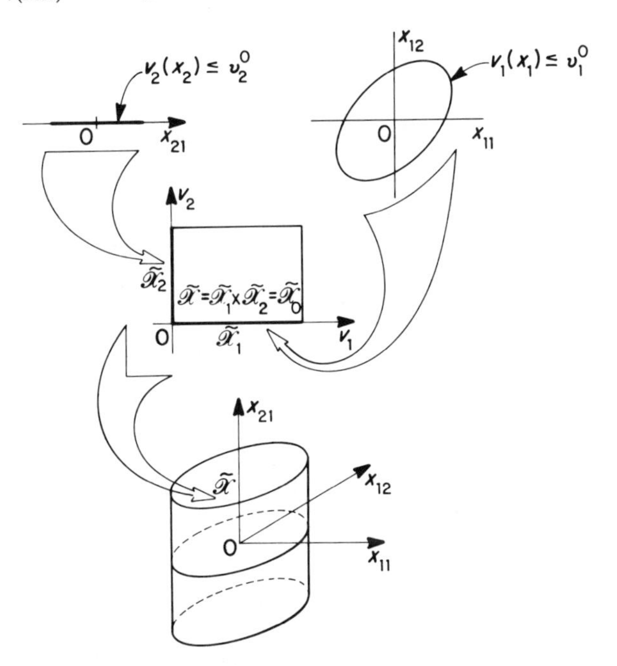

FIGURE 2.8. Stability regions.

that for each positive definite matrix G, there is a positive definite $s \times s$ matrix H as a unique solution of the Liapunov matrix equation

$$\overline{W}^T H + H\overline{W} = -G. \tag{2.301}$$

Once H is determined, we construct the function $\nu_3(v) = v^T H v$ as in (2.288). According to (2.282), the indicated minimization can be carried out as

$$\gamma_3 = \min_{1 \leq i \leq s} \{\hat{h}_{ii}^{-1}(v_i^0)^2\}, \tag{2.302}$$

where $\hat{h}_{ii}$ is a diagonal element of H^{-1}, or as

$$\gamma_3 = (\det H) \min_{1 \leq i \leq s} \{\mu_i^{-1}(v_i^0)^2\}, \tag{2.303}$$

where μ_i is the ith leading principal minor of the matrix H.

Further improvements of the above estimates of the connective stability regions and their comparison, as well as applications, can be obtained by following the original work of Weissenberger (1973). Several results outlined in this section will be used in the local stabilization of the Large Space Telescope outlined in Chapter 6.

REFERENCES

Antosiewicz, H. A. (1962), "An Inequality for Approximate Solutions of Ordinary Differential Equations", *Mathematische Zeitschrift*, 78, 44–52.

Antosiewicz, H. A. (1965), "Recent Contributions to Liapunov's Second Method", *Proceedings of the NATO Advanced Study Institute*, Padua, Italy, Publisher Edizioni "Oderisi", Gubbio, 1966, 57–60.

Araki, M., Ando, K., and Kondo, B. (1971), "Stability of Sampled-Data Composite Systems with Many Nonlinearities", *IEEE Transactions*, AC-16, 22–27.

Araki, M., and Kondo, B. (1972), "Stability and Transient Behavior of Composite Nonlinear Systems", *IEEE Transactions*, AC-17, 537–541.

Araki, M. (1975), "Application of $\mathfrak{M}$-Matrices to the Stability Problems of Composite Dynamical Systems", *Journal of Mathematical Analysis and Applications*, 52, 309–321.

Arrow, K. J. (1966), "Price-Quantity Adjustments in Multiple Markets with Rising Demands", *Proceedings of a Symposium on the Mathematical Methods in the Social Sciences*, K. J. Arrow, S. Karlin, and P. Suppes (eds.), Stanford University Press, Palo Alto, California, 3–15.

Athans, M., Sandell, N. R., Jr., and Varaiya, P. (1975), "Stability of Interconnected Systems", *Proceedings of the 1975 IEEE Conference on Decision and Control*, Houston, Texas, 456–462.

Bailey, F. N. (1966), "The Application of Lyapunov's Second Method to Interconnected Systems", *SIAM Journal of Control*, 3, 443–462.

Barbashin, E. A. (1970), *Liapunov's Functions* (in Russian), Nauka, Moscow.

Barbashin, E. A., and Krassovskii, N. N. (1952), "On Stability in the Large" (in Russian), *Dokladi Akademii Nauk SSSR*, 86, 454–456.

Barnett, S., and Šiljak, D. D. (1977), "Routh's Algorithm: A Centennial Survey", *SIAM Review*, 19, 472–489.

Bellman, R. (1962), "Vector Lyapunov Functions", *SIAM Journal of Control*, 1, 32–34.

Berztiss, A. T. (1975), *Data Structures: Theory and Practice*, Academic, New York.

Busacker, R. G., and Saaty, T. L. (1965), *Finite Graphs and Networks: An Introduction with Applications*, McGraw-Hill, New York.

Bitsoris, G., and Burgat, C. (1976), "Stability Conditions and Estimates of the Stability Region of Complex Systems", *International Journal of Systems Science*, 7, 911–928.

Bitsoris, G., and Burgat, C. (1977), "Stability Analysis of Complex Discrete Systems with Locally and Globally Stable Subsystems", *International Journal of Control*, 25, 413–424.

Brauer, F. (1965), "The Use of Comparison Theorems for Ordinary Differential Equations", *Proceedings of the NATO Advanced Study Institute*, Padua, Italy, Publisher Edizioni "Oderisi", Gubbio, 1966, 29–50.

Callier, F. M., Chan, W. S., and Desoer, C. A. (1976), "Input-Output Stability Theory of Interconnected Systems Using Decomposition Techniques", *IEEE Transactions*, CAS-23, 714–729.

Coddington, E. A., and Levinson, N. (1955), *Theory of Ordinary Differential Equations*, McGraw-Hill, New York.

Conti, R. (1956), "Sulla prolungbilita delle soluzioni di un sistema di equazioni differenziali ordinarie", *Bollogna Unione Matematica Italiana*, 11, 510–514.

Cook, P. A. (1974), "On the Stability of Interconnected Systems", *International Journal of Control*, 20, 407–415.

Coppel, W. A. (1965), *Stability and Asymptotic Behavior of Differential Equations*, Heath, Boston, Massachusetts.

Cordeanu, C. (1960), "An Application of Differential Inequalities to the Theory of Stability" (in Russian), *Analele Ştiintifice ale Universitătii ≪ Al. I. Çuza≫ din Iaşi*, 6, 47–58.

Cordeanu, C. (1961), "An Application of Differential Inequalities to the Theory of Stability" (in Russian), *Analele Ştiintifice ale Universitătii ≪ Al. I. Çuza≫ din Iaşi*, 7, 241–252.

Diamond, P. (1975), "Stochastic Exponential Stability Concepts and Large-Scale Discrete Systems", *International Journal of Control*, 22, 141–145.

Duhem, M. P. (1902), "Sur les conditions nécessaires pour la stabilité de l'équilibre d'un sisteme visqueux", *Comptes Rendus*, 126, 939–941.

Ertegov, V. D. (1970), "On Stability of Solutions of Difference Equations" (in Russian), *Transactions of the Kazan Aviation Institute*, 125, 14–19.

Fiedler, M., and Pták, V. (1962), "On Matrices with Nonpositive Off-Diagonal Elements and Principal Minors", *Czechoslovakian Mathematical Journal*, 12, 382–400.

Franklin, T. N. (1968), *Matrix Theory*, Prentice-Hall, Englewood Cliffs, New Jersey.

Gantmacher, F. R. (1960), *The Theory of Matrices*, Vols. I and II, Chelsea, New York.

Grujić, Lj. T. (1973), "Uniform Asymptotic Stability of Discrete Large-Scale Systems", *IEEE Transactions*, SMC-3, 636–643.

Grujić, Lj. T. (1974), "Stability Analysis of Large-Scale Systems with Stable and Unstable Subsystems", *International Journal of Control*, 20, 453–463.

Grujić, Lj. T. (1975a), "Uniform Practical and Finite-Time Stability of Large-Scale Systems", *International Journal of Systems Science*, 6, 181–195.

Grujić, Lj. T. (1975b), "Non-Lyapunov Stability Analysis of Large-Scale Systems on Time-Varying Sets", *International Journal of Control*, 21, 401–415.

Grujić, Lj. T., Gentina, J. C., and Borne, P. (1976), "General Aggregation of Large-Scale Systems by Vector Lyapunov Functions and Vector Norms", *International Journal of Control*, 24, 529–550.

Grujić, Lj. T., and Šiljak, D. D. (1973a), "On Stability of Discrete Composite Systems" *IEEE Transactions*, AC-18, 522–524.

Grujić, Lj. T., and Šiljak, D. D. (1973b), "Asymptotic Stability and Instability of Large-Scale Systems", *IEEE Transactions*, AC-18, 636–645.

Grujić, Lj. T., and Šiljak, D. D. (1974), "Exponential Stability of Large-Scale Discrete Systems", *International Journal of Control*, 19, 481–491.

Gunderson, R. W. (1970), "On a Stability Property of Krassovskii", *International Journal of Non-Linear Mechanics*, 5, 507–512.

Gunderson, R. W. (1971), "A Stability Condition for Linear Comparison Systems", *Quarterly of Applied Mathematics*, 29, 327–328.

Hahn, W. (1967), *Stability of Motion*, Springer, New York.

Harrary, F., Norman, R. Z., and Cartwright, D. (1965), *Structural Models: An Introduction to the Theory of Directed Graphs*, Wiley, New York.

Hale, J. K. (1969), *Ordinary Differential Equations*, Wiley, New York.

Kamke, E. (1930), *Differentialgleichungen reeler Funktionen*, Akademische Verlagsgeselshaft, Leipzig, Germany.

Kamke, E. (1932), "Zur Theorie der Systeme gewönlicher Differentialgleichungen. II", *Acta mathematica*, 58, 57–85.

Kloeden, P. E. (1975), "Aggregation-Decomposition and Ultimate Boundedness", *The Journal of the Australian Mathematical Society*, 19, 249–258.

Kloeden, P. E., Diamond, P. (1977), "Converse Theorems for Stochastic Exponential Stability", *International Journal of Control*, 25, 507–512.

Krassovskii, N. N. (1959), *Some Problems of the Theory of Stability of Motion* (in Russian), Fizmatgiz, Moscow (English Translation: Stanford University Press, Palo Alto, California, 1963).

Ladde, G. S. (1975a), "Systems of Differential Inequalities and Stochastic Differential Equations. II", *Journal of Mathematical Physics*, 16, 894–900.

Ladde, G. S. (1975b), "Variational Comparison Theorem and Perturbations of Nonlinear Systems", *Proceedings of the American Mathematical Society*, 52, 181–187.

Ladde, G. S. (1976), "Stability of Large-Scale Hereditary Systems Under Structural Perturbations", *Proceedings of the IFAC Symposium on Large Scale Systems Theory and Applications*, G. Guardabassi and A. Locatelli (eds.), Udine, Italy, 215–226.

Ladde, G. S. (1977), "Stability of Large-Scale Functional Systems Under Structural Perturbations", (to appear).

Ladde, G. S., and Šiljak, D. D. (1975), "Connective Stability of Large-Scale Stochastic Systems", *International Journal of Systems Science*, 6, 713–721.

Lakshmikantham, V. (1974), "On the Method of Vector Lyapunov Functions", *Proceedings of the Twelfth Annual Allerton Conference on Circuit and System Theory*, University of Illinois, Urbana, Illinois, 71–76.

Lakshmikantham, V., and Leela, S. (1969), "*Differential and Integral Inequalities*", Vols. I and II, Academic, New York.

Lakshmikantham, V., and Leela, S. (1977), "Cone-Valued Lyapunov Functions", *Nonlinear Analysis, Theory, Methods, and Applications*, 1, 215–222.

Liapunov, A. M. (1907), "Problème général de la stabilité du mouvement", *Annales de la faculté des Sciences de l'université de Toulouse*, 9, 203–474 (Reprinted in Annals of Matematics Studies, Princeton University Press, Princeton, New Jersey, Vol. 17, 1949.)

LaSalle, J. P. (1975), "Vector Lyapunov Functions", *Bulletin of the Institute of Mathematics Academia Sinica*, 3, 139–150.

Lur'e, A. I. (1951), *Some Nonlinear Problems in the Theory of Automatic Control* (in Russian), GOSTEHIZDAT, Moscow (English Translation: Her Majesty's Stationery Office, London, 1957).

Matrosov, V. M. (1962), "On the Theory of Stability of Motion" (in Russian), *Prikladnaya Matematika i Mekhanika*, 26, 992–100.

Matrosov, V. M. (1963), "On the Theory of Stability of Motion" (in Russian), *Transactions of the Kazan Aviation Institute*, 80, 22–33.

Matrosov, V. M. (1965), "Development of the Method of Liapunov Functions in Stability Theory" (in Russian), *Proceedings of the Second All-Union Conference in Theoretical and Applied Mechanics*, 112–125.

Matrosov, V. M. (1967), "On Differential Equations and Inequalities with Discontinuous Right-Hand Sides" (in Russian), *Differentsial'nie Uravnenya*, Part I:395-409, Part II: 839–848.

Matrosov, V. M. (1971), "Vector Liapunov Functions in the Analysis of Nonlinear Interconnected Systems", *Instituto Nazionale di Alta Matematica: Symposia Mathematica*, 6, 209–242.

Matrosov, V. M. (1972a), "Method of Vector Liapunov Functions of Interconnected Systems with Distributed Parameters (Survey)" (in Russian), *Avtomatika i Telemekhanika*, 33, 63–75.

Matrosov, V. M. (1972b), "Method of Vector Liapunov Functions in Feedback Systems" (in Russian), *Automatika i Telemekhanika*, 33, 63–75.

McClamroch, N. H., and Ianculescu, G. D. (1975), "Global Stability of Two Interconnected Nonlinear Systems", *IEEE Transactions*, AC-20, 678–642.

Michel, A. N. (1970a), "Quantitative Analysis of Simple and Interconnected Systems: Stability, Boundedness, and Trajectory Behavior", *IEEE Transactions*, CT-17, 292–301.

Michel, A. N. (1970b), "Stability, Transient Behavior, and Trajectory Bounds of Interconnected Systems", *International Journal of Control*, 11, 703–715.

Michel, A. N. (1974), "Stability Analysis of Interconnected Systems", *SIAM Journal of Control*, 12, 554–579.

Michel, A. N. (1975a), "Stability Analysis of Stochastic Composite Systems", *IEEE Transactions*, AC-20, 246–250.

Michel, A. N. (1975b), "Stability and Trajectory Behavior of Composite Systems", *IEEE Transactions*, CAS-22, 305–312.

Michel, A. N., and Porter, D. W. (1972), "Stability Analysis of Composite Systems", *IEEE Transactions*, AC-17, 222–226.

Michel, A. N., and Rasmussen, R. D. (1976), "Stability of Stochastic Composite Systems", *IEEE Transactions*, AC-21, 89–94.

Montemayor, J. J., and Womack, B. F. (1975), "On a Conjecture by Šiljak", *IEEE Transactions*, AC-20, 572–573.

Moylan, P. J., and Hill, D. J. (1978), "Stability Criteria for Large-Scale Systems", *IEEE Transactions*, AC-23 (to appear).

Müler, M. (1926), "Über das Fundamentaltheorem in der Theorie der gewönlichen Differentialgleichungen", *Mathematische Zeitschrift*, 26, 619–645.

Newman, P. K. (1959), "Some Notes on Stability Conditions", *Review of Economic Studies*, 72, 1–9.

Nyquist, H. (1932), "Regeneration Theory", *Bell System Technical Journal*, 11, 126–147.

Piontkovskii, A. A., and Rutkovskaia, L. D. (1967), "Investigation of Certain Stability-Theory by the Liapunov Function Method" (in Russian), *Avtomatika i Telemekhanika*, 28, 23–31.

Popov, V. M. (1973), *Hyperstability of Control Systems*, Springer, New York.

Porter, D. W., and Michel, A. N. (1974), "Input-Output Stability of Time-Varying Nonlinear Multiloop Feedback Systems", *IEEE Transactions*, AC-19, 422–427.

Rasmussen, R. D., and Michel, A. N. (1976a), "On Vector Lyapunov Functions for Stochastic Dynamical Systems", *IEEE Transactions*, AC-21, 250–254.

Rasmussen, R. D., and Michel, A. N. (1976b), "Stability of Interconnected Dynamical Systems Described on Banach Spaces", *IEEE Transactions*, AC-21, 464–471.

Šiljak, D. D. (1969), *Nonlinear Systems*, Wiley, New York.

Šiljak, D. D. (1971), "On Large-Scale System Stability", *Proceedings of the Ninth Annual Allerton Conference on Circuit and System Theory*, University of Illinois, Monticello, Illinois, 731–740.

Šiljak, D. D. (1972a), "Stability of Large-Scale Systems Under Structural Perturbations", *IEEE Transactions*, SMC-2, 657–663.

Šiljak, D. D. (1972b), "Stability of Large-Scale Systems", *Proceedings of the Fifth IFAC Congress*, Paris, C-32:1–11.
Šiljak, D. D. (1973), "On Stability of Large-Scale Systems Under Structural Perturbations", *IEEE Transactions*, SMC-3, 415–417.
Šiljak, D. D. (1975a), "Connective Stability of Competitive Equilibrium", *Automatica*, 11, 389–400.
Šiljak, D. D. (1975b), "Large-Scale Systems: Stability, Complexity, Reliability", *Proceedings of the Utah State University–NASA Ames Research Center Seminar Workshop on Large-Scale Dynamic Systems*, NASA SP-371, Washington D.C., 147–162.
Šiljak, D. D. (1977a), "On Pure Structure of Dynamic Systems", *Nonlinear Analysis, Theory, Methods, and Applications*, 1, 397–413.
Šiljak, D. D. (1977b), "Vulnerability of Dynamic Systems", *Proceedings of the IFAC Workshop on Control and Management of Integrated Industrial Complexes*, Toulouse, France, 133–144.
Šiljak, D. D., and Sun, C. K. (1971), "Exponential Absolute Stability of Discrete Systems", *Zeitschrift für Angewandte Mathematik und Mechanik*, 51, 271–275.
Šiljak, D. D., and Sun, C. K. (1972), "On Exponential Absolute Stability", *International Journal of Control*, 16, 1003–1008.
Šiljak, D. D., and Vukčević, M. B. (1976), "Large-Scale Systems: Stability, Complexity, Reliability", *Journal of the Franklin Institute*, 301, 49–69.
Šiljak, D. D., and Weissenberger, S. (1970), "A Construction of the Lur'e-Liapunov Function", *Regelungstechnik und Process-Dataverarbeitung*, 10, 455–456.
Sundareshan, M. K., and Vidyasagar, M. (1977), "L_2-Stability of Large-Scale Dynamical Systems: Criteria via Positive Operator Theory", *IEEE Transactions*, AC-22, 396–399.
Szaraki, J. (1965), *Differential Inequalities*, Monographie Matematyczne, Vol. 43, PWN, Warszawa, Poland.
Thompson, W. E. (1970), "Exponential Stability of Interconnected Systems", *IEEE Transactions*, AC-15, 504–506.
Thompson, W. E., and Koenig, H. E. (1972), "Stability of a Class of Interconnected Systems", *International Journal of Control*, 15, 751–763.
Vakhonina, G. S., Zemliakov, A. S., and Matrosov, V. M. (1972), "Methods of Construction of Quadratic Vector Liapunov Functions for Linear Systems" (in Russian), *Avtomatika i Telemekhanika*, 33, 5–16.
Vidyasagar, M. (1977a), "L_2-Instability Criteria for Interconnected Systems", *SIAM Journal of Control and Optimization*, 15, 312–328.
Vidyasagar, M. (1977b), "On the Instability of Large-Scale Systems", *IEEE Transactions*, AC-22, 267–269.
Walter, W. (1970), *Differential and Integral Inequalities*, Springer, New York.
Ważewski, T. (1950), "Systèmes des équations et des inégalites différentielles ordinaires aux deuxièmes membres monotones et leurs applicationes", *Annales de la Societé Polonaise de mathematiques*, 23, 112–166.
Weissenberger, S. (1966), "Stability-Boundary Approximations for Relay-Control Systems via a Steepest-Ascent Construction of Liapunov Functions", *ASME Transactions: Journal of Basic Engineering*, 88, 419–428.

Weissenberger, S. (1973), "Stability Regions of Large-Scale Systems", *Automatica*, 9, 653–663.

Willems, J. C. (1976a), "Stability of Large Scale Interconnected Systems", *Directions in Large-Scale Systems, Many Person Optimization, and Decentralized Control*, Y. C. Ho and S. K. Mitter (eds.), Plenum Press, New York, 401–410.

Willems, J. C. (1976b), "Lyapunov Functions for Diagonally Dominant Systems", *Automatica*, 12, 519–523.

Yakubovich, V. A. (1962), "Solution of Certain Special Matrix Inequalities Occurring in the Theory of Automatic Control" (in Russian), *Dokladi Akademii Nauk SSSR*, 143, 1304–1307.

Yakubovich, V. A. (1964), "The Method of Matrix Inequalities in the Stability Theory of Nonlinear Control Systems" (in Russian), *Avtomatika i Telemekhanika*, 25, 1017–1029.

Zemliakov, A. S. (1972), "On the Problem of Comparison System Construction" (in Russian), *Transactions of the Kazan Aviation Institute*, 144, 46–54.

3

SYNTHESIS

Decentralized Control

Now that we have derived the conditions for connective stability of large-scale systems, we may ask: Can we synthesize reliable complex systems by using feedback? A positive answer to this question is provided in this chapter, and we will present a decentralized multilevel scheme for synthesizing large-scale systems which are stable under structural perturbations, that is, connectively stable.

By relatively simple examples we demonstrated in Section 1.10 that Simon's *intuitive* arguments about reliability of hierarchic structures are true. The objective of this chapter is to show *rigorously* that Simon's intuitive recipe can be used to construct dynamically reliable large-scale systems by hierarchic feedback control. Local feedback controllers are used to stabilize each subsystem when isolated from the rest of the system. Then, regarding the interactions among the subsystems as perturbations, a global controller is utilized to minimize the coupling effect of subsystem interconnections. Finally, connective stability of the overall system is established by testing for stability of the aggregate model as proposed in the preceding chapter.

Prior to applying a decentralized feedback control, we have to make sure that the available inputs can influence (reach) each part of a large system and thus alter its performance according to the requirements. Since we propose to feed back the states and outputs, we also need to check that the state of each subsystem can be estimated from the outputs. These preliminary considerations are structural in nature and therefore are formulated as input and output reachability to be examined in the framework of directed graphs. We will also develop decentralization schemes which can be used to transform multivariable systems into input or output decentralized representations. These schemes are more general than simple permutations of binary matrices of directed graphs, but they are restricted to linear systems.

Decentralization arose in a natural way in economic systems (see Chapter 4), and it was there that decentralized control and decision strategies were proposed for teams and large organizations (Arrow, 1964; Marschak and Radner, 1971). Such strategies are based upon the fact that the system information pattern and control structure are restricted (for either physical or economical reasons) in such a way that each subsystem is controlled by its own input only. A problem of stabilizing a linear dynamic system using a simple decentralized control scheme with underlying economic features was first studied by McFadden (1969). This problem was later given much more general treatment in control literature by Lau, Persiano, and Varaiya (1972), Aoki (1972), Wang and Davison (1973), and Corfmat and Morse (1976). A survey of these and many other results concerning decentralized control was provided by Sandell, Varaiya, and Athans (1975).

As distinct from previous decentralized control schemes, the scheme proposed by Šiljak and Vukčević (1976a, c) and presented in this chapter attempts to stabilize a large linear system by manipulating only subsystem matrices. Thus, the dimensionality of control and estimation problems is reduced in much the same way as it is by the classical decomposition techniques for solving a large number of algebraic equations (e.g. Himmelblau, 1973) and mathematical programming problems involving a large number of variables (e.g. Dantzig and Wolfe, 1960). Besides a considerable saving in numerical aspects of control, the presented scheme produces systems which are dynamically reliable with respect to structural perturbations and can tolerate a wide class of nonlinearities in the interactions among the subsystems. In fact, by this scheme, systems can be stabilized in cases where we have no information about the actual shape of the nonlinear interactions among the subsystems, and only their bounds are available to the designer. Its reliability and its robustness in accommodating imprecise knowledge of interactions are two additional features of decentralized control which must be taken into account whenever a question of centralized vs. decentralized strategy appears in controlling a large-scale dynamic system.

3.1. REACHABILITY, VULNERABILITY, AND CONDENSATIONS

The behavior of a physical system can be altered efficiently by feedback control without changing the system itself. The principle of feedback is to choose inputs to the system as functions of its outputs so that the closed-loop system accomplishes a desired controlled behavior. Before we can use this simple but powerful principle in a system design, we have to make sure that the inputs can "reach" each part (state) of the system, and that all parts of the system are "represented" by the outputs. These two inherent properties of dynamic systems were defined as *input* and *output reachability* (Šiljak, 1977a, b). In this section, we use these results to formulate, study, and partially solve the problem of input and output reachability in the control and estimation of large-scale systems which are considered in the rest of this chapter. The material of this section will also be used in the following chapters for model building in such diverse fields as economics, space flight, ecology, and power systems.

The formalization and study of input and output reachability, decentralized control and estimation schemes, canonical structures, and structural perturbations are carried out in the natural framework of directed graphs (digraphs) and interconnection matrices. Only a bare minimum of notions and concepts from digraph theory are defined here. For a deeper understanding of the structural analysis of dynamic systems outlined in this section, the books of Harary, Norman, and Cartwright (1965), Harary (1969), Deo (1974), and Berztiss (1975) are recommended.

Let us consider a system $\mathbb{S}$ which is described by the equations

$$\begin{aligned} \dot{x} &= f(t, x, u), \\ y &= g(t, x), \end{aligned} \tag{3.1}$$

where $x(t) \in \mathfrak{R}^n$ is the state, $u(t) \in \mathfrak{R}^m$ is the input, and $y(t) \in \mathfrak{R}^l$ is the output of $\mathbb{S}$. The functions f: $\mathfrak{R} \times \mathfrak{R}^n \times \mathfrak{R}^m \to \mathfrak{R}^n$ and g: $\mathfrak{R} \times \mathfrak{R}^n \to \mathfrak{R}^l$ are sufficiently smooth so that $\mathbb{S}$ represents a *dynamic system* (Kalman, Falb, and Arbib, 1969).

With the dynamic system $\mathbb{S}$ we associate a *directed graph* (*digraph*) (Šiljak, 1977b) as the ordered pair $\mathfrak{D} = (V, R)$, where $V = U \cup X \cup Y$ and $U = \{u_1, u_2, \ldots, u_m\}$, $X = \{x_1, x_2, \ldots, x_n\}$, $Y = \{y_1, y_2, \ldots, y_l\}$ are nonempty sets of input, state, and output points, respectively. R is a relation in V, that is, R is a set of ordered pairs which are the lines (u_j, x_i), (x_j, x_i), or (x_j, y_i) joining the points of $\mathfrak{D}$. We make an important assumption about $\mathbb{S}$ by requiring that $\mathfrak{D}$ does not contain lines of the type (u_j, u_i), (u_j, y_i), (x_j, u_i),

(y_j, x_i), (y_j, u_i), and (y_j, y_i). This requirement may seem to be overrestrictive, but in fact it is not, since it reflects the structure of what we ordinarily consider as a dynamic system $\mathbb{S}$ described by the equations (3.1). We merely assume that there are no lines joining the input points, no lines from the input points to the output points, etc.

A convenient way to represent a digraph $\mathfrak{D}$ associated with $\mathbb{S}$ is to use interconnection matrices. We propose that the $p \times p$ interconnection matrix $M = (m_{ij})$, which we define as a composite matrix

$$M = \begin{bmatrix} E & L & 0 \\ 0 & 0 & 0 \\ F & 0 & 0 \end{bmatrix} \tag{3.2}$$

such that $i, j = 1, 2, \ldots, p$ and $p = n + m + l$, be used to describe the *basic structure* of $\mathbb{S}$. In (3.2), the $n \times n$ state connection matrix $E = (e_{ij})$ is defined as a binary matrix with elements e_{ij} specified by

$$e_{ij} = \begin{cases} 1, & (x_j, x_i) \in R, \\ 0, & (x_j, x_i) \notin R, \end{cases} \tag{3.3}$$

where $i, j = 1, 2, \ldots, n$. That is, $e_{ij} = 1$ if x_j occurs in $f_i(t, x, u)$, and $e_{ij} = 0$ if x_j does not occur in $f_i(t, x, u)$. Similarly, we define the $n \times m$ input connection matrix $L = (l_{ij})$ as

$$l_{ij} = \begin{cases} 1, & (u_j, x_i) \in R, \\ 0, & (u_j, x_i) \notin R, \end{cases} \tag{3.4}$$

where $i = 1, 2, \ldots, n$ and $j = 1, 2, \ldots, m$. In other words, $l_{ij} = 1$ if u_j occurs in $f_i(t, x, u)$, and $l_{ij} = 0$ if u_j does not occur in $f_i(t, x, u)$. Finally, the $l \times n$ output connection matrix $F = (f_{ij})$ is defined by

$$f_{ij} = \begin{cases} 1, & (x_j, y_i) \in R, \\ 0, & (x_j, y_i) \notin R, \end{cases} \tag{3.5}$$

where $i = 1, 2, \ldots, l$ and $j = 1, 2, \ldots, n$. Again, $f_{ij} = 1$ if x_j occurs in $g_i(t, x)$, and $f_{ij} = 0$ if x_j does not occur in $g_i(t, x)$.

If no component u_j of the input vector u can influence a state x_i either directly or via other states of $\mathbb{S}$, then there is no way to alter the behavior of $\mathbb{S}$ associated with the state x_i. Similarly, if a state x_j does not influence any component y_i of the output vector y either directly or via other states of $\mathbb{S}$, then it is impossible to estimate the state x_j. In order to express these facts in the graph-theoretic terms, we replace the word "influence" by "reach" and rely on the reachability concept of digraphs (Harary, Norman, and Cartwright, 1965).

To define input and output reachability in terms of graphs, we need several well-known notions from the theory of directed graphs. We consider again the digraph $\mathcal{D} = (V, R)$, where the set $V = \{v_1, v_2, \ldots, v_k\}$ is specified as $V = U \cup X \cup Y$ and $p = m + n + l$. If a collection of distinct points $v_1, v_2, \ldots, v_k$ together with the lines $(v_1, v_2), (v_2, v_3), \ldots, (v_{k-1}, v_k)$ are placed in sequence, then the ordered set $\{(v_1, v_2), (v_2, v_3), \ldots, (v_{k-1}, v_k)\}$ is a *(directed) path* from v_1 to v_k. Then v_i is *reachable* from v_j if there is a path from v_j to v_i. A *reachable set* $V_i(v_j)$ *of a point* v_j is a set of points v_i reachable from v_j. Carrying this a step further, we define a *reachable set* $V_i(V_j)$ *of a set* V_j as a set of points v_i reachable from any point $v_j \in V_j$. Therefore, $V_i(V_j)$ is the union of the sets $V_i(v_j)$ for $v_j \in V_j$. An *antecedent set* $V_j(v_i)$ *of* v_i consists of points v_j from which v_i is reachable. Similarly, an *antecedent set* $V_j(V_i)$ *of a set* V_i is a set of points v_j from which some point v_i of V_i is reachable.

Now, we can state the following (Šiljak, 1977a,b):

Definition 3.1. *A system $\mathcal{S}$ with a digraph $\mathcal{D} = (U \cup X \cup Y, R)$ is input-reachable if X is a reachable set of U.*

The "directional dual" of Definition 1 is

Definition 3.2. *A system $\mathcal{S}$ with a digraph $\mathcal{D} = (U \cup X \cup Y, R)$ is output-reachable if X is an antecedent set of Y.*

By imitating Definitions 3.1 and 3.2, but otherwise using different sets, we can formulate the following:

Definition 3.3. *A system $\mathcal{S}$ with a digraph $\mathcal{D} = (U \cup X \cup Y, R)$ is input-output-reachable if Y is a reachable set of U and U is an antecedent set of Y.*

In Definitions 3.1–3, we ignore the trivial fact that every point of a digraph belongs to its reachable set. We also do not require that a reachable set includes all corresponding reachable points. Furthermore, from the above definitions we can conclude that if a system is both input- and output-reachable, then it is also input-output-reachable, but the converse is not true in general.

We consider reachability of $\mathcal{S}$ in terms of the $p \times p$ *path matrix* $P = (p_{ij})$ which corresponds to the digraph $\mathcal{D} = (V, R)$ and is defined as

$$p_{ij} = \begin{cases} 1, & \text{there is a path from } v_j \text{ to } v_i, \\ 0, & \text{there is no path from } v_j \text{ to } v_i. \end{cases} \tag{3.6}$$

In this definition of P, the trivial paths of zero length are excluded (the length of a path being the number of lines in the path).

To determine the path matrix for a given digraph, we need the following result which was obtained by Festinger, Schachter, and Back (1950):

Theorem 3.1. *Let $M = (m_{ij})$ be the $p \times p$ interconnection matrix corresponding to a digraph $\mathcal{D} = (V, R)$, and let $N = (n_{ij})$ be the $p \times p$ matrix such that $N = M^d$, where $d \in \{1, 2, \ldots, p\}$. Then n_{ij} is the total number of distinct sequences $(v_j, \ldots), \ldots, (\cdots, v_i)$ of length d in $\mathcal{D}$.*

Proof. We can prove this theorem by induction, following Berztiss (1975). We show that the theorem is true for $d = 1$, and then show that it is true for $d + 1$ whenever it is true for d. Then the theorem is true for any d. For $d = 1$, the theorem is actually the definition of the matrix M. For $d + 1$, $m_{ik} n_{kj} = n_{kj}$ if (v_k, v_i) is a line, and $m_{ik} n_{kj} = 0$ if (v_k, v_i) is not a line. The total number of sequences of length $d + 1$ having the form $(v_j, \ldots), \ldots, (\cdots, v_k), (v_k, v_i)$ is equal to $\sum_{k=1}^{p} m_{ik} n_{kj}$, which is the ijth element of M^{d+1}. This proves Theorem 3.1.

The path matrix is calculated using the following:

Corollary 3.1. *Let $P = (p_{ij})$ be a path matrix of $\mathcal{D} = (V, R)$, and $Q = (q_{ij})$ be a matrix defined as*

$$Q = M + M^2 + \cdots + M^p. \tag{3.7}$$

Then $p_{ij} = 1$ if and only if $q_{ij} \neq 0$.

Once reachability is formulated in terms of the path matrix P, we can calculate P from a given M to determine the input and output reachability of $\mathcal{S}$.

Let us note that M^d can be written as

$$M^d = \begin{bmatrix} E^d & E^{d-1}L & 0 \\ 0 & 0 & 0 \\ FE^{d-1} & FE^{d-2}L & 0 \end{bmatrix}, \tag{3.8}$$

where again $d \in \{1, 2, \ldots, p\}$. The matrix Q of (3.7) has the form

$$Q = \begin{bmatrix} E^0 & L^0 & 0 \\ 0 & 0 & 0 \\ F^0 & H^0 & 0 \end{bmatrix}, \tag{3.9}$$

where

$$
\begin{aligned}
E^0 &= E + E^2 + \cdots + E^p, \\
L^0 &= (I + E + \cdots + E^{p-1})L, \\
F^0 &= F(I + E + \cdots + E^{p-1}), \\
H^0 &= F(I + E + \cdots + E^{p-2})L.
\end{aligned}
\tag{3.10}
$$

We arrive immediately at the following (Šiljak, 1977a,b):

Theorem 3.2. *A system $\mathbb{S}$ with an input-output connection matrix M defined in (3.2) is input-reachable if and only if the matrix L^0 of (3.10) has no zero rows, it is output-reachable if and only if the matrix F^0 of (3.10) has no zero columns, and it is input-output-reachable if and only if H^0 of (3.10) has neither zero rows nor zero columns.*

Proof. By constructing the path matrix P using Q of (3.9) and Corollary 3.1, the proof of Theorem 3.2 is automatic.

To illustrate the application of Theorem 3.2, let us consider a system $\mathbb{S}$ with the digraph shown in Figure 3.1. The interconnection matrix M of (3.2) is given as

$$
M = \begin{array}{c} \begin{array}{cccccc} x_1 & x_2 & x_3 & x_4 & u & y \end{array} \\ \begin{bmatrix} 0 & 1 & 0 & 1 & 1 & 0 \\ 1 & 0 & 0 & 1 & 0 & 0 \\ 0 & 0 & 0 & 1 & 0 & 0 \\ 0 & 0 & 1 & 0 & 0 & 0 \\ 0 & 0 & 0 & 0 & 0 & 0 \\ 0 & 0 & 1 & 0 & 0 & 0 \end{bmatrix} \end{array} \begin{array}{c} x_1 \\ x_2 \\ x_3 \\ x_4 \\ u \\ y \end{array}, \tag{3.11}
$$

and the corresponding path matrix P is calculated via the matrix Q of (3.9) as

$$
P = \begin{array}{c} \begin{array}{cccccc} x_1 & x_2 & x_3 & x_4 & u & y \end{array} \\ \begin{bmatrix} 1 & 1 & 1 & 1 & 1 & 0 \\ 1 & 1 & 1 & 1 & 1 & 0 \\ 0 & 0 & 1 & 1 & 0 & 0 \\ 0 & 0 & 1 & 0 & 0 & 0 \\ 0 & 0 & 0 & 0 & 0 & 0 \\ 0 & 0 & 1 & 1 & 0 & 0 \end{bmatrix} \end{array} \begin{array}{c} x_1 \\ x_2 \\ x_3 \\ x_4 \\ u \\ y \end{array}. \tag{3.12}
$$

By applying Theorem 3.2 to the matrix P of (3.12), we conclude that the

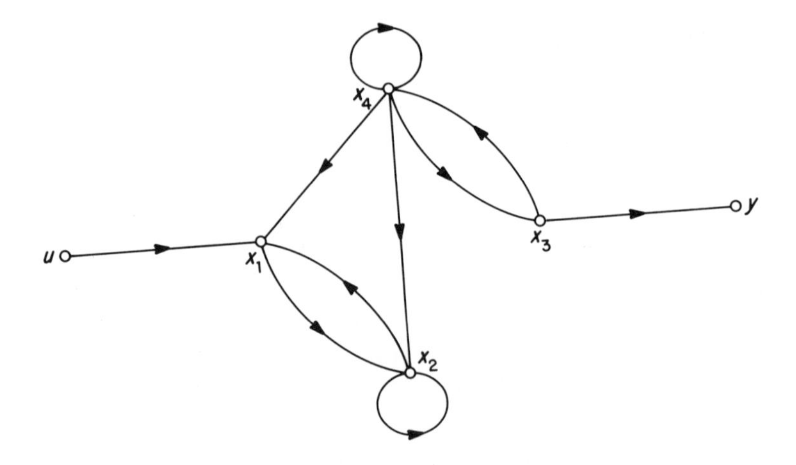

FIGURE 3.1. Input- and output-unreachable system.

system $\mathcal{S}$ is neither input- nor output-reachable. Therefore, it cannot be input-output-reachable, which is confirmed by P.

If we interchange the input and output in the digraph of Figure 3.1 to get the digraph shown in Figure 3.2, then we obtain a system which is both input- and output-reachable. The matrix M corresponding to the digraph of Figure 3.2 is

$$M = \begin{array}{c} \begin{matrix} x_1 & x_2 & x_3 & x_4 & u & y \end{matrix} \\ \begin{bmatrix} 0 & 1 & 0 & 1 & 0 & 0 \\ 1 & 0 & 0 & 1 & 0 & 0 \\ 0 & 0 & 0 & 1 & 1 & 0 \\ 0 & 0 & 1 & 0 & 0 & 0 \\ 0 & 0 & 0 & 0 & 0 & 0 \\ 1 & 0 & 0 & 0 & 0 & 0 \end{bmatrix} \end{array} \begin{matrix} x_1 \\ x_2 \\ x_3 \\ x_4 \\ u \\ y \end{matrix}, \tag{3.13}$$

which produces the path matrix

$$P = \begin{array}{c} \begin{matrix} x_1 & x_2 & x_3 & x_4 & u & y \end{matrix} \\ \begin{bmatrix} 1 & 1 & 1 & 1 & 1 & 0 \\ 1 & 1 & 1 & 1 & 1 & 0 \\ 0 & 0 & 1 & 1 & 1 & 0 \\ 0 & 0 & 1 & 0 & 1 & 0 \\ 0 & 0 & 0 & 0 & 0 & 0 \\ 1 & 1 & 1 & 1 & 1 & 0 \end{bmatrix} \end{array} \begin{matrix} x_1 \\ x_2 \\ x_3 \\ x_4 \\ u \\ y \end{matrix}, \tag{3.14}$$

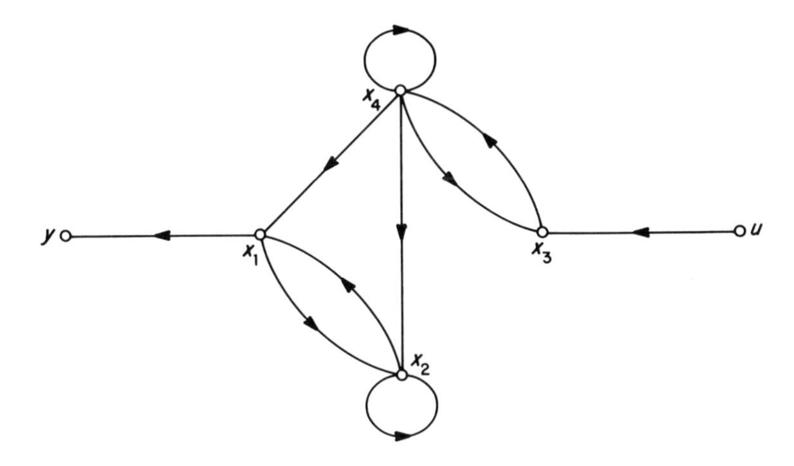

FIGURE 3.2. Input- and output-reachable system.

confirming the statement about the digraph of Figure 3.2. Since $\mathbb{S}$ is both input- and output-reachable, it is input-output-reachable, which is also confirmed by P in (3.14) and verified by inspection of Figure 3.2.

The computation of the path matrix by generating powers of the interconnection matrix, as performed above, is not a numerically attractive procedure for verifying reachability properties of the corresponding dynamic system. There are numerous algorithms developed to avoid various numerical difficulties in identifying the paths of a digraph, which started with the well-known Boolean representation algorithm of Warshall (1962) and culminated recently in the depth-first search method of Tarjan (1972). A survey of these algorithms is given by Bowie (1976).

Now, we turn our attention to *partitions* and *condensations* of digraphs. For either conceptual or numerical reasons, large-scale dynamic models in ecology, economics, and engineering may be considered as dynamic systems partitioned into interconnected subsystems. In order to gain insight into the structure of such models we propose to investigate their condensation digraphs. These digraphs are obtained from the original ones by replacing the subgraphs corresponding to the subsystems by points, and joining the new points by lines which represent the interconnections among the subsystems.

Let us consider again a system $\mathbb{S}$ described by the equations

$$\begin{aligned} \dot{x} &= f(t, x, u), \\ y &= g(t, x), \end{aligned} \tag{3.1}$$

where $x(t) \in \mathfrak{R}^n$ is the state, $u(t) \in \mathfrak{R}^m$ is the input, and $y(t) \in \mathfrak{R}^l$ is the output of $\mathcal{S}$. We assume that $\mathcal{S}$ is decomposed into s interconnected subsystems $\mathcal{S}_i$ described by the equations

$$\begin{aligned} \dot{x}_i &= f_i(t, x, u), \\ y_i &= g_i(t, x), \qquad i = 1, 2, \ldots, s, \end{aligned} \tag{3.15}$$

where $x_i(t) \in \mathcal{R}^{n_i}$ is the state, $u_i(t) \in \mathcal{R}^{m_i}$ is the input, and $y_i \in \mathcal{R}^{l_i}$ is the output of $\mathcal{S}_i$. We have

$$\begin{aligned} \mathcal{R}^n &= \mathcal{R}^{n_1} \times \mathcal{R}^{n_2} \times \cdots \times \mathcal{R}^{n_s}, \\ \mathcal{R}^m &= \mathcal{R}^{m_1} \times \mathcal{R}^{m_2} \times \cdots \times \mathcal{R}^{m_r}, \\ \mathcal{R}^l &= \mathcal{R}^{l_1} \times \mathcal{R}^{l_2} \times \cdots \times \mathcal{R}^{l_q}, \end{aligned} \tag{3.16}$$

so that

$$\begin{aligned} x &= (x_1^T, x_2^T, \ldots, x_s^T)^T, \\ u &= (u_1^T, u_2^T, \ldots, u_r^T)^T, \\ y &= (y_1^T, y_2^T, \ldots, y_q^T)^T \end{aligned} \tag{3.17}$$

are the state, input, and output vectors of $\mathcal{S}$, respectively.

We associate again a digraph $\mathcal{D} = (V, R)$ with the dynamic system $\mathcal{S}$ described by (3.1). Then we partition each set U, X, Y of V into disjoint subsets $u_1, u_2, \ldots, u_r$; $x_1, x_2, \ldots, x_s$; $y_1, y_2, \ldots, y_q$ whose union is all of V. Thus, each element of V is in exactly one subset. The condensation $\mathcal{D}^* = (V^*, R^*)$ of $\mathcal{D} = (V, R)$ with respect to this partition is the digraph whose points V^* are those subsets of V, that is, $V^* = U^* \cup X^* \cup Y^*$ and $U^* = \{u_1, u_2, \ldots, u_r\}$, $X^* = \{x_1, x_2, \ldots, x_s\}$, $Y^* = \{y_1, y_2, \ldots, y_q\}$. The lines of the condensation $\mathcal{D}^*$ are determined by the following rule: There is a line (u_j, x_i) in $\mathcal{R}^*$ if and only if there is at least one line in R from a point of the subset u_j of U to a point of the subset x_i of X. Similarly, the rule holds for the lines (x_j, x_i) and (x_j, y_i) of $\mathcal{D}^*$. In this way, the condensation $\mathcal{D}^*$ represents (uniquely) the structure of the composite system $\mathcal{S}$ described by (3.15) with points of $\mathcal{D}^*$ standing for the subsystems and the lines of $\mathcal{D}^*$ standing for the interconnections among them.

Another way to represent a partition of $\mathcal{S}$ is to use interconnection matrices in the same was as they were used to describe the original system $\mathcal{S}$. Rewriting the matrix M of (3.2), but using different submatrices, we can define the $p^* \times p^*$ matrix

$$M^* = \begin{bmatrix} E^* & L^* & 0 \\ 0 & 0 & 0 \\ F^* & 0 & 0 \end{bmatrix}, \tag{3.18}$$

which we associate with the condensation $\mathcal{D}^*$ in an obvious way. In (3.18), E^*, L^*, F^* are $s \times s$, $s \times r$, $q \times s$ matrices, respectively, and $p^* = s + r + q$. Now, input and output reachability of the condensation $\mathcal{D}^*$ can be determined by applying Theorem 3.2 to the matrix M^* of (3.18).

There are many different ways in which a dynamic system $\mathbb{S}$ and the corresponding digraph $\mathcal{D}$ may be partitioned. In the pure theory of structures (Harary, Norman, and Cartwright, 1965), it is common to partition a digraph into its *strong components* so that each point of the condensation $\mathcal{D}^*$ corresponds to one and only one strong component of $\mathcal{D}$—a strong component of a digraph $\mathcal{D}$ being a subgraph of $\mathcal{D}$ in which every two points are mutually reachable. To be able to conclude input and output reachability of such a special condensation, let us denote by $\mathcal{D}_x = (X, R_x)$ the *state-truncation* of the digraph $\mathcal{D} = (V, R)$, which is obtained from $\mathcal{D}$ by removing all the input points U and the output points Y as well as all the lines in R connected to the points of U and Y. By $\mathcal{D}_x^*$ we denote the truncated condensation of $\mathcal{D}^*$ which corresponds to the truncation $\mathcal{D}_x$. Now, we can prove the following:

Theorem 3.3. *Let the condensation $\mathcal{D}_x^*$ be constructed with respect to the strong components of $\mathcal{D}_x$. Then the digraph $\mathcal{D}$ is input- (output-) reachable if and only if the condensation $\mathcal{D}^*$ is input- (output-) reachable.*

Proof. We prove only the input-reachability part of the theorem, since the output-reachability part is its directional dual. If $\mathcal{D}^*$ is input-reachable, then X^* is a reachable set of U^* and there is a path to each point of X^* from a point of U^*. Since $\mathcal{D}_x^*$ is a condensation with respect to strong components of $\mathcal{D}_x$, it follows from the reachability of components that there is a path to each point of X from a point of U. That is, X is a reachable set of U, and $\mathcal{D}$ is input-reachable. Conversely, if $\mathcal{D}^*$ is not input-reachable, then X^* is not a reachable set of U^*, and there are points of X^* that cannot be reached by any point of U^*. Obviously, by the definition of condensation, those points of X that correspond to the unreachable points of X^* cannot be reached by any point of U, and $\mathcal{D}$ is not input-reachable. This proves Theorem 3.3.

With an abuse of Definitions 3.1 and 3.2, in the above proof we referred to input (output) reachability of the digraphs $\mathcal{D}$ and $\mathcal{D}^*$. This was done to avoid ambiguity arising from the fact that both the digraph $\mathcal{D}$ and its condensation $\mathcal{D}^*$ are related to the same dynamic system $\mathbb{S}$. The change in terminology should create no confusion, since input (output) reachability is defined unambiguously in terms of digraphs. Still another abuse of notation is committed in (3.17), where it would be appropriate to use starred notation on all components x_i, u_i, y_i of the vectors x, u, y, since they represent a partition and condensation rather than the individual states, inputs, and outputs of the system $\mathbb{S}$. This again should not pose any difficulty, since in the rest of this chapter, if not obvious or indicated to the contrary, all components of state, input, and output vectors correspond to partitions and condensations.

Algorithms of Purdom (1970), Munro (1971), and Kevkorian (1975) can be used to compute the strongly connected components of the digraph $\mathcal{D}_x$. Then the input and output reachability of a system $\mathbb{S}$ can be determined by means of the appropriate condensation $\mathcal{D}_x^*$ and Theorem 3.3.

Partitions of mathematical models of dynamic processes in ecology, economics, and engineering are most often guided by the special structural properties of the models. Therefore, in general, partitions of the digraphs are not performed with respect to their strong components, and there is no reason why the corresponding condensations have to be constituted that way. Furthermore, it is also possible to alter our notion of subsystems and allow "overlapping" of subspaces $\mathcal{R}^{n_i}$ in (3.16) so that, for example, the state vector x_k of the subsystem $\mathbb{S}_k$ is formed pairwise as $x_k = (x_i^T, x_j^T)^T$, where x_i and x_j are defined in (3.17). For purposes of control and estimation, it is desirable to obtain still another partition of $\mathbb{S}$ into input- and output-reachable subsystems $\mathbb{S}_i$; that is, it is advantageous to partition the corresponding digraph $\mathcal{D}$ into input- and output-reachable components. After a short historical note, this topic is considered next in the context of Kalman's (1963) canonical structure of linear systems.

E. F. Moore (1956) introduced the concept of reachability in the context of finite-state systems by formulating the notion of a strongly connected sequential machine. He wrote:

> A machine $\mathbb{S}$ will be said to be strongly connected if for any ordered pair (q_i, q_j) of states of $\mathbb{S}$, there exists a sequence of inputs which will take the machine from the state q_i to state q_j,

and thus laid a foundation for the concept of controllability of dynamic systems introduced later by Kalman (1963). While binary interconnection relationships in sequential machines lead naturally to connectedness and reachability considerations of the corresponding directed graphs, the general matrices of linear dynamic systems imposed a stronger rank condition for controllability which includes that of reachability. The same is true for Kalman's notion of observability, which is the dynamic system analog of Moore's notion of distinguishability of sequential machines.

Although controllability and observability conditions are indispensable in certain basic problems of control and estimation of dynamic systems (Kalman, Falb, and Arbib, 1969), there are at least three good reasons why they may be replaced by the weaker reachability tests of the interconnection matrices as proposed above. First, the conceptual significance of the difference between the controllability-observability rank condition and its reachability counterpart can be disputed on physical grounds. Cases where reachability tests succeed and the rank conditions fail can be dismissed as unrealistic, since a slight perturbation of system parameters can restore the controllability and observability properties of the system. For example, consider a linear constant system

$$\begin{aligned} \dot{x} &= Ax + bu, \\ y &= c^T x, \end{aligned} \tag{3.19}$$

where

$$A = \begin{bmatrix} 1 & 1 \\ 1 & 1 \end{bmatrix}, \quad b = \begin{bmatrix} 1 \\ 1 \end{bmatrix}, \quad c = \begin{bmatrix} 1 \\ 1 \end{bmatrix}. \tag{3.20}$$

It is a well-known fact (Kalman, 1963) that the system (3.19) is both uncontrollable and unobservable, because both matrices

$$[b \,|\, Ab] = \left[\begin{array}{c|c} 1 & 2 \\ 1 & 2 \end{array}\right], \quad [c^T \,|\, A^T c^T] = \left[\begin{array}{c|c} 1 & 2 \\ 1 & 2 \end{array}\right] \tag{3.21}$$

have rank less than two. Now, from the digraph of the system shown in Figure 3.3, it can be concluded by inspection that both states x_1 and x_2 are directly accessible to the input u and the output y. It is also obvious that if the elements of the triple $\{A, b, c\}$ are slightly perturbed, it is possible to recover both controllability and observability of the system (3.19). Therefore, looking from the side of the rank conditions for controllability and

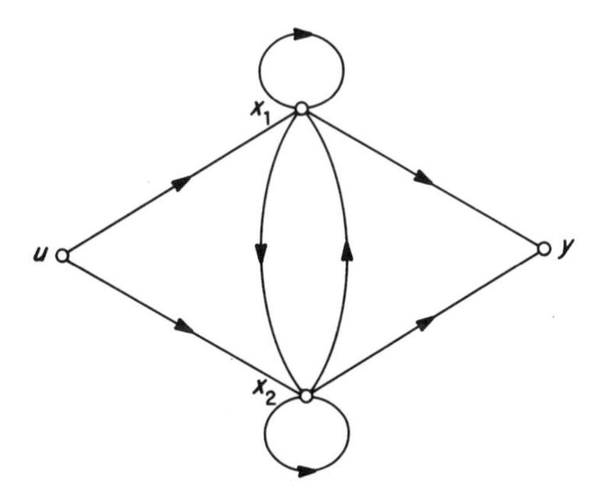

FIGURE 3.3. System structure for the example.

observability, we cannot tell whether the failure of the system to satisfy the conditions is for structural reasons, or is due to a special choice of system parameters. To avoid this ambiguity in large-scale systems where elimination of trivial cases can be quite costly, it is advantageous to use the "pure" structural considerations of input and output reachability.

The second reason for preferring reachability considerations is that the rank conditions are ill-posed numerical problems as compared to purely structural manipulations using binary interconnection matrices. For this reason, Lin (1974), Kevkorian (1975), and Shields and Pierson (1976) recommended a return to connectedness investigations via a new term of "structural controllability". It is in the spirit of Moore's original investigations that we formulated the concepts of input and output reachability above and opened the possibility of using effective computing schemes (Bowie, 1976) from the pure theory of structures in determining the input and output reachability of dynamic systems.

Finally, the third reason for considering the reachability conditions more appealing is that they apply to linear and nonlinear systems alike. This provided a real possibility of formulating for the first time the *pure canonical structure* of dynamic systems (Šiljak, 1977b). By imitating the canonical structure of linear systems defined by Kalman (1963) in the context of controllability and observability, but otherwise using input and output reachability, we present next the canonical structure of nonlinear dynamic systems.

Let us consider again the dynamic system $\mathbb{S}$ described by (3.1), together with its interconnection matrix M of (3.2). By permutation of rows and columns of M, it can be transformed into a matrix $\tilde{M}$ which has the following form:

$$\tilde{M} = \begin{array}{c} \begin{matrix} \tilde{x}_{io} & \tilde{x}_{\not{i}o} & \tilde{x}_{i\emptyset} & x_{\not{i}\emptyset} & u & y \end{matrix} \\ \begin{bmatrix} E_{11} & E_{12} & 0 & 0 & L_1 & 0 \\ 0 & E_{22} & 0 & 0 & 0 & 0 \\ E_{31} & E_{32} & E_{33} & E_{34} & L_3 & 0 \\ 0 & E_{42} & 0 & E_{44} & 0 & 0 \\ 0 & 0 & 0 & 0 & 0 & 0 \\ F_1 & F_2 & 0 & 0 & 0 & 0 \end{bmatrix} \end{array} \begin{matrix} \tilde{x}_{io} \\ \tilde{x}_{\not{i}o} \\ \tilde{x}_{i\emptyset} \\ \tilde{x}_{\not{i}\emptyset} \\ u \\ y \end{matrix} \tag{3.22}$$

The partition of the state vector $\tilde{x}$ of the transformed system $\tilde{\mathbb{S}}$,

$$\tilde{x} = (\tilde{x}_{io}^T, \tilde{x}_{\not{i}o}^T, \tilde{x}_{i\emptyset}^T, \tilde{x}_{\not{i}\emptyset}^T)^T \tag{3.23}$$

into four components, which can be identified from $\tilde{M}$ in (3.22), represents the four subsystems with the following properties:

$\tilde{\mathbb{S}}_{io}$, input-reachable and output-reachable,
$\tilde{\mathbb{S}}_{\not{i}o}$, input-unreachable and output-reachable,
$\tilde{\mathbb{S}}_{i\not{o}}$, input-reachable and output-unreachable,
$\tilde{\mathbb{S}}_{\not{i}\not{o}}$, input-unreachable and output-unreachable .

By using condensation, we can represent each component of $\tilde{x}$ in (3.23) as a point of the condensation $\tilde{\mathcal{D}}^*$ as shown in Figure 3.4. This digraph represents the *pure canonical structure* of the dynamic system $\mathbb{S}$.

Now, we turn our attention to the *decentralized systems*, which are the main subject of this chapter. Decentralization is an effective way to cope with complexity and gross changes in the interactions of large-scale dynamic systems; thus decentralized systems, as well as control and estimation schemes which exploit decentralization in one form or another, have attracted rapidly increasing interest (see Sandell, Varaiya, and Athans, 1975). Nevertheless, it is only recently (Šiljak and Vukčević, 1976c) that attempts have been made to formulate an appropriate definition of a decentralized system and produce effective decentralization schemes for linear dynamic systems. The definitions and reachability properties of decentralized systems are outlined here; a presentation of the decentralization techniques is postponed till the next section.

Whether a system is decentralized or not is essentially a matter of its structure, and therefore decentrality of dynamic systems can be effectively defined in terms of digraphs. Intuitively, a system $\mathbb{S}$ is an *input-decentralized system* if each subsystem $\mathbb{S}_i$ has its own input u_i. In order to put this intuitive notion of input decentralization into digraph-theoretic terms, we recall that

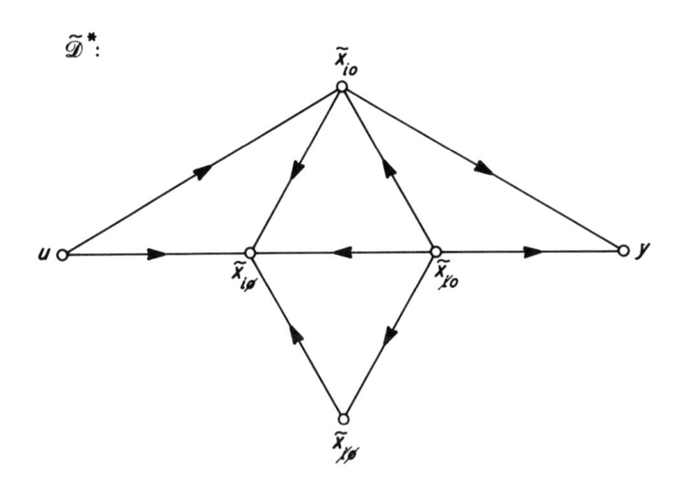

FIGURE 3.4. Canonical structure.

the *point basis* B of a digraph $\mathcal{D} = (V, R)$ is a minimal subset of V from which all points of $\mathcal{D}$ are reachable. Furthermore, a 1-*basis* is a minimal collection B_1 of mutually nonadjacent points in $\mathcal{D}$ such that every point of $\mathcal{D}$ is either in B_1 or adjacent to a point of B_1. To avoid trivial cases which are of no interest to us, we assume that no points in the subsets U and Y of $V = U \cup X \cup Y$ are disconnected, that is, $\mathrm{od}(u_k) \neq 0$ for all $k = 1, 2, \ldots, m$, and $\mathrm{id}(y_k) \neq 0$ for all $k = 1, 2, \ldots, l$, where "od" and "id" stand for *outdegree* and *indegree* (Harary, Norman, and Cartwright, 1965).

To be able to use the 1-basis in formulating definitions of decentralized systems, we form the *input-truncated condensation* $\mathcal{D}_u^* = (U^* \cup X^*, R_u^*)$ from the condensation $\mathcal{D}^*$ of $\mathcal{D}$ by removing all the output points y_i of Y^*. Now we state

Definition 3.4. *A system* $\mathcal{S}$ *with an input-truncated condensation* $\mathcal{D}_u^* = (U^* \cup X^*, R_u^*)$ *is input-decentralized if and only if* $r = s$ *and the set* U^* *is a* 1-*basis of* $\mathcal{D}_u^*$ *such that* $\mathrm{od}(u_i) = 1$ *for all* $i = 1, 2, \ldots, r$.

To provide a definition of an *output-decentralized system*, we need the notion of *point contrabasis* C of $\mathcal{D} = (V, R)$, which is a minimal subset of V such that V is a set of points from which some point of C is reachable (Harary, Norman, and Cartwright, 1965). We also recall that 1-*contrabasis* of $\mathcal{D}$ is a minimal collection C_1 of mutually nonadjacent points such that every point of $\mathcal{D}$ is either in C_1 or adjacent to a point of C_1. By $\mathcal{D}_y^* = (X^* \cup Y^*, R_y^*)$ we denote the *output-truncated condensation* of $\mathcal{D}$. Then we have

Definition 3.5. *A system* $\mathcal{S}$ *with an output-truncated condensation* $\mathcal{D}_y^* = (X^* \cup Y^*, R_y^*)$ *is output-decentralized if and only if* $q = s$ *and the set* Y *is a* 1-*contrabasis of* $\mathcal{D}_y^*$ *such that* $\mathrm{id}(y_i) = 1$ *for all* $i = 1, 2, \ldots, q$.

It is simple to combine Definitions 3.4 and 3.5 and formulate a definition of *input-output-decentralized systems*.

The digraph $\mathcal{D}_u^*$ of an input-centralized system is shown in Figure 3.5(a). The input-decentralized version of the same system is given in Figure 3.5(b). Outputs are omitted in the digraphs, since they have no effect on the input-decentralization property. Output-centralized and -decentralized systems can be represented by directional duals of the digraphs shown in Figure 3.5.

A convenient way to characterize input- and output-decentralized systems is to use the interconnection matrix M^* defined in (3.18). Then a system $\mathcal{S}$ with a condensation $\mathcal{D}^*$ is input-decentralized if and only if in M^* of (3.18), $L^* = I_s$, where I_s is the $s \times s$ identity matrix. Similarly, $\mathcal{S}$ is an output-decentralized system if and only if $F^* = I_s$.

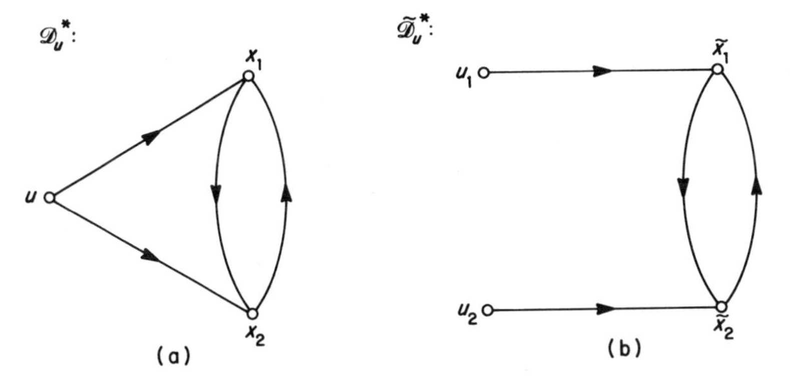

FIGURE 3.5. (a) Input-centralized system. (b) Input-decentralized system.

By using the methods of Kevkorian (1975), it is possible to permute the states of a nonlinear system $\mathbb{S}$ and choose an appropriate condensation $\mathcal{D}^*$ of the corresponding digraph $\mathcal{D}$ so that the new transformed system is in either input- or output-decentralized form or both. As shown in Section 3.2, in linear systems it is possible to do better (Šiljak and Vukčević, 1976c) and use similarity transformations to produce input- and output-decentralized systems. Such decentralization procedures may be required preliminary steps in the stabilization, optimization, and estimation of linear dynamic systems by decentralized feedback, which are considered in the rest of this chapter.

By applying Theorem 3.2 to M^* of (3.18), we conclude that if $L^* = F^* = I_s$, then the input-output-decentralized system $\mathbb{S}$ is input-output-reachable whenever each subsystem $\mathbb{S}_i$ is input-output-reachable regardless of the form of the state interconnection matrix E^* of $\mathcal{D}^*$. This somewhat trivial result leads to an interesting conclusion: in decentralized systems input-output reachability is invariant under perturbations of the interconnection structure among the subsystems. This is the *vulnerability* aspect of dynamic systems, which is considered next.

We argued in the previous chapter that large-scale dynamic systems, which are composed of interconnected subsystems, quite commonly either by design or fault do not stay "in one piece" during operation. They are subject to structural perturbations whereby groups of subsystems are disconnected from and again connected to each other in an unpredictable way. In Chapter 1 and in the rest of this chapter, we consider the effects of structural perturbations on the stability and optimality of dynamic systems. In the rest of this section, we complement these results by investigating reachability under structural perturbations.

In terms of digraphs (Harary, Norman, and Cartwright, 1965), disconnecting subsystems from each other is equivalent to "line removals", and disconnecting subsystems from the overall system is equivalent to "point removals". These perturbations can be conveniently described by interconnection matrices (Šiljak, 1975). For this purpose, the $s \times s$ fundamental interconnection matrix $\bar{E}^* = (\bar{e}_{ij}^*)$ is associated with a condensation $\bar{\mathfrak{D}}^* = (U^* \cup X^* \cup Y^*, \bar{R}^*)$ of $\mathfrak{S}$ in (3.15) as follows:

$$\bar{e}_{ij}^* = \begin{cases} 1, & (x_j, x_i) \in \bar{R}^*, \\ 0, & (x_j, x_i) \notin \bar{R}^*. \end{cases} \tag{3.24}$$

That is, $\bar{e}_{ij}^* = 1$ if x_j occurs in $f_i(t, x, u)$, and $\bar{e}_{ij}^* = 0$ if x_j does not occur in $f_i(t, x, u)$. Now, a structural perturbation is represented by the removal of a line (or a number of lines) of the condensation $\bar{\mathfrak{D}}^* = (U^* \cup X^* \cup Y^*, \bar{R}^*)$ between points of X^*. That results in a spanning subgraph $\mathfrak{D}^*$ of $\bar{\mathfrak{D}}^*$, that is, a subgraph with the same set of points as $\bar{\mathfrak{D}}^*$. All spanning subgraphs of $\bar{\mathfrak{D}}^*$ obtained this way can be represented uniquely by an interconnection matrix E^* which is obtained from $\bar{E}^*$ as follows: $\bar{e}_{ij}^* = 0$ in $\bar{E}^*$ implies $e_{ij}^* = 0$ in E^* for all $i, j = 1, 2, \ldots, s$; and a removal of a line (x_j, x_i) of $\bar{\mathfrak{D}}^*$ implies that $\bar{e}_{ij}^* = 1$ in $\bar{E}^*$ is replaced by $e_{ij}^* = 0$ in E^*. The fact that an interconnection matrix E^* is generated in this way by the fundamental interconnection matrix $\bar{E}^*$ is denoted by $E^* \in \bar{E}^*$. Finally, without loss of generality, a point removal can be treated in the same way, as a special case of line removals. If a kth point of $\bar{\mathfrak{D}}^*$ is removed, then $e_{ik}^* = e_{kj}^* = 0$ for all $i, j = 1, 2, \ldots, s$.

Structural perturbations are illustrated in Figure 3.6. The top digraph $\bar{\mathfrak{D}}_y^*$ is the output truncation which represents the basic structure of the composite system $\mathfrak{S}$ when all the inputs are removed, and which corresponds to the fundamental interconnection matrices

$$\bar{E}^* = \begin{bmatrix} 0 & 0 & 1 \\ 1 & 0 & 0 \\ 0 & 1 & 0 \end{bmatrix}, \qquad \bar{F}^* = [0 \quad 1 \quad 1]. \tag{3.25}$$

The structural perturbations formed by line removals are represented by the digraphs $\mathfrak{D}_y^*$ below the digraph $\bar{\mathfrak{D}}_y^*$. The digraph at the bottom of Figure 3.6 corresponds to $E^* = 0$ and thus to the total disconnection of states x_1, x_2, x_3. Inputs can be added to $\mathfrak{D}_y^*$ in an obvious way if input reachability is investigated in the presence of structural perturbations.

On the basis of the above considerations, we introduce the following:

Definition 3.6. *A system $\mathfrak{S}$ with a condensation $\bar{\mathfrak{D}}^* = (U^* \cup X^* \cup Y^*, \bar{R}^*)$ is connectively input- (output-) reachable if and only if it is input- (output-) reachable for all interconnection matrices $E^* \in \bar{E}^*$.*

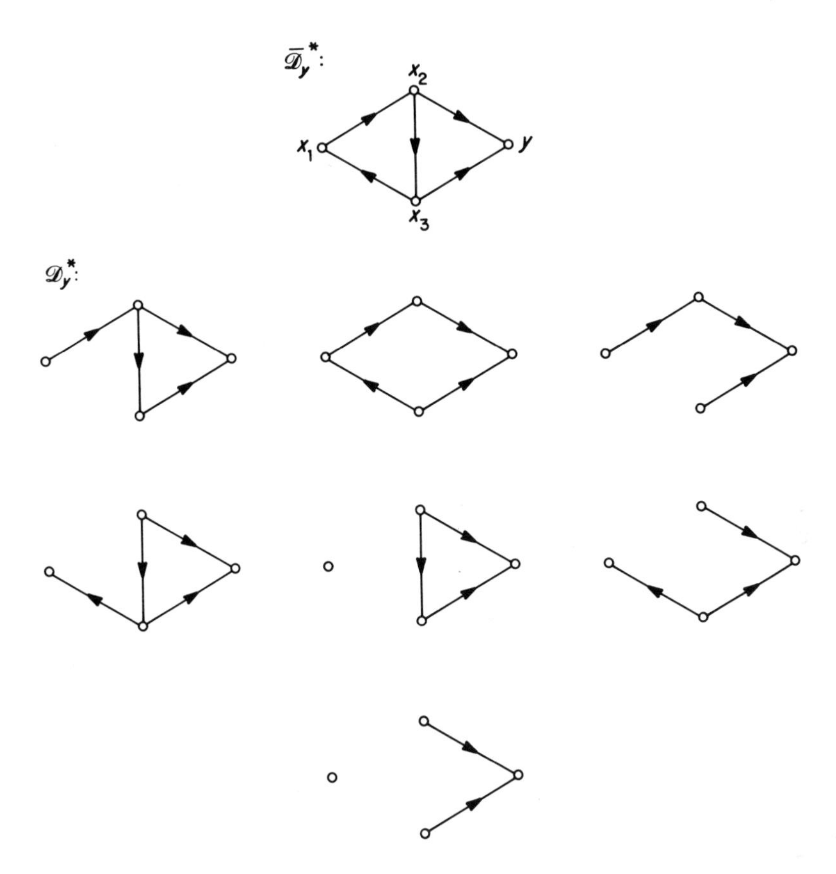

FIGURE 3.6. Structural perturbations.

Definition 3.6 parallels those of connective stability (Šiljak, 1975) and connective suboptimality (Šiljak and Sundareshan, 1976a, b). In the case of control systems, the three definitions complement each other in an important way discussed in the latter part of this chapter.

Now we prove the following:

Theorem 3.4. *A system* $\mathbb{S}$ *with an input-truncated condensation* $\bar{\mathfrak{D}}_u^* = (U^* \cup X^*, \bar{R}_u^*)$ *is connectively input-reachable if and only if the set* U^* *is a* 1*-basis of* $\bar{\mathfrak{D}}_u^*$ *for* $E^* = 0$.

Proof. The "if" part for $E^* = 0$ follows directly from the definition of the 1-basis. For $E^* \neq 0$, we get the corresponding digraph by adding lines to the one that corresponds to $E^* = 0$. But it is obvious that there are no lines in any digraph whose addition can destroy its reachability property. For if (x_j, x_i) is any line of $\mathfrak{D}_u^*$, any path in $\mathfrak{D}_u^* - (x_j, x_i)$ is also in $\mathfrak{D}_u^*$.

On the other hand, if U^* is not a 1-basis for $E^* = 0$, then there must be points of X^* that are not reachable from U^* for all $E^* \in \bar{E}^*$, and $\mathbb{S}$ is not connectively input-reachable. This proves the "only if" part and thus Theorem 3.4.

By the principle of duality of digraphs (Harary, 1969), it is possible to obtain an analog to Theorem 3.4 for connective output reachability of $\mathbb{S}$:

Theorem 3.5. *A system $\mathbb{S}$ with an output truncated condensation $\bar{\mathfrak{D}}_y^* = (X^* \cup Y^*, \bar{R}_y^*)$ is connectively output-reachable if and only if the set Y^* is a* 1*-contrabasis of $\bar{\mathfrak{D}}_y^*$ for $E^* = 0$.*

Proof. The theorem is an obvious dual of Theorem 3.4.

The result of Theorems 3.4 and 3.5 is intuitively clear: If we want reachability to be preserved under structural perturbations, we should check the "worst case", that of $E^* = 0$. It is interesting to compare this statement with the connective stability of competitive dynamic systems considered in Chapter 2, where the worst case was $E^* = \bar{E}^*$. That is, stability is improved by reducing the number of interconnections among the subsystems, but reachability may be destroyed.

Another straightforward but nevertheless important conclusion coming from Theorems 3.4 and 3.5 is that for reachability to be invulnerable to structural perturbations, each subsystem should have its own input and output or share them with other subsystems. In other words, if a subsystem is reachable from the input only through another subsystem, then it is liable to become input-unreachable due to structural perturbations. Similarly, it can become output-unreachable if its state is represented at the output only by an adjacent subsystem. If a system is under structural perturbations, then to preserve reachability, each subsystem when isolated should be input-output-reachable. This is a symmetric situation to that of the connective stability of competitive dynamic systems, where each subsystem was stable when isolated. The symmetry is not complete, since reachability of each isolated subsystem is a necessary and sufficient condition for connective reachability, while stability of each isolated subsystem was only a necessary condition for connective stability of large-scale systems.

In Figure 3.6, the digraph $\bar{\mathfrak{D}}_y^*$ represents an output-reachable system, and so do the three digraphs immediately below $\bar{\mathfrak{D}}_y^*$. The remaining four digraphs describe output-unreachable systems. By the principle of duality, input reachability under structural perturbations can be visualized using the same Figure 3.6.

Finally, one can generalize the above connective-reachability concept by defining *partial connective reachability* with respect to a pair of interconnection matrices $(\bar{E}^*, \hat{E}^*)$, where a fixed matrix $\hat{E}^* \in \bar{E}^*$ takes the role of

$E^* = 0$. Again, it is intuitively clear that if U^* is a 1-basis for $\hat{E}^*$, this being the worst case, then $\mathbb{S}$ is partially connectively input-state-reachable for all $E^* \in \bar{E}^* - \hat{E}^*$, where by the difference $\bar{E}^* - \hat{E}^*$ we mean all interconnection matrices generated by $\bar{E}^*$ which have the unit elements corresponding to $\hat{E}^*$.

It is of interest to observe that input- and output-decentralized systems are invulnerable with respect to structural perturbations. That is, we have the following corollary to the Theorems 4 and 5:

Corollary 3.2. *A system $\mathbb{S}$ is connectively input- (output-) reachable if it is input- (output-) decentralized.*

This corollary announces a strong advantage of decentralized systems over centralized ones, where common inputs are shared by the subsystems. This fact has not been stated yet in the open literature on control systems, but is probably appreciated intuitively by the authors of the many recent new results on decentralized control (Sandell, Varaiya, and Athans, 1975).

Further use of digraphs in studying vulnerability of dynamic systems is a wide open field. One way to approach problems that arise in this context, is to apply the results obtained in vulnerability studies of communication nets (Boesch and Thomas, 1970). These results, however, would have to be modified in an essential way in order to reflect the inherent structural properties of dynamic systems as defined above. Another possible aid in reliability studies of large control systems and breakdown phenomena, is the fault tree analysis (see Barlow, Fussell, and Singpurwalla, 1975). This other approach could shed some light on the intricate interplay between the components reliability and the reliability of the overall dynamic system.

3.2. DECENTRALIZATION

Despite the effective solution of many important problems in system theory by decentralized control, our knowledge of how to decentralize a dynamic system has remained superficial. Recently, an effective decentralization procedure was proposed (Šiljak and Vukčević, 1976a, c), which is a preliminary step in a multilevel control and estimation scheme for large-scale linear systems. The procedure yields a number of subsystems that have either decentralized inputs, or decentralized outputs, or both. The process of decentralization is carried out on the subsystem level, and it does not require a test for controllability and observability of the overall system.

Let us consider a system $\mathbb{S}$ described by the linear differential equation

$$\dot{z} = \hat{A}z + \hat{B}u, \tag{3.26}$$

where $z(t) \in \mathcal{R}^n$ is the state of the system, $u(t) \in \mathcal{R}^s$ is the input to the system, and $\hat{A}$ and $\hat{B}$ are constant $n \times n$ and $n \times s$ matrices.

We decompose $\mathcal{S}$ into r dynamic elements

$$\dot{z}_p = \hat{A}_p z_p + \sum_{\substack{q=1 \\ q \neq p}}^{r} \hat{A}_{pq} z_q + \hat{B}_p u, \qquad p = 1, 2, \ldots, r, \tag{3.27}$$

where $z_p(t) \in \mathcal{R}^{l_p}$ and $z = (z_1^T, z_2^T, \ldots, z_r^T)^T$, $n = \sum_{p=1}^{n} l_p$, such that all pairs $(\hat{A}_p, \hat{B}_p)$ are controllable, that is, the $l_p \times l_p s$ matrix

$$[\hat{B}_p \mid \hat{A}_p \hat{B}_p \mid \hat{A}_p^2 \hat{B}_p \mid \cdots \mid \hat{A}_p^{p-1} \hat{B}_p] \tag{3.28}$$

has rank equal to l_p (for this well-known result see Chen, 1970).

By using the linear transformation proposed by Luenberger (1967) (see also Chen, 1970), we can write the elements (3.27) as

$$\dot{\bar{z}}_p = \bar{A}_p \bar{z}_p + \sum_{\substack{q=1 \\ q \neq p}}^{r} \bar{A}_{pq} \bar{z}_q + \bar{B}_p u, \qquad p = 1, 2, \ldots, r, \tag{3.29}$$

such that the matrices $\bar{B}_p$ have the following form:

$$\bar{B}_p = \begin{bmatrix} \bar{b}_1^p & & & 0 \\ & \bar{b}_2^p & & \\ & & \ddots & \\ 0 & & & \bar{b}_s^p \end{bmatrix}, \qquad \bar{b}_i^p = \begin{bmatrix} 1 \\ 0 \\ \vdots \\ 0 \end{bmatrix}, \tag{3.30}$$

and $\bar{b}_i^p \in \mathcal{R}^{n_{pi}}$, $l_p = \sum_{i=1}^{s} n_{pi}$. The linear nonsingular transformation

$$\bar{z}_p = Q_p^{-1} z_p \tag{3.31}$$

is defined by

$$Q_p = \left[\hat{b}_1^p, \ldots, \hat{A}_p^{n_{p1}-1} \hat{b}_1^p; \ldots; \hat{b}_s^p, \ldots, \hat{A}_p^{n_{ps}-1} \hat{b}_s^p\right], \tag{3.32}$$

where $\hat{b}_i^p \in \mathcal{R}^{l_p}$, $i = 1, 2, \ldots, s$, are the columns of the matrix $\hat{B}_p$. Then in (3.29) we have

$$\bar{A}_p = Q_p^{-1} \hat{A}_p Q_p, \qquad \bar{A}_{pq} = Q_p^{-1} \hat{A}_{pq} Q_q, \qquad \bar{B}_p = Q_p^{-1} \hat{B}_p. \tag{3.33}$$

Due to (3.30), we can decompose the state $\bar{z}_p$ of each transformed element (3.29) as $\bar{z}_p = (\bar{z}_{p1}^T, \bar{z}_{p2}^T, \ldots, \bar{z}_{ps}^T)^T$, $\bar{z}_{pi}(t) \in \mathcal{R}^{n_{pi}}$, so that with each $\bar{z}_{pi}$ we associate the vector $\bar{b}_i^p$ and the component $u_i \in \mathcal{R}$ of the input vector $u \in \mathcal{R}^s$. Now, we group the $\bar{z}_{pi}$'s of each of p elements which correspond to the same input u_i, and form the ith subsystem with the state $x_i(t) \in \mathcal{R}^{n_i}$, such that $x_i = (\bar{z}_{1i}^T, \bar{z}_{2i}^T, \ldots, \bar{z}_{ri}^T)^T$ and $n_i = \sum_{p=1}^r n_{pi}$. This process of grouping yields finally the representation of the system $\mathbb{S}$ as composed of s interconnected subsystems $\mathbb{S}_i$ described by equations

$$\dot{x}_i = A_i x_i + \sum_{\substack{j=1 \\ j \neq i}}^{s} A_{ij} x_j + b_i u_i, \qquad i = 1, 2, \ldots, s. \tag{3.34}$$

To compute the matrices A_i, A_{ij}, and the vector b_i from $\bar{A}_p$, $\bar{A}_{pq}$, $\bar{B}_p$, let us denote by $\bar{z} = (\bar{z}_1^T, \bar{z}_2^T, \ldots, \bar{z}_r^T)^T$ and $x = (x_1^T, x_2^T, \ldots, x_s^T)^T$ the state vectors of the overall systems corresponding to (3.29) and (3.34), respectively. Then the grouping process described above is carried out by the nonsingular linear transformation

$$x = P\bar{z}, \tag{3.35}$$

where the permutation matrix P has the block form $P = (P_1^T, P_2^T, \ldots, P_s^T)^T$, and the ith block $P_i \in \mathcal{R}^{n_i \times n}$ is defined by

$$P_i = \left.\begin{bmatrix} 0 & \cdots & 0 & I_{1i} & \cdots & \cdots & 0 & 0 & 0 & \cdots & 0 \\ 0 & \cdots & 0 & 0 & \cdots & \cdots & I_{2i} & 0 & 0 & \cdots & 0 \\ \cdots & \cdots & \cdots & \cdots & \cdots & \cdots & \cdots & \cdots & \cdots & \cdots & \cdots \\ 0 & \cdots & 0 & 0 & \cdots & \cdots & 0 & I_{ri} & 0 & \cdots & 0 \end{bmatrix}\right\} r, \tag{3.36}$$

$$\underbrace{\qquad\qquad}_{i-1} \qquad \underbrace{\qquad\qquad\qquad}_{(r-1)s+1} \qquad \underbrace{\qquad\qquad}_{s-i}$$

where I_{pi} is the $n_{pi} \times n_{pi}$ identity matrix, and the zero matrices in (3.36) have the appropriate dimensions.

Now we write (3.29) as

$$\dot{\bar{z}} = \bar{A}\bar{z} + \bar{B}u, \tag{3.37}$$

where

$$\bar{A} = \begin{bmatrix} \bar{A}_1 & \bar{A}_{12} & \cdots & \bar{A}_{1r} \\ \bar{A}_{21} & \bar{A}_2 & \cdots & \bar{A}_{2r} \\ \cdots & \cdots & \ddots & \cdots \\ \bar{A}_{r1} & \bar{A}_{r2} & \cdots & \bar{A}_r \end{bmatrix}, \qquad \bar{B} = \begin{bmatrix} \bar{B}_1 \\ \bar{B}_2 \\ \vdots \\ \bar{B}_r \end{bmatrix}, \tag{3.38}$$

and apply (3.35) to (3.37), to get

$$\dot{x} = Ax + Bu, \tag{3.39}$$

with

$$A = P\bar{A}P^T, \qquad B = P\bar{B}, \tag{3.40}$$

and

$$B = \begin{bmatrix} b_1 & & & 0 \\ & b_2 & & \\ & & \ddots & \\ 0 & & & b_s \end{bmatrix}, \qquad b_i = \begin{bmatrix} \bar{b}_i^1 \\ \bar{b}_i^2 \\ \vdots \\ \bar{b}_i^r \end{bmatrix}, \tag{3.41}$$

where $b_i \in \mathcal{R}^{n_i}$ and $n = \sum_{i=1}^{s} n_i$. Finally, we identify the overall system (3.39) as s interconnected input-decentralized subsystems described by the equations (3.34).

It is important to note that all steps but the last in the input-decentralization scheme are performed on the subsystem level. The last step, which involves the transformation (3.35) and is performed on the overall system level, consists of regrouping the components of the state vector and does not require the matrix inversion. That is, for the permutation matrix P in (3.35), $P^{-1} = P^T$.

For the output decentralization, we consider the linear system $\mathbb{S}$ as

$$\begin{aligned} \dot{z} &= \hat{A}z + \hat{B}u, \\ y &= \hat{C}z, \end{aligned} \tag{3.42}$$

where the output $y(t) \in \mathcal{R}^m$ and $\hat{C}$ is an $m \times n$ constant matrix. By the output-decentralization scheme, we get the system (3.42) as m interconnected subsystems

$$\begin{aligned} \dot{x}_i &= A_i x_i + \sum_{\substack{j=1 \\ j \neq i}}^{m} A_{ij} x_j + B_i u, \qquad i = 1, 2, \ldots, m, \\ y_i &= c_i^T x_i, \end{aligned} \tag{3.43}$$

where $y_i(t) \in \mathcal{R}$ is a scalar output of the ith subsystem, and c_i is a constant n_i-vector.

To obtain (3.43), we form the dual of (3.42) as

$$\begin{aligned} \dot{z} &= \hat{A}^T z + \hat{C}^T u, \\ y &= \hat{B}^T z, \end{aligned} \tag{3.44}$$

where for convenience we have kept the same notation $z(t)$, $u(t)$, $y(t)$ but recall that now $z(t) \in \mathcal{R}^n$, $u(t) \in \mathcal{R}^m$, $y(t) \in \mathcal{R}^s$. Applying the input-decentralization scheme to (3.44), we get

$$\begin{aligned} \dot{x}_i &= A_i^T x_i + \sum_{\substack{j=1 \\ j \neq i}}^{m} A_{ji}^T x_j + c_i u_i, \\ & \qquad\qquad\qquad\qquad i = 1, 2, \ldots, m, \\ y &= \sum_{i=1}^{m} B_i^T x_i, \end{aligned} \tag{3.45}$$

where $c_i(t) \in \mathcal{R}^{n_i}$, and the B_i's are $s \times n_i$ matrices obtained from $\hat{B}$ in the course of the input decentralization of (3.44). It remains to note that (3.45) is the dual of (3.43), and conclude that (3.43) is the output-decentralization version of (3.42).

To illustrate the decentralization procedure, let us consider the system

$$\dot{z} = \begin{bmatrix} 1 & 3 & 4 & 5 & 6 \\ 0 & 5 & 2 & 4 & 1 \\ 2 & 4 & 0 & 0 & 6 \\ 2 & 0 & 2 & 1 & 6 \\ 2 & 2 & 3 & 0 & 1 \end{bmatrix} z + \begin{bmatrix} 1 & 2 \\ 0 & 4 \\ 2 & 0 \\ 0 & 2 \\ 1 & 3 \end{bmatrix} u, \tag{3.46}$$

which is of the form (3.26) with $n = 5$, $s = 2$. The system (3.46) is decomposed into two ($r = 2$) dynamic elements

$$\begin{aligned} \dot{z}_1 &= \begin{bmatrix} 1 & 3 \\ 0 & 5 \end{bmatrix} z_1 + \begin{bmatrix} 4 & 5 & 6 \\ 2 & 4 & 1 \end{bmatrix} z_2 + \begin{bmatrix} 1 & 2 \\ 0 & 4 \end{bmatrix} u, \\ \dot{z}_2 &= \begin{bmatrix} 0 & 0 & 6 \\ 2 & 1 & 6 \\ 3 & 0 & 1 \end{bmatrix} z_2 + \begin{bmatrix} 2 & 4 \\ 2 & 0 \\ 2 & 2 \end{bmatrix} z_1 + \begin{bmatrix} 2 & 0 \\ 0 & 2 \\ 1 & 3 \end{bmatrix} u, \end{aligned} \tag{3.47}$$

so that $l_1 = 2$, $l_2 = 3$. The labeled digraph of the decomposed system (3.47) is shown in Figure 3.7.

Since both pairs $(\hat{A}_1, \hat{B}_1)$ and $(\hat{A}_2, \hat{B}_2)$ in (3.47) are controllable, we can use the transformation (3.31):

$$Q_1 = \begin{bmatrix} 1 & 2 \\ 0 & 4 \end{bmatrix}, \qquad Q_2 = \begin{bmatrix} 2 & 6 & 0 \\ 0 & 10 & 2 \\ 1 & 7 & 3 \end{bmatrix} \tag{3.48}$$

to get (3.42) as

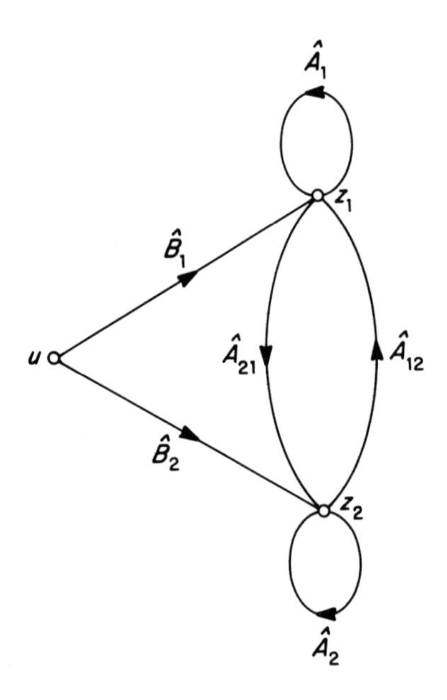

FIGURE 3.7. Input-centralized system.

$$\begin{bmatrix} \dot{\bar{z}}_{11} \\ \dot{\bar{z}}_{12} \end{bmatrix} = \begin{bmatrix} 1 & 4 \\ 0 & 5 \end{bmatrix} \begin{bmatrix} \bar{z}_{11} \\ \bar{z}_{12} \end{bmatrix} + \begin{bmatrix} 11.50 & 86.50 & 22.50 \\ 1.25 & 14.75 & 2.75 \end{bmatrix} \begin{bmatrix} \bar{z}_{21} \\ \bar{z}_{22} \end{bmatrix} + \begin{bmatrix} 1 & 0 \\ 0 & 1 \end{bmatrix} \begin{bmatrix} u_1 \\ u_2 \end{bmatrix},$$

$$\begin{bmatrix} \dot{\bar{z}}_{21} \\ \dot{\bar{z}}_{22} \end{bmatrix} = \begin{bmatrix} 0 & -4.09 & -0.82 \\ 1 & 8.36 & 3.27 \\ 0 & -9.82 & -6.36 \end{bmatrix} \begin{bmatrix} \bar{z}_{21} \\ \bar{z}_{22} \end{bmatrix} + \begin{bmatrix} 0.45 & 8.91 \\ 0.18 & 0.36 \\ 0.09 & 0.18 \end{bmatrix} \begin{bmatrix} \bar{z}_{11} \\ \bar{z}_{12} \end{bmatrix} + \begin{bmatrix} 1 & 0 \\ 0 & 0 \\ 0 & 1 \end{bmatrix} \begin{bmatrix} u_1 \\ u_2 \end{bmatrix}. \tag{3.49}$$

The matrices $\bar{B}_1$ and $\bar{B}_2$ have the form (3.30) with

$$\bar{b}_1^1 = 1, \qquad \bar{b}_2^1 = 1, \qquad \bar{b}_1^2 = (1, 0)^T, \qquad \bar{b}_2^2 = 1 \tag{3.50}$$

and $n_{11} = 1$, $n_{12} = 1$, $n_{21} = 2$, $n_{22} = 1$. Therefore, $n_1 = n_{11} + n_{21} = 3$, $n_2 = n_{12} + n_{22} = 2$, and the input-decentralization scheme produces two subsystems of second and third order associated with the input components u_1 and u_2, respectively. The two subsystems are formed by regrouping the components of

$$\bar{z}_1 = (\bar{z}_{11}^T, \bar{z}_{12}^T)^T, \qquad \bar{z}_2 = (\bar{z}_{21}^T, \bar{z}_{22}^T)^T \tag{3.51}$$

in (3.49) using the permutation matrices P_1 and P_2 defined by (3.36) as

$$P_1 = \begin{bmatrix} I_{11} & 0 & 0 & 0 \\ 0 & 0 & I_{21} & 0 \end{bmatrix}; \qquad I_{11} = 1, \qquad I_{21} = \begin{bmatrix} 1 & 0 \\ 0 & 1 \end{bmatrix};$$
$$P_2 = \begin{bmatrix} 0 & I_{12} & 0 & 0 \\ 0 & 0 & 0 & I_{22} \end{bmatrix}; \qquad I_{12} = 1, \qquad I_{22} = 1. \tag{3.52}$$

The two subsystems have the states

$$x_1 = (\bar{z}_{11}^T, \bar{z}_{21}^T), \qquad x_2 = (\bar{z}_{12}^T, \bar{z}_{22}^T)^T, \tag{3.53}$$

and have the representation (3.34), which is

$$\dot{x}_1 = \begin{bmatrix} 1 & 11.50 & 86.50 \\ 0.45 & 0 & -4.09 \\ 0.18 & 1 & 8.36 \end{bmatrix} x_1 + \begin{bmatrix} 4 & 22.50 \\ 8.91 & -0.82 \\ 0.36 & 3.27 \end{bmatrix} x_2 + \begin{bmatrix} 1 \\ 1 \\ 0 \end{bmatrix} u_1,$$
$$\dot{x}_2 = \begin{bmatrix} 5 & 2.75 \\ 0.18 & -6.36 \end{bmatrix} x_2 + \begin{bmatrix} 0 & 1.25 & 14.75 \\ 0.18 & 0 & -9.82 \end{bmatrix} x_1 + \begin{bmatrix} 1 \\ 1 \end{bmatrix} u_2. \tag{3.54}$$

The two interconnected subsystems (3.54) have an input-decentralized digraph as shown in Figure 3.8.

Now, several remarks concerning the decentralization scheme are in order. Although to perform the decentralization procedure we do not need controllability and observability tests of the overall systems, the procedure implicitly depends on the system controllability and observability. This fact is particularly important in the next section when we propose to stabilize a linear composite system by decentralized control after testing for controllability of the subsystems only. To avoid futile attempts to stabilize a system which is controllable piece by piece but is uncontrollable as a whole, we should test for input reachability of the system first, using the results of Section 3.1. This way, we still avoid testing for controllability of the overall system, which is an ill-posed numerical problem, especially when too many variables are involved, but differs from the input-reachability property only in "physically unrealistic" cases.

3.3. STABILIZATION

Once a large-scale system is given in the input-decentralized form, either as a result of input decentralization or by being identified as such through physical considerations, we propose here to stabilize the system by a

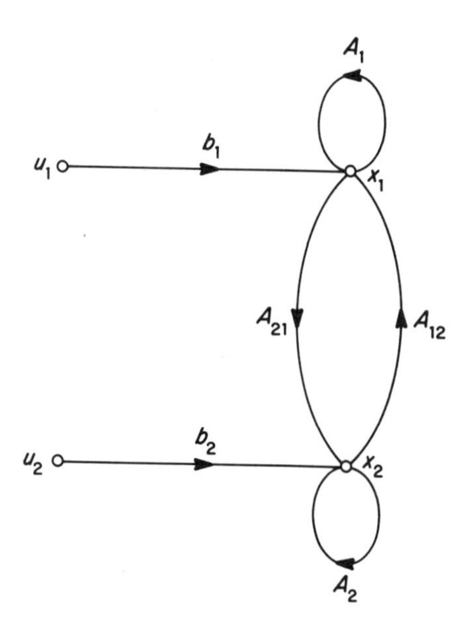

FIGURE 3.8. Input-decentralized system.

multilevel control scheme based upon the decomposition-aggregation stability analysis presented in Chapter 2. In the scheme, local controllers are used to stabilize each subsystem when decoupled, while global controllers are applied to reduce the effect of interconnections among the subsystems. The local controllers provide a desired degree of stability for each subsystem separately, and can be designed by any of the classical techniques, such as pole shifting by state feedback (Chen, 1970), root-locus method (Thaler and Brown, 1960), parameter plane method (Mitrović, 1959; Šiljak, 1969), etc. After the subsystems are stabilized, an aggregate model is used to deduce stability of the whole system.

To stabilize the system $\mathcal{S}$ given as

$$\dot{x}_i = A_i x_i + \sum_{j=1}^{s} A_{ij} x_j + b_i u_i, \qquad i = 1, 2, \ldots, s, \tag{3.34}$$

we apply the decentralized multilevel control

$$u_i(t) = u_i^l(t) + u_i^g(t), \tag{3.55}$$

where $u_i^l(t)$ is the local control law chosen as

$$u_i^l = -k_i^T x_i, \tag{3.56}$$

with a constant vector $k_i \in \mathcal{R}^{n_i}$, and $u_i^g(t)$ is the global control law chosen as

$$u_i^g = -\sum_{j=1}^{s} k_{ij}^T x_j, \tag{3.57}$$

where $k_{ij} \in \mathcal{R}^{n_j}$ are constant vectors.

By substituting the control (3.55) into (3.34), we get the closed-loop system as

$$\dot{x}_i = (A_i - b_i k_i^T)x_i + \sum_{j=1}^{s} (A_{ij} - b_i k_{ij}^T)x_j, \qquad i = 1, 2, \ldots, s. \tag{3.58}$$

Since each pair (A_i, b_i) is controllable, a simple choice of k_i can be always made to place the eigenvalues of $A_i - b_i k_i^T$ at any desired distinct locations $-\sigma_1^i \pm j\omega_1^i, \ldots, -\sigma_p^i \pm j\omega_p^i, -\sigma_{p+1}^i, \ldots, -\sigma_{n_i-p}^i$ ($\sigma_q^i > 0$, $q = 1, 2, \ldots, n_i - p$, and $0 \leqslant p \leqslant [n_i/2]$). Then each uncoupled subsystem

$$\dot{x}_i = (A_i - b_i k_i^T)x_i, \qquad i = 1, 2, \ldots, s, \tag{3.59}$$

is stabilized with a degree of exponential stability

$$\pi_i = \min_q \sigma_q^i, \qquad q = 1, 2, \ldots, n_i - p. \tag{3.60}$$

It is assumed that the subsystems (3.59) are of low order, and numerous classical techniques can be used to achieve a suitable set of subsystem eigenvalues, which is all that is needed to aggregate the subsystem stability properties into a single scalar Liapunov function. This is remarkable, since the entire arsenal of standard design techniques can be used on the subsystem level to get an appropriate aggregate model and deduce the stability of the overall system. We will come back to this point later and show how the classical techniques are used for stabilization of the subsystems. Now we turn to the construction of an aggregate model for the interconnected system (3.58).

To provide a Liapunov function with the exact estimate of π_i for each decoupled subsystem, we apply to (3.59) the linear nonsingular transformation

$$x_i = T_i \tilde{x}_i \tag{3.61}$$

to get the system (3.59) as

$$\dot{\tilde{x}}_i = \Lambda_i \tilde{x}_i, \tag{3.62}$$

where $\Lambda_i = T_i^{-1}(A_i - b_i k_i^T)T_i$ has the quasidiagonal form

$$\Lambda_i = \text{diag}\left\{\begin{bmatrix} -\sigma_1^i & \omega_1^i \\ -\omega_1^i & -\sigma_1^i \end{bmatrix}, \ldots, \begin{bmatrix} -\sigma_p^i & \omega_p^i \\ -\omega_p^i & -\sigma_p^i \end{bmatrix}, -\sigma_{p+1}^i, \ldots, -\sigma_{n_i-p}^i\right\}. \tag{3.63}$$

For the system (3.62) we choose the Liapunov function $v_i\colon \mathscr{R}^{n_i} \to \mathscr{R}_+$,

$$v_i(\tilde{x}_i) = (\tilde{x}_i^T \tilde{H}_i \tilde{x}_i)^{1/2}, \tag{3.64}$$

where

$$\Lambda_i^T \tilde{H}_i + \tilde{H}_i \Lambda_i = -\tilde{G}_i, \tag{3.65}$$

and

$$\begin{aligned} \tilde{H}_i &= \theta_i I_i, \\ \tilde{G}_i &= 2\theta_i \text{diag}\left\{\sigma_1^i, \sigma_1^i, \ldots, \sigma_p^i, \sigma_p^i, \sigma_{p+1}^i, \ldots, \sigma_{n_i-p}^i\right\}. \end{aligned} \tag{3.66}$$

In (3.66), $\theta_i > 0$ is an arbitrary constant and I_i is the $n_i \times n_i$ identity matrix. That the chosen Liapunov function (3.66) provides the exact estimate of π_i defined by (3.60) is shown in Section 6.5.

The aggregate comparison system involving the vector Liapunov function $v\colon \mathscr{R}^n \to \mathscr{R}_+^s$,

$$v = (v_1, v_2, \ldots, v_s)^T, \tag{3.67}$$

is obtained for the transformed system (3.58):

$$\dot{\tilde{x}}_i = \Lambda_i \tilde{x}_i + \sum_{j=1}^{s} (\tilde{A}_{ij} - \tilde{b}_i \tilde{k}_{ij}^T)\tilde{x}_j, \qquad i = 1, 2, \ldots, s, \tag{3.68}$$

where $\tilde{A}_{ij} = T_i^{-1} A_{ij} T_j$, $\tilde{b}_i = T_i^{-1} b_i$, $\tilde{k}_{ij}^T = k_{ij}^T T_j$, and using the Liapunov functions $v_i(\tilde{x}_i)$ defined in (3.64). Using the aggregation method presented in Chapter 2, we construct the aggregate model $\mathscr{Q}$ as

$$\dot{v} \leqslant \tilde{W} v, \tag{3.69}$$

where the constant $s \times s$ matrix $\tilde{W} = (\tilde{w}_{ij})$ has the elements defined as

$$\tilde{w}_{ij} = -\delta_{ij} \pi_i + \tilde{\xi}_{ij}, \tag{3.70}$$

in which δ_{ij} is the Kronecker symbol, π_i is defined in (3.60), and

$$\tilde{\xi}_{ij} = \lambda_M^{1/2}[(\tilde{A}_{ij} - \tilde{b}_i \tilde{k}_{ij}^T)^T (\tilde{A}_{ij} - \tilde{b}_i \tilde{k}_{ij}^T)]. \tag{3.71}$$

Here λ_M is the maximum eigenvalue of the indicated matrix.

Now we apply Theorem 2.15 from the preceeding chapter and deduce the stability of the system $\mathbb{S}$ by the Sevastyanov-Kotelyanskii conditions (2.132), which for $\tilde{W} = (\tilde{w}_{ij})$ defined by (3.70) and (3.71) have the following form:

$$(-1)^k \begin{vmatrix} -\pi_1 + \tilde{\xi}_{11} & \tilde{\xi}_{12} & \cdots & \tilde{\xi}_{1k} \\ \tilde{\xi}_{21} & -\pi_2 + \tilde{\xi}_{22} & \cdots & \tilde{\xi}_{2k} \\ \cdots & \cdots & \cdots & \cdots \\ \tilde{\xi}_{k1} & \tilde{\xi}_{k2} & \cdots & -\pi_k + \tilde{\xi}_{kk} \end{vmatrix} > 0, \qquad k = 1, 2, \ldots, s. \tag{3.72}$$

To satisfy these conditions, we choose the vectors $\tilde{k}_{ij}$ in (3.68) so as to minimize the nonnegative numbers $\tilde{\xi}_{ij}$ which reflect the strengths of the interconnections among the subsystems in (3.68). Such choice is provided by

$$\tilde{k}_{ij}^0 = [(\tilde{b}_i^T \tilde{b}_i)^{-1} \tilde{b}_i \tilde{A}_{ij}]^T, \tag{3.73}$$

where $(\tilde{b}_i^T \tilde{b}_i)^{-1} \tilde{b}_i^T$ is the Moore-Penrose generalized inverse of $\tilde{b}_i$ (Langenhop, 1967). The choice of $\tilde{k}_{ij}^0$ in (3.73) produces the optimal aggregate matrix $\tilde{W}^0$ in the sense that $\tilde{W}^0 \leqslant \tilde{W}$ (that is, $\tilde{W}^0 - \tilde{W} \leqslant 0$) is valid for all $\tilde{k}_{ij}$. That is equivalent to saying (see Appendix) that $\lambda_M(\tilde{W}^0) \leqslant \lambda_M(\tilde{W})$ for all $\tilde{k}_{ij}$. Since the conditions (3.72) are necessary and sufficient for $\lambda_M(\tilde{W}) < 0$ (that is, for stability of $\tilde{W}$), the choice $\tilde{k}_{ij} = \tilde{k}_{ij}^0$ is justified.

To deduce the stability of the overall system (3.68) with the optimal choice $\tilde{k}_{ij} = \tilde{k}_{ij}^0$, which is

$$\dot{\tilde{x}}_i = \Lambda_i \tilde{x}_i + [I_i - \tilde{b}_i(\tilde{b}_i^T \tilde{b}_i)^{-1} \tilde{b}_i^T] \sum_{j=1}^{s} \tilde{A}_{ij} \tilde{x}_j, \qquad i = 1, 2, \ldots, s, \tag{3.74}$$

we apply the determinantal inequalities (3.72) to the optimal aggregate matrix $\tilde{W}^0 = (\tilde{w}_{ij}^0)$ defined by (3.70) and $\tilde{\xi}_{ij} = \tilde{\xi}_{ij}^0 = \lambda_M^{1/2}\{\tilde{A}_{ij}^T[I_i - \tilde{b}_i(\tilde{b}_i^T \tilde{b}_i)^{-1} \tilde{b}_i^T]\tilde{A}_{ij}\}$.

If the outlined stabilization procedure is applied to the system $\mathbb{S}$, the feedback control function $u(t) \in \mathcal{R}^s$ can be calculated using k_i and $\tilde{k}_{ij}^0$ obtained on the subsystem level. The control function $u(t)$ is given by

$$u = -\hat{K}z, \tag{3.75}$$

where

$$\hat{K} = KPQ^{-1}, \tag{3.76}$$

in which

$$K = \begin{bmatrix} k_1^T + k_{11}^T & k_{12}^T & \cdots & k_{1s}^T \\ k_{21}^T & k_2^T + k_{22}^T & \cdots & k_{2s}^T \\ \cdots & \cdots & \cdots & \cdots \\ k_{s1}^T & k_{s2}^T & \cdots & k_s^T + k_{ss}^T \end{bmatrix}, \tag{3.77}$$

the matrix P is defined in (3.36), and $Q = \text{diag}(Q_1, Q_2, \ldots, Q_r)$. Here the matrices Q_p, $p = 1, 2, \ldots, r$, are given in (3.32), and $k_{ij}^T = \tilde{k}_{ij}^{0T} T_j^{-1}$. Therefore, we arrive at the following:

Theorem 3.6. *A linear constant system* $\mathbb{S}$ *defined by* (3.26) *is stabilized by the linear control law* $u = -\hat{K}z$, *where the matrix* $\hat{K}$ *is given by* (3.76), *if the corresponding aggregate matrix* $\tilde{W}^0 = [-\delta_{ij}\pi_i + \tilde{\xi}_{ij}^0]$ *satisfies the conditions* (3.72).

Successful application of the above theorem depends on appropriate choice of the eigenvalues for the decoupled subsystems (3.59). Once the subsystem eigenvalues are prescribed, the matrix K in (3.77) and thus the matrix $\hat{K}$ in (3.76) are computed uniquely using the proposed algorithm. Therefore, if for a computed matrix K the conditions (3.72) are not met, a reassignment of the subsystems eigenvalues is required. The search for an appropriate set of eigenvalues for each subsystem can proceed in many different ways, and we will now describe some of them.

Let us start with the standard *state feedback* and show that if the pair (A, b) is completely controllable, that is

$$\text{rank}[b \mid Ab \mid \cdots \mid A^{n-1}b] = n, \tag{3.78}$$

then it is a well-known fact (e.g. Chen, 1970) that a constant vector k can always be found so that the eigenvalues $\lambda_i(A - bk^T)$ corresponding to the closed-loop uncoupled subsystem

$$\dot{x} = (A - bk^T)x \tag{3.79}$$

can be placed at arbitrary locations

$$\lambda_1 = -\sigma_1 + j\omega_1, \ldots, \lambda_{2p} = -\sigma_p - j\omega_p, \lambda_{2p+1} = -\sigma_{p+1}, \ldots, \lambda_n = -\sigma_{n-p}, \tag{3.80}$$

where $\sigma_q > 0$, $q = 1, 2, \ldots, n - p$, and $0 \leqslant p \leqslant [n/2]$. For convenience, we have dropped the subscript i used to identify the subsystem.

To compute k which produces the eigenvalues λ_i in (3.80) of the matrix $A - bk^T$, we note that they prescribe the characteristic polynomial

$$\beta(s) = s^n + \beta_1 s^{n-1} + \cdots + \beta_n, \tag{3.81}$$

defined as $\beta(s) = \det(sI - A + bk^T)$, by the indentity

$$s^n + \beta_1 s^{n-1} + \cdots + \beta_n \equiv \prod_{i=1}^{n} (s - \lambda_i). \tag{3.82}$$

Now, the problem is to compute k so that the equation (3.81) is established. That is, we need the following (Kalman, Falb, and Arbib, 1969):

Theorem 3.7. *Let the pair (A, b) be completely controllable, and $s^n + \beta_1 s^{n-1} + \cdots + \beta_n$ be an arbitrary polynomial. Then there is an n-vector k such that* (3.81) *is satisfied.*

Proof. Since the characteristic polynomial is invariant under a similarity transformation, we can choose any convenient basis for the vector space $\mathcal{R}^n$ when computing k. Therefore, we choose the set of vectors

$$\begin{aligned} p_1 &= A^{n-1}b + \alpha_1 A^{n-2}b + \cdots + \alpha_{n-1}b, \\ p_2 &= A^{n-2}b + \alpha_1 A^{n-3}b + \cdots + \alpha_{n-2}b, \\ &\vdots \\ p_n &= b \end{aligned} \tag{3.83}$$

for such a basis. The set of vectors $\{p_1, p_2, \ldots, p_n\}$ is a triangular linear combination of the vectors $\{b, Ab, \ldots, A^{n-1}b\}$, which are linearily independent because the pair (A, b) is completely controllable, that is, the rank condition (3.78) is satisfied. So $\{p_1, p_2, \ldots, p_n\}$ form a basis for $\mathcal{R}^n$, because the same is true for $\{b, Ab, \ldots, A^{n-1}b\}$. In the basis $\{p_1, p_2, \ldots, p_n\}$, A and b have the new representation

$$A^c = \begin{bmatrix} 0 & 1 & 0 & \cdots & 0 \\ 0 & 0 & 1 & \cdots & 0 \\ \cdots & \cdots & \cdots & \cdots & \cdots \\ 0 & 0 & 0 & \cdots & 1 \\ -\alpha_n & -\alpha_{n-1} & -\alpha_{n-2} & \cdots & -\alpha_1 \end{bmatrix}, \qquad b^c = \begin{bmatrix} 0 \\ 0 \\ \vdots \\ 0 \\ 1 \end{bmatrix}, \tag{3.84}$$

where A^c is called a *companion matrix*. The representation (3.84) can be verified by direct computation of Ap_n, Ap_{n-1}, ..., Ap_2, and concluding that $Ap_n = -\alpha_n p_n$ from $\beta(A) = 0$, which is the well-known Cayley-Hamilton theorem (Zadeh and Desoer, 1963). Now, because of the special form of b^c, the matrix $A^c - b^c k^{cT}$ is also a companion matrix with the characteristic polynomial

$$\beta(s) = s^n + (\alpha_1 + k_n^c)s^{n-1} + \cdots + (\alpha_n + k_1^c), \tag{3.85}$$

where $k^c = (k_1^c, k_2^c, \dots, k_n^c)^T$. Comparing (3.85) with (3.82), we generate the formulas

$$k_{n-i+1}^c = \beta_i - \alpha_i, \qquad i = 1, 2, \dots, n, \tag{3.86}$$

for calculating k^c. Finally, we compute the original vector k from k^c by the equation

$$k = k^c P^{-1}, \tag{3.87}$$

where $P = [p_1 \,|\, p_2 \,|\, \cdots \,|\, p_n]$. Equation (3.87) follows from the relation $P(A - bk^T)P^{-1} = A^c - b^c(k^c)^T$. This proves Theorem 3.7.

A slight modification of Theorem 3.7 applies to a multivariable case when each subsystem has more than one input. The details can be found in Chen (1970). We will consider the multivariable subsystems in Section 3.5 on multilevel optimization, which can in turn be used to stabilize the large-scale systems which have more inputs than subsystems.

To illustrate the use of state feedback and application of Theorems 3.6 and 3.7, let us again consider the fifth-order system described by Equation (3.46). The eigenvalues of the system matrix A in (3.46) are $\lambda_{1,2} = 0.76 \pm j1.83$, $\lambda_3 = 11.54$, $\lambda_4 = -3.89$, $\lambda_5 = -1.16$, and the system is unstable.

To stabilize the system (3.46) we start with its input-decentralized representation (3.51) and transform each uncoupled subsystem into its companion form to get

$$\begin{aligned} \dot{x}_1^c &= \begin{bmatrix} 0 & 1 & 0 \\ 0 & 0 & 1 \\ -8.86 & 8.50 & 9.36 \end{bmatrix} x_1^c + \begin{bmatrix} 3.20 & 1.98 \\ -14.72 & 0.49 \\ -7.92 & 36.01 \end{bmatrix} x_2^c + \begin{bmatrix} 0 \\ 0 \\ 1 \end{bmatrix} u_1, \\ \dot{x}_2^c &= \begin{bmatrix} 0 & 1 \\ 32.32 & -1.36 \end{bmatrix} x_2^c + \begin{bmatrix} 1.69 & 1.26 & 0.08 \\ -7.52 & -5.23 & 0.49 \end{bmatrix} x_1^c + \begin{bmatrix} 0 \\ 1 \end{bmatrix} u_2, \end{aligned} \tag{3.88}$$

where $x_1 = Q_1^c x_1^c$, $x_2 = Q_2^c x_2^c$, and the transformation matrices Q_1^c, Q_2^c are constructed on the subsystem level as shown by Chen (1970). The transformation to the companion form is of no conceptual significance, and is performed for two practical reasons. First, it is convenient for subsystem stabilization by pole assignment utilizing the state feedback, and secondly, the diagonal form (3.68) with no complex roots can be obtained from the companion form (3.88) using the Vandermonde matrix T_i in (3.61), where x_i is replaced by x_i^c.

Now, by using the feedback law (3.55) and vectors $k_1^T = (1791.14, 458.50, 46.36)$, $k_2^T = (33.82, 1.14)$, we relocate the eigenvalues of the uncoupled subsystems (3.88) from

$$\lambda_1^1 = 0.63, \qquad \lambda_1^2 = 5.04,$$
$$\lambda_2^1 = -1.39, \qquad \lambda_2^2 = -6.41, \tag{3.89}$$
$$\lambda_3^1 = 10.12$$

to

$$\lambda_1^1 = -10, \qquad \lambda_1^2 = -1,$$
$$\lambda_2^1 = -12, \qquad \lambda_2^2 = -1.5, \tag{3.90}$$
$$\lambda_3^1 = -15.$$

After the local stabilization, the interconnected subsystems have the quasidiagonal form

$$\dot{\tilde{x}}_1 = \begin{bmatrix} -10 & 0 & 0 \\ 0 & -12 & 0 \\ 0 & 0 & -15 \end{bmatrix} \tilde{x}_1 + \begin{bmatrix} -23.52 & -43.78 \\ 40.23 & 68.97 \\ -15.49 & -24.96 \end{bmatrix} \tilde{x}_2 + \begin{bmatrix} 0.1 \\ -0.17 \\ 0.07 \end{bmatrix} u_1^g,$$
$$\dot{\tilde{x}}_2 = \begin{bmatrix} -1 & 0 \\ 0 & -1.5 \end{bmatrix} \tilde{x}_2 + \begin{bmatrix} 179.95 & 247.53 & 367.40 \\ -182.58 & -249.03 & -365.95 \end{bmatrix} \tilde{x}_1 + \begin{bmatrix} 2 \\ -2 \end{bmatrix} u_2^g, \tag{3.91}$$

which is not identical to (3.68). For the moment, we have not made use of the global control u_1^g, u_2^g in (3.91). In order to demonstrate the effect of the global controllers, we set $\tilde{k}_{12} = \tilde{k}_{21} = 0$.

From (3.60) and (3.91), we have $\pi_1 = 10$, $\pi_2 = 1$. Using (3.71) and (3.91), we compute $\tilde{\xi}_{12} = 98.51$, $\tilde{\xi}_{21} = 676.68$. The aggregate matrix $\tilde{W}$ in (3.69) is obtained as

$$\tilde{W} = \begin{bmatrix} -10 & 98.51 \\ 676.68 & -1 \end{bmatrix}, \tag{3.92}$$

which does not satisfy the conditions (3.72). Therefore, we cannot infer the stability of the overall system.

Let us use now the global control specified by (3.73) and computed as $\tilde{k}_{12}^{0T} = (-238.95, -415.34)$, $\tilde{k}_{21}^{0T} = (90.63, 124.14, 183.33)$, which yields the subsystems (3.91) as

$$\dot{\tilde{x}}_1 = \begin{bmatrix} -10 & 0 & 0 \\ 0 & -12 & 0 \\ 0 & 0 & -15 \end{bmatrix} \tilde{x}_1 + \begin{bmatrix} 0.37 & -2.25 \\ 0.40 & -0.25 \\ 0.44 & 2.73 \end{bmatrix} \tilde{x}_2,$$
$$\dot{\tilde{x}}_2 = \begin{bmatrix} -1 & 0 \\ 0 & -1.5 \end{bmatrix} \tilde{x}_2 + \begin{bmatrix} -1.31 & -0.75 & 0.72 \\ -1.31 & -0.75 & 0.72 \end{bmatrix} \tilde{x}_1, \tag{3.93}$$

and the aggregate matrix

$$\tilde{W}^0 = \begin{bmatrix} -10 & 3.55 \\ 2.37 & -1 \end{bmatrix}, \tag{3.94}$$

which satisfies the conditions (3.72). The overall closed-loop system

$$\dot{z} = (\hat{A} - \hat{B}\hat{K})z \tag{3.95}$$

is stable with eigenvalues $\lambda_{1,2} = -1.03 \pm j0.16$, $\lambda_3 = -10.27$, $\lambda_4 = -11.99$, $\lambda_5 = -15.17$, and the matrix $\hat{K}$ defined in (3.76) and computed as

$$\hat{K} = \begin{bmatrix} -127.58 & 96.33 & 105.29 & 166.25 & 36.63 \\ 0.05 & 0.68 & 0.62 & 0.20 & -0.70 \end{bmatrix}. \tag{3.96}$$

It is also interesting to note that an upper estimate of the degree π of exponential stability of the system $\mathbb{S}$ is provided by the aggregate matrix $\tilde{W}^0$, since in general $\pi \leqslant \min_i \pi_i$. In other words, the degree of exponential stability of the overall system π stabilized by the proposed method is smaller than the degree of exponential stability of each decoupled subsystem.

If only outputs of the subsystems are available for control, then state feedback cannot be used unless state estimators are constructed. In cases when we have one-input, one-output subsystems, we can avoid building the state estimators and attempt to design each subsystem by the *output feedback* and *root-locus method.*

To sketch the root-locus application, let us consider a system $\mathbb{S}$ described as s interconnected subsystems

$$\begin{aligned} \dot{x}_i &= A_i x_i + \sum_{j=1}^{s} d_{ij} y_j + b_i u_i, \\ y_i &= c_i^T x_i, \end{aligned} \qquad i = 1, 2, \dots, s \tag{3.97}$$

where $x_i \in \mathcal{R}^{n_i}$, $u_i \in \mathcal{R}$, and $y_i \in \mathcal{R}$. To stabilize the system (3.97), we can use again the multilevel control

$$u_i(t) = u_i^l(t) + u_i^g(t), \tag{3.55}$$

but choose as the output control

$$u_i^l = \kappa_i(r_i - y_i), \qquad u_i^g = -\sum_{j=1}^{s} \kappa_{ij} y_j, \tag{3.98}$$

where r_i is the reference input at each subsystem $\mathbb{S}_i$, and κ_i, κ_{ij} are constants to be determined in the stabilization process. The closed-loop system is given as

$$\dot{x}_i = (A_i - \kappa_i b_i c_i^T)x_i + \kappa_i b_i r_i + \sum_{j=1}^{s} (d_{ij} - \kappa_{ij} b_j)c_j^T x_j, \qquad i = 1, 2, \ldots, s.$$
$$y_i = c_i^T x_i, \tag{3.99}$$

We again drop the subscript i and proceed to find the constant κ such that uncoupled closed-loop subsystem

$$\dot{x} = (A - \kappa bc^T)x \tag{3.100}$$

has a desired set of eigenvalues (3.80). It is a well-known fact in control engineering (e.g. Chen, 1970) that if for the open-loop system

$$\dot{x} = Ax + bu,$$
$$y = c^T x, \tag{3.101}$$

the transfer function $y(s)/u(s) = c^T(sI - A)^{-1}b$ has the factored form

$$\frac{y(s)}{u(s)} = \frac{(s - z_1)(s - z_2)\cdots(s - z_m)}{(s - p_1)(s - p_2)\cdots(s - p_n)}, \tag{3.102}$$

where $p_1, p_2, \ldots, p_n$ are open-loop poles, $z_1, z_2, \ldots, z_m$ are open-loop zeros, and $n > m$, then the closed-loop transfer function $y(s)/r(s)$ has the form

$$\frac{y(s)}{r(s)} = \frac{\kappa(s - z_1)(s - z_2)\cdots(s - z_m)}{(s - p_1)(s - p_2)\cdots(s - p_n) + \kappa(s - z_1)(s - z_2)\cdots(s - z_m)}. \tag{3.103}$$

The denominator on the right-hand side of (3.103) is the characteristic polynomial $\beta(s)$ of the matrix $A - \kappa bc^T$, and we can apply the classical root-locus technique (Thaler and Brown, 1960) to select κ so that

$$\beta(s) = (s - p_1)(s - p_2)\cdots(s - p_n) + \kappa(s - z_1)(s - z_2)\cdots(s - z_m) \tag{3.104}$$

has zeros at desired locations (3.80). Once the zeros (3.80) are computed numerically, one can proceed to aggregate the subsystems by Liapunov functions. Finally, the gains κ_{ij} of the global controllers can be chosen as in (3.73),

$$\tilde{\kappa}_{ij}^0 = (\tilde{b}_i^T \tilde{b}_i)^{-1} \tilde{b}_i^T \tilde{d}_{ij}, \tag{3.105}$$

to minimize the effect of interactions among the subsystems, and deduce the stability of the overall system using the aggregate model and Theorem 3.6.

To illustrate the application of the root-locus method, let us consider a linear constant system described by the equations

$$\dot{x} = \begin{bmatrix} 1 & 2 & 3 & 0.5 & 1 \\ 0 & 1 & 3 & 0 & 0.5 \\ 2 & 1 & 1 & 0 & 0.5 \\ -2 & -1 & 0.5 & 0 & 0.5 \\ 1 & -1 & 0.5 & 0.5 & 0 \end{bmatrix} x + \begin{bmatrix} 1 & 0 \\ 0 & 0 \\ 1 & 0 \\ 0 & 1 \\ 0 & 1 \end{bmatrix} u, \tag{3.106}$$

$$y = \begin{bmatrix} 2 & 1 & 1 & 0 & 0 \\ 0 & 0 & 0 & 1 & 2 \end{bmatrix} x.$$

The eigenvalues of the system are $\lambda_{1,2} = -0.831 \pm j0.391$, $\lambda_{3,4} = -1.505 \pm j0.845$, $\lambda_5 = 4.672$, and the system is unstable.

We consider the system to be composed of two interconnected subsystems

$$\begin{aligned} \dot{x}_1 &= \begin{bmatrix} 1 & 2 & 3 \\ 0 & 1 & 3 \\ 2 & 1 & 1 \end{bmatrix} x_1 + \begin{bmatrix} 0.5 & 1 \\ 0 & 0.5 \\ 0 & 0.5 \end{bmatrix} x_2 + \begin{bmatrix} 1 \\ 0 \\ 1 \end{bmatrix} u_1, \\ \dot{x}_2 &= \begin{bmatrix} -2 & -1 & 0.5 \\ 1 & -1 & 0.5 \end{bmatrix} x_1 + \begin{bmatrix} 0 & 0.5 \\ 0.5 & 0 \end{bmatrix} x_2 + \begin{bmatrix} 1 \\ 1 \end{bmatrix} u_2. \end{aligned} \tag{3.107}$$

By using the output feedback around each subsystem and plotting the root loci shown in Figure 3.9, the open-loop poles of the free subsystems are relocated from $s^1_{1,2} = -0.761 \pm j0.552$, $s^1_3 = 4.522$, $s^2_{1,2} = -1.5 \pm j0.866$ to $s^1_{1,2} = -0.844 \pm j0.871$, $s^1_3 = -16.312$, $s^2_1 = -2.127$, $s^2_2 = -9.872$. The gains of the local feedback controls are $\kappa_1 = -7$, $\kappa_2 = -3$. No global control is used. The corresponding aggregate matrix is

$$\tilde{W} = \begin{bmatrix} -0.843 & 0.350 \\ 4.038 & -2.127 \end{bmatrix}, \tag{3.108}$$

which satisfies the conditions (3.72), and the overall (transformed) system

$$\dot{\tilde{x}} = \left[\begin{array}{ccc|cc} -0.843 & -0.871 & 0 & 0.164 & 0.237 \\ 0.871 & -0.843 & 0 & -0.037 & -0.208 \\ 0 & 0 & -16.312 & -0.038 & -1.167 \\ \hline 1.002 & 3.717 & 0.278 & -2.127 & 0 \\ -0.530 & -1.120 & 0.751 & 0 & -9.873 \end{array}\right] \tilde{x} \tag{3.109}$$

is stable. The eigenvalues corresponding to (3.109) are $\lambda_{1,2} = -0.989 \pm j0.666$, $\lambda_3 = -1.822$, $\lambda_4 = -9.755$, $\lambda_5 = -16.445$, which confirms our conclusion based upon stability of the aggregate matrix (3.108).

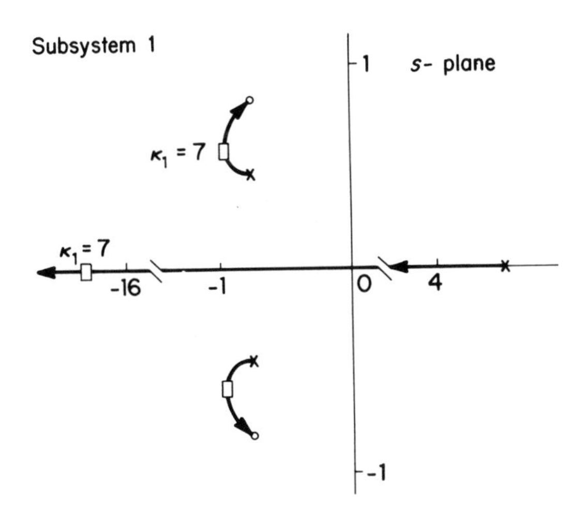

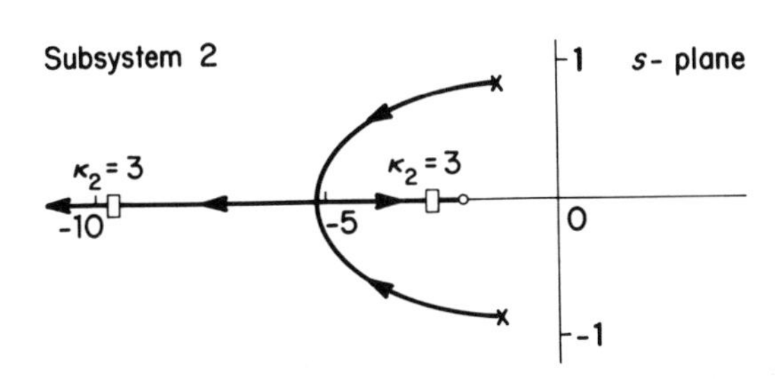

FIGURE 3.9. Root-loci.

The stabilized linear system (3.74) is connectively stable and tolerates nonlinearities in coupling. That is, stability of (3.74) implies connective stability of the nonlinear time-varying system

$$\dot{\tilde{x}}_i = \Lambda_i \tilde{x}_i + \sum_{j=1}^{s} e_{ij}(t)[\tilde{h}_{ij}(t, \tilde{x}) - \tilde{b}_i \tilde{k}_{ij}^{0T} \tilde{x}_j], \qquad i = 1, 2, \ldots, s, \qquad (3.110)$$

where $e_{ij}\colon \mathcal{T} \to [0, 1]$ are elements of the $s \times s$ interconnection matrix $E(t)$; $\tilde{h}_{ij}\colon \mathcal{T} \times \mathcal{R}^n \to \mathcal{R}^{n_i}$ are interconnection functions which are sufficiently smooth and satisfy the constraints

$$\|\tilde{h}_{ij}(t, \tilde{x})\| \leqslant (\tilde{\xi}_{ij}^0 - \tilde{\eta}_{ij})\|\tilde{x}_j\| \qquad \forall (t, x) \in \mathcal{R}^{n+1}, \quad i, j = 1, 2, \ldots, s, \quad (3.111)$$

where $\tilde{\eta}_{ij} = \lambda_M^{1/2}(\tilde{k}_{ij}^0 \tilde{b}_i^T \tilde{b}_i \tilde{k}_{ij}^{0T}) = (\tilde{b}_i^T \tilde{b}_i \tilde{k}_{ij}^{0T} \tilde{k}_{ij}^0)^{1/2}$; and $\tilde{k}_{ij}^0$ is given in (3.73). By using the transformation (3.61) we can interpret readily the constraints on nonlinearities, (3.111), in terms of the original system (3.58), where $A_{ij} x_j$ is replaced by a nonlinear function $h_{ij}(t, x)$.

Therefore, recalling Theorem 2.15, the following result is automatic:

Theorem 3.8. *The nonlinear time-varying system* (3.110) *is connectively stabilized by the linear control law* $u = -Kx$, *where the matrix* K *is defined in* (3.77), *if the corresponding aggregate matrix* $\tilde{W}^0 = [-\delta_{ij}\pi_i + \tilde{\xi}_{ij}^0]$ *associated with the fundamental interconnection matrix* $\bar{E}$ *satisfies the conditions* (3.72).

Stability of the system (3.110) is understood as global exponential connective stability of the equilibrium $x^* = 0$ as in Definition 2.7.

To demonstrate the dynamic reliability of systems stabilized by the hierarchic method outlined, we shall compare the method with the usual "one shot" stabilization by state feedback presented by Chen (1970). For this purpose, we consider the system (3.98) in the form

$$\dot{x}^c = \begin{bmatrix} 0 & 1 & 0 & 3.2 & 1.98 \\ 0 & 0 & 1 & -14.72 & 0.49 \\ -8.86 & 8.5 & 9.36 & -7.92 & 36.01 \\ 1.69 & 1.26 & 0.08 & 0 & 1 \\ -7.52 & -5.23 & 0.49 & 32.32 & -1.36 \end{bmatrix} x^c + \begin{bmatrix} 0 & 0 \\ 0 & 0 \\ 1 & 0 \\ 0 & 0 \\ 0 & 1 \end{bmatrix} u. \tag{3.112}$$

Applying the method of Chen (1970) to (3.112), we use the state feedback

$$u = \begin{bmatrix} u_1 \\ u_2 \end{bmatrix} = -\begin{bmatrix} M_1^T \\ M_2^T \end{bmatrix} x^c, \qquad x^c = \begin{bmatrix} x_1^c \\ x_2^c \end{bmatrix}, \tag{3.113}$$

where $M_1, M_2 \in \mathcal{R}^n$. Since the system (3.112) is controllable by the component u_1 alone, following Chen (1970) we choose $M_2 = 0$ and compute

$$M_1 = (268.69, \quad 216.26, \quad 47.50, \quad 1650.05, \quad 327.46)^T \tag{3.114}$$

to have the same eigenvalues of the closed-loop system as those of (3.95). The closed-loop system is

$$\dot{x}^c = \begin{bmatrix} 0 & 1 & 0 & 3.20 & 1.98 \\ 0 & 0 & 1 & -14.72 & 0.49 \\ -277.56 & -207.76 & -38.14 & -1657.97 & -291.45 \\ 1.69 & 1.26 & 0.08 & 0 & 1 \\ -7.52 & -5.23 & 0.49 & 32.32 & -1.36 \end{bmatrix} x^c. \tag{3.115}$$

The system (3.115) can now be rewritten as two subsystems,

$$\begin{aligned} \dot{x}_1^c &= (A_1^c - b_1^c m_1^T)x_1^c + e_{12}(t)(A_{12}^c - b_1^c m_{12}^T)x_2^c, \\ \dot{x}_2^c &= A_2^c x_2^c + e_{21}(t)A_{21}^c x_1^c, \end{aligned} \tag{3.116}$$

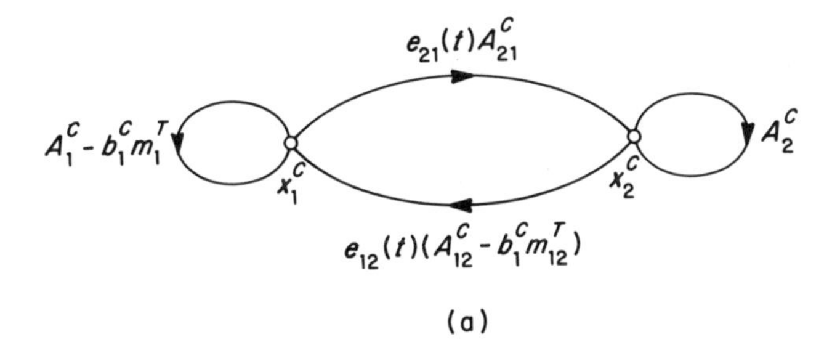

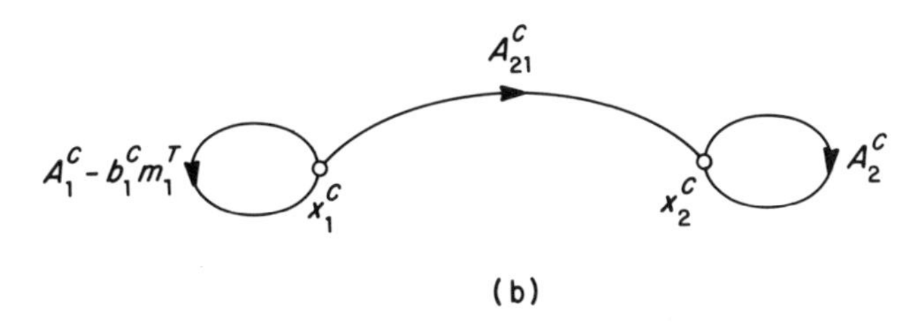

FIGURE 3.10. "One shot" stabilization.

which constitute the closed-loop system corresponding to (3.88) and (3.114) with $M_1 = (m_1^T,\ m_{12}^T)^T$. In (3.116), $e_{12}(t)$ and $e_{21}(t)$ are elements of the interconnection matrix $E(t)$ which corresponds to the fundamental interconnection matrix $\bar{E}$ as

$$E(t) = \begin{bmatrix} 0 & e_{12}(t) \\ e_{21}(t) & 0 \end{bmatrix}, \qquad \bar{E} = \begin{bmatrix} 0 & 1 \\ 1 & 0 \end{bmatrix}. \tag{3.117}$$

The digraph of the system (3.116) is shown in Figure 3.10(a).

Now if, for some time t_1, the interconnection matrix becomes

$$E(t_1) = \begin{bmatrix} 0 & 0 \\ 1 & 0 \end{bmatrix}, \tag{3.118}$$

the system (3.116) reduces to

$$\begin{aligned} \dot{x}_1^c &= (A_1^c - b_1^c m_1^T)x_1^c, \\ \dot{x}_2^c &= A_2^c x_2^c + A_{21}^c x_1^c, \end{aligned} \tag{3.119}$$

as shown in Figure 3.10(b). The eigenvalues of the overall system corresponding to (3.119) formed under the structural perturbation (3.118) are

$$\lambda_1 = -2.10, \quad \lambda_2 = -4.14, \quad \lambda_3 = -31.89, \quad \lambda_4 = 5.04, \quad \lambda_5 = -6.41, \tag{3.120}$$

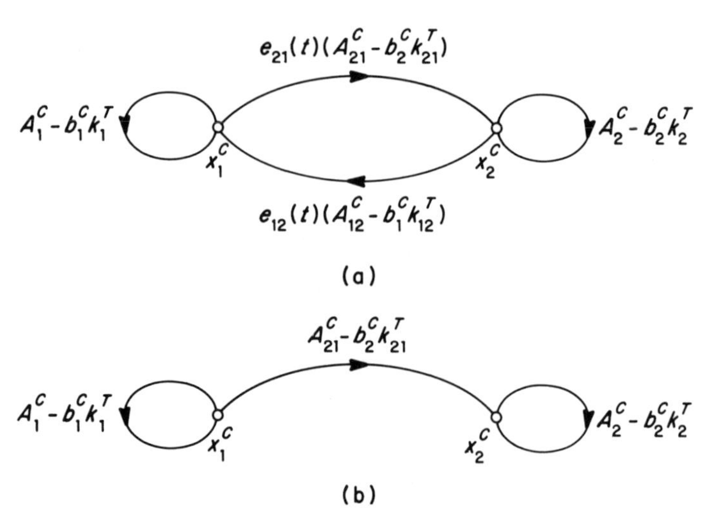

FIGURE 3.11. Multilevel stabilization.

which shows that the overall system (3.120) is unstable.

The same two subsystems (3.88) were stabilized by the multilevel method to get the closed-loop system (3.93), which has the form

$$\begin{aligned} \dot{x}_1^c &= (A_1^c - b_1^c k_1^T)x_1^c + e_{12}(t)(A_{12}^c - b_1^c k_{12}^T)x_2^c, \\ \dot{x}_2^c &= (A_2^c - b_2^c k_2^T)x_2^c + e_{21}(t)(A_{21}^c - b_2^c k_{21}^T)x_1^c \end{aligned} \tag{3.121}$$

and the digraph shown in Figure 3.11(a). Under the same structural perturbation described by (3.118), the system (3.121) becomes

$$\begin{aligned} \dot{x}_1^c &= (A_1^c - b_1^c k_1^T)x_1^c, \\ \dot{x}_2^c &= (A_2^c - b_2^c k_2^T)x_2^c + (A_{21}^c - b_2^c k_{21}^T)x_1^c, \end{aligned} \tag{3.122}$$

as shown in Figure 3.11(b). The eigenvalues of the overall system corresponding to (3.122) are those of (3.90). The system (3.121) under the structural perturbation (3.118) remains stable.

The crucial step in the proposed multilevel stabilization scheme is the search for an appropriate set of subsystem eigenvalues. The search can be aided by an interactive computer program outlined by Šiljak and Vukčević (1976a). The efficiency of the computer program relies on the low order of the subsystems, and on the simplicity in testing the Sevastyanov-Kotelyanskii conditions (3.72) and computing the matrix $\hat{K}$ in (3.76). Furthermore, the computerized procedure provides considerable freedom to the designer to apply his understanding of the system and familiarity with the method to come up with a successful design.

In the remainder of this section, we will consider a class of interconnected systems which can always be stabilized by the proposed method using only *local controllers*. This result was obtained by Vukčević (1975), relying on a similar result obtained by Davison (1974) for a different class of systems.

A system of the class considered by Vukčević is described by

$$\dot{x}_i = A_i x_i + \sum_{\substack{j=1 \\ j \neq i}}^{s} A_{ij} x_j + b_i u_i, \qquad i = 1, 2, \ldots, s, \tag{3.34}$$

where the $n_i \times n_i$ matrix A_i and the n_i-vector b_i are

$$A_i = \begin{bmatrix} 0 & 1 & \cdots & 0 \\ 0 & 0 & \cdots & 0 \\ \cdots & \cdots & \cdots & \cdots \\ 0 & 0 & \cdots & 1 \\ -a_1^i & -a_2^i & \cdots & -a_{n_i}^i \end{bmatrix}, \qquad b_i = \begin{bmatrix} 0 \\ 0 \\ \vdots \\ 0 \\ 1 \end{bmatrix}, \tag{3.123}$$

and the $n_i \times n_j$ matrices $A_{ij} = (a_{pq}^{ij})$ are such that

$$a_{pq}^{ij} = 0, \qquad p < q, \tag{3.124}$$

where $p = 1, 2, \ldots, n_i$ and $q = 1, 2, \ldots, n_j$.

In order to stabilize the system (3.34) characterized by (3.123) and (3.124), we apply the local control

$$u_i^l = -k_i^T x_i \tag{3.56}$$

and get (3.34) as

$$\dot{x}_i = (A_i - b_i k_i^T) x_i + \sum_{\substack{j=1 \\ j \neq i}}^{s} A_{ij} x_j, \qquad i = 1, 2, \ldots, s. \tag{3.125}$$

Gain vectors k_i are chosen so that each matrix $A_i - b_i k_i^T$ has a set $\mathfrak{L}_i$ of distinct real eigenvalues λ_p^i defined by

$$\mathfrak{L}_i = \{\lambda_p^i \colon \lambda_p^i = -\alpha \sigma_p^i;\ \alpha \geqslant 1,\ \sigma_p^i > 0,\ p = 1, 2, \ldots, n_i\}, \quad i = 1, 2, \ldots, s. \tag{3.126}$$

The positive constant α is to be determined so that the overall system (3.34) is stabilized.

Following the development of the multilevel stabilization, we transform (3.125) into

$$\dot{\tilde{x}}_i = \Lambda_i \tilde{x}_i + \sum_{\substack{j=1 \\ j \neq i}}^{s} \tilde{A}_{ij} \tilde{x}_j, \qquad i = 1, 2, \ldots, s, \tag{3.127}$$

where the transformation (3.61) is used to get

$$\Lambda_i = T_i^{-1}(A_i - b_i k_i^T) T_i, \qquad \tilde{A}_{ij} = T_i^{-1} A_{ij} T_j, \tag{3.128}$$

with Λ_i in the diagonal form

$$\Lambda_i = \text{diag}\{-\alpha\sigma_1^i, -\alpha\sigma_2^i, \ldots, -\alpha\sigma_{n_i}^i\}. \tag{3.129}$$

In this case, the transformation matrix T_i can be factorized as

$$T_i = R_i \hat{T}_i, \tag{3.130}$$

where

$$R_i = \text{diag}\{1, \alpha, \ldots, \alpha^{n_i - 1}\}, \tag{3.131}$$

and $\hat{T}_i$ is the Vandermonde matrix

$$\hat{T}_i = \begin{bmatrix} 1 & 1 & \cdots & 1 \\ -\sigma_1^i & -\sigma_2^i & \cdots & -\sigma_{n_i}^i \\ \vdots & \vdots & & \vdots \\ (-\sigma_1^i)^{n_i-1} & (-\sigma_2^i)^{n_i-1} & \cdots & (-\sigma_{n_i}^i)^{n_i-1} \end{bmatrix}. \tag{3.132}$$

For the moment, we consider the free uncoupled subsystems

$$\dot{\tilde{x}}_i = \Lambda_i \tilde{x}_i, \qquad i = 1, 2, \ldots, s. \tag{3.133}$$

Each subsystem (3.133) is stabilized with a degree of exponential stability

$$\pi_i = \alpha \hat{\pi}_i, \tag{3.134}$$

where

$$\hat{\pi}_i = \min_p \sigma_p^i. \tag{3.135}$$

Now we choose again the Liapunov function $v\colon \mathcal{R}^{n_i} \to \mathcal{R}_+$,

$$v_i(\tilde{x}_i) = (\tilde{x}_i^T \tilde{H}_i \tilde{x}_i)^{1/2}, \tag{3.64}$$

where

$$\Lambda_i \tilde{H}_i + \tilde{H}_i \Lambda_i = -\tilde{G}_i \tag{3.65}$$

and

$$\tilde{G}_i = 2\theta_i \operatorname{diag}\left\{\alpha\sigma_1^i, \alpha\sigma_2^i, \ldots, \alpha\sigma_{n_i}^i\right\}, \qquad \tilde{H}_i = \theta_i I_i. \tag{3.66}$$

The aggregate model $\tilde{\mathscr{A}}$ given as

$$\dot{v} \leqslant \tilde{W}v \tag{3.69}$$

is formed as above by computing the elements $\tilde{w}_{ij}$ of the aggregate matrix $\tilde{W}$ with

$$\tilde{\xi}_{ij} = \lambda_M^{1/2}(\tilde{A}_{ij}^T \tilde{A}_{ij}) \tag{3.136}$$

and

$$\tilde{A}_{ij} = \hat{T}_i^{-1} R_i^{-1} A_{ij} R_j \hat{T}_j. \tag{3.137}$$

Our ability to stabilize the system depends ultimately on satisfying the Sevastyanov-Kotelyanskii conditions (3.72) by the aggregate matrix $\tilde{W} = (\tilde{w}_{ij})$ defined by

$$\tilde{w}_{ij} = -\delta_{ij}\pi_i + (1 - \delta_{ij})\tilde{\xi}_{ij}. \tag{3.138}$$

The conditions (3.72) are equivalent to the quasidominant diagonal property of $\tilde{W}$ (see Appendix),

$$d_j|\tilde{w}_{jj}| > \sum_{\substack{i=1 \\ i \neq j}}^{s} d_i |\tilde{w}_{ij}|, \qquad j = 1, 2, \ldots, s, \tag{3.139}$$

where the d_i's are positive numbers. Apparently, we can make the matrix $\tilde{W}$ satisfy the conditions (3.139) if we can increase the diagonal elements $\tilde{w}_{jj}$ sufficiently while keeping the off-diagonal elements $\tilde{w}_{ij}$ bounded. This is exactly the case with the class of systems under consideration. We notice that the diagonal elements $(i = j)$,

$$\tilde{w}_{ii} = -\alpha\hat{\pi}_i, \tag{3.140}$$

depend linearly on the adjustable parameter α. The off-diagonal elements $(i \neq j)$,

$$\tilde{w}_{ij} = \tilde{\xi}_{ij}(\alpha), \tag{3.141}$$

are bounded functions of α. To see this, we note that the elements $\alpha^{q-p} a_{pq}^{ij}$ of the matrices $R_i^{-1} A_{ij} R_j$ either are zero for $p < q$ due to (3.124), or are bounded for $p \geqslant q$ due to nonpositive powers of α. We have

$$\lim_{\alpha \to +\infty} R_i^{-1} A_{ij} R_j = D_{ij}, \tag{3.142}$$

where the matrix $D_{ij} = (d_{pq}^{ij})$ is defined by $d_{pq}^{ij} = a_{pq}^{ij}$ when $p = q$, and $d_{pq}^{ij} = 0$ when $p \neq q$. From (3.137) and (3.142), we define $\tilde{D}_{ij} = \hat{T}_i^{-1} D_{ij} \hat{T}_j$ and conclude from

$$\lim_{\alpha \to +\infty} \tilde{\xi}_{ij}(\alpha) = \lambda_M^{1/2}(\tilde{D}_{ij}^T \tilde{D}_{ij}) \tag{3.143}$$

that the off-diagonal elements $\tilde{w}_{ij}$ are bounded in α.

Therefore, for the selected class of dynamic systems we can always choose a sufficiently large parameter α, and use local linear feedback control to stabilize the systems. From (3.126), we see that by increasing the value of α, we move the subsystem eigenvalues away from the origin, thus increasing the degree of exponential stability of each subsystem. This, however, requires an increase of the local feedback gains in the course of stabilization.

Let us illustrate the local stabilization procedure using the following example:

$$\dot{x} = \begin{bmatrix} 0 & 1 & 0 & 2 & 0 \\ 0 & 0 & 1 & 3 & 4 \\ -2 & -1 & -1 & 2 & 1 \\ 4 & 0 & 0 & 0 & 1 \\ 5 & 6 & 0 & -3 & -2 \end{bmatrix} x + \begin{bmatrix} 0 & 0 \\ 0 & 0 \\ 1 & 0 \\ 0 & 0 \\ 0 & 1 \end{bmatrix} u. \tag{3.144}$$

The eigenvalues of the system matrix A corresponding to (3.144) are

$$\lambda_1 = 1.7244, \quad \lambda_2 = 5.1042, \quad \lambda_3 = -1.2633, \quad \lambda_{4,5} = -4.2826 \pm j1.7755, \tag{3.145}$$

and the system (3.144) is unstable.

The system (3.144) can be decomposed as

$$\dot{x}_1 = \begin{bmatrix} 0 & 1 & 0 \\ 0 & 0 & 1 \\ -2 & -1 & -1 \end{bmatrix} x_1 + \begin{bmatrix} 2 & 0 \\ 3 & 4 \\ 2 & 1 \end{bmatrix} x_2 + \begin{bmatrix} 0 \\ 0 \\ 1 \end{bmatrix} u_1, \tag{3.146a}$$

$$\dot{x}_2 = \begin{bmatrix} 0 & 1 \\ -3 & -2 \end{bmatrix} x_2 + \begin{bmatrix} 4 & 0 & 0 \\ 5 & 6 & 0 \end{bmatrix} x_1 + \begin{bmatrix} 0 \\ 1 \end{bmatrix} u_2. \tag{3.146b}$$

The eigenvalues of the subsystem (3.146a) are moved from

$$\lambda_1^1 = -1.3532, \qquad \lambda_{2,3}^1 = 0.1766 \pm j1.2028 \tag{3.147}$$

to the new locations

$$\lambda_1^1 = -\sigma_1^1 = -1, \qquad \lambda_2^1 = -\sigma_2^1 = -2, \qquad \lambda_3^1 = -\sigma_3^1 = -3 \quad (3.148)$$

by applying the local control (3.34) and

$$k_1^T = (4, 10, 5). \quad (3.149)$$

Similarly, the eigenvalues of the subsystem (3.146b) are changed from

$$\lambda_{1,2}^2 = -1 + j1.4142 \quad (3.150)$$

to

$$\lambda_1^2 = -\sigma_1^2 = -1, \qquad \lambda_2^2 = -\sigma_2^2 = -2 \quad (3.151)$$

by applying the local control (3.34) and

$$k_2^T = (-1, 1). \quad (3.152)$$

Referring to (3.129), we see that in (3.148) and (3.151), the parameter $\alpha = 1$.

We construct the transformation matrices R_1, R_2, $\hat{T}_1$, $\hat{T}_2$ for $\alpha > 1$ as

$$R_1 = \begin{bmatrix} 1 & 0 & 0 \\ 0 & \alpha & 0 \\ 0 & 0 & \alpha^2 \end{bmatrix}, \qquad \hat{T}_1 = \begin{bmatrix} 1 & 1 & 1 \\ -1 & -2 & -3 \\ 1 & 4 & 9 \end{bmatrix},$$
$$R_2 = \begin{bmatrix} 1 & 0 \\ 0 & \alpha \end{bmatrix}, \qquad \hat{T}_2 = \begin{bmatrix} 1 & 1 \\ -1 & -2 \end{bmatrix}. \quad (3.153)$$

The numbers $\hat{\pi}_1$, $\hat{\pi}_2$ are both set equal to one. Then, the aggregation matrix of (3.69) defined by (3.140) is given as

$$\tilde{W} = \begin{bmatrix} -\alpha & \tilde{\xi}_{12} \\ \tilde{\xi}_{21} & -\alpha \end{bmatrix}, \quad (3.154)$$

which for $\alpha = 1$ takes the form

$$\tilde{W} = \begin{bmatrix} -1 & 17.0011 \\ 12.2936 & -1 \end{bmatrix}, \quad (3.155)$$

where

$$\tilde{\xi}_{12} = \lambda_M^{1/2}(\hat{T}_1^{-1} A_{12} \hat{T}_2), \qquad \tilde{\xi}_{21} = \lambda_M^{1/2}(\hat{T}_2^{-1} A_{21} \hat{T}_1),$$

and A_{12}, A_{21} are specified in (3.146).

It is obvious that the matrix $\tilde{W}$ in (3.155) does not satisfy the inequalities (3.72).

From (3.146) and (3.142), we find that

$$D_{12} = \begin{bmatrix} 2 & 0 \\ 0 & 4 \\ 0 & 0 \end{bmatrix}, \qquad D_{21} = \begin{bmatrix} 4 & 0 & 0 \\ 0 & 6 & 0 \end{bmatrix}, \tag{3.156}$$

and for $\alpha > 15$, we have $\tilde{\xi}_{12} \approx 32.55$, $\tilde{\xi}_{21} \approx 18.98$. Thus, for $\alpha = 25$, we have the aggregate matrix

$$\tilde{W} = \begin{bmatrix} -25 & 32.55 \\ 18.98 & -25 \end{bmatrix}, \tag{3.157}$$

which satisfies the conditions (3.72) and the overall system is stable. The corresponding eigenvalues of the overall closed-loop system are

$$\begin{aligned} \lambda_1 &= -36.0364, \qquad \lambda_{2,3} = -25.9599 \pm j3.5219, \\ \lambda_{4,5} &= -68.5213 \pm j6.0474. \end{aligned} \tag{3.158}$$

For the chosen value of $\alpha = 25$, we have the eigenvalue sets $\mathfrak{L}_1$ and $\mathfrak{L}_2$ defined in (3.126) given as

$$\begin{aligned} \mathfrak{L}_1 &= \{-\alpha\sigma_1^1, -\alpha\sigma_2^1, -\alpha\sigma_3^1\} = \{-25, -50, -75\}, \\ \mathfrak{L}_2 &= \{-\alpha\sigma_1^2, -\alpha\sigma_2^2\} = \{-25, -50\}. \end{aligned} \tag{3.159}$$

The locations of the subsystem eigenvalues specified by $\mathfrak{L}_1$, $\mathfrak{L}_2$ of (3.159) are achieved by the local state-variable feedback defined by (3.34) and

$$\begin{aligned} k_1^T &= (93748, 6874, 149), \\ k_2^T &= (1247, 73). \end{aligned} \tag{3.160}$$

The gains in (3.160) are relatively high, which is due to the use of local controllers only. The gains can be considerably reduced by applying global controllers in the multilevel scheme outlined at the beginning of this section.

Applications of the decentralized stabilization scheme presented in this section to spacecraft control systems are outlined in Chapter 6, and applications to interconnected power systems are given in Chapter 7.

3.4. ESTIMATION

The scheme for multilevel stabilization by state feedback which was presented in the preceding section was based upon the assumption that each state of all the subsystems can be read out as outputs. In large dynamic systems, this assumption cannot be expected to hold even if the subsystems are simple. Therefore, we should be able to build the state estimator whose task it is to use the knowledge of each subsystem (its input-output equations) and its actual input and output, and produce a good estimate of the unknown present state of the overall system. If a system is stabilized as "one piece", then standard design procedures can be used to build an asymptotic state estimator for "one shot" determination of the system state, as reviewed by Chen (1970). Such procedures require that the observability test be applied to the overall system, which for high-dimensional systems can be a costly, complicated endeavor whose final outcome is unreliable due to errors in computations. The question then is: Can we build a state estimator for a large dynamic system in the spirit of decomposition principle by building low-order estimators for each subsystem separately? An affirmative answer to this question has been provided only recently (Šiljak and Vukčević, 1977c, 1978), and a multilevel estimation scheme is available for state estimation of large-scale systems. One of the most pleasing facts about the scheme is that the stabilization of the error between the real state and the estimated state is accomplished by the same decentralized control method developed in the preceding section.

Let us consider the large-scale linear system

$$\dot{z} = \hat{A}z + \hat{B}u,$$
$$y = \hat{C}z, \tag{3.42}$$

where $z \in \mathfrak{R}^n$, $u \in \mathfrak{R}^s$, $y \in \mathfrak{R}^m$. We assume that the system (3.42) is given in the output-decentralized form either as a product of the output-decentralized scheme, or by being recognized as such directly from physical considerations. That is, we consider (3.42) as

$$\dot{x}_i = A_i x_i + \sum_{\substack{j=1 \\ j \neq i}}^{m} A_{ij} x_j + B_i u, \qquad i = 1, 2, \ldots, m, \tag{3.43}$$
$$y_i = c_i^T x_i,$$

where each pair (A_i, c_i^T) is observable (Chen, 1970), that is

$$\operatorname{rank}[c_i \mid A_i^T c_i^T \mid \cdots \mid (A_i^T)^{n_i - 1} c_i^T] = n_i. \tag{3.161}$$

Comparing (3.161) with (3.78), we conclude that (A_i, c_i^T) is observable if and only if the pair (A_i^T, c_i) is controllable. This fact is needed below.

In order to estimate the states $x_i(t)$, we construct subsystem observers of the form

$$\dot{\hat{x}}_i = F_i\hat{x}_i + g_i y_i + \sum_{\substack{j=1 \\ j\neq i}}^{m} F_{ij}\hat{x}_j + \sum_{\substack{j=1 \\ j\neq i}}^{m} g_{ij} y_j + B_i u, \qquad i = 1, 2, \ldots, m, \tag{3.162}$$

where the matrices and vectors F_i, F_{ij}, g_i, g_{ij} are to be determined so that (3.162) are the identity observers for the subsystems (3.43).

For the error of estimation,

$$w_i = x_i - \hat{x}_i, \tag{3.163}$$

we subtract (3.162) from (3.43) to obtain

$$\dot{w}_i = (A_i - g_i c_i^T)x_i - F_i\hat{x}_i + \sum_{\substack{j=1 \\ j\neq i}}^{m} (A_{ij} - g_{ij}c_j^T)x_j - \sum_{\substack{j=1 \\ j\neq i}}^{m} F_{ij}\hat{x}_j. \tag{3.164}$$

We choose

$$F_i = A_i - g_i c_i^T, \qquad F_{ij} = A_{ij} - g_{ij}c_j^T, \tag{3.165}$$

and from (3.164) we get the equations describing the error between the real state and the estimated state as

$$\dot{w}_i = F_i w_i + \sum_{\substack{j=1 \\ j\neq i}}^{m} F_{ij} w_j, \qquad i = 1, 2, \ldots, m. \tag{3.166}$$

In order to obtain the asymptotic estimator from (3.166), we stabilize the dual of (3.166),

$$\dot{\tilde{w}}_i = F_i^T \tilde{w}_i + \sum_{\substack{j=1 \\ j\neq i}}^{m} F_{ji}^T \tilde{w}_j, \qquad i = 1, 2, \ldots, m. \tag{3.167}$$

Using (3.165) and (3.167), we write

$$\dot{\tilde{w}}_i = (A_i^T - c_i g_i^T)\tilde{w}_i + \sum_{\substack{j=1 \\ j\neq i}}^{m} (A_{ji}^T - c_i g_{ji}^T)\tilde{w}_j, \qquad i = 1, 2, \ldots, m. \tag{3.168}$$

We recognize immediately the important fact that the equations (3.168) have the same form as (3.58) and the pairs (A_i^T, c_i) are all controllable. Therefore, we can select g_i, g_{ij} by the hierarchic scheme to stabilize (3.168) as we used to select k_i, k_{ij} to stabilize (3.58). Then, all $\tilde{w}_i$ will approach zero exponentially; and regardless of how large is the discrepancy between $\hat{x}_{i0}$ and x_{i0}, each estimate $\hat{x}_i$ will rapidly approach the corresponding state vector x_i.

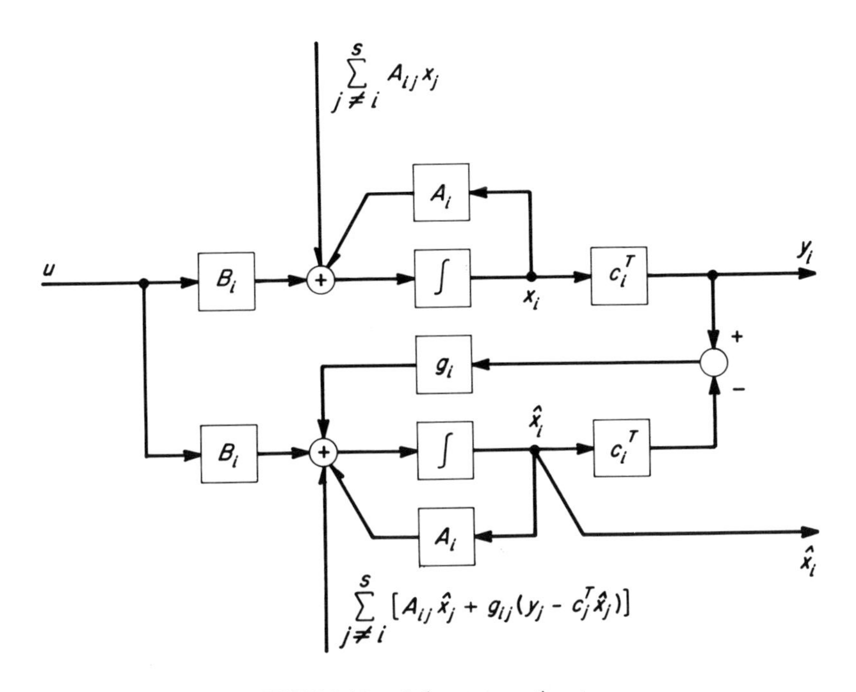

FIGURE 3.12. Subsystem estimator.

The final form of the estimator obtained after the appropriate choice of g_i, g_{ij}, can be given as

$$\dot{\hat{x}}_i = A_i \hat{x}_i + g_i(y_i - c_i^T \hat{x}_i) + \sum_{\substack{j=1 \\ j \neq i}}^{m} [A_{ij}\hat{x}_j + g_{ij}(y_j - c_j^T \hat{x}_j)] + B_i u, \tag{3.169}$$

$$i = 1, 2, \ldots, m.$$

The "wiring diagram" of the ith estimator for the ith plant (3.43) can be drawn by inspection as shown in Figure 3.12.

It is important to note that the outlined hierarchic construction of estimators, which results into the interconnected estimators (3.169), reduces the dimensionally of the estimation problem for the overall system (3.42). Instead of constructing one estimator for the nth-order system (3.42), we propose to design m estimators for n_ith-order subsystems.

Now we are in a position to specify the regulator for each subsystem, as a system comprising the state estimator and the control law. When the plant of ith subsystem is described by the equations

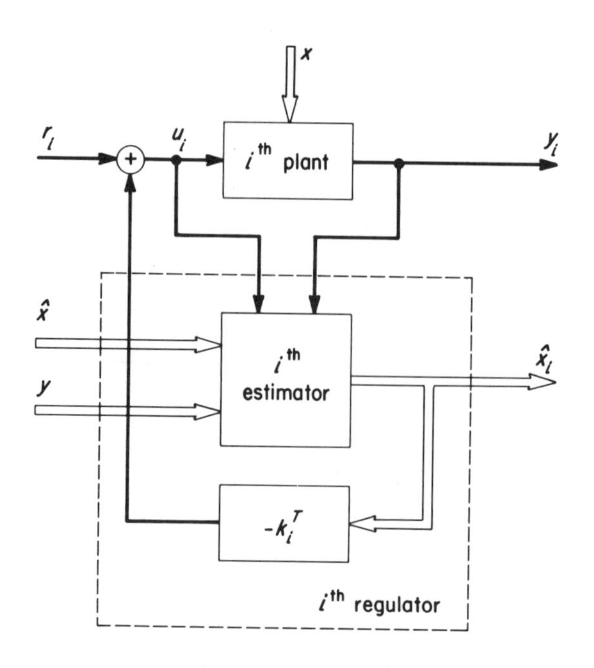

FIGURE 3.13. Subsystem regulator.

$$\dot{x}_i = A_i x_i + \sum_{\substack{j=1 \\ j \neq i}}^{s} A_{ij} x_j + b_i u_i, \qquad i = 1, 2, \ldots, s, \tag{3.170}$$

$$y_i = c_i^T x_i,$$

the block diagram of the ith regulator is as shown in Figure 3.13, where a suitable notation is introduced for the input and output signals (see Figure 3.12). In the scheme of Figure 3.13, only the local control law

$$u_i = r_i - k_i^T x_i \tag{3.171}$$

is used, where $r_i(t)$ is a reference input for the ith subsystem.

It is of interest to demonstrate the fact that the separation property (Chen, 1970) holds for the arrangement of Figure 3.13, and that we can use the control

$$u_i = r_i - k_i^T \hat{x}_i, \tag{3.172}$$

where we have replaced the real state x_i by the estimated state $\hat{x}_i$, instead of (3.171).

We assume that stabilization of the estimator is accomplished by choosing the g_i's only, and that the g_{ij}'s in (3.164) are all zero. Now, by substituting (3.172) into (3.170), and using (3.164), the subsystem of Figure 3.13 can be described by

$$\begin{bmatrix} \dot{x}_1 \\ \cdots \\ \dot{x}_s \\ \hline \dot{\hat{x}}_1 \\ \cdots \\ \dot{\hat{x}}_s \end{bmatrix} = \left[\begin{array}{ccc|ccc} A_1 & \cdots & A_{1s} & -b_1 k_1^T & \cdots & 0 \\ \cdots & \cdots & \cdots & \cdots & \cdots & \cdots \\ A_{1s} & \cdots & A_s & 0 & \cdots & -b_s k_s^T \\ \hline g_1 c_1^T & \cdots & 0 & A_1 - g_1 c_1^T + b_1 k_1^T & \cdots & A_{1s} \\ \cdots & \cdots & \cdots & \cdots & \cdots & \cdots \\ 0 & \cdots & g_s c_s^T & A_{s1} & \cdots & A_s - g_s c_s^T + b_s k_s^T \end{array}\right] \begin{bmatrix} x_1 \\ \cdots \\ x_s \\ \hline \hat{x}_1 \\ \cdots \\ \hat{x}_s \end{bmatrix} + \begin{bmatrix} b_1 r_1 \\ \cdots \\ b_s r_s \\ \hline b_1 r_1 \\ \cdots \\ b_s r_s \end{bmatrix}. \tag{3.173}$$

By applying the following transformation to Equation (3.173),

$$\begin{bmatrix} x_1 \\ \cdots \\ x_s \\ \hline w_1 \\ \cdots \\ w_s \end{bmatrix} = \left[\begin{array}{c|c} I & 0 \\ \hline I & -I \end{array}\right] \begin{bmatrix} x_1 \\ \cdots \\ x_s \\ \hline \hat{x}_1 \\ \cdots \\ \hat{x}_s \end{bmatrix}, \tag{3.174}$$

we get

$$
\begin{bmatrix} \dot{x}_1 \\ \cdots \\ \dot{x}_s \\ \hline \dot{w}_1 \\ \cdots \\ \dot{w}_s \end{bmatrix}
= \left[\begin{array}{ccc|ccc} A_1 - b_1 k_1^T & \cdots & A_{1s} & b_1 k_1^T & \cdots & 0 \\ \cdots & \cdots & \cdots & \cdots & \cdots & \cdots \\ A_{s1} & \cdots & A_s - b_s k_s^T & 0 & \cdots & b_s k_s^T \\ \hline & & & A_1 - g_1 c_1^T & \cdots & A_{1s} \\ & 0 & & \cdots & \cdots & \cdots \\ & & & A_{s1} & \cdots & A_s - g_s c_s^T \end{array}\right]
\begin{bmatrix} x_1 \\ \cdots \\ x_s \\ \hline w_1 \\ \cdots \\ w_s \end{bmatrix}
+ \begin{bmatrix} b_1 r_1 \\ \cdots \\ b_s r_s \\ \hline 0 \end{bmatrix}. \tag{3.175}
$$

Since (3.175) can be split into two independent sets of equations, the separation property is established. From (3.175) we see that as far as stability is concerned there is no difference between using the estimated state $\hat{x}$ and the real state x. The stabilization scheme of the previous section can be used twice to independently stabilize the two matrices in the diagonal blocks on the right-hand side of (3.175), by applying the state feedback. This fact about the multilevel estimation is of interest in a multilevel optimization proposed in the next section, for which the availability of all the states is absolutely essential.

Some comments can be made about the multilevel estimation scheme presented in this section. First, it is immediately possible to reduce the order of the estimator for each subsystem as outlined by Chen (1970). Secondly, as in the case of multilevel stabilization, it is possible to select the class of large-scale systems for which the subsystem asymptotic estimators can always be built by local design only. As expected, this class is the dual of the system defined by (3.34) and (3.123). Thirdly, it is obvious that the separation property, established above for local stabilization only, can be directly extended to include the global controllers. Fourthly, we have considered systems without noise. If there is noise associated with subsystems, subsystem estimators can be suitably modified (e.g., Anderson and Moore, 1971) to include noise in the decentralized estimator for the overall system. Finally, it should be mentioned that the asymptotic estimators produced by the proposed multilevel scheme are highly reliable under structural perturbations and can tolerate a wide range of nonlinearities in coupling among the subsystems. This important robustness of decentralized estimators was first observed by Weissenberger (1976) and later used by Šiljak and Vukčević (1978).

3.5. OPTIMIZATION

By a simple example in Section 1.10, we demonstrated a possibility that optimal systems may become unstable if subjected to structural perturbations. Intuitively, the higher the degree of cooperation among different parts of the system (subsystems), the higher the efficiency of the overall system. Increased cooperation, however, means increased interdependence among the subsystems, which in turn may jeopardize the functioning of the overall system when some of the subsystems cease to participate. Since for a proper operation of large-scale systems, it is essential that structural changes do not cause a breakdown, a trade-off between optimality and dynamic reliability should be established. Therefore, the control schemes for large systems should be designed to ensure dynamic reliability despite a possible deterioration of the optimal performance.

In this section, we will outline a multilevel control scheme proposed by Šiljak and Sundareshan (1976a, b), which inherently incorporates the desired trade-off. It is assumed that a large system is decomposed into a number of subsystems which are optimized by local feedback controllers with respect to a local performance index when ignoring the interactions among the subsystems. That is, each subsystem optimizes its performance as if it were decoupled from the rest of the system. This strategy is shown to result in a reliable system when the interconnections are suitably limited.

When the locally optimal subsystems are interconnected, the interconnections act as perturbations causing a degradation of the system performance and thus a suboptimal performance of the overall system. For this reason, a suboptimality index is defined which measures the performance deterioration and represents the trade-off between optimality and reliability in large-scale dynamic systems.

Unless the subsystems are weakly coupled (Kokotović, Perkins, Cruz, D'Ans, 1969; Kokotović and Singh, 1971), the local optimizations may be in conflict with each other, producing a poor or even unstable overall system. To reduce the conflict among the goals of the subsystems, we introduce the global controllers, whose function is to decrease the effect of some (or all) interactions among the subsystems using the partial (or total) information available on the subsystem level. Therefore, the interactions among the subsystems are necessarily treated as perturbation terms, thus deliberately ignoring their possible beneficial effects. Such an approach, however, opens a real possibility for treating nonlinearities in the interconnections as well as in the subsystems and establishing the robustness of the overall system. It should be noted, however, that at present we do not know how to attach cost to global control. If this aspect is critical, global control being optional may be entirely excluded, and the proposed decentralized optimization can be carried out by using only local feedback control associated with each subsystem.

The suboptimality design due to Šiljak and Sundareshan (1976a), which will be presented in this section, uses the results of Popov (1960) and Rissanen (1966) obtained in the context of perturbed optimal systems. A similar approach has also been used by Bailey and Ramapriyan (1973) and by Weissenberger (1974). Neutralization of the interconnection effects by the global control is motivated by the work of Johnson (1971) concerning disturbance rejection in linear systems.

Besides the reliability aspect, the proposed multilevel scheme offers advantages in the following situations:

(1) When the individual subsystems have no information about the actual shape of interactions except that they are bounded, suboptimality and stability of the overall system can be accomplished using local controllers only.

(2) When a system is too large, a straightforward optimization is either uneconomical because excessive computer time is required, or impossible because excessive computer storage is needed to complete the optimization.

(3) When the state of the overall system is not accessible for direct measurement, and a single observer is not feasible, the proposed multilevel optimization scheme can accommodate the decentralized estimators described in the preceding section.

Let us consider a system $\mathfrak{S}$ described by the equation

$$\dot{x} = f(t, x, u), \tag{3.1}$$

where again $x(t) \in \mathfrak{R}^n$ is the state of the system $\mathfrak{S}$ and $u(t) \in \mathfrak{R}^m$ is the input at time $t \in \mathfrak{T}$. The function $f\colon \mathfrak{T} \times \mathfrak{R}^n \times \mathfrak{R}^m \rightarrow \mathfrak{R}^n$ is continuous on a bounded domain $\mathfrak{D}$ in $\mathfrak{T} \times \mathfrak{R}^n$ and is locally Lipschitzian with respect to x in $\mathfrak{D}$, so that for every fixed input function $u(t)$, a unique solution $x(t; t_0, x_0)$ exists for all initial conditions $(t_0, x_0) \in \mathfrak{D}$ and all $t \in \mathfrak{T}_1$. The symbol $\mathfrak{T}_1$ represents the closed time interval $[t_0, t_1]$, and $\mathfrak{D} = \{(t, x) \in \mathfrak{T} \times \mathfrak{R}^n: t_0 \leqslant t \leqslant t_1, \|x\| \leqslant \rho\}$, where $0 < \rho < +\infty$.

We assume that the system $\mathbb{S}$ can be decomposed into s interconnected subsystems $\mathbb{S}_i$ described by the equations

$$\dot{x}_i = g_i(t, x_i, u_i^l) + h_i(t, x), \qquad i = 1, 2, \ldots, s, \tag{3.176}$$

where $x_i(t) \in \mathfrak{R}^{n_i}$ is the state of the ith subsystem, so that $\mathfrak{R}^n = \mathfrak{R}^{n_1} \times \mathfrak{R}^{n_2} \times \cdots \times \mathfrak{R}^{n_s}$. In (3.176), $u_i^l(t) \in \mathfrak{R}^{m_i}$ is the control function available locally for controlling the ith subsystem, and $\mathfrak{R}^m = \mathfrak{R}^{m_1} \times \mathfrak{R}^{m_2} \times \cdots \times \mathfrak{R}^{m_s}$. Further, $g_i: \mathfrak{R} \times \mathfrak{R}^{n_i} \times \mathfrak{R}^{m_i} \to \mathfrak{R}^{n_i}$ represents the dynamics of the decoupled subsystems

$$\dot{x}_i = g_i(t, x_i, u_i^l), \qquad i = 1, 2, \ldots, s, \tag{3.177}$$

which are all completely locally controllable about any admissible solution $x_i(t; t_0, x_{i0})$; and $h: \mathfrak{R} \times \mathfrak{R}^n \to \mathfrak{R}^{n_i}$ is the function which represents the interconnection of the ith subsystem inside the overall system $\mathbb{S}$.

By using the local control law

$$u_i^l(t) = k_i^l(t, x_i), \tag{3.178}$$

each isolated subsystem (3.177) is optimized with respect to the performance index

$$J_i(t_0, x_{i0}, u_i^l) = v_i[t_1, x_i(t_1)] + \int_{t_0}^{t_1} L_i[t, x_i(t), u_i^l(t)]\, dt, \tag{3.179}$$

where $v_i: \mathfrak{R} \times \mathfrak{R}^{n_i} \to \mathfrak{R}_+$, $L_i: \mathfrak{R} \times \mathfrak{R}^{n_i} \times \mathfrak{R}^{m_i} \to \mathfrak{R}_+$ are functions of the class C^2 in all arguments, and $x_i(t)$ denotes the solution $x_i(t; t_0, x_{i0})$ of (3.177) for the fixed control function $u_i^l(t)$. Implicit in this optimization is the fact that each decoupled subsystem (3.177) is of low order and has a simple structure, so that it is relatively easy to find the optimal control law

$$u_i^{l0}(t) = k_i^l(t, x_i), \tag{3.180}$$

which produces the optimal cost

$$J_i^0(t_0, x_{i0}) = J_i[t_0, x_{i0}, k_i^l(t_0, x_{i0})]. \tag{3.181}$$

We assume that all points of the optimal trajectories $x_i^0(t; t_0, x_{i0})$ belong to $\mathfrak{D}$ for all $(t_0, x_{i0}) \in \mathfrak{D}$.

In addition to the complexities arising from the size of the system, there are certain structural considerations that should be taken into account when designing controllers for large systems. A system which is composed of interconnected subsystems may undergo structural perturbations whereby subsystems are disconnected (and again connected) in various ways during the operation. In order to guarantee a satisfactory performance of the system despite the on-off participation of the subsystems, we should preserve as much as possible the autonomy of each isolated subsystem (3.177). Therefore, our performance index for the overall system $\mathbb{S}$ is simply

$$J(t_0, x_0, u^l) = \sum_{i=1}^{s} J_i(t_0, x_{i0}, u_i^l), \tag{3.182}$$

where $u^l(t) \in \mathcal{R}^m$ is defined above as $u^l = [(u_1^l)^T, (u_2^l)^T, \ldots, (u_s^l)^T]^T$. That is, each subsystem (3.177) is optimized with respect to its own performance index (3.179) regardless of the behavior of the other subsystems. Therefore, the interconnections among the subsystems are regarded as perturbation terms, thus ignoring their possible beneficial effects. This is justified because any use of the beneficial effects will naturally increase the dependence among the subsystems and, hence, increase vulnerability to structural perturbations. This structural aspect of the present optimization scheme will be treated in detail later.

The optimization strategy chosen, however, cannot achieve the optimal performance index

$$J^0(t_0, x_0) = \sum_{i=1}^{s} J_i^0(t_0, x_{i0}) \tag{3.183}$$

by using only the local control $u^l(t)$, unless all the interactions are absent ($h_i \equiv 0$, $i = 1, 2, \ldots, s$). The value of the performance index for the overall system $\mathbb{S}$ when the interactions are present is given by

$$\tilde{J}(t_0, x_0) = \sum_{i=1}^{s} v_i[t_1, \tilde{x}_i(t_1)] + \int_{t_0}^{t_1} L_i[t, \tilde{x}_i(t), k_i^l(t, \tilde{x}_i)]\, dt, \tag{3.184}$$

where $\tilde{x}_i(t)$ denotes the solution $\tilde{x}_i(t; t_0, x_{i0})$ of system $\mathbb{S}$, and generally

$$\tilde{J}(t_0, x_0) \leqslant J^0(t_0, x_0) \qquad (t_0, x_0) \in \mathfrak{D}. \tag{3.185}$$

Therefore, the local control law $u^l(t)$ is chosen as a suboptimal policy with an index of suboptimality $\varepsilon > 0$ defined by the inequality

$$\tilde{J}(t_0, x_0) \geqslant (1 + \varepsilon) J^0(t_0, x_0) \qquad \forall (t_0, x_0) \in \mathfrak{D}. \tag{3.186}$$

The suboptimality index ε for the system with the optimal local control,

$$\dot{x}_i = g_i[t, x_i, k_i^l(t, x_i)] + h_i(t, x), \qquad i = 1, 2, \ldots, s, \tag{3.187}$$

depends on the size of the interactions $h_i(t, x)$ among the subsystems, and therefore it is a measure of the deterioration of the performance index $J^0(t_0, x_0)$ due to interactions.

We introduce the following:

Definition 3.7. *The system* (3.187) *with the local control law* (3.180) *is said to be suboptimal with index* ε *if there exists a number* $\varepsilon > 0$ *for which the inequality* (3.186) *is satisfied.*

Since suboptimality in the system is a result of the presence of interconnections, the index ε depends on the size of $h_i(t, x)$, and the following problem is of interest:

Problem 3.1. *Establish conditions on* $h_i(t, x)$ *to guarantee a prescribed value of the suboptimality index* ε.

A solution to Problem 3.1, which will be given later, involves only bounds on the norms of the interconnection functions $h_i(t, x)$. Therefore, the results obtained are valid for a class of $h_i(t, x)$ and thus do not depend on the actual form of these nonlinear functions. This robustness is of major importance for modeling uncertainties and possible variations in the shape of nonlinear interconnections during operation (Weissenberger, 1976).

It is important to observe now that the suboptimal performance of the system has resulted from the use only of controllers associated locally with each individual subsystem. The suboptimality index ε is a measure of the performance degradation, which is directly proportional to the size of interconnections. Hence, an improvement in the system performance is possible if $\|h_i(t, x)\|$ can be reduced. We propose to accomplish this by using additional control functions that neutralize the effect of interconnections. These functions are generated by a global controller on a higher hierarchical level using the states of the subsystems. If, however, some of the subsystem states are not available to the global controller, it can perform its function partially or be *excluded entirely*.

Since the global control functions are introduced to modify the existing interconnections $h_i(t, x)$, the effective interconnections can be represented by $h_i(t, x, u_i^g)$, where $u_i^g(t) \in \mathcal{R}^{m_i}$ is the global control component applied to the subsystem $\mathcal{S}_i$. Now, the interconnected subsystems are described by

$$\dot{x}_i = g_i(t, x_i, u_i^l) + h_i(t, x, u_i^g), \qquad i = 1, 2, \ldots, s, \tag{3.188}$$

which is an obvious modification of the equations (3.176).

With this modification, the index ε becomes a function of $\|h(t, x, u^g)\|$, where $h\colon \mathcal{R} \times \mathcal{R}^n \times \mathcal{R}^m \to \mathcal{R}^n$ is $h = (h_1^T, h_2^T, \ldots, h_s^T)^T$ and $u^g \in \mathcal{R}^m$ is $u^g = [(u_1^g)^T, (u_2^g)^T, \ldots, (u_s^g)^T]^T$, and we solve the following:

Problem 3.2. *Find a control law*

$$u^g(t) = k^g(t, x) \tag{3.189}$$

for which

$$\varepsilon^0 = \inf_{u^g(t)} \{\varepsilon[\|h(t, x, u^g)\|]\} \qquad \forall (t, x) \in \mathfrak{D}. \tag{3.190}$$

It is important to note that the choice of global control to neutralize the interconnections using available subsystem states does not disturb the reliability of the system accomplished by the proposed use of local controllers. Therefore, a solution to Problem 3.2, outlined subsequently, mitigates suboptimality and at the same time preserves the reliability of the closed-loop system.

Now that Problems 3.1 and 3.2 are precisely formulated, we proceed to solve them by decentralized optimization proposed by Šiljak and Sundareshan (1976a).

A solution to Problem 3.1 may be obtained by using the classical Hamilton-Jacobi theory. Since in our optimization procedure we chose the local control laws to optimize the decoupled subsystems, the optimal indices satisfy the corresponding Hamilton-Jacobi equations. When the subsystems are interconnected, the equations are not satisfied by the respective performance indices, and the overall system is not optimal. However, a majorization procedure is possible to provide an estimate of the performance deviation from the optimum due to interactions.

We assume that the optimal index $V_i(t, x_i)$ is a function $V_i\colon \mathfrak{R} \times \mathfrak{R}^n \to \mathfrak{R}_+$ which belongs to the class C^2 in both arguments, and satisfies the Hamilton-Jacobi equation (e.g. Anderson and Moore, 1971)

$$\frac{\partial V_i(t, x_i)}{\partial t} + [\text{grad } V_i(t, x_i)]^T g_i[t, x_i, k_i^l(t, x_i)] + L_i[t, x_i, k_i^l(t, x_i)] = 0,$$

$$i = 1, 2, \dots, s, \qquad \forall (t, x) \in \mathfrak{D}, \tag{3.191}$$

and $V_i[t_1, x_i(t_1)] = \nu[t_1, x_i(t_1)]$, $[t_1, x_i(t_1)] \in \mathfrak{D}$.

Let us define the functions $V\colon \mathfrak{R} \times \mathfrak{R}^n \to \mathfrak{R}_+$, $\nu\colon \mathfrak{R} \times \mathfrak{R}^n \to \mathfrak{R}_+$, $L\colon \mathfrak{R} \times \mathfrak{R}^n \times \mathfrak{R}^m \to \mathfrak{R}_+$ as

$$\begin{aligned} V(t, x) &= \sum_{i=1}^{s} V_i(t, x_i), \\ \nu(t, x) &= \sum_{i=1}^{s} \nu_i(t, x_i), \\ L(t, x, u^l) &= \sum_{i=1}^{s} L_i(t, x_i, u_i^l), \end{aligned} \tag{3.192}$$

and k^l: $\mathcal{R} \times \mathcal{R}^n \to \mathcal{R}^m$ as $k^l = [(k_1^l)^T, (k_2^l)^T, \ldots, (k_s^l)^T]^T$, where the $k_i^l(t, x_i)$'s are defined in (3.178).

Now we provide a solution to Problem 3.1 by the following:

Theorem 3.9. *Let the interactions $h_i(t, x)$ in (3.187) satisfy the constraint*

$$[\text{grad } V(t,x)]^T h(t,x) \leqslant \frac{\varepsilon}{1+\varepsilon} L[t, x, k^l(t,x)] \qquad \forall (t,x) \in \mathcal{D}. \tag{3.193}$$

Then the composite system (3.187) is suboptimal with index ε.

Proof. From (3.191) and (3.192) we get the Hamilton-Jacobi equation

$$\frac{\partial V(t,x)}{\partial t} + [\text{grad } V(t,x)]^T g[t, x, k^l(t,x)] + L[t, x, k^l(t,x)] = 0 \qquad \forall (t,x) \in \mathcal{D}, \tag{3.194}$$

with $V[t_1, x(t_1)] = \nu[t_1, x(t_1)]$, for the optimal system

$$\dot{x} = g[t, x, k^l(t,x)], \tag{3.195}$$

where g: $\mathcal{R} \times \mathcal{R}^n \times \mathcal{R}^m \to \mathcal{R}^n$ is defined by $g = (g_1^T, g_2^T, \ldots, g_s^T)^T$ and the g_i's are as in (3.177).

The total time derivative $\dot{V}(t,x)$ of the function $V(t,x)$ is calculated along the trajectories $\tilde{x}(t)$ of the composite system (3.187) as

$$\dot{V}[t, \tilde{x}(t)] = \frac{\partial V[t, \tilde{x}(t)]}{\partial t} + \{\text{grad } V[t, \tilde{x}(t)]\}^T \{g(t, \tilde{x}(t), k^l[t, \tilde{x}(t)]) + h[t, \tilde{x}(t)]\}. \tag{3.196}$$

Using (3.194), we can rewrite (3.196) as

$$-\dot{V}(t,\tilde{x}) + [\text{grad } V(t,\tilde{x})]^T h(t,\tilde{x}) = L[t, \tilde{x}, k^l(t,\tilde{x})] \qquad \forall (t,\tilde{x}) \in \mathcal{D}, \tag{3.197}$$

where $\tilde{x} = \tilde{x}(t)$. This is equivalent to

$$L[t, \tilde{x}, k^l(t,\tilde{x})] = -(1+\varepsilon)\dot{V}(t,\tilde{x}) + (1+\varepsilon)[\text{grad } V(t,\tilde{x})]^T h(t,\tilde{x}) - \varepsilon L[t, \tilde{x}, k^l(t,\tilde{x})] \qquad \forall (t,\tilde{x}) \in \mathcal{D}. \tag{3.198}$$

By integrating (3.198) from t_0 to t_1, we get

$$\tilde{J}(t_0, x_0) = (1+\varepsilon) J^0(t_0, x_0) + (1+\varepsilon) \int_{t_0}^{t_1} \left\{ [\text{grad } V(t,\tilde{x})]^T h(t,\tilde{x}) - \frac{\varepsilon}{1+\varepsilon} L[t, \tilde{x}, k^l(t,\tilde{x})] \right\} dt, \tag{3.199}$$

where we use $V[t_1, \tilde{x}(t_1)] = v[t_1, \tilde{x}(t_1)]$.

By using the condition (3.193) of the theorem, from (3.199) we obtain the inequality (3.186), and the proof of Theorem 3.9 is complete.

In the context of Theorem 3.9, the inequality (3.193) may be interpreted in two ways. It places constraints on the interactions $h_i(t, x)$ so as to guarantee a prescribed suboptimality index ε. Alternatively, starting with specified interactions $h_i(t, x)$, the inequality (3.193) enables one to calculate the suboptimality index ε. The latter interpretation suggests that we may be able to choose the global control function $u^g(t)$ and solve the optimization Problem 3.2.

Let us observe that when the global control is present, the constraint on the interactions $h_i(t, x, u_i^g)$ given by

$$\|\text{grad } V(t,x)\| \, \|h(t,x,u^g)\| \leqslant \frac{\varepsilon}{1+\varepsilon} L[t, x, k^I(t,x)] \qquad \forall (t,x) \in \mathfrak{D} \quad (3.200)$$

implies the inequality (3.193). If the expressions for the functions V and L are available, (3.200) can be used to get an explicit relationship between ε and $\|h(t, x, u^g)\|$, which can be used to suitably choose $u^g(t)$ and solve Problem 3.2. This is precisely the case when the free subsystems are linear and optimized with respect to quadratic costs. We shall demonstrate this in the latter part of this section. Now we turn our attention to connective aspects of the proposed optimization scheme.

The multilevel optimization of large-scale systems, outlined above, implies an important structural property for the optimized systems: The suboptimality index is invariant under structural perturbations whereby subsystems are disconnected from each other and again connected together in various ways during operation. The notion of connective suboptimality introduced here is, therefore, an extension of the dynamic reliability concept that originated in stability studies of large-scale systems (see Chapters 1 and 2).

Let us consider the system $\mathbb{S}$ given by (3.176), and let us assume that the interactions $h_i(t, x)$ are of the form

$$h_i(t,x) = h_i(t, e_{i1}x_1, e_{i2}x_2, \ldots, e_{is}x_s), \qquad i = 1, 2, \ldots, s, \quad (3.201)$$

where $e_{ij} = e_{ij}(t)$ are elements of the $s \times s$ interconnection matrix $E(t) = [e_{ij}(t)]$, such that $0 \leqslant e_{ij}(t) \leqslant 1$ $\forall t \in \mathfrak{T}$. The elements $e_{ij}(t)$ are arbitrary continuous functions of time t.

Now, relying on Definition 3.7, we state the following:

Definition 3.8. *The system* (3.176) *with local control law* (3.180) *is said to be connectively suboptimal with index* ε *if it is suboptimal with index* ε *for all interconnection matrices* $E(t) \in \bar{E}$.

It is of interest to obtain conditions for connective suboptimality, which would be expressed in terms of the constant fundamental interconnection matrix $\bar{E} = (\bar{e}_{ij})$, but are valid for all interconnection matrices $E(t) = [e_{ij}(t)]$. Let us show how this goal can be accomplished when $\mathbb{S}$ is a linear system described by

$$\dot{x} = Ax + Bu, \tag{3.202}$$

where $x(t) \in \mathfrak{R}^n$ is the state of the system and $u(t) \in \mathfrak{R}^m$ is the control. Here $A = A(t)$ and $B = B(t)$ are $n \times n$ and $n \times m$ matrices continuous in t on the interval $\mathfrak{T} = (\tau, +\infty)$, where τ is a number or the symbol $-\infty$.

We assume that the system (3.202) can be decomposed into s interconnected subsystems $\mathbb{S}_i$ described by

$$\dot{x}_1 = A_i x_i + B_i u_i + \sum_{j=1}^{s} e_{ij} A_{ij} x_j, \qquad i = 1, 2, \ldots, s, \tag{3.203}$$

where $x_i(t) \in \mathfrak{R}^{n_i}$ is the state of the subsystem $\mathbb{S}_i$, $u_i(t) \in \mathfrak{R}^{m_i}$ is the control, and $A_i = A_i(t)$, $B_i = B_i(t)$, and $A_{ij} = A_{ij}(t)$ are matrices of proper dimensions.

In the proposed multilevel optimization scheme, the control $u_i(t)$ is considered as consisting of two parts, the local control $u_i^l(t)$ and the global control $u_i^g(t)$,

$$u_i(t) = u_i^l(t) + u_i^g(t). \tag{3.204}$$

The local control $u_i^l(t)$ is chosen as a linear control law

$$u_i^l = -K_i x_i, \tag{3.205}$$

to optimize isolated subsystems, and the global control law

$$u_i^g = -\sum_{j=1}^{s} e_{ij} K_{ij} x_j, \tag{3.206}$$

to minimize degradation of performance due to interconnections among the subsystems.

When $E = 0$, the interconnected system (3.203) breaks down into s decoupled subsystems $\mathbb{S}_i$ described by

$$\dot{x}_i = A_i x_i + B_i u_i^l, \qquad i = 1, 2, \ldots, s. \tag{3.207}$$

We assume that all s pairs (A_i, B_i) are completely controllable, and that with each isolated subsystem (3.207) a quadratic performance index

$$J_i(t_0, x_{i0}, u_i^l) = \int_{t_0}^{\infty} \left[\|x_i(t)\|_{Q_i}^2 + \|u_i^l(t)\|_{R_i}^2 \right] dt \tag{3.208}$$

is associated. In (3.208), $Q_i = Q_i(t)$ is an $n_i \times n_i$ symmetric, nonnegative definite matrix, $R_i = R_i(t)$ is an $m_i \times m_i$ symmetric, positive definite matrix, and $\|x_i\|^2_{Q_i}$ stands for $x_i^T Q_i x_i$.

The local control $u_i^l(t)$ in (3.205) can now be chosen to minimize the performance index $J_i(t_0, x_{i0}, u_i^l)$ in (3.208). From linear regulator theory (e.g. Anderson and Moore, 1971) the optimal control u_i^{l0} is given by (3.205) where

$$K_i = R_i^{-1} B_i^T P_i. \tag{3.209}$$

In (3.209), $P_i = P_i(t)$ is an $n_i \times n_i$ symmetric, positive definite matrix which is the solution of the Riccati equation

$$\dot{P}_i + P_i A_i + A_i^T P_i - P_i B_i R_i^{-1} B_i^T P_i + Q_i = 0. \tag{3.210}$$

The optimal cost $J_i^0(t_0, x_{i0}) = J_i(t_0, x_{i0}, u_i^{l0})$ is

$$J_i^0(t_0, x_{i0}) = \|x_{i0}\|^2_{P_i}. \tag{3.211}$$

Assume that Q_i can be factored as $Q_i = C_i C_i^T$, where C_i is an $n_i \times n_i$ constant matrix, so that the pair (A_i, C_i) is completely observable. Then each closed-loop subsystem

$$\dot{x}_i = (A_i - B_i R_i^{-1} B_i^T P_i) x_i, \qquad i = 1, 2, \ldots, s, \tag{3.212}$$

is globally asymptotically stable.

The control $u_i^{l0}(t)$, $i = 1, 2, \ldots, s$, will not generally be optimal for the composite system (3.203), and it would not result in the optimal cost $J^0(t_0, x_0)$ defined by (3.183) and (3.211) unless $E(t) \equiv 0$. When $E(t) \not\equiv 0$, the controls $u_i^{l0}(t)$ can only be suboptimal, and that takes place if a number $\varepsilon > 0$ exists such that

$$\tilde{J}(t_0, x_0) \leqslant (1 + \varepsilon) J^0(t_0, x_0) \qquad \forall (t_0, x_0) \in \mathcal{T} \times \mathcal{R}^n, \tag{3.213}$$

as discussed above. To establish the existence of ε for the composite closed-loop system

$$\dot{x}_i = (A_i - B_i K_i) x_i + \sum_{j=1}^{s} e_{ij} (A_{ij} - B_i K_{ij}) x_j, \qquad i = 1, 2, \ldots, s \tag{3.214}$$

we prove the following:

Theorem 3.10. *The system* (3.214) *is connectively suboptimal with index* ε *if*

$$\sum_{i=1}^{s} \sum_{j=1}^{s} \bar{e}_{ij} \xi_{ij} \leqslant \frac{1}{2} \frac{\varepsilon}{1 + \varepsilon} \frac{\min_i \lambda_m[W_i(t)]}{\max_i \lambda_M[P_i(t)]} \qquad \forall t \in \mathcal{T}, \tag{3.215}$$

where $\xi_{ij} = \lambda_M^{1/2}[H_{ij}^T(t)H_{ij}(t)]$, *the matrices* $H_{ij}(t)$ *and* $W_i(t)$ *are given as*

$$H_{ij} = A_{ij} - B_i K_{ij}, \qquad W_i = P_i B_i R_i^{-1} B_i^T P_i + Q_i, \tag{3.216}$$

and λ_m, λ_M *are the minimum and the maximum eigenvalues of the respective matrices at time* t.

Proof. Since the decoupled subsystems (3.212) are optimal, the functions $V_i(t, x_i) = \|x_i\|_{P_i}^2$, $i = 1, 2, \ldots, s$, satisfy individually the Hamilton-Jacobi equations

$$\frac{\partial V_i(t, x_i)}{\partial t} + [\text{grad } V_i(t, x_i)]^T[(A_i - B_i K_i)x_i] = -\|x_i\|_{Q_i}^2 + \|K_i x_i\|_{R_i}^2 \\ \forall (t, x_i) \in \mathcal{T} \times \mathcal{R}^{n_i}, \\ i = 1, 2, \ldots, s. \tag{3.217}$$

Now, reproducing the proof of Theorem 3.11, but paying attention to the connectivity aspect of the system $\mathbb{S}$, we compute the total time derivative $\dot{V}_i(t, x_i)$ along the trajectories $\tilde{x}_i(t) \in \mathcal{R}^{n_i}$ of the composite system (3.214) for the fundamental interconnection matrix $\bar{E}$ as

$$\dot{V}_i(t, \tilde{x}_i) = \frac{\partial V_i(t, \tilde{x}_i)}{\partial t} + [\text{grad } V_i(t, \tilde{x}_i)]^T[(A_i - B_i K_i)\tilde{x}_i] + \sum_{j=1}^{s} \bar{e}_{ij} H_{ij} \tilde{x}_j. \tag{3.218}$$

By substituting (3.218) into (3.217) and rearranging the terms, we get

$$\|\tilde{x}_i\|_{W_i}^2 = -(1 + \varepsilon)\dot{V}_i(t, \tilde{x}_i) + (1 + \varepsilon)[\text{grad } V_i(t, \tilde{x}_i)]^T \sum_{j=1}^{s} \bar{e}_{ij} H_{ij} \tilde{x}_j - \varepsilon\|x_i\|_{W_i}^2, \tag{3.219}$$

where $x = (x_1^T, x_2^T, \ldots, x_s^T)^T$.

By summing the equations (3.219) and integrating from t_0 to ∞, we obtain

$$\tilde{J}(t_0, x_0) = (1 + \varepsilon)J^0(t_0, x_0) \\ + (1 + \varepsilon)\int_{t_0}^{\infty} \sum_{i=1}^{s} \left\{[\text{grad } V_i(t, \tilde{x}_i)]^T \sum_{j=1}^{s} \bar{e}_{ij} H_{ij} \tilde{x}_j - \frac{\varepsilon}{1 + \varepsilon}\|\tilde{x}_i\|_{W_i}^2\right\} dt, \tag{3.220}$$

where $\tilde{J}$ and J^0 are defined in (3.184) and (3.183).

We use the inequalities

$$\|H_{ij}\tilde{x}_j\| \leqslant \xi_{ij}\|\tilde{x}_j\| \tag{3.221}$$

and

$$\left\| [\text{grad } V_i(t, \tilde{x}_i)]^T \sum_{j=1}^{s} \bar{e}_{ij} H_{ij} \tilde{x}_j \right\| \leqslant 2\lambda_M(P_i) \sum_{j=1}^{s} \bar{e}_{ij} \xi_{ij} \|\tilde{x}_j\|^2, \quad (3.222)$$

together with the inequality (3.215) of the theorem, to establish the inequality (3.186) from (3.220). Since $E(t) \leqslant \bar{E}$ element by element for all $t \in \mathfrak{T}$, the inequality (3.222) holds for all $E(t) \in \bar{E}$, and the closed-loop system (3.214) is connectively suboptimal with index ε. The proof of Theorem 3.10 is complete.

We can now show that a closed-loop composite system which is connectively suboptimal is also connectively asymptotically stable in the large. To demonstrate this fact for the system (3.214), we choose the function

$$V(t, \tilde{x}) = \sum_{i=1}^{s} V_i(t, \tilde{x}_i) = \sum_{i=1}^{s} \|\tilde{x}_i\|_{P_i}^2 \quad (3.223)$$

as a Liapunov function. Since each P_i is a positive definite solution of the Riccati equation (3.210), $V(t, \tilde{x})$ is a positive definite quadratic form. Taking the total time derivative of $\dot{V}(t, \tilde{x})$ along the trajectories of the system (3.214) for the case $E(t) \equiv \bar{E}$, and using (3.215) and (3.221), we get

$$\dot{V}(t, \tilde{x}) = -\sum_{i=1}^{s} \|\tilde{x}_i\|_{W_i}^2 + 2 \sum_{i=1}^{s} \tilde{x}_i^T P_i \sum_{j=1}^{s} \bar{e}_{ij} H_{ij} \tilde{x}_j \leqslant 0, \qquad \forall (t, x) \in \mathfrak{T} \times \mathfrak{R}^n. \quad (3.224)$$

Therefore, the equilibrium $x^* = 0$ of the system (3.214) for $E(t) = \bar{E}$ is asymptotically stable in the large. Since $E(t) \leqslant \bar{E}$ element by element, the equilibrium $x^* = 0$ is asymptotically stable in the large for all interconnection matrices $E(t)$, that is, it is connectively asymptotically stable in the large.

So far we have not commented on the role of the global controls $u_i^g(t)$ except to modify the original interconnections $A_{ij} x_j$ to obtain the effective interconnections $(A_{ij} - B_i K_{ij}) x_j$. It is obvious that when the original interconnections $A_{ij} x_j$ satisfy the inequality (3.228), then the connective suboptimality with index ε is achieved by local controls $u_i^l(t)$ only (corresponding to the trivial case $K_{ij} \equiv 0$; $i, j = 1, 2, \ldots, s$).

The role of the global controls $u_i^g(t)$ is to reduce the effect of interconnections. This is useful when the size of the existing interconnections is such that no value of ε can be found to satisfy the inequality (3.215). Even when an appropriate value of ε could be found, the global control can be implemented to reduce the value of ε and hence the deterioration of the optimal performance. It should be noted, however, that it is not clear how the cost of global control can be included in the proposed optimization scheme.

From (3.215), we see that the index ε is a function of the norm ξ_{ij} of the matrix $H_{ij} = A_{ij} - B_i K_{ij}$ defined in (3.216). To minimize ξ_{ij}, we choose $K_{ij} = B_i^\dagger A_{ij}$, where $B_i^\dagger$ is the Moore-Penrose generalized inverse of B_i (Langenhop, 1967). When rank $B_i = m_i$, we have $B_i^\dagger = (B_i^T B_i)^{-1} B_i^T$ and

$$K_{ij} = (B_i^T B_i)^{-1} B_i^T A_{ij}, \tag{3.225}$$

which represents a solution to Problem 3.2 above. In the particular case when the rank of the composite matrix $[B_i | A_{ij}]$ is equal to the rank of the matrix B_i itself, the choice (3.225) of K_{ij} produces $H_{ij} = 0$. That means that the effective interconnection from the jth subsystem to the ith subsystem in the closed-loop optimized system is nullified.

It is now of interest to show that linear systems optimized by the decentralized schemes are robust, that is, they can tolerate a wide range of nonlinearities in the interactions, a fact first pointed out by Weissenberger (1974). Let us consider a simple modification of the equations (3.203),

$$\dot{x}_i = A_i x_i + B_i u_i^l + h_i(t, x), \qquad i = 1, 2, \ldots, s, \tag{3.226}$$

where $h_i(t, x)$ represent the interactions and have the form shown in (3.201). We exclude the global control and consider the closed-loop system as

$$\dot{x}_i = (A_i - B_i R_i^{-1} B_i^T P_i) x_i + h_i(t, x), \qquad i = 1, 2, \ldots, s. \tag{3.227}$$

Now, if there exist numbers $\xi_{ij} \geqslant 0$ such that the inequalities

$$\|h_i(t, x)\| \leqslant \sum_{j=1}^{s} \bar{e}_{ij} \xi_{ij} \|x_j\| \qquad \forall (t, x) \in \mathcal{T} \times \mathcal{R}^n, \quad \forall i = 1, 2, \ldots, s \tag{3.228}$$

and (3.215) are satisfied, then from the proof of Theorem 3.10, it follows immediately that the composite system (3.227) is connectively suboptimal with index ε.

To illustrate the usefulness of Theorem 3.10 and, in particular, emphasize its structural aspect, let us consider a system of equations

$$\begin{aligned} ml^2 \ddot{\theta}_1 &= -ka^2(\theta_1 - \theta_2) - mgl\theta_1 - u_1, \\ ml^2 \ddot{\theta}_2 &= -ka^2(\theta_2 - \theta_1) - mgl\theta_2 - u_2, \end{aligned} \tag{3.229}$$

which describe the motion of two identical pendulums coupled by a spring and subject to two distinct inputs as shown on Figure 3.14.

We consider the case when the position $a(t)$ of the spring can change along the full length of the pendulums, that is

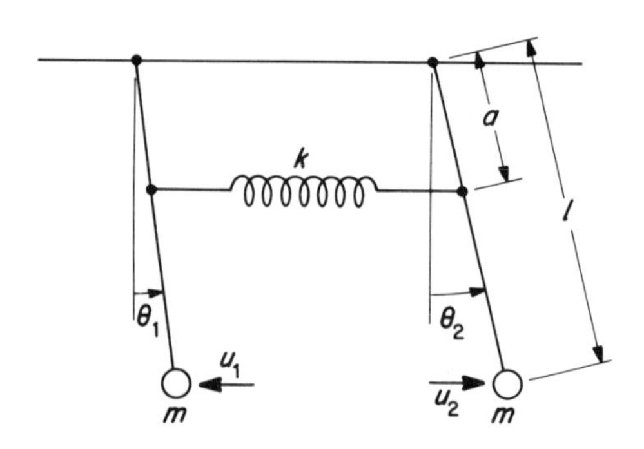

FIGURE 3.14. Interconnected pendulums.

$$0 \leqslant a(t) \leqslant l, \tag{3.230}$$

for all t. The problem of interest is to optimally regulate the system (3.229) for any unspecified function $a(t)$. In view of the arbitrariness of the function $a(t)$, the standard "one shot" optimization of the system is difficult to apply. Furthermore, if optimization is conducted for any particular position of the spring, the computed optimal control may be inappropriate for another position. This problem can be resolved by using the proposed multilevel control scheme.

Choosing the state vector as $x = (\theta_1, \dot{\theta}_1, \theta_2, \dot{\theta}_2)^T$, the equations (3.229) can be rewritten as

$$\dot{x} = \begin{bmatrix} 0 & 1 & 0 & 0 \\ -\frac{g}{l} - \frac{ka^2}{ml^2} & 0 & \frac{ka^2}{ml^2} & 0 \\ 0 & 0 & 0 & 1 \\ \frac{ka^2}{ml^2} & 0 & -\frac{g}{l} - \frac{ka^2}{ml^2} & 0 \end{bmatrix} x + \begin{bmatrix} 0 & 0 \\ -\frac{1}{ml^2} & 0 \\ 0 & 0 \\ 0 & -\frac{1}{ml^2} \end{bmatrix} u, \tag{3.231}$$

where $u = (u_1, u_2)^T$. This system can be decomposed into two subsystems described by

$$\begin{aligned} \dot{x}_1 &= \begin{bmatrix} 0 & 1 \\ -\alpha & 0 \end{bmatrix} x_1 + \begin{bmatrix} 0 \\ -\beta \end{bmatrix} u_1 + e_{11} \begin{bmatrix} 0 & 0 \\ -\gamma & 0 \end{bmatrix} x_1 + e_{12} \begin{bmatrix} 0 & 0 \\ \gamma & 0 \end{bmatrix} x_2, \\ \dot{x}_2 &= \begin{bmatrix} 0 & 1 \\ -\alpha & 0 \end{bmatrix} x_2 + \begin{bmatrix} 0 \\ -\beta \end{bmatrix} u_2 + e_{21} \begin{bmatrix} 0 & 0 \\ \gamma & 0 \end{bmatrix} x_1 + e_{22} \begin{bmatrix} 0 & 0 \\ -\gamma & 0 \end{bmatrix} x_2, \end{aligned} \tag{3.232}$$

where $x_1 = (\theta_1, \dot{\theta}_1)^T$, $x_2 = (\theta_2, \dot{\theta}_2)^T$, $\alpha = g/l$, $\beta = 1/ml^2$, $\gamma = k/m$, and $e_{11}(t) = e_{12}(t) = e_{21}(t) = e_{22}(t) = a^2(t)/l^2$.

We note that $0 \leqslant e_{ij}(t) \leqslant 1$ for all t and $i, j = 1, 2$, and the disconnection of the two subsystems occurs when $a(t) = 0$, that is, $e_{ij}(t) = 0$. In that case, the spring is moved all the way to the support, and the two pendulums are decoupled from each other.

The decoupled pendulums (subsystems) are optimized with respect to the costs specified by $Q_1 = Q_2 = I_{2\times 2}$, $R_1 = R_2 = 1$, by the local controls

$$u_1^l = -\beta[p_{12} \quad p_{22}]x_1, \qquad u_2^l = -\beta[p_{12} \quad p_{22}]x_2, \tag{3.233}$$

where $p_{12} = (1/\beta^2)[(\alpha^2 + \beta^2)^{1/2} - \alpha]$ and $p_{22} = (1/\beta)(1 + 2p_{12})^{1/2}$ are elements of the 2×2 symmetric positive definite solutions P_1, P_2 ($P_1 = P_2$) of the Riccati equations.

Due to the presence of interconnections, the computed local controls are suboptimal with index ε given by the inequality (3.215) as

$$\gamma \leqslant \frac{1}{4} \frac{\varepsilon}{\varepsilon + 1} \frac{\lambda_m(W_1)}{\lambda_M(P_1)}, \qquad t \in \mathfrak{T}, \tag{3.234}$$

where $W_1 (W_1 = W_2)$ is calculated from (3.216).

Now, by choosing the global controls

$$\begin{aligned} u_1^g &= -\frac{1}{\beta}[-\gamma e_{11} \quad 0 \quad \gamma e_{12} \quad 0]x, \\ u_2^g &= -\frac{1}{\beta}[\gamma e_{21} \quad 0 \quad -\gamma e_{22} \quad 0]x, \end{aligned} \tag{3.235}$$

we can reduce the suboptimality index to zero. The global controls (3.235), together with the local controls (3.233), result in a two-level control strategy that optimizes the system with respect to the chosen performance indices, and at the same time preserves dynamic reliability to structural perturbations caused by the changes in the position of the spring that couples the two pendulums.

The multilevel scheme can easily accommodate the nonlinearities in the interconnections among the subsystems. In the context of the two pendulums, we can include a nonlinear spring and consider a modification of the equations (3.231) according to (3.227), where k is replaced by $k(t, x)$. Then the condition

$$|k(t, x)| \leqslant m\gamma, \qquad (t, x) \in \mathfrak{T} \times \mathfrak{R}^n \tag{3.236}$$

and the inequality (3.230) guarantee suboptimality of the system with the same suboptimality index as in the case of the linear model. It should be emphasized that we do not need to know the actual shape of the function $k(t, x)$, but only the constraint on its size. The robustness of the proposed suboptimal control should be appreciated when the proposed multilevel scheme is contrasted with the conventional optimal control.

An application of the decentralized optimization technique outlined in this section will be given in Chapter 6 where we design a control system for the NASA Large Space Telescope, as well as in Chapter 7 where we consider automatic generation control of interconnected power systems.

REFERENCES

Anderson, B. D. O., and Moore, J. B. (1971), *Linear Optimal Control*, Prentice-Hall, Englewood Cliffs, New Jersey.

Aoki, M. (1972), "On Feedback Stabilizability of Decentralized Dynamic Systems", *Automatica*, 8, 163–173.

Aoki, M., and Li, M. T. (1973), "Partial Reconstruction of state Vectors in Decentralized Dynamic Systems", *IEEE Transactions*, AC-18, 289–294.

Arrow, K. J. (1964), "Control in Large Organizations", *Management Science*, 10, 397–408.

Arrow, K. J., and Hurwicz, L. (1963), "Decentralization and Computation in Resource Allocation", *Essays in Economics and Econometrics*, R. W. Pfouts (ed.), University of North Carolina Press, Chapel Hill, North Carolina, 34–104.

Bailey, F. N., and Ramapriyan, H. K. (1973), "Bounds on Suboptimality in the Control of Linear Dynamic Systems", *IEEE Transactions*, AC-18, 532–534.

Barlow, R. E., Fussell, J. B., and Singpurwalla, N. D. (eds.) (1975), *Reliability and Fault Tree Analysis*, SIAM, Philadelphia, Pennsylvania.

Berge, C. (1962), *The Theory of Graphs and Their Applications*, Wiley, New York.

Berztiss, A. T. (1975), *Data Structures: Theory and Practice*, Academic, New York.

Boesch, F. T., and Thomas, R. E. (1970), "On Graphs of Invulnerable Communication Nets", *IEEE Transactions*, CT-17, 183–192.

Bowie, W. S. (1976), "Applications of Graph Theory in Computer Systems", *International Journal of Computer and Information Sciences*, 5, 9–31.

Callier, F. M., Chan, W. S., and Desoer, C. A. (1976), "Input-Output Stability of Interconnected Systems Using Decomposition Techniques", *IEEE Transactions*, CAS-23, 714–729.

Chen, C. T. (1970), *Introduction to Linear System Theory*, Holt, Rinehart, and Winston, New York.

Chu, K. C. (1972), "Team Decesion Theory and Information Structures in Optimal Control Problems-Part II", *IEEE Transactions*, AC-17, 22–28.

Corfmat, J. P., and Morse, A. S. (1973), "Stabilization with Decentralized Feedback Control", *IEEE Transactions*, AC-18, 673–682.

Corfmat, J. P., and Morse, A. S. (1976), "Decentralized Control of Linear Multivariable Systems", *Automatica*, 12, 479–495.

Dantzig, G. B., and Wolfe, P. (1960), "Decomposition Principle for Linear Programs", *Operations Research*, 8, 101–111.

Davison, E. J. (1974), "The Decentralized Stabilization and Control of a Class of Unknown Non-Linear Time-Varying Systems", *Automatica*, 10, 309–316.

Davison, E. J. (1977), "Connectability and Structural Controllability of Composite Systems", *Automatica*, 13, 109–123.

Davison, E. J., and Rau, N. S. (1973), "The Optimal Decentralized Control of a Power System Consisting of a Number of Interconnected Synchronous Machines", *International Journal of Control*, 18, 1313–1328.

Deo, N. (1974), *Graph Theory with Applications to Engineering and Computer Science*, Prentice-Hall, Englewood Cliffs, New Jersey.

Festinger, L., Schachter, S., and Back, K. (1950), *Social Pressures in Informal Groups*, Harper, New York.

Harary, F. (1969), *Graph Theory*, Addison-Wesley, Reading, Massachusetts.

Harary, F., Norman, R. Z., and Cartwright, D. (1965), *Structural Models: An Introduction to the Theory of Directed Graphs*, Wiley, New York.

Himelblau, M. (ed.) (1973), *Decomposition of Large Scale Problems*, North-Holland, Amsterdam, Holland.

Ho, Y. C., and Chu, K. C. (1972), "Team Decision Theory and Information Structures in Optimal Control Problems—Part I", IEEE Transactions, AC-17, 15–22.

Ho, Y. C., and Chu, K. C. (1974), "Information Structure in Many Person Optimization Problems", *Automatica*, 10, 341–351.

Ho, Y. C., and Mitter, S. K. (eds.) (1976), *Directions in Large Scale Systems, Many Person Optimization, and Decentralized Control*, Plenum, New York.

Johnson, D. C. (1971), "Accomodation of External Disturbances in Linear Regulator and Servomechanism Problems", *IEEE Transactions*, AC-16, 552–554.

Jury, E. I. (1974), *Inners and Stability of Dynamic Systems*, Wiley, New York.

Kalman, R. E. (1963), "Mathematical Description of Dynamical Systems", *SIAM Journal of Control*, 1, 152–192.

Kalman, R. E., Falb, P. L., and Arbib, M. A. (1969), *Topics in Mathematical System Theory*, McGraw-Hill, New York.

Kevkorian, A. K. (1975), "Structural Aspects of Large Dynamic Systems", *Proceedings of the Sixth IFAC Congress*, Boston, Massachusetts, 19.3:1–7.

Kokotović, P., Perkins, W. R., Cruz, J. B., D'Ans, G. (1969), "ε—Coupling Method for Near-Optimum Design of Large-Scale Linear Systems", *Proceedings of IEE*, 116, 889–892.

Kokotović, P., and Singh, G. (1971), "Optimization of Coupled Non-Linear Systems", *International Journal of Control*, 14, 51–64.

Kornai, J., and Simonovits, A. (1977), "Decentralized Control Problems in Neumann-Economies", *Journal of Economic Theory*, 14, 44–67.

Langenhop, C. E. (1967), "On Generalized Inverses of Matrices", *SIAM Journal of Applied Mathematics*, 15, 1239–1246.

Lau, R., Persiano, R. C. M., and Varaiya, P. P. (1972), "Decentralized Information and Control: A Network Example", *IEEE Transactions*, AC-17, 446–473.

Lin, C. T. (1974), "Structural Controllability", *IEEE Transactions*, AC-19, 201–208.

Luenberger, D. G. (1964), "Observing the State of a Linear System", *IEEE Transactions*, MIL-8, 190–197.

Luenberger, D. G. (1967), "Canonical Forms for Linear Multivariable Systems", *IEEE Transactions*, AC-12, 290–293.

Mahmoud, M. S. (1977), "Multileve Systems Control and Applications", *IEEE Transactions*, SMC-7, 125–143.

Marschak, J., and Radner, R. (1971), *The Economic Theory of Teams*, Yale University Press, New Haven, Connecticut.

McFadden, D. (1969), "On the Controllability of Decentralized Macro-Economic System: The Assignement Problem", *Mathematical System Theory and Economics I*, H. W. Kuhn and G. P. Szegö (eds.), Springer, New York, 221–239.

Mesarović, M. D., Macko, M., and Takahara, Y. (1970), *Theory of Hierarchical Multilevel Systems*, Academic, New York.

Mitrović, D. (1959), "Graphical Analysis and Synthesis of Feedback Control Systems", *AIEE Transactions*, AI-77, 476–503.

Moore, E. F. (1956), "Gedanken-Experiments on Sequential Machines", *Automata Studies*, C. Shannon and J. McCarthy (eds.), Princeton University Press, Princeton, New Jersey, 34, 129–153.

Munro, I. (1971), "Efficient Determination of the Transitive Closure of a Directed Graph", *Information Proceeding Letters*, 1, 56–58.

Özgüner, Ü. (1975), "On the Weak Interconnection of Composite Dynamic Systems", *Proceedings of the IEEE Conference of Decision and Control*, Houston, Texas, 810–814.

Özgüner, Ü., and Perkins, W. R. (1976a), "Structural Properties of Large-Scale Composite Systems", *Large-Scale Dynamical Systems*, R. Saeks (ed.), Point Lobos Press, Los Angeles, California, 5–32.

Özgüner, Ü., and Perkins, W. R. (1976b), "A Series Solution to the Nash Strategy for Large-Scale Interconnected Systems", *Proceedings of the IFAC Symposium on Large Scale Systems Theory and Applications*, G. Guardabassi and A. Locatelli (eds.), Udine, Italy, 435–440.

Popov, V. M. (1960), "Criterion of Quality for Nonlinear Controlled Systems", *Proceedings of the First IFAC Congress*, Moscow, 1, 173–176.

Purdom, P. (1970), "A Transitive Closure Algorithm", *Nordisk Tidskrift for Informationbehandlung*, 10, 76–94.

Rissanen, J. J. (1966), "Performance Deterioration of Optimum Systems", *IEEE Transactions*, AC-11, 530–532.

Sage, A. P. (1977), *Methodology for Large-Scale Systems*, McGraw-Hill, New York.

Sandell, N. R., Varaiya, P., and Athans, M. (1975), "A Survey of Decentralized Control Methods for Large Scale Systems", *Proceedings of the ERDA Conference on System Engineering for Power: Status and Prospects*, Henniker, New Hampshire, 1–19. (Also in: *IEEE Transactions*, AC-23, to appear).

Shields, R. W., and Pearson, J. B. (1976), "Structural Controllability of MultiInput Linear Systems", *IEEE Transactions*, AC-21, 203–212.

Šiljak, D. D. (1969), *Nonlinear Systems*, Wiley, New York.

Šiljak, D. D. (1975), "Connective Stability of Competitive Equilibrium", *Automatica*, 11, 389–400.

Šiljak, D. D. (1976a), "Competitive Economic Systems: Stability, Decomposition, and Aggregation", *IEEE Transactions*, AC-21, 149–160.

Šiljak, D. D. (1976b), "Stabilization of Large-Scale Systems: A Spinning Flexible Spacecraft", *Automatica*, 12, 309–320.
Šiljak, D. D. (1977a), "On Reachability of Dynamic Systems", *International Journal of Systems Science*, 8, 321–338.
Šiljak, D. D. (1977b), "On Pure Structure of Dynamic Systems", *Nonlinear Analysis, Theory, Methods, and Applications*, 1, 397–413.
Šiljak, D. D. (1977c), "Vulnerability of Dynamic Systems", *Proceedings of the IFAC Workshop on Control and Management of Integrated Industrial Complexes*, Toulouse, France, 133–144.
Šiljak, D. D. (1977d), "On Decentralized Control of Large-Scale Systems", *Proceedings of the IUTAM Symposium on Dynamics of Multibody Systems, Munchen*, Germany (to be published in 1978 by Springer, New York).
Šiljak, D. D., and Sundareshan, M. K. (1974), "On Hierarchic Optimal Control of Large-Scale Systems", *Proceedings of the Eight Asilomar Conference on Circuits, Systems, and Computers*, Pacific Grove, California, 495–502.
Šiljak, D. D., and Sundareshan, M. K. (1976a), "A Multilevel Optimization of Large-Scale Dynamic Systems", *IEEE Transactions*, AC-21, 79–84.
Šiljak, D. D., and Sundareshan, M. K. (1976b), "Large-Scale Systems: Optimality Vs. Reliability", *Directions in Large-Scale Systems, Many Person Optimization, and Decentralized Control*, Y. C. Ho and S. K. Mitter (eds.), Plenum, New York, 411–425.
Šiljak, D. D., Sundareshan, M. K., and Vukčević, M. B. (1975), "A Multilevel Control System for the Large Space Telescope", *Final Report on NASA Contract NAS 8-27799*, University of Santa Clara, Santa Clara, California.
Šiljak, D. D., and Vukčević, M. B. (1976a), "Multilevel Control of Large-Scale Systems: Decentralization, Stabilization, Estimation, and Reliability", *Large-Scale Dynamical Systems*, R. Saeks (ed.), Point Lobos Press, Los Angeles, California, 34–57.
Šiljak, D. D., and Vukčević, M. B. (1976b), "Large-Scale Systems: Stability, Complexity, Reliability", *Journal of the Franklin Institute*, 301, 49–69.
Šiljak, D. D., and Vukčević, M. B. (1976c), "Decentralization, Stabilization, and Estimation of Large-Scale Linear Systems", *IEEE Transactions*, AC-21, 363–366.
Šiljak, D. D., and Vukčević, M. B. (1977), "Decentrally Stabilizable Linear and Bilinear Large-Scale Systems", *International Journal of Control*, 26, 289–305.
Šiljak, D. D., and Vukčević, M. B. (1978), "On Decentralized Estimation", *International Journal of Control*, to appear.
Simon, H. A. (1962), "The Architecture of Complexity", *Proceedings of the American Philosophical Society*, 106, 467–482.
Singh, M. G., Hassan, M. F., and Titli, A. (1976), "Multilevel Feedback Control for Interconnected Systems Using The Prediction Principle", *IEEE Transactions*, SMC-6, 233–239.
Sundareshan, M. K. (1976), "Large-Scale Discrete Systems: A Two-Level Optimization Scheme", *International Journal of Control*, 7, 901–909.
Sundareshan, M. K. (1977), "Generation of Multilevel Control and Estimation Schemes for Large-Scale Systems: A Perturbation Approach", *IEEE Transactions*, SMC-7, 144–152.

Tarjan, R. E. (1972), "Depth-First Search and Linear Graph Algorithm", *SIAM Journal of Computing*, 1, 146–160.

Thaler, G. T., and Brown, R. G. (1960), *Analysis and Design of Feedback Control Systems*, McGraw-Hill, New York.

Varaiya, P. (1972), "Trends in the Theory of Decision Making in Large Systems", *Annals of Economic and Social Measurement*, 1, 493–499.

Varaiya, P., and Warland, J. (1977), "Decentralized Stochastic Control", *Proceedings of the IFAC Workshop on Control and Management of Integrated Industrial Complexes*, Toulouse, France, 97–105.

Vukčević, M. B. (1975), "Locally Stabilizable Large-Scale Systems", *Proceedings of the Ninth Asilomar Conference on Circuits, Systems, and Computers*, Pacific Grove, California, 97–100.

Wang, S. H., and Davison, E. J. (1973), "On the Stabilization of Decentralized Control Systems", *IEEE Transactions*, AC-18, 473–478.

Warfield, J. N. (1974), "Toward Interpretation of Complex Structural Models", *IEEE Transactions*, SMC-4, 405–417.

Warshall, S. (1962), "A Theorem on Boolean Matrices", *Journal of the Association for Computing Machinery*, 9, 11–12.

Weissenberger, S. (1973), "Stability Regions of Large-Scale Systems", *Automatica*, 9, 653–663.

Weissenberger, S. (1974), "Tolerance of Decentrally Optimal Controllers to Nonlinearity and Coupling", *Proceedings of the Twelfth Annual Allerton Conference on Circuit and System Theory*, University of Illinois, Monticello, Illinois, 87–95.

Weissenberger, S. (1976), "On the Sensitivity of Large, Decentrally Optimal Control Systems to Modeling Errors", *Large-Scale Dynamical Systems*, R. Saeks (ed.), Point Lobos Press, Los Angeles, California, 168–181.

Wismer, D. A. (ed.) (1963), *Optimization Methods for Large-Scale Systems*, McGraw-Hill, New York.

Zadeh, L. A., and Desoer, C. A. (1963), *Linear System Theory*, McGraw-Hill, New York.

4

ECONOMICS

Competitive Equilibrium

Ever since Adam Smith argued in 1776 that there is an "invisible hand" that brings tranquility to the market place, there has been a long line of eminent economists who have endeavored to show rigorously that such a state of a competitive economy *is* or *could be* a reality. The central question is: Under what conditions in a fully decentralized exchange economy can a large number of economic agents who are guided only by their self-interest reach the state of the economy in which their decisions are mutually compatible? The underlying assumption in this context is that in an economy with given tastes, endowments, and technology, the economic agents make decisions on the basis of the prices of commodities they transact, but they have no control over these prices and take them as given. For each commodity a quantity demanded and a quantity supplied are established by the decisions of individual agents, and if the two quantities happen to be equal, the economy is in the desired state of equilibrium and the corresponding prices are equilibrium prices. This economic situation is the classical *competitive equilibrium.* Its major social significance lies in the fact that it brings about a rational allocation of resources which is optimal in some definite (Pareto-efficient) sense.

The concept of competitive equilibrium has two inseparable parts: existence and stability. It is of primary importance to find out what are the conditions under which competitive equilibrium is possible, and once that is established, it is necessary to demonstrate that the equilibrium can be reached by following a set of rational rules and procedures. Since it is highly improbable that an exchange economy will be at equilibrium by chance, the mere knowledge of the existence of such a state is useless if ability of the economy to achieve it is not established. The roots of this problem can be traced back to the classical English economists and in particular to John Stuart Mill, who recognized the role of demand conditions and prices on the formation of equilibrium situations. But, it was Léon Walras, the founder of the Lausanne school, who in 1874 proposed a rigorous formulation of a competitive economy and brought about the full recognition of general equilibrium theory.

The Walrasian competitive economy is made up of consumers and producers. Consumers own the factors of production, such as labor and capital assets, the sale of which provides their income. With this income, consumers choose commodities at the given prices, which are supplied by the producers operating under fixed coefficients (constant returns to scale). Therefore, the total demand and supply for each commodity, which is obtained by aggregation over all individuals, are functions of prices of both consumer commodities and resources. Now, transactions of commodities take place if and only if all the (generally conflicting) individual plans of consumers are compatible. That is, when the aggregate supply and demand on all markets are equal, and the corresponding prices are equilibrium prices.

If we establish (or assume) that a competitive equilibrium exists, we have to show that if a set of prices is chosen at random (Walras's "*prix criés par hasard*"), then the competitive system can reach the equilibrium through decentralized decisions by the consumers, who behave solely as price-takers. Walras was the first to recognize fully that this capability is essential for the workability of a competitive system, and for this purpose proposed a price adjustment scheme which he termed *tâtonnement*. The scheme is conceived by Walras as a trial-and-error procedure by which the market, like a huge computer, locates an equilibrium by solving an enormous number of supply-equals-demand equations. If, at the given set of prices, the quantity demanded and the quantity supplied of each commodity are not equal, the law of supply and demand comes into play. Prices of those commodities for which demand exceeds supply rise, whereas prices of those commodities for which supply exceeds demand fall. If for the new set of prices are not equilibrium prices, the prices are altered again using the same rule. This iterative process goes on until an equilibrium is reached.

Walras was aware of the fact that the *tâtonnement*, as described above, need not ever converge to any equilibrium price. To assure convergence he argued that the supply and demand for any given commodity are more influenced by changes in its own price than by changes in all other prices. Therefore, during the adjustment process the dominance of the price in its own market overrides the cross-coupling effects from other markets, and after each step the prices get closer to their equilibrium values.

Although *tâtonnement* as described by Walras represents a discrete dynamic process, it was not until the work of Paul Samuelson in 1941 that the competitive market was explicitly treated as a dynamic system. With this new description, the convergence problem of *tâtonnement* was recognized as a stability problem, and a whole host of strong results from the theory of differential equations were made available to the economists. Numerous contributions in mathematical economics made along these lines are surveyed in the books by Quirk and Saposnik (1968), Nikaido (1969), and Arrow and Hahn (1971).

Our main objective in this chapter is to show that the Walras-Samuelson dynamic model of competitive equilibrium, involving price adjustment by *tâtonnement*, is an inherently robust dynamic system. Once stability of the equilibrium is established by using the well-known diagonal dominance condition (which resembles the dominance condition of Walras), it remains invariant under a wide class of structural, nonlinear, nonstationary and stochastic perturbations. In view of the existing stability analysis of competitive equilibrium (Arrow and Hahn, 1971), the robustness property of stability is a relatively simple and straightforward implication. It is nevertheless believed that it is an important one, since it reinforces the credibility of competitive-equilibrium models by showing that certain inaccuracies in data or economic parameters, nonlinearities, random fluctuations, etc., cannot alter the models in so essential a way as to render them useless as descriptions of the real world (Šiljak, 1976a).

Nonstationary models of competitive equilibrium are introduced to consider a reduction of the commodity space, nonlinear saturation phenomena, the change of one commodity from a substitute to a complement for another over a time interval, time-varying shifts in excess demand, etc. Both continuous and discrete models will be considered, and in both cases it will be shown that the diagonal dominance of system matrices ensures connective stability.

Unpredictable changes in supply and demand schedules, due to random changes in economic environment as well as to inaccuracies of the *tâtonnement* process, introduce stochastic disturbances of competitive equilibrium. In the framework of Itô differential equations and stochastic stability in the mean, we will establish a trade-off between the degree of stability and the size of the random shocks that can be absorbed by stable nonlinear and nonstationary equilibrium models.

By using the modern mathematical methods of the comparison principle and the vector Liapunov function outlined in Chapter 2, a hierarchic model is proposed for consideration of large competitive systems. A commodity can be split up into a number of subcommodities, or several commodities can be combined into one composite commodity. Then, a low-order linear aggregate market model can be formed which represents the price-adjustment mechanism for composite commodities. Stability of the aggregate model implies stability of the original nonlinear and nonstationary market, and the stability is connective. That is, stability of the aggregate model implies stability of each subset of markets for composite commodities. This is interesting in that it shows a wide tolerance of stable market models with composite commodities to nonlinear nonstationary phenomena, and therefore an inherent robustness of large equilibrium models.

Stability results presented in this chapter, motivate the structural and robustness considerations throughout this book. They also lay solid ground for the reliable decentralized control scheme of Chapter 3, which mimics the decentralized structure of competitive market system.

Throughout this chapter considerable use is made of the properties of matrices with nonnegative off-diagonal elements, called Metzler matrices. These properties are given by Newman (1959), and can also be obtained directly from the survey by Fiedler and Pták (1962). A collection of the properties of Metzler matrices is provided in the Appendix.

4.1. COMMODITIES AND PRICES: DEMAND FUNCTIONS

Consumers and producers meet in the market place to exchange commodities. The exchange is based upon the prices of commodities, which are determined on the market. Demand and supply decisions concerning commodities are taken by both kinds of economic agents (consumers and producers), and the market demand and supply functions are defined by aggregating the quantities demanded and the quantities supplied by the individual economic agents. The demand and supply functions of each economic agent are functions of the prices of the commodities, and therefore the market demand and supply functions also have prices as independent variables. The notion of market equilibrium is essential in this context, since it is concerned with the compatibility of the decisions made by the economic agents. At equilibrium prices the demand and supply of each commodity are equal, and every agent can achieve what he wants to achieve.

In the stability analysis of market equilibrium, it is convenient to define the *excess demand function* as the difference between the quantity demanded and the quantity supplied on the market at each price. Then an equilibrium price for a commodity is a price at which the excess demand function is zero. We derive this function next.

We assume that there are $n + 1$ commodities on the market, and by P_i, $i = 0, 1, \ldots, n$, we denote their respective prices. Under the assumption of perfect competition, economic agents decide the terms on which they exchange commodities, and their decisions consist of the prices of various commodities that they take as given. Furthermore, under pure exchange we assume that the jth consumer has a utility function

$$U_j(Q_{0j}, Q_{1j}, \ldots, Q_{nj}), \tag{4.1}$$

which he maximizes over the quantities Q_{0j}, Q_{1j}, $\ldots$, Q_{nj} of all commodities subject to his budget constraints

$$\sum_{i=0}^{n} P_i Q_{ij} = \sum_{i=0}^{n} P_i \overline{Q}_{ij}, \tag{4.2}$$

where Q_{ij} denotes the quantity of the ith commodity finally obtained (consumed) by the jth agent, and $\overline{Q}_{ij}$ is the quantity of the ith commodity initially held by the jth agent. The jth agent chooses the optimal quantities Q_{0j}^0, Q_{1j}^0, $\ldots$, Q_{nj}^0 such that

$$U_j(Q_{0j}^0, Q_{1j}^0, \ldots, Q_{nj}^0) \geqslant U_j(Q_{0j}, Q_{1j}, \ldots, Q_{nj}) \tag{4.3}$$

for all Q_{0j}, Q_{1j}, $\ldots$, Q_{nj} satisfying the budget constraint (4.2). Under the appropriate assumptions (the utility function is sufficiently smooth and strictly quasiconcave, the consumer's indifference curves do not intersect any axis, and no consumer has a bliss point) this utility maximization problem can be solved by standard optimization techniques (Intriligator, 1971).

The excess demand of the jth consumer for the ith commodity is defined as

$$F_{ij} = Q_{ij} - \overline{Q}_{ij}. \tag{4.4}$$

Then the utility function U_j of (4.1) can be expressed in terms of F_{ij} as $U_j(F_{0j} + \overline{Q}_{0j}, F_{1j} + \overline{Q}_{1j}, \ldots, F_{nj} + \overline{Q}_{nj})$, and the Lagrangian for this problem can be formulated as

$$V_j = U_j(F_{0j} + \overline{Q}_{0j}, F_{1j} + \overline{Q}_{1j}, \ldots, F_{nj} + \overline{Q}_{nj}) + \lambda \sum_{i=0}^{n} P_i F_{ij}, \tag{4.5}$$

where λ is the Lagrange multiplier for the jth consumer. Now the necessary conditions for maximizing consumer's utility subject to the budget constraint are

$$\left.\begin{aligned} \frac{\partial V_j}{\partial F_{ij}} &= \frac{\partial U_j}{\partial F_{ij}} + \lambda P_j = 0, \\ \frac{\partial V_j}{\partial \lambda} &= \sum_{i=0}^{n} P_i F_{ij} = 0, \end{aligned}\right\} \quad i = 0, 1, \ldots, n. \tag{4.6}$$

Under certain well-known conditions [the corresponding bordered Hessian determinants alternate in sign (Quirk and Saposnik, 1968)], we can apply the implicit-function theorem to the equations (4.6), and assert that there is a regular constrained maximum of the utility function. In a sufficiently small neighborhood of the utility-maximizing point, we can use one equation of (4.6) to eliminate λ and solve the remaining n for the excess demands as functions of prices:

$$F_{ij} = F_{ij}(P_0, P_1, \ldots, P_n), \qquad j = 1, 2, \ldots, n. \tag{4.7}$$

Now, the aggregate excess demand for the ith commodity is defined as

$$F_i(P_0, P_1, \ldots, P_n) = \sum_{j=1}^{m} F_{ij}(P_0, P_1, \ldots, P_n), \tag{4.8}$$

where m is the number of consumers. That is, the aggregate excess demand function for the ith commodity is obtained by summing up the individual excess demand functions of the m consumers.

The fact that we eliminated λ from the first n equations of (4.6) means that the implicitly defined quantities are invariant under the multiplication of prices by a positive constant. That is,

$$F_i(\kappa P_0, \kappa P_1, \ldots, \kappa P_n) = \kappa F_i(P_0, P_1, \ldots, P_n), \qquad i = 1, 2, \ldots, n, \tag{4.9}$$

for any positive number κ, and the aggregate excess demand functions are homogeneous of degree zero in prices. Therefore, we can choose arbitrarily a commodity (say commodity 0) to act as a medium of exchange, and we refer to that commodity as the *numéraire*. Prices of commodities other than the *numéraire* can be considered as "normalized prices" $p_i = P_i/P_0$, $i = 1, 2, \ldots, n$. Then the aggregate excess demand function for the ith commodity in terms of the normalized prices is defined as

$$f_i(p_1, p_2, \ldots, p_n) = \sum_{j=1}^{m} f_{ij}(p_1, p_2, \ldots, p_n), \qquad i = 0, 1, \ldots, n, \tag{4.10}$$

where $f_{ij}(p_1, p_2, \ldots, p_n)$ is obviously the excess-demand function of the jth consumer for the ith commodity with normalized prices as independent variables.

The market system described by the aggregate excess demand functions (4.10) is truly a decentralized system. Each economic agent takes prices as given and maximizes his utility with respect to his own budget constraint. The question of fundamental interest is: Are the desired actions of economic agents mutually compatible for some set of admissible prices?

This is the question of the existence of equilibrium prices $p_1^*, p_2^*, \ldots, p_n^*$ at which the aggregate excess demand functions are equal to zero, and the independent decisions of economic agents can be carried out simultaneously.

Since our interest is centered on the stability problem of competitive equilibrium, we shall assume that the system satisfies the conditions which ensure the existence of the equilibrium prices. That is, we assume that there is a set of (normalized) prices $p_1^*, p_2^*, \ldots, p_n^*$ such that the aggregate excess demand functions become simultaneously zero, that is,

$$f_i(p_1^*, p_2^*, \ldots, p_n^*) = 0, \qquad i = 1, 2, \ldots, n. \tag{4.11}$$

The aggregate excess demand functions and the existence of the equilibrium prices are two basic ingredients in setting up a stage for the analysis of price adjustment processes in dynamic models of the market systems. Our sketchy description of these two entities is all we can do here. For a detailed treatment of both subjects, the reader is referred to the books by Quirk and Saposnik (1968), Nikaido (1969), and Arrow and Hahn (1971). We should also note that we have used the neoclassical Hicks-Samuelson analysis to derive the excess demand functions of a market. The reason for this choice of economic environment is that it is more suitable for stability considerations of competitive equilibrium than the alternative set-theoretic approach, which is more convenient for equilibrium-existence problems.

4.2. A DYNAMIC MODEL

The ability of a competitive economy to allocate resources efficiently to producers and distribute the products among consumers would be of little use if the competitive equilibrium were not stable. The mere existence (and even uniqueness) of competitive equilibrium explains nothing about how the equilibrium is achieved if the market is not already in it. Even if the market is at equilibrium, any changes in conditions and parameters (such as tastes of consumers, technology, resources, weather, etc.) would move the market away from the equilibrium. Being aware of this problem, Walras (1874) suggested that the market competitive mechanism works as a large computing machine which solves the numerous market equations for the equilibrium on the basis of the *law of supply and demand* and an iterative procedure which he called *tâtonnement*. Assume, as Walras did, that an initial set of prices is not at the equilibrium and that for some commodities demand exceeds supplies, whereas for others supply exceeds demand. Considering commodities in a definite order, we can adjust the price of the first commodity so that supply equals demand. This we do on the basis of the law of supply and demand: We raise the price if demand exceeds supply and lower it if supply exceeds demand. When we repeat the procedure for the second commodity, we generally destroy the partial equilibrium of the first commodity market. Therefore, at the time we adjust the last commodity market, all prices of the preceeding markets will be at nonequilibrium values. Walras argued, however, that adjustments of partial equilibria are most significant, since the demand and supply for any commodity depend on its own price much more than on the prices of other commodities. Therefore, adjusting partial equilibria recursively, we move the overall market closer and closer to its equilibrium.

To aid in understanding the iterative process of the price adjustments, we may imagine an auctioneer who at the beginning of the marketing day announces a set of prices to the market participants and then collects transaction offers from them. If the offers do not match, the announced prices are not the equilibrium prices, and he calls another set according to a rule based upon the law of supply and demand. The auctioneer raises the price of a commodity for which demand exceeds supply and decreases it in the opposite case. No transactions are allowed to take place at the nonequilibrium set of prices. Following the arguments of Walras, we conclude that the auctioneer may never be able to get the transaction offers to match, but applying his rule he will drive the economy closer and closer to the equilibrium.

Although intuitively appealing, the arguments of Walras concerning the "price mechanism" in the working of a decentralized economy cannot be accepted as rigorous. This fact motivated Hicks (1939) to use the shapes of the supply and demand functions in the neighborhood of equilibrium and derive stability conditions for convergence of the price adjustment process to the equilibrium. He proposed a set of determinantal inequalities as conditions for a linear market model to be stable, and those conditions will be used throughout this exposition.

The stability analyses carried out by Walras, Hicks, and some others (see Negishi, 1962) were purely static and did not correspond to an explicit dynamic model of the market. This situation was rectified by Samuelson (1947), who introduced time into competitive-equilibrium models in an essential way by identifying *tâtonnements* as solutions of the system of differential equations

$$\frac{dp_i}{dt} = k_i f_i(p_1, p_2, \ldots, p_n), \qquad i = 1, 2, \ldots, n, \tag{4.12}$$

where $p_i = p_i(t)$, $f_i(p_1, p_2, \ldots, p_n)$, and $k_i > 0$ are the price, the excess demand function, and the coefficient of price adjustment of the ith commodity, respectively. We note immediately that the system of differential equations (4.12) is a "dynamic representation" of the classical law of supply and demand, since $f_i(p_1, p_2, \ldots, p_n) > 0$ implies a rise in the price p_i, and $f_i(p_1, p_2, \ldots, p_n) < 0$ implies a fall in the price p_i. Therefore, we expect the solutions $p_i(t)$, $i = 1, 2, \ldots, n$, to mimic the price adjustment process which we believe goes on in actual markets.

There are several comments that we should make at this time. First, we should mention that the existence of solutions of (4.12) for all times is obviously a prerequisite for the possibility of *tâtonnement* and its stability properties. Furthermore, it would be equally awkward if the uniqueness of solutions were to fail, so that for given initial prices $p_{i0} = p_i(t_0)$, $i = 1, 2, \ldots, n$, there were a number of possible solutions $p_i(t)$, $i = 1, 2, \ldots, n$, for $t \geqslant t_0$. It is logical, therefore, to assume that excess demand functions $f_i(p_1, p_2, \ldots, p_n)$ are sufficiently smooth to guarantee the existence and uniqueness of solutions to the system of equations (4.12). This assumption would be in agreement with the utility-maximization process which produced the excess demand functions in the previous section, because the implicit-function theorem asserts the continuity of these functions and the existence of continuous first partial derivatives. This fact allows us also to take for granted that the solutions of (4.12) are continuous with respect to initial conditions t_0, p_{i0}, $i = 1, 2, \ldots, n$.

It is also possible (Arrow and Hahn, 1971) to make an analysis in the case when prices are announced by the auctioneer in terms of the *numéraire* which is used as a unit of accounting and transaction. Then we consider a dynamic model given as

$$\begin{aligned} \frac{dP_0}{dt} &\equiv 0, \\ \frac{dP_i}{dt} &= K_i F_i(P_0, P_1, \ldots, P_n), \qquad i = 1, 2, \ldots, n, \end{aligned} \tag{4.13}$$

where the functions $F_i(P_0, P_1, \ldots, P_n)$ are defined by (4.8).

Finally, we should make some comments on the role of time in the model (4.12). Obviously, in postulating the model and auctioneer's rules, we assumed that when prices change they do so in continuous time. This assumption is made for convenience, and it is fairly clear that a *tâtonnement* process can also be described in terms of difference equations and other types of functional equations. It is only for descriptive reasons that the *tâtonnement* process was formulated as if it were taking place in real time. The process could equally well be envisioned as an iterative one, whose steps have no direct relation to real time.

Our primary interest in the following consideration of competitive equilibrium is to remove the assumptions that tastes of consumers, technology, weather, etc., remain constant during the price adjustment process, and thus treat cases when, in general, the basic underlying nature of the economy is not assumed to be stationary. This generalization of competitive-equilibrium models requires that time appear explicitly in aggregate excess demand functions on the right side of Equation (4.12). That is, we consider a model of a competitive market economy to be given as a system of differential equations

$$\frac{dp_i}{dt} = f_i(t, p_1, p_2, \ldots, p_n), \qquad i = 1, 2, \ldots, n, \tag{4.14}$$

where again $p_i = p_i(t)$, $i = 1, 2, \ldots, n$, are prices of the n commodities. For convenience, the coefficients of the speeds of adjustment are assumed equal to unity for all commodities (that is, $k_i = 1$, $i = 1, 2, \ldots, n$). The time t in the excess-demand functions $f_i(t, p_1, p_2, \ldots, p_n)$, which is the same as that appearing implicitly in the derivative dp_i/dt, will be used to describe various nonstationary phenomena that may take place in the economy, such as shifts in excess-demand schedules, inaccuracy of the *tâtonnement* process, reduction in the number of commodities during adjustment of prices, etc. All these situations will be discussed separately in the course of this exposition.

It should be noted that nonstationary models—that is, models which involve time explicitly as that of (4.14)—were introduced in economic analysis by Samuelson (1947) to discuss the *moving equilibrium* for price. Considerations of nonstationary linear models were made later by Arrow (1966) to study multiple markets with steadily increasing demands. An extensive study of nonstationary nonlinear models of the general type (4.14) were initiated recently by Šiljak (1975c, 1976a). These results will be outlined in the following sections, except for the competitive analysis of nonnormalized processes, which can be found in Šiljak (1976b).

4.3. CONNECTIVE STABILITY: LINEAR CASE

The objective of this section is to introduce the concept of connective stability in competitive equilibrium analysis via linear constant models. These models can be obtained from the general nonlinear stationary models formulated in the preceeding section by applying linearization.

Let us consider a stationary version of the market model (4.14) in the vector form

$$\dot{p} = f(p), \tag{4.15}$$

where $p \in \mathfrak{R}^n$ is the price vector and $f\colon \mathfrak{R}^n \to \mathfrak{R}^n$ is the vector of aggregate excess-demand functions. We assume that $f(p)$ has continuous first derivatives and that $f(p^*) \equiv 0$, so that p^* is an equilibrium price of the market (4.15). Expanding $f(p)$ about p^* in a Taylor series and substituting $p = p^* + q$, we obtain (4.15) as

$$\dot{q} = Aq + b + g(q), \tag{4.16}$$

where $A = [\partial f_i(p^*)/\partial p_j]$ is a constant $n \times n$ Jacobian matrix, $b = f(p^*)$, and $g(q)$ denotes the higher-order terms in the Taylor series expansion. To study the local properties of the nonlinear model (4.15) in the neighborhood of the equilibrium p^*, we can use the linear model

$$\dot{p} = Ap + b, \tag{4.17}$$

which is obtained from (4.16) when the higher-order terms of the Taylor expansion are neglected and q is replaced by p. We use the model (4.17) to introduce the basic idea of structural perturbations and connective stability.

We assume that all the commodities are substitutes, so that rise in the price on any market decreases the demand in that market, but increases the demand in some (or all) other markets. This amounts to saying that the elements of the matrix $A = (a_{ij})$ in (4.17) are such that

$$a_{ij} \begin{cases} < 0, & i = j, \\ \geqslant 0, & i \neq j. \end{cases} \tag{4.18}$$

Since A has nonnegative off-diagonal elements, it is a Metzler matrix (Arrow, 1966; Newman, 1959). The gross-substitute case is of special importance in the context of the model (4.17), and various rich properties of the related Metzler matrices, which we are going to use throughout this exposition, are listed in the Appendix. We also assume for a moment that the vector b in (4.17) is nonnegative ($b \geqslant 0$), by which we mean $b_i \geqslant 0$, $i = 1, 2, \ldots, n$.

As shown by Arrow (1966), solutions $p(t; t_0, p_0)$ are nonnegative whenever they "start nonnegative". That is, for all initial conditions $t_0, p_0 \geqslant 0$, we have

$$p(t; t_0, p_0) \geqslant 0, \qquad t \geqslant t_0. \tag{4.19}$$

The equilibrium price of the market is a constant solution of Equation (4.17) determined by the algebraic equation

$$Ap + b = 0. \tag{4.20}$$

If $\det A \neq 0$, the equilibrium p^* is a constant vector given as

$$p^* = -A^{-1}b, \tag{4.21}$$

which is the unique solution of (4.20).

By using a simple transformation

$$q = p - p^*, \tag{4.22}$$

(4.17) can be rewritten as

$$\dot{q} = Aq, \tag{4.23}$$

and stability of the price equilibrium p^* of the market model (4.17) implies and is implied by stability of the equilibrium $q^* = 0$ of (4.23). Stability of $q^* = 0$ in (4.23) is equivalent to stability of A (eigenvalues of A all have negative real parts). It is well known (Newman, 1959) that a Metzler matrix is stable if and only if it is a Hicks matrix, that is, all its odd-order principal minors are negative and all its even-order principal minors are positive (Theorem A.2). That we can use only leading principal minors to test for stability of a Metzler matrix is also well known (Gantmacher, 1960), and the conditions

$$(-1)^k \begin{vmatrix} a_{11} & a_{12} & \cdots & a_{1k} \\ a_{21} & a_{22} & \cdots & a_{2k} \\ \cdots & \cdots & \cdots & \cdots \\ a_{k1} & a_{k2} & \cdots & a_{kk} \end{vmatrix} > 0, \qquad k = 1, 2, \ldots, n, \tag{4.24}$$

are both necessary and sufficient for stability of A (Theorem A.9). These conditions were obtained independently by Kotelyanskii (1952), using the results of Sevastyanov (1951), and by Hawkins and Simon (1949).

For a Metzler matrix A with negative diagonal elements as in (4.18), the conditions (4.24) are equivalent (Newman, 1959) to its quasidominant diagonal property, which amounts to saying that there exist numbers $d_j > 0, j = 1, 2, \ldots, n$, such that

$$d_j |a_{jj}| > \sum_{\substack{i=1 \\ i \neq j}}^{n} d_i |a_{ij}|, \qquad j = 1, 2, \ldots, n. \tag{4.25}$$

The last inequality in (4.24) implies that $\det A \neq 0$, and stability of A implies uniqueness of the equilibrium price p^* given by (4.21). Furthermore, it can also be shown that stability of A is equivalent to saying that A^{-1} is nonpositive element by element. Since A^{-1} cannot have a zero row (it is nonsingular), nonpositivity of A^{-1} in (4.21) means that for any positive b (that is, $b > 0$) we have a positive equilibrium p^*.

Now, if $A = (a_{ij})$ satisfies (4.25) and is a stable matrix, then the system (4.17) would remain stable even if some of the nonzero off-diagonal elements a_{ij} $(i \neq j)$ were to go to zero. This fact is obvious from (4.25).

Furthermore, since the perturbed matrix is still a stable Metzler matrix, the price adjustment process is nonnegative, and so is the equilibrium price of the system (4.17). A systematic investigation of such structural perturbations of the matrix A in (4.17) can be carried out in the framework of connective stability, which is considered next.

Let us reconsider the system (4.17) in the form

$$\dot{p} = \bar{A}p + b, \tag{4.26}$$

where $p(t)$ is again the price vector, and $\bar{A} = (\bar{a}_{ij})$ is the $n \times n$ constant matrix with elements defined as

$$\bar{a}_{ij} = \begin{cases} -\alpha_i + \bar{e}_{ii}\,\alpha_{ii}, & i = j, \\ \bar{e}_{ij}\,\alpha_{ij}, & i \neq j. \end{cases} \tag{4.27}$$

Here α_i, $\alpha_{ij} \geqslant 0$ are numbers, and $\bar{e}_{ij}$ are binary elements of the $n \times n$ fundamental interconnection matrix $\bar{E} = (\bar{e}_{ij})$ defined as

$$\bar{e}_{ij} = \begin{cases} 1, & \alpha_{ij} \neq 0, \\ 0, & \alpha_{ij} = 0. \end{cases} \tag{4.28}$$

By $E = (e_{ij})$ we denote a constant $n \times n$ interconnection matrix which is obtained from the fundamental interconnection matrix $\bar{E} = (\bar{e}_{ij})$ by replacing the unit elements $\bar{e}_{ij}$ with numbers e_{ij} such that

$$0 \leqslant e_{ij} \leqslant 1, \qquad i, j = 1, 2, \ldots, n, \tag{4.29}$$

while the zero elements $\bar{e}_{ij}$ remain the zero elements e_{ij} in the matrix E. As in the preceeding chapters, by $E \in \bar{E}$ we denote the fact that E is generated by $\bar{E}$. We also recall that $\bar{E} \in \bar{E}$, since we can choose $E = \bar{E}$.

We note immediately that for each $E \in \bar{E}$ we have a different system (4.17), which we denote by $\mathbb{S}_{(4.17)}$, since the constant $n \times n$ matrix $A = (a_{ij})$ is now defined as

$$a_{ij} = \begin{cases} -\alpha_i + e_{ii}\,\alpha_{ii}, & i = j, \\ e_{ij}\,\alpha_{ij}, & i \neq j. \end{cases} \tag{4.30}$$

The underlying idea of the concept of connective stability is the possibility of establishing the stability of a class of systems $\mathbb{S}_{(4.17)}$ obtained for $E \in \bar{E}$, by proving the stability of one member of that class which corresponds to $\bar{E}$, namely, the system represented by (4.26). Thus, we need the following:

Definition 4.1. *The system $\mathbb{S}_{(4.17)}$ is connectively stable if the equilibrium $p^* = -A^{-1}b$ is stable in the sense of Liapunov for all $E \in \bar{E}$.*

Comparing Definition 4.1 with the definitions of connective stability formulated in Chapter 2, we note that in Definition 4.1 we say that "the system $\mathbb{S}_{(4.17)}$ is connectively stable" instead of "the equilibrium p^* is connectively stable". This modification is necessary because in the system (4.17), for each $E \in \bar{E}$ we have a distinct equilibrium p^* which is, in general, different from that of (4.26),

$$\bar{p}^* = -\bar{A}^{-1}b. \tag{4.31}$$

However, our stability conditions will establish the stability of each equilibrium p^* by showing the stability of $\bar{p}^*$ only.

Before we engage in analyzing the connective stability of $\mathbb{S}_{(4.17)}$, let us consider the effect on the equilibrium price p^* of changing the strength of interactions between individual markets measured by the elements e_{ij} of the interconnection matrix E. We note first that from (4.27), (4.29), and (4.30), we have

$$A \leqslant \bar{A}, \qquad E \in \bar{E}, \tag{4.32}$$

where the inequality is taken element by element ($A - \bar{A} \leqslant 0$). It is a known fact that for Metzler matrices A, $\bar{A}$, the inequality (4.32) means that if $\bar{A}$ satisfies the conditions (4.24), so does A. This is obvious from the equivalent conditions (4.25). The last inequality in (4.24) implies that $\det A \neq 0$ and $p^* = -A^{-1}b$ exists for all $E \in \bar{E}$. We can go a step further and actually show that for stable market system $\mathbb{S}_{(4.17)}$, the weaker the interactions among individual markets, the smaller the equilibrium price. In other words, we have the following:

Theorem 4.1. *If $\bar{p}^*$ is stable, then*

$$p^* \leqslant \bar{p}^*, \qquad E \in \bar{E}. \tag{4.33}$$

Proof. We show that $\bar{p}^*$ is stable if and only if the Metzler matrix $\bar{A}$ is a Hicks matrix. This and the inequality (4.32) imply that (Theorem A.4)

$$\bar{A}^{-1} \leqslant A^{-1} \leqslant 0, \qquad E \in \bar{E}, \tag{4.34}$$

where again the inequalities are taken element by element. Since $\bar{A}^{-1}$, A^{-1} are nonsingular, they cannot have a row of zeros, and from (4.21), (4.31), and (4.34) we conclude (4.33). This proves Theorem 4.1.

The Metzlerian structure of the matrices A, $\bar{A}$ allows us to establish an even stronger result than that of Theorem 4.1:

Theorem 4.2. *If $\bar{p}^*$ is stable, then*

$$p(t; t_0, p_0) \leqslant \bar{p}(t; t_0, \bar{p}_0), \qquad t \geqslant t_0, \quad E \in \bar{E}, \tag{4.35}$$

whenever $p_0 \leqslant \bar{p}_0$.

Proof. By using the simple transformation

$$q = \bar{p} - p, \tag{4.36}$$

from (4.17) and (4.26) we obtain the equation

$$\dot{q} = Bq, \tag{4.37}$$

where $B = \bar{A} - A$ is again a Metzler matrix due to (4.32). Therefore, $p_0 \leqslant \bar{p}_0$ implies $q_0 \geqslant 0$, and we have $q(t; t_0, q_0) \geqslant 0$, $t \geqslant t_0$, as in (4.19). Nonnegativity of $q(t; t_0, q_0)$ for all $t \geqslant t_0$ and $E \in \bar{E}$ implies (4.35) whenever $p_0 \leqslant \bar{p}_0$. This proves Theorem 4.2.

An interesting interpretation of Theorems 4.1 and 4.2 can be given in the case where the kth commodity "disappears" from the market. When the price of a commodity becomes negative, then the agent selling that commodity has to give up at the same time units of some other commodity. If the agent can "throw away" commodities without using up units of other commodities, then he will prefer this option of "free disposal", and negative prices cannot arise. A reduction of the number of commodities was considered by Arrow and Hurwicz (1962) in the context of the nonlinear model (4.15). Under certain conditions imposed on the excess demand functions of free commodities, they concluded that properties of the whole market remains valid for the reduced market. Our interest here is in considering linear models in which a reduction of the commodity space can take place and extending the conclusion of Arrow and Hurwicz to include stability.

By using interconnection matrices, we can describe a structural perturbation caused by the disappearance of the kth commodity by setting

$$e_{ik} = e_{kj} = 0, \qquad i, j = 1, 2, \ldots, n. \tag{4.38}$$

From Theorem 4.1, we conclude that the equilibrium price p^* of the reduced market will be always smaller than the equilibrium price $\bar{p}^*$ of the original stable market system. Theorem 4.2 tells us that price adjustment process $p(t; t_0, p_0)$ will have the same property for the two market systems. The immediate question is then: can we prove a similar result for stability properties of the whole and the reduced markets? To this effect, we prove the following:

Theorem 4.3. *The system $\mathbb{S}_{(4.17)}$ is connectively asymptotically stable in the large if and only if the Metzler matrix $\bar{A}$ is a Hicks matrix.*

Proof. As in the proof of Theorem 4.1, we use the fact (see Appendix) that for Metzler matrices A, $\bar{A}$ the inequality (4.32) implies that A is a stable matrix (that is, a Hicks matrix) if and only if $\bar{A}$ is. This proves the "if" part of the theorem. To establish the "only if" part, we notice that if $\bar{A}$ is not a Hicks matrix, then it is not a stable matrix, and therefore the system (4.26) corresponding to $\bar{E} \in \bar{E}$ is not a stable system. Therefore, $\mathbb{S}_{(4.17)}$ is not a stable system for all $E \in \bar{E}$, and the proof of Theorem 4.3 is complete.

An immediate conclusion is that in case of the structural perturbation (4.38), stability of the original market system implies stability of the reduced market system. It is obvious that this conclusion holds even when a number of commodities disappear from a given market system. This fact was observed by Lange (1945) when he established the notion of total stability. In the connective-stability framework, we are not restricted to the case (4.38), but can treat any case described by $E \in \bar{E}$. Furthermore, we can extend our results to nonlinear models as well, thus giving considerable generality to the structural perturbations of the multiple market systems.

4.4. CONNECTIVE STABILITY: NONLINEAR CASE

In this section, we show that a stable competitive equilibrium is robust and can tolerate not only the structural perturbations, but also a wide range of nonlinearities in the interactions among the individual markets. This task however, requires a more refined analysis.

Let us consider a market described by nonlinear differential equations (4.14) written in the vector form

$$\dot{p} = f(p), \tag{4.39}$$

where $p(t) \in \mathfrak{R}^n$ is the price vector; the excess-demand function f: $\mathfrak{R}^n \to \mathfrak{R}^n$ is defined, bounded, and continuous on the domain $\mathfrak{R}^n$, so that solutions $p(t; t_0, p_0)$ of (4.39) exist for all initial conditions $(t_0, p_0) \in \mathfrak{T} \times \mathfrak{R}^n$ and $t \in \mathfrak{T}_0$. We recall that $\mathfrak{T}$ is the time interval $(\tau, +\infty)$, where τ is a number or the symbol $-\infty$, and $\mathfrak{T}_0$ is the semi-infinite time value $[t_0, +\infty)$.

Since we would like to use the effective diagonal-dominance conditions (4.25) for nonlinear models, it is convenient to consider (4.39) in the form

$$\dot{p} = A(p)p, \tag{4.40}$$

where $p(t) \in \mathfrak{R}^n$, and in the excess-demand function $A(p)p$ the matrix function A: $\mathfrak{R}^n \to \mathfrak{R}^{n^2}$ is defined, bounded, and continuous on $\mathfrak{R}^n$, so that solutions $p(t; t_0, p_0)$ of (4.40) exist for all $(t_0, p_0) \in \mathcal{T} \times \mathcal{R}^n$ and all $t \in \mathcal{T}_0$. It is obvious that we can always choose the $n \times n$ matrix $A\cdot = [a_{ij}(p)]$ as a diagonal matrix, $A(p) = \text{diag}\{f_1(p)/p_1, f_2(p)/p_2, \ldots, f_n(p)/p_n\}$, but this choice is by no means unique. For example,

$$
\begin{aligned}
f(p) &= \begin{bmatrix} -2p_1|p_2| \\ -3p_2|p_1| \end{bmatrix} = \begin{bmatrix} -2|p_2| & 0 \\ 0 & -3|p_1| \end{bmatrix} \begin{bmatrix} p_1 \\ p_2 \end{bmatrix} \\
&= \begin{bmatrix} -|p_2| & -p_1 \operatorname{sgn} p_2 \\ -2p_2 \operatorname{sgn} p_1 & -|p_1| \end{bmatrix} \begin{bmatrix} p_1 \\ p_2 \end{bmatrix},
\end{aligned} \tag{4.41}
$$

where sgn $0 = 0$.

We assume that there exists an equilibrium price $p^* > 0$ of (4.39), so that $f(p^*) = 0$ for all $t \in \mathcal{T}$. To consider a model with the equilibrium at the origin, we use the transformation $q = p - p^*$ in (4.39), and get $\dot{q} = f(q + p^*)$. Denoting $g(q) \equiv f(q + p^*)$, we get the system (4.39) as $\dot{q} = g(q)$, where the equilibrium $p^* > 0$ of (4.39) is represented by the equilibrium $q^* = 0$, since $g(0) \equiv f(p^*) \equiv 0$. With the system $\dot{q} = g(q)$ we associate the system $\dot{q} = B(q)q$ in the same way as before. For convenience, we replace q by p and B by A, and consider again the system (4.40). It is important to note, however, that in this new interpretation of (4.40) we have to allow the price p to be negative, with the understanding that in the original system this means that $p(t; t_0, p_0)$ is below the equilibrium price $p^* > 0$.

In order to establish conditions for asymptotic connective stability of the market (4.40), we define the coefficients $a_{ij}(p)$ of the matrix $A(p)$ as

$$
a_{ij}(p) = \begin{cases} -\varphi_i(p) + e_{ii}\varphi_{ii}(p), & i = j, \\ e_{ij}\varphi_{ij}(p), & i \neq j, \end{cases} \tag{4.42}
$$

where $\varphi_i(p), \varphi_{ij}(p) \in C(\mathcal{R}^n)$ are nonlinear functions which represent the nonlinear interdependence among the individual markets. In (4.42), the elements e_{ij} of the interconnections matrices $E \in \bar{E}$ are assumed to be constant.

We further assume that there exist numbers $\alpha_{ij} \geqslant 0$, $\alpha_i > \alpha_{ii} \geqslant 0$ such that

$$
\varphi_i(p)|p_i| \geqslant \alpha_i \phi_i(|p_i|), \qquad \varphi_{ij}(p)p_j \leqslant \alpha_{ij}\phi_j(|p_j|), \qquad i, j = 1, 2, \ldots, n, \qquad \forall p \in \mathcal{R}^n. \tag{4.43}
$$

Here $\phi_i\colon \mathcal{R}_+ \to \mathcal{R}_+$ are comparison functions of the class $\mathcal{K}$: $\phi_i(\zeta) \in C(\mathcal{R}_+)$; $\phi_i(0) = 0$; and $\phi_i(\zeta_1) < \phi_i(\zeta_2)$ for all $\zeta_1, \zeta_2\colon 0 \leqslant \zeta_1 < \zeta_2 < +\infty$ (see Definition 2.11).

Since in the case of the system (4.40) we consider the equilibrium price $p^* = 0$ fixed during the adjustment process, we need a modification of Definition 4.1, which is stated as follows:

Definition 4.2. *The equilibrium price* $p^* = 0$ *of the market* (4.40) *is asymptotically connectively stable in the large if it is asymptotically stable in the large for all* $E \in \bar{E}$.

To establish this kind of stability for a nonlinear nonstationary market (4.40), we can use the following (Šiljak, 1976a):

Theorem 4.4. *The equilibrium price* $p^* = 0$ *of the market* (4.40) *is asymptotically connectively stable in the large if the* $n \times n$ *constant Metzler matrix* $\bar{A} = (\bar{a}_{ij})$ *defined by* (4.27) *is a Hicks matrix.*

Proof. Consider the function $\nu: \mathcal{R}^n \to \mathcal{R}_+$,

$$\nu(p) = \sum_{i=1}^{n} d_i |p_i|, \tag{4.44}$$

as a candidate for Liapunov's function, where $d_i > 0$, $i = 1, 2, \ldots, n$, are yet unspecified numbers. As proposed by Rosenbrock (1963), we define the functional σ_i as

$$\sigma_i = \begin{cases} 1 & \text{if } p_i(t) > 0 \text{ or if } p_i(t) = 0 \text{ and } \dot{p}_i(t) > 0, \\ 0 & \text{if } p_i(t) = 0 \text{ and } \dot{p}(t) = 0, \\ -1 & \text{if } p_i(t) < 0 \text{ or if } p_i(t) = 0 \text{ and } \dot{p}_i(t) < 0, \end{cases} \tag{4.45}$$

and calculate the right-hand derivative $D^+\nu(p)$ with respect to (4.40) to get

$$\begin{aligned} D^+\nu(p)_{(4.40)} &= \sum_{i=1}^{n} d_i \sigma_i \dot{p}_i \\ &= \sum_{i=1}^{n} d_i \sigma_i \sum_{j=1}^{n} a_{ij} p_j \\ &= \sum_{j=1}^{n} d_j \sigma_j p_j a_{jj} + \sum_{j=1}^{n} p_j \sum_{\substack{i=1 \\ i \neq j}}^{n} d_i \sigma_i a_{ij} \\ &\leqslant - \sum_{j=1}^{n} d_j \phi_j(|p_j|)\, |\bar{a}_{jj}| + \sum_{j=1}^{n} \phi_j(|p_j|) \sum_{\substack{i=1 \\ i \neq j}}^{n} d_i |\bar{a}_{ij}| \\ &\leqslant - \sum_{j=1}^{n} d_j \Big(|\bar{a}_{jj}| - d_j^{-1} \sum_{\substack{i=1 \\ i \neq j}}^{n} d_i |\bar{a}_{ij}|\Big) \phi_j(|p_j|) \qquad \forall (t,p) \in \mathcal{T} \times \mathcal{R}^n, \quad \forall E \in \bar{E}. \end{aligned} \tag{4.46}$$

Since $\bar{A}$ is a Metzler matrix, the fact that it is also a Hicks matrix is equivalent to saying that it is also a quasidominant diagonal matrix. Hence there exist positive numbers d_i, π such that

$$|\bar{a}_{jj}| - d_j^{-1} \sum_{\substack{i=1 \\ i \neq j}}^{n} d_i |\bar{a}_{ij}| \geqslant \pi, \qquad j = 1, 2, \ldots, n. \tag{4.47}$$

From (4.46) and (4.47), we get the differential inequality

$$D^+ \nu(p)_{(4.40)} \leqslant -\pi \sum_{j=1}^{n} d_j \phi_j(|p_j|) \qquad \forall (t,p) \in \mathcal{T} \times \mathcal{R}^n, \quad \forall E \in \bar{E}, \tag{4.48}$$

and write

$$\begin{aligned} \phi_1(\|p\|) \leqslant \nu(p) \leqslant \phi_2(\|p\|), \\ D^+ \nu(p)_{(4.40)} \leqslant -\phi_3(\|p\|), \end{aligned} \qquad \forall (t,p) \in \mathcal{T} \times \mathcal{R}^n, \quad \forall E \in \bar{E}, \tag{4.49}$$

where

$$\phi_1(\|p\|) = d_m \|p\|, \; \phi_2(\|p\|) = n^{1/2} d_M \|p\|, \; \phi_3(\|p\|) = \pi \textstyle\sum_{j=1}^{n} d_j \phi_j(|p_j|),$$

and $d_m = \min_i d_i$, $d_M = \max_i d_i$. In deriving (4.49), we have used the well-known relationship $\|p\| \leqslant |p| \leqslant n^{1/2} \|p\|$ between the Euclidean norm $\|p\| = \sum_{i=1}^{n} p_i^2$ and the absolute-value norm $|p| = \sum_{i=1}^{n} |p_i|$.

It is a well-known fact (see Theorem 2.7) that the inequalities (4.49) imply global asymptotic stability of $p^* = 0$. Since the inequalities (4.49) are valid for all $E \in \bar{E}$, the stability is also connective. This proves Theorem 4.4.

In connection with Theorem 4.4, we should mention the fact that diagonally dominant market models which are described by the equation (4.40) were already considered by Arrow, Block, and Hurwicz (1959) and Arrow and Hahn (1971) by interpreting the matrix $A(p)$ as the Jacobian corresponding to the system (4.15). Arrow and Hahn wrote: "The kind of result that we need here is one that would allow us to deduce global stability from the postulate that the Jacobian of excess supplies has everywhere DD [diagonal dominance]. No such result is available, but neither are counter-examples". Theorem 4.4 confirms this conjecture. To see this, we assume that $f(p) \in (C^1 \mathcal{R}^n)$ and recall (see Demidovich, 1967) that by setting $q = p + \mu h$ $(0 \leqslant \mu \leqslant 1)$, we can write

$$f(p+h) - f(p) = \int_0^1 \frac{d}{d\mu}[f(q)]\, d\mu = \int_0^1 J(q) h \, d\mu, \tag{4.50}$$

where $J = (\partial f_i / \partial p_j)$ is the $n \times n$ Jacobian matrix of $f(p)$. When $p^* = 0$ and $f(0) \equiv 0$, we can choose $A(p) = \int_0^1 J(\mu p)\, d\mu$ to represent (4.39) by (4.40). Now if $J(p)$ is negative quasidominant diagonal for all p, so also is the matrix $A(p)$, and by Theorem 4.4 the equilibrium price $p^* = 0$ is globally asymptotically stable. Therefore, we arrive at the following:

Corollary 4.1. *The equilibrium price $p^* = 0$ of the market* (4.39) *is asymptotically stable in the large if the $n \times n$ Jacobian matrix $J = (\partial f_i/\partial p_j)$ of the excess demand function $f(p)$ is negative quasidominant diagonal for all $p \neq 0$.*

We should also note that the representation of $f(p)$ by $A(p)p$ using the Jacobian $J(p)$ is neither unique nor best. We can form representations $A(p)p$ even in cases when $f(p)$ is not a continuously differentiable function in p and the Jacobian is undefined. This is illustrated by the example (4.41).

An alternative to the result of Corollary 4.1 is provided by the classical theorem of Krassovskii (1959). By using $\|f(p)\|$ as a Liapunov function, Krassovskii established global asymptotic stability of $p^* = 0$ by negative definiteness of the Jacobian $J(p)$ for all p. For extensions of Krassovskii's result to the so called "systems with convergence" introduced by Pliss (1964), one can use the book of Demidovich (1967). These extensions can be readily used to establish results similar to that of Corollary 4.1 for nonlinear nonstationary market models. It should be noted, however, that the diagonal-dominance condition is more appealing than positive definiteness, due to its straightforward economic interpretation.

We can improve the result of Theorem 4.4 in several different ways. By imitating Definition 2.8, but otherwise using a different system, we can prove both necessary and sufficient conditions for stability of (4.40) which is absolute, connective, and exponential. In order to do this, let us assume that the nonlinear functions φ_i, φ_{ij} in (4.42) belong to the classes of functions

$$\begin{aligned} \Phi_i &= \{\varphi_i(p)\colon \varphi_i(p) \geqslant \alpha_i\}, \\ \Phi_{ij} &= \{\varphi_{ij}(p)\colon |\varphi_{ij}(p)| \leqslant \alpha_{ij}\}, \qquad i, j = 1, 2, \ldots, n, \qquad \forall p \in \mathscr{R}^n, \end{aligned} \tag{4.51}$$

where again $\alpha_i > \alpha_{ii} \geqslant 0$, $\alpha_{ij} \geqslant 0$. The conditions (4.51) follow from those of (4.43) when the comparison functions $\phi_j(|p_j|)$ are chosen as $|p_j|$.

We want to show that stability of the constant Metzler matrix $\bar{A}$ is both necessary and sufficient for stability of the equilibrium price $p^* = 0$ of the market (4.40) for arbitrary functions $\varphi_i(p)$, $\varphi_{ij}(p)$ which belong to the classes Φ_i, Φ_{ij}, and for arbitrary matrices E generated by the matrix $\bar{E}$. The fact that the system is asymptotically stable in the large for all $\varphi_i \in \Phi_i$, $\varphi_{ij} \in \Phi_{ij}$ we acknowledge by the term "absolutely stable" (see Definition 2.8). We can do even better, as follows:

Definition 4.3. *The equilibrium price $p^* = 0$ of the market* (4.40) *is absolutely, exponentially, and connectively stable in the large if there exist two positive numbers Π and π independent of the initial conditions (t_0, p_0), such that*

$$\|p(t; t_0, p_0)\| \leqslant \Pi \|p_0\| \exp[-\pi(t - t_0)] \qquad \forall t \in \mathscr{T}_0 \tag{4.52}$$

for all $(t_0, p_0) \in \mathfrak{T} \times \mathfrak{R}^n$, *all* $\varphi_i \in \Phi_i$, $\varphi_{ij} \in \Phi_{ij}$, *and all* $E \in \bar{E}$.

To establish this kind of stability, we can use the following:

Theorem 4.5. *The equilibrium price* $p^* = 0$ *of the market* (4.40) *is absolutely, exponentially, and connectively stable in the large if and only if the* $n \times n$ *constant Metzler matrix* $\bar{A} = (\bar{a}_{ij})$ *is a Hicks matrix.*

Proof. The proof of the "if" part of the theorem follows the proof of Theorem 4.4. We again choose $\nu(p)$ as in (4.43), and as in (4.45) obtain the expression

$$D^+\nu(p)_{(4.40)} \leqslant -\sum_{j=1}^{n} d_j\Big(|\bar{a}_{jj}| - d_j^{-1}\sum_{\substack{i=1\\ i\neq j}}^{n} d_i|\bar{a}_{ij}|\Big)|p_j| \qquad \forall(t,p) \in \mathfrak{T}\times\mathfrak{R}^n, \qquad \forall E \in \bar{E}. \tag{4.53}$$

Since $\bar{A}$ is a Hicks matrix, using (4.46) we rewrite (4.53) as

$$D^+\nu(p)_{(4.40)} \leqslant -\pi\nu(p) \qquad \forall(t,p) \in \mathfrak{T}\times\mathfrak{R}^n, \quad \forall E \in \bar{E}. \tag{4.54}$$

By integrating (4.53), we obtain

$$\nu[p(t)] \leqslant \nu_0(p_0)\exp[-\pi(t - t_0)] \qquad \forall t \in \mathfrak{T}_0 \quad \forall(t,p) \in \mathfrak{T}\times\mathfrak{R}^n, \quad \forall E \in \bar{E}. \tag{4.55}$$

Using the well-known relationship

$$\|p\| \leqslant |p| \leqslant n^{1/2}\|p\| \tag{4.56}$$

between the Euclidean norm $\|p\| = \sum_{i=1}^{n} p_i^2$ and the absolute-value norm $|p| = \sum_{i=1}^{n} |p_i|$, we can rewrite (4.46) as (4.52) with

$$\Pi = n^{1/2} d_M d_m^{-1}, \; d_M = \max_i d_i, \; d_m = \min_i d_i. \tag{4.57}$$

This establishes the "if" part of Theorem 4.5.

To prove the "only if" part of the theorem, we need only to notice that in Definition 4.3 we require stability for all $E \in \bar{E}$, and thus for $\bar{E}$ too. If the Metzler matrix $\bar{A}$ is not a Hicks matrix, then it is not a stable matrix, and the system (4.40) corresponding to

$$\varphi_i(p) = \alpha_i, \; \varphi_{ij}(p) = \alpha_{ij}, \tag{4.58}$$

and $\bar{E}$ is unstable. This proves Theorem 4.5.

Several comments are now in order:

First, we notice that Theorem 4.5 applies to a "mixed" market of substitute and complement commodities, and, more importantly, it allows for a change of a commodity from a substitute to a complement for another commodity. Moreover, the validity of Theorem 4.5 does not rest explicitly on Walras's law (Arrow and Hahn, 1971) expressed by the budget constraint (4.2), or any other assumption concerning market excess demand.

Second, constraints on the nonlinear interactions $\varphi_{ij}(p)$ can represent a "leveling off" of the influence of the price p_j of the jth commodity on the price p_i of the ith commodity, when the price p_j is far from its equilibrium value p_j^*.

Third, we note that under the conditions of Theorem 4.5, the equilibrium price $p^* = 0$ of the system (4.40) is connectively and exponentially stable in the large even if the matrix function $A(p)$ is replaced by $K(p)A(p)$, where

$$K = \text{diag}\{k_{11}, k_{22}, \ldots, k_{nn}\} \tag{4.59}$$

is a diagonal matrix with elements $k_{ii}(p)$ which are functions of p bounded by arbitrary positive (but finite) numbers $\bar{k}_{ii}$. This is a consequence of the fact (see Appendix) that if a constant Metzler matrix $\bar{A} = (\bar{a}_{ij})$ with a negative diagonal is a Hicks matrix (that is, a quasidominant diagonal matrix), so is the matrix

$$\bar{K}\bar{A} = (\bar{k}_{ii}\bar{a}_{ij}), \tag{4.60}$$

which is obvious from (4.47). Therefore, when $\bar{A}$ is a Hicks matrix, the stability property assured by Theorem 4.5 holds for all speeds of adjustment of prices (as was established for the linear case by Metzler, 1945), but with possibly different degree π of exponential stability for different K-matrices.

Fourth, we recall that for Metzler matrices with negative diagonal the Hicksian conditions (4.24) are equivalent to the quasidominancy conditions (4.25) or (4.47). The latter conditions are more convenient for economic interpretation, especially when we have the special case in which all $d_i = 1$, so that

$$|a_{jj}| > \sum_{\substack{i=1 \\ i \neq j}}^{n} |a_{ij}|, \tag{4.61}$$

which can be accomplished by altering suitably the measurement of commodities. From (4.61), we see that stability of the market is assured when the price of any given commodity is more affected by the changes in its own price than by total absolute change in prices of other commodities. It is believed that this justifies the intuitive argument that Walras (1874) used to establish convergence of the *tâtonnement* process in his original investigations.

A number of extensions and applications of the results presented in this section are possible in various other areas and models. A competitive analysis of Richardson's (1960) model of the arms race was developed by Šiljak (1976b, 1977b) to study how formations of alliances and neutral countries affect the equilibrium and stability of the armament processes. Applications of the obtained results to pharmacokinetics models (Bellman, 1962) and compartmental systems have been proposed by Ladde (1976b, c). Further possibilities of using these results are in studying interactions in social groups along the lines of Simon (1957) and Sandberg (1974), as well as in the analysis of certain electronic circuits initiated by Sandberg (1969).

4.5. NONSTATIONARY MODELS: MOVING EQUILIBRIUM

In initiating the dynamic analysis of competitive equilibrium, Samuelson (1947) assumed that both supply and demand functions are explicit functions of time. Consequently, the price at which supply equals demand and the excess demand is zero becomes a function of time. For this price to be an equilibrium, it must be constant (equilibria are constant solutions of the corresponding differential equations). This case takes place only if the effect of time in the market is restricted to changes in the slopes of supply and demand characteristics. The situation in which shifts in demand cause a time variation in the price of zero excess demand may be termed a "moving equilibrium for price" (Samuelson, 1947). Samuelson used some simple examples to examine whether the adjustment process diverges from, follows, or reaches moving equilibrium, and thus initiated a stability analysis of competitive equilibrium under shifts in excess demand.

Steady upward shifts in time of demand functions on some or all of the interrelated markets were considered by Arrow (1966). He assumed that the demand is shifting upward in time and that the supply curve may do the same, but never more rapidly than the demand; he then showed that the prices rise at a rate that approaches a limiting value. The positive linear time function was chosen to represent trends in an otherwise stable, linear, and constant market model.

In this section, we will consider the shifts in demand and supply functions which have no specified sign or form except that they are bounded. We will show that in stable market systems under bounded shifts the role of the equilibrium is played by a compact region, and prices on all the markets are ultimately bounded globally with respect to that region. That is, all prices reach the region in a finite time, and once in the region, they stay there for all future times. This property of the price adjustment process will be established for a nonlinear and time-varying model studied in the preceeding section. We will provide an upper estimate of the above mentioned region by means of the same Liapunov function used to determine global stability properties of the model. The estimate is directly proportional to the size of the shifts in the excess-demand functions. Furthermore, the estimate of the region is invariant under structural perturbations, and the adjustment process is again exponential—prices on all the markets reach the region faster than exponentially despite structural changes in the models.

We continue to consider the price adjustment model of type (4.40),

$$\dot{p} = A(t,p)p + b(t,p), \tag{4.62}$$

where a function $b\colon \mathfrak{T} \times \mathfrak{R}^n \to \mathfrak{R}^n$ is added on the right side of (4.62). The function $b(t,p)$ has components of the form

$$b_i(t,p) = l_i(t)\psi_i(t,p), \tag{4.63}$$

where $l_i(t)$ are components of the interconnection vector $l(t) = [l_1(t), l_2(t), \ldots, l_n(t)]^T$, such that $l_i(t) \in [0,1]$ for all $t \in \mathfrak{T}$. Similarly, as in the case of the matrix $\bar{E}$, we define the binary vector $\bar{l} \in \mathfrak{R}_+^n$ as

$$\bar{l}_i = \begin{cases} 1, & \text{there is a demand shift on the } i\text{th market: } \psi_i(t,p) \not\equiv 0, \\ 0, & \text{there is no demand shift on the } i\text{th market: } \psi_i(t,p) \equiv 0, \end{cases} \tag{4.64}$$

and denote by $l(t) \in \bar{l}$ all vectors obtained from $\bar{l}$ by replacing unit elements with the corresponding functions $l_i(t)$.

In (4.63), the functions $\psi_i(t,p) \in C(\mathfrak{T} \times \mathfrak{R}^n)$ satisfy the conditions

$$|\psi_i(t,p)| \leqslant \beta_i, \tag{4.65}$$

where the β_i's are nonnegative numbers. Furthermore, we define a constant vector $\bar{b} \in \mathfrak{R}_+^n$ as

$$\bar{b}_i = \bar{l}_i \beta_i. \tag{4.66}$$

With the system (4.62) we associate a compact region

$$\mathfrak{P} = \{p \in \mathfrak{R}^n\colon \|p\| \leqslant \Gamma\}, \tag{4.67}$$

where Γ is a nonnegative number. In the absence of shifts, $b(t,p) \equiv 0$ and the region $\mathfrak{P}$ is reduced to the equilibrium $p^* = 0$, that is, $\Gamma = 0$. By $\mathfrak{P}^c$ we denote the complement of the region $\mathfrak{P}$.

Now we state a connective version (Šiljak, 1975b) of ultimate boundedness (Yoshizava, 1966) as

Definition 4.4. *The price adjustment process $p(t; t_0, p_0)$ of the market* (4.62) *is connectively exponentially and ultimately bounded in the large with respect*

to the region $\mathcal{P} = \{p \in \mathcal{R}^n\colon \|p\| \leqslant \Gamma\}$ *if and only if there exist positive numbers* $\gamma < \Gamma$, Π, *and* π *independent of the initial conditions* (t_0, p_0) *such that*

$$\|p(t; t_0, p_0)\| \leqslant \gamma + \Pi \|p_0\| \exp[-\pi(t - t_0)] \qquad \forall t \in \mathcal{T}_0 \tag{4.68}$$

for all $t_0 \in \mathcal{T}$, $p_0 \in \mathcal{P}^c$, *all interconnection matrices* $E(t) \in \bar{E}$, *and all vectors* $l(t) \in \bar{l}$.

As in the work of Arrow (1966), we shall assume that the system (4.62) would be stable in the absence of shifts, that is, if $b(t,p) \equiv 0$, so that (4.24) is valid for the matrix $\bar{A}$ defined in (4.27). We can prove the following:

Theorem 4.6. *The adjustment process* $p(t; t_0, p_0)$ *of the market* (4.62) *is connectively exponentially and ultimately bounded with respect to the region* $\mathcal{P} = \{p \in \mathcal{R}^n\colon \|p\| \leqslant \Gamma\}$, *where*

$$\Gamma = n^{-1/2} d_M^{-1} \pi^{-1} (\beta + \varepsilon), \tag{4.69}$$

$d_M = \max_i d_i$, $\beta = \sum_{i=1}^n d_i \bar{b}_i$, *and* $\varepsilon > 0$ *is an arbitrarily small number, if the* $n \times n$ *constant Metzler matrix* $\bar{A} = (\bar{a}_{ij})$ *defined in* (4.27) *is a Hicks matrix.*

Proof. We use again the function $\nu(p)$ defined in (4.44), and calculate $D^+\nu(p)$ with respect to the system (4.62) as

$$\begin{aligned} D^+\nu(p)_{(4.62)} &= \sum_{i=1}^n d_i \sigma_i \left(\sum_{j=1}^n a_{ij} p_j + b_i \right) \\ &\leqslant -\sum_{j=1}^n d_j |p_j| |\bar{a}_{jj}| + \sum_{j=1}^n |p_j| \sum_{\substack{i=1 \\ i \neq j}}^n d_i |\bar{a}_{ij}| + \sum_{i=1}^n d_i \bar{b}_i . \end{aligned} \tag{4.70}$$

The Hicksian property of $\bar{A}$ is equivalent to saying that $\bar{A}$ is a quasidominant diagonal matrix, as stated in (4.47). Using (4.47), we can rewrite (4.70) as

$$D^+\nu(p)_{(4.62)} \leqslant -\pi\nu(p) + \beta \qquad \forall (t,p) \in \mathcal{T} \times \mathcal{R}^n. \tag{4.71}$$

From (4.71), we conclude that the region $\mathcal{N}$ is reached in finite time by every solution $p(t; t_0, p_0)$ which starts in its complement $\mathcal{N}^c = \{p \in \mathcal{R}^n\colon \nu(p) \geqslant \pi^{-1}(\beta + \varepsilon)\}$, where $\varepsilon > 0$ can be chosen as an arbitrarily small number. This is because from (4.71) we have $D^+\nu(p)_{(4.62)} \leqslant -\varepsilon$ in $\mathcal{N}^c$. By integrating (4.71), we get

$$\nu[p(t)] \leqslant \pi^{-1}\beta + \nu(p_0) \exp[-\pi(t - t_0)] \qquad \forall t \in \mathcal{T}_0, \tag{4.72}$$

which is valid for all $t_0 \in \mathcal{T}$, $p_0 \in \mathcal{N}^c$, and all $E(t)$, $l(t)$. As in the case of (4.55), we obtain (4.68) where

$$\gamma = d_m^{-1}\pi^{-1}\beta, \tag{4.73}$$

and Π, π are the same as before. The region $\mathcal{P}$ is now determined by (4.69). Since (4.72) is valid for all $E(t) \in \overline{E}, l(t) \in \bar{l}$, so are Π, π, γ, Γ, and the region $\mathcal{P}$. This proves Theorem 4.6.

It is now simple to compute from (4.72) an upper estimate $t_1 = t_0 + \pi^{-1}\ln[\varepsilon^{-1}(\pi\nu_0 - \beta)]$ of the time necessary for the price adjustment process to enter the computed region $\mathcal{P}$.

By the above analysis of ultimate boundedness, we conclude that the price adjustment process enters a compact region, but we do not have any information about what the process does once in the region except that it stays there for all future times. Since our model has a forcing function to represent the demand shifts in time, we expect that the process under certain conditions converges to a curve in the price space which plays the role of equilibrium. At the cost of more refined analysis we can actually show that this intuitive proposition is true under pretty much the same conditions used to show stability of a single equilibrium price in the previous section.

We need first Demidovich's (1967) generalization of the notion of a convergent system

$$\dot{p} = f(t,p), \tag{4.74}$$

introduced by Pliss (1964):

Definition 4.5. *A system (4.74) is a system with convergence if:*

(1) *Solutions $p(t; t_0, p_0)$ are defined on $\mathcal{T} \times \mathcal{R}^n$;*

(2) *there exists a unique solution $p^*(t)$ which is bounded for all $t \in \mathcal{T}$, that is,*

$$\sup_{t\in\mathcal{T}} p^*(t) < +\infty; \tag{4.75}$$

(3) *the solution $p^*(t)$ is asymptotically stable in the large and*

$$\lim_{t\to+\infty}\{p(t; t_0, p_0) - p^*(t)\} = 0. \tag{4.76}$$

In the context of competitive systems, the solution $p^*(t)$ represents a limiting regime that is approached by all adjustment processes. In order to establish this fact by Liapunov's direct method, we need first the following Lemma due to Demidovich (1967):

Lemma 4.1. *Let $f(p) \in C^1(\mathcal{R}^n)$, and let $J_s = (s_{ij})$ be the corresponding symmetric $n \times n$ Jacobian matrix defined as*

$$s_{ij} = \frac{1}{2}\left(\frac{\partial f_i}{\partial p_j} + \frac{\partial f_j}{\partial p_i}\right) \tag{4.77}$$

with $\lambda(p)$ and $\Lambda(p)$ as the minimum and maximum eigenvalues of $J_s(p)$. Then

$$\lambda_m \|h\|^2 \leqslant [f(p+h) - f(p)]^T h \leqslant \Lambda_M \|h\|^2, \tag{4.78}$$

where

$$\lambda_m = \inf_{\mu \in [0,1]} \lambda(p + \mu h), \qquad \Lambda_M = \sup_{\mu \in [0,1]} \Lambda(p + \mu h). \tag{4.79}$$

Proof. By setting $q = p + \mu h$, $0 \leqslant \mu \leqslant 1$, we obtain

$$f(p+h) - f(p) = \int_0^1 \frac{d}{d\mu}[f(q)]\, d\mu = \int_0^1 J(q) h \, d\mu, \tag{4.80}$$

where $J = (\partial f_i / \partial p_j)$ is the $n \times n$ Jacobian matrix corresponding to $f(p)$. From (4.80) we get

$$[f(p+h) - f(p)]^T h = \int_0^1 h^T J(q) h \, d\mu. \tag{4.81}$$

Since $h^T J(q) h = h^T J_s(q) h$, (4.81) implies (4.78), and Lemma 4 is established.

We also need the following result of Luzin (see Demidovich, 1967):

Lemma 4.2. *Assume that in the system (4.74), $f(t,p) \in C^{(0,1)}(\mathfrak{T} \times \mathfrak{R}^n)$ and all solutions $p(t; t_0, p_0)$ of (4.74) exist and are unique on $\mathfrak{T} \times \mathfrak{R}^n$. Furthermore, $\|p_0\| = \rho$ implies $p(t; t_0, p_0) \in \mathfrak{R}_\rho = \{p \in \mathfrak{R}^n : \|p\| \leqslant \rho\}$ for all $t \in \mathfrak{T}_0$. Then there is at least one solution $p^*(t)$ of (4.74) defined and such that $\|p^*(t)\| < \rho$ for all $t \in \mathfrak{T}$.*

Proof. Let $\mathfrak{R}_\rho = \mathfrak{R}_0$, $t_0 = 0$, and define $\mathfrak{R}_1 = \{p_0 \in \mathfrak{R}^n : p(t; -1, p_0) \in \mathfrak{R}_\rho\}$. That is, $\mathfrak{R}_1$ is a set of all $p \in \mathfrak{R}^n$ with the property that for all solutions $p(t; 0, p_0)$ of (4.74), $p = p_0$ implies $p(t; -1, p_0) \in \mathfrak{R}_0$. Similarly define $\mathfrak{R}_k = \{p_0 \in \mathfrak{R}^n : p(t; -k, p_0) \in \mathfrak{R}_0\}$, $k = 1, 2, \ldots$. From the conditions of the Lemma, $\mathfrak{R}_0$ is an invariant set and $\mathfrak{R}_1 \subset \mathfrak{R}_0$. Consequently,

$$\mathfrak{R}_0 \supset \mathfrak{R}_1 \supset \mathfrak{R}_2 \supset \cdots. \tag{4.82}$$

Since the $\mathfrak{R}_k$'s are closed, it follows from (4.82) that there is a point

$$p_0^* = \bigcap_{k=0}^{\infty} \mathfrak{R}_k, \tag{4.83}$$

where $p_0^* \in \mathfrak{R}_0$.

We consider now the solution $p^*(t) = p(t; 0, p_0^*)$. Since $p_0^* \in \mathfrak{R}_k$, $k = 1, 2, \ldots$, there exists a solution $p(t; -k, p_k^*)$ for $t \geqslant -k$, $\|p_k^*\| \leqslant \rho$, such that

$p(0;-k,p_k^*) = p_0$. The uniqueness of solutions implies $p(t;0,p_0^*) \equiv p(t;-k, p_k^*)$, and therefore $p(t;0,p_0^*)$ is defined for all $-k \leqslant t < +\infty$. Since k is arbitrary, we conclude that the solution $p^*(t) = p(t;0,p_0^*)$ is defined for all $t \in \mathfrak{T}$, and $\sup_{t\in\mathfrak{T}}\|p^*(t)\| \leqslant \rho$. The proof of Lemma 4.2 is completed.

With Lemmas 4.1 and 4.2, we are in a position to prove the following result of Demidovich (1967):

Theorem 4.7. *Let us assume that in the system* (4.74), $f(t,p) \in C^{(0,1)}(\mathfrak{T} \times \mathfrak{R}^n)$,

$$\sup_{t\in\mathfrak{T}}\|f(t,0)\| = \zeta < +\infty, \tag{4.84}$$

and the maximum eigenvalue $\Lambda(t,p)$ *of the symmetric* $n \times n$ *Jacobian matrix* $J_s = (s_{ij})$ *with coefficients* $s_{ij} = s_{ij}(t,p)$ *defined by* (4.77) *satisfies the inequality*

$$\Lambda(t,p) \leqslant -\pi \qquad \forall (t,p) \in \mathfrak{R} \times \mathfrak{T}^n, \tag{4.85}$$

where π *is a positive number. Then the system* (4.74) *is a convergent system.*

Proof. Consider the function ν: $\mathfrak{R}^n \to \mathfrak{R}_+$ defined as

$$\nu(p) = \tfrac{1}{2}\|p\|^2, \tag{4.86}$$

and obtain

$$\dot{\nu}(p) = p^T f(t,p) \tag{4.87}$$

using (4.74). Then

$$\dot{\nu}(p) = p^T[f(t,p) - f(t,0)] + p^T f(t,0). \tag{4.88}$$

Using Lemma 4.1 and (4.85), we get

$$\dot{\nu}(p) \leqslant -\pi\|p\|^2 + |p^T f(t,0)| \qquad \forall (t,p) \in \mathfrak{T} \times \mathfrak{R}^n. \tag{4.89}$$

Since $|p^T f(t,0)| \leqslant \|f(t,0)\|\,\|p\| \leqslant \zeta\|p\|$, we obtain (4.89) as

$$\dot{\nu}(p) \leqslant -\pi\|p\|^2 + \zeta\|p\| \leqslant 0 \qquad \forall (t,p) \in \mathfrak{T} \times \mathfrak{R}_\rho, \tag{4.90}$$

where $\rho = \zeta\pi^{-1}$. By Lemma 4.2 we conclude that there exists a solution $p^*(t)$ of (4.74) which is defined and bounded on $\mathfrak{T}$ so that $\sup_{t\in\mathfrak{T}}\|p^*(t)\| \leqslant \rho$.

To show that $p^*(t)$ is globally asymptotically stable, we set

$$q(t) = p(t) - p^*(t). \tag{4.91}$$

For $\nu(q) = \tfrac{1}{2}\|q\|^2$ and $\dot{q} = f(t,p) - f(t,p^*)$, we compute

$$\begin{aligned}\dot{\nu}(q) &= q^T[f(t,p) - f(t,p^*)] \\ &\leqslant -2\pi\nu(q),\end{aligned} \tag{4.92}$$

so that

$$\nu[q(t)] \leqslant \nu(q_0)\exp[-2\pi(t - t_0)] \qquad \forall t \in \mathcal{T}_0 \tag{4.93}$$

and

$$\|p(t) - p^*(t)\| \leqslant \|p_0 - p_0^*\|\exp[-\pi(t - t_0)] \qquad \forall t \in \mathcal{T}_0. \tag{4.94}$$

The solution $p^*(t)$ has the required stability property. Furthermore, (4.94) implies the boundedness and uniqueness of $p^*(t)$ for all $t \in \mathcal{T}$. Therefore the system (4.74) is a convergent system and Theorem 4.7 is established.

With the help of Theorem 4.7, we can now consider a market model of the type (4.62),

$$\dot{p} = A(t,p)p + b(t), \tag{4.95}$$

and derive conditions for the model to be a convergent system. For this purpose, we define the quasidominant diagonal property of the $n \times n$ functional matrix $A = (a_{ij})$. We say that A is a negative quasidominant diagonal matrix if there exist positive numbers d_i, $i = 1, 2, \ldots, n$, and π such that

$$a_{jj}(t,p) < 0, \qquad |a_{jj}(t,p)| - d_j^{-1}\sum_{\substack{i=1 \\ i\neq j}}^{n} d_i a_{ij}(t,p) \geqslant \pi, \qquad j = 1, 2, \ldots, n, \tag{4.96}$$

$$\forall (t,p) \in \mathcal{T} \times \mathcal{R}^n.$$

We shall also assume that $b(t)$ is a bounded function, that is, $\sup_{t\in\mathcal{T}}\|b(t)\| = \zeta$, where ζ is a positive number. Now we establish the following:

Theorem 4.8. *A system* (4.95) *is convergent if* $A(t,p)$ *is everywhere a negative quasidominant diagonal matrix.*

Proof. The proof of Theorem 4.8 follows directly from the proofs of Theorems 4.6 and 4.7.

In the context of competitive equilibrium, Theorem 4.8 states that if the matrix $A(t,p)$ is dominant diagonal and if the shifts in excess demand $b(t)$ are bounded, then there is a price adjustment process $p^*(t)$ which is bounded and globally stable. That is, all prices tend to that process as the time progresses.

To express the above result in terms of the Jacobian, we can use either

Corollary 4.2. *A system* (4.95) *is convergent if* $J(t,p)$ *is everywhere a negative quasidominant diagonal matrix.*

or

Corollary 4.3. *A system* (4.95) *is convergent if* $J_s(t,p)$ *satisfies the inequality* (4.85).

By choosing

$$A(t,p) = \int_0^1 J(t,\mu p)\,d\mu \tag{4.97}$$

and using the Liapunov function $\nu(p) = \sum_{i=1}^n d_i|p_i|$, Corollary 4.2 follows from Theorem 4.8. By using the function $\nu(p) = \frac{1}{2}\|p\|^2$, Corollary 4.3 is a consequence of Theorem 4.7.

We can now turn our attention to the case of a linear constant market with time-dependent shifts,

$$\dot{p} = Ap + b(t), \tag{4.98}$$

which was considered by Arrow (1966). We assumed that A is an $n \times n$ constant Metzler matrix and $b(t) = ct$, where $c \in \mathfrak{R}^n$ is a constant vector. Under the assumption that A is a stable matrix, Arrow showed that the actual price $p(t)$ is always under the market clearing price ("moving equilibrium" in previous terminology of Samuelson, 1947) $p^0(t)$ defined by $0 = Ap^0 + b(t)$, that is, $p^0 = -A^{-1}b(t)$. Furthermore, the difference $p^0 - p$ approaches a limit which decreases as the speeds of adjustments on the different markets increase.

If we consider bounded shifts such that $b(t) \in C(\mathfrak{T})$ and $\sup_{t\in\mathfrak{T}}\|b(t)\| = \rho$, where ρ is a positive number, we conclude from Corollary 4.2 that stability of A implies that the market (4.98) is convergent. This is because for Metzler matrices stability implies and is implied by the negative quasidominant diagonal property of A. Furthermore, the limiting process $p^*(t)$ is determined by

$$p^*(t) = \int_{t_0}^t e^{A(t-\tau)}b(\tau)\,d\tau. \tag{4.99}$$

In the context of the market model (4.98) it would be of interest to determine conditions under which prices on all markets approach the market clearing price $p^0(t)$. That would take place if the difference

$$q(t) = p^*(t) - p^0(t) \tag{4.100}$$

approached zero as the time progresses. To obtain the desired conditions, let us use the fact that $\dot{p}^* = Ap^* + b(t)$ and $p^0(t) = -A^{-1}b(t)$ to get from

$\dot{q}(t) = \dot{p}^*(t) - \dot{p}^0(t)$ the following equation:

$$\dot{q} = Aq + c(t), \tag{4.101}$$

where $c(t) = A^{-1}\dot{b}(t)$. Now we prove the following:

Theorem 4.9. *The actual price* $p(t; t_0, p_0)$ *converges to the clearing price* $p^0(t) = -A^{-1}b(t)$ *of the market* (4.98) *if* A *is a negative quasidominant diagonal matrix,* $c(t)$ *is bounded, and* $\lim_{t\to\infty} c(t) = 0$.

Proof. Applying a result of Hahn (1967) to (4.101) and having in mind that $\|c(t)\| \leqslant \zeta$, we can say that for each $\varepsilon > 0$ we can find a t_1 such that $\|b(t)\| < \varepsilon$ for $t > t_1$. Also, for $t_0 < t \leqslant t_1$, $\|c(t)\| \leqslant \zeta$. Then we recall that

$$q(t; t_0, q_0) = e^{A(t-t_0)}q_0 + \int_{t_0}^{t} e^{A(t-\tau)}c(\tau)\, d\tau, \tag{4.102}$$

and for $t > t_1$,

$$\begin{aligned}\left\| \int_{t_0}^{t} e^{A(t-\tau)}c(\tau)\, d\tau \right\| &\leqslant \int_{t_0}^{t} \|e^{A(t-\tau)}\|\, \|c(\tau)\|\, d\tau + \int_{t_1}^{t} \|e^{A(t-\tau)}\|\, \|c(\tau)\|\, d\tau \\ &\leqslant \Pi e^{-\pi(t-t_0)}\|q_0\| + \Pi\zeta \int_{t_0}^{t_1} e^{-\pi(t-\tau)}\, d\tau + \Pi\varepsilon \int_{t_0}^{t} e^{-\pi(t-\tau)}\, d\tau \\ &= \Pi e^{-\pi(t-t_0)}\|q_0\| + \zeta\frac{\Pi}{\pi}e^{-\pi t}\left(e^{\pi t_1} - e^{\pi t_0}\right) + \varepsilon\frac{\Pi}{\pi}\left[1 - e^{\pi(t_1 - t)}\right]\end{aligned} \tag{4.103}$$

Therefore, $\|q(t; t_0, q_0)\|$ becomes arbitrarily small as t increases and ε decreases, and Theorem 4.9 follows from (4.100).

From Theorem 4.9, it follows that if the market is stable without the shifts and if the shifts are slow and tend to a constant value, then the actual price can get arbitrarily close to a market clearing price. If, in particular, for some $t_1 \in \mathfrak{T}$,

$$b(t) = c \qquad \forall t \geqslant t_1, \tag{4.104}$$

where $c \in \mathfrak{R}_+^n$ is a constant vector, then stability of A implies

$$\lim p(t; t_0, p_0) = -A^{-1}c \qquad \forall (t_0, p_0) \in \mathfrak{T} \times \mathfrak{R}^n, \tag{4.105}$$

and we can show that the convergence is exponential.

By introducing time and nonlinearities in the multiple market models considered in this and the preceeding section, we considerably widen the scope of the general equilibrium analysis in the Hicks-Metzler algebraic setting. The concept of connective stability and structural perturbations has made explicit the powerful implications of the Hicks and dominant diagonal conditions, and made the analysis of relevent models closer to the nonstationary and nonlinear reality of competitive market systems.

4.6. A STOCHASTIC MODEL

There are at least two strong reasons for considering stochastic rather than deterministic models of competitive equilibrium. First, due to unpredictable changes in the tastes of consumers, technology, weather, etc., it is quite realistic to consider fluctuations of supply and demand schedules as external random disturbances of the equilibrium. Secondly, inaccuracies of the *tâtonnement* process are likely to produce random perturbations of the internal adjustment mechanism of a competitive-equilibrium model. Both external and internal random disturbances introduce destabilizing effects into the market, which cannot be estimated satisfactorily from deterministic models. For these reasons, Turnovsky and Weintraub (1971) initiated a stochastic stability analysis of competitive equilibrium and derived explicit conditions under which linear market models are stable with probability one. For a list of general references and a short discussion concerning stochastic stability as used in this context, see the beginning of Section 5.8.

The objective of this section is to consider nonlinear stochastic models of competitive equilibrium, which were introduced (Šiljak, 1977a), as natural extensions of the nonlinear matrix models studied in Sections 4.4 and 4.5. This approach enables us to use the convenient diagonal-dominance condition and derive explicit conditions for stochastic stability and instability of nonlinear equilibrium models under random disturbances. The condition is ideally suited for establishing a trade-off between the degree of stability of the deterministic part of the model and the size of random disturbances that can be tolerated by a stable equilibrium. Furthermore, under the diagonal-dominance condition, stability is both connective and exponential in the mean.

Let us consider again n interrelated markets of n commodities, and let us assume that the price adjustment process is governed by a stochastic equation of the Itô type,

$$dp = A(t,p)p\,dt + B(t,p)p\,dz, \tag{4.106}$$

where $p(t) \in \mathfrak{R}^n$ is the price vector, and $z(t) \in \mathfrak{R}$ is a random variable which is a normalized Wiener process with

$$\mathcal{E}\{[z(t_1) - z(t_2)]^2\} = |t_1 - t_2|, \tag{4.107}$$

where $\mathcal{E}$ denotes expectation. In (4.106), the $n \times n$ functional matrix $A: \mathcal{T} \times \mathcal{R}^n \to \mathcal{R}^{n^2}$ represents the deterministic interaction among prices, whereas the $n \times n$ functional diffusion matrix $B: \mathcal{T} \times \mathcal{R}^n \to \mathcal{R}^{n^2}$ describes the influence of stochastic disturbances on the price adjustment process. The functional matrices $A(t,p)$, $B(t,p)$ are sufficiently smooth that the solution process $p(t; t_0, p_0)$ of (4.106) exists for all initial conditions $(t_0, p_0) \in \mathcal{T} \times \mathcal{R}^n$ and all $t \in \mathcal{T}_0$.

In the following analysis we consider stochastic stability of the equilibrium price $p^* = 0$ of the market model (4.106). We remember, however, that it is of interest to study the case when $A(t,p^*) = B(t,p^*) = 0$ for all $t \in \mathcal{T}$, but $p^* \neq 0$. Then we define the nonlinear functional matrices $\hat{A}(t,q)q \equiv A(t,q+p^*)(q+p^*)$, $\hat{B}(t,q) \equiv B(t,q+p^*)$, and consider the equation $dq = \hat{A}(t,q)q\,dt + \hat{B}(t,q)q\,dz$, which is (4.106) with p replaced by q and with $p^* = 0$.

In order to include the connective property of stochastic stability, we write the elements $a_{ij}(t,p)$, $b_{ij}(t,p)$ of the matrices $A(t,p)$, $B(t,p)$ as

$$a_{ij}(t,p) = \begin{cases} -\varphi_i(t,p) + e_{ii}(t)\varphi_{ii}(t,p), & i = j, \\ e_{ij}(t)\varphi_{ij}(t,p), & i \neq j, \end{cases} \tag{4.108}$$

$$b_{ij}(t,p) = l_{ij}(t)\psi_{ij}(t,p).$$

As in the preceding section, we assume

$$\varphi_i(t,p) \geqslant \alpha_i, \quad |\varphi_{ij}(t,p)| \leqslant \alpha_{ij}, \quad |\psi_{ij}(t,p)| \leqslant \beta_{ij} \qquad \forall (t,p) \in \mathcal{T} \times \mathcal{R}^n. \tag{4.109}$$

In (4.109), the functions e_{ij}, l_{ij}: $\mathcal{T} \to [0,1]$ are elements of the $n \times n$ interconnection matrices $E = (e_{ij})$, $L = (l_{ij})$. The elements e_{ij} measure the strength of the deterministic interactions among prices in the market, whereas the elements l_{ij} measure the influence of the external stochastic disturbance on the market.

Again, we need the notion of the fundamental $n \times n$ interconnection matrices $\bar{E} = (\bar{e}_{ij})$ and $\bar{L} = (\bar{l}_{ij})$ defined by

$$\bar{e}_{ij} = \begin{cases} 1, & \varphi_{ij}(t,p) \not\equiv 0, \\ 0, & \varphi_{ij}(t,p) \equiv 0, \end{cases} \qquad \bar{l}_{ij} = \begin{cases} 1, & \psi_{ij}(t,p) \not\equiv 0, \\ 0, & \psi_{ij}(t,p) \equiv 0. \end{cases} \tag{4.110}$$

Therefore, the matrix pair $(\bar{E}, \bar{L})$ represent the basic structure of the market system (4.106), and any pair of interconnection matrices (E, L) can be generated from the pair $(\bar{E}, \bar{L})$ by replacing the unit elements of $(\bar{E}, \bar{L})$ by corresponding elements e_{ij}, l_{ij} of (E, L). This fact is denoted by $(E, L) \in (\bar{E}, \bar{L})$.

Stochastic stability of the equilibrium price $p^* = 0$ of the market (4.106) is a convergence of the solution process $p(t; t_0, p_0)$ starting at time t_0 and an initial price $p_0 = p(t_0)$ toward the equilibrium. The convergence is measured in terms of "stochastic closeness" (e.g., in the mean, almost sure, in probability, etc.), which in turn generates various notions of stochastic stability. In considerations of the market model (4.106), we are interested in establishing conditions for globally exponential and connective stability in the mean (Šiljak, 1977a)—that is, conditions under which the expected value of the distance between the price adjustment process $p(t; t_0, p_0)$ and the equilibrium price $p^* = 0$, which is denoted by $\mathcal{E}\{\|p(t; t_0, p_0)\|\}$, tends to zero exponentially as time increases for all initial data $(t_0, p_0) \in \mathcal{T} \times \mathcal{R}^n$ and all interconnection matrices $(E, L) \in (\overline{E}, \overline{L})$. More precisely, we state the following:

Definition 4.6. *The equilibrium price $p^* = 0$ of the market* (4.106) *is globally and exponentially connectively stable in the mean if there exist two positive numbers Π and π, independent of the initial conditions (t_0, p_0), such that*

$$\mathcal{E}\{\|p(t; t_0, p_0)\|\} \leqslant \Pi \|p_0\| \exp[-\pi(t - t_0)] \qquad \forall t \in \mathcal{T}_0 \tag{4.111}$$

for all $(t_0, p_0) \in \mathcal{T} \times \mathcal{R}^n$ and all $(E, L) \in (\overline{E}, \overline{L})$.

Conditions for this kind of stability can be expressed again in terms of diagonal dominance of system matrices. We recall that an $n \times n$ matrix $C = (c_{ij})$ is a *negative dominant diagonal matrix* if

$$\begin{aligned} &c_{jj} < 0, \\ &|c_{jj}| > \sum_{\substack{i=1 \\ i \neq j}}^{n} |c_{ij}|, \qquad j = 1, 2, \ldots, n. \end{aligned} \tag{4.112}$$

Now we define a matrix $\overline{C}$ as

$$\overline{C} = \overline{A} + \overline{A}^T + \overline{B}, \tag{4.113}$$

and prove the following:

Theorem 4.10. *The equilibrium price $p^* = 0$ of the market* (4.106) *is globally and exponentially connectively stable in the mean if the $n \times n$ matrix $\overline{C}$ defined by* (4.113) *is a negative dominant diagonal matrix.*

Proof. Consider a decrescent, positive definite, and radially unbounded function $\nu: \mathcal{R}^n \to \mathcal{R}_+$ defined by

$$\nu(p) = \sum_{i=1}^{n} p_i^2 \tag{4.114}$$

as a candidate for Liapunov's function for the market system (4.106). Using

Itô's calculus (Gikhman and Skorokhod, 1969), we examine the expression

$$\mathfrak{L}v(p) = \frac{\partial v(p)}{\partial p} A(t,p)p + \frac{1}{2} \sum_{i,j=1}^{n} \frac{\partial^2 v(p)}{\partial p_i \partial p_j} s_{ij}(t,p), \tag{4.115}$$

where $\partial v/\partial p = (\partial v/\partial p_1, \partial v/\partial p_2, \ldots, \partial v/\partial p_n)$ is the gradient of $v(p)$, $\partial^2 v/\partial p_i \partial p_j$ is the (i,j)th element of the Hessian matrix related to $v(p)$, and the s_{ij}'s are the elements of the $n \times n$ matrix $S(t,p) = B(t,p)pp^T B(t,p)$. To establish the stability of the equilibrium price $p^* = 0$, we observe that $v(p)$ is a positive definite function, and demonstrate that $\mathfrak{L}v(p)_{(4.106)}$ is negative definite.

Let us calculate $\mathfrak{L}v(p)_{(4.106)}$ as

$$\begin{aligned} \mathfrak{L}v(p)_{(4.106)} &= \sum_{i=1}^{n} 2p_i \left(\sum_{j=1}^{n} a_{ij} p_j \right) + \sum_{i=1}^{n} \left(\sum_{j=1}^{n} b_{ij} p_j \right)^2 \\ &= \sum_{j=1}^{n} 2a_{jj} p_j^2 + \sum_{j=1}^{n} p_j \sum_{\substack{i=1 \\ i \neq j}}^{n} 2a_{ij} p_i + \sum_{i=1}^{n} \left(\sum_{j=1}^{n} b_{ij} p_j \right)^2. \end{aligned} \tag{4.116}$$

The negative dominant diagonal property of $\overline{C}$ in (4.113) is equivalent to

$$\begin{gathered} 2\bar{a}_{jj} + \bar{b}_{jj} < 0, \\ 2\bar{a}_{jj} + \bar{b}_{jj} + \sum_{\substack{i=1 \\ i \neq j}}^{n} (\bar{a}_{ij} + \bar{a}_{ji} + \bar{b}_{ij}) \leqslant -\bar{\pi}, \qquad j = 1, 2, \ldots, n, \end{gathered} \tag{4.117}$$

where $\bar{\pi}$ is a positive number. From (4.116) and (4.117), we get the inequality

$$\mathfrak{L}v(p)_{(4.106)} \leqslant -\bar{\pi} v(p) \qquad \forall (t,p) \in \mathfrak{T} \times \mathfrak{R}^n, \quad \forall (E,L) \in (\overline{E}, \overline{L}.) \tag{4.118}$$

By applying Ladde's (1975) stochastic comparison principle to (4.118), we obtain

$$\begin{gathered} \mathcal{E}\{v[p(t; t_0, p_0)]\} \leqslant v(p_0) \exp[-\bar{\pi}(t - t_0)] \qquad \forall t \in \mathfrak{T}_0, \quad \forall (t_0, p_0) \in \mathfrak{T} \times \mathfrak{R}^n, \\ \forall (E,L) \in (\overline{E}, \overline{L}.) \end{gathered} \tag{4.119}$$

From (4.114) and (4.119), we get (4.111) with $\Pi = 1$, $\pi = \frac{1}{2}\bar{\pi}$. The proof of Theorem 4.10 is complete.

We note immediately that due to the nonnegativity of the matrix $\overline{B}$, it follows from (4.117) that the "random shocks are likely to introduce instability into the system", as observed by Turnovsky and Weintraub (1971) on a linear constant model of multiple markets. Theorem 4.10 extends this observation to nonlinear nonstationary market models. From

(4.117) we conclude that the matrix $\bar{A} + \bar{A}^T$ should be sufficiently negative dominant diagonal to offset positivity of the matrix $\bar{B}$.

We also mention the fact that the negative dominant diagonal property is more conservative than the usual quasidominant diagonal property used in the previous sections to establish the stability of deterministic market models. This is so because we were limited in choice of Liapunov functions that are twice differentiable, and had to choose the function (4.114) instead of (4.44).

Stochastic-instability considerations of competitive equilibrium were carried out by Šiljak (1977a) using the stability framework of Theorem 4.10 and instability analysis of deterministic market models of (Šiljak, 1976a). The stochastic-instability conditions are important in that they can provide a necessity part which is missing in our sufficient stability conditions. Two distinct sources of instability were exposed (Šiljak, 1977a): The existence of strongly inferior goods for which the law of supply and demand does not hold [this is known as the Giffen paradox (Arrow and Hahn, 1971)], and the presence of random disturbances. Similar instability results have been obtained for stochastic ecological models and are presented in Section 5.9.

The results outlined in this section can be improved in a number of different ways. It is possible to relax the constraints on the interactions among the individual markets and establish the weaker global asymptotic property of stochastic stability as shown in Section 5.10. In the same section, it will be shown how to construct stochastic hierarchic models of ecosystems which in turn can be used to represent markets of composite commodities in stochastic environment along the lines of deterministic hierarchic models considered in Section 4.8 of this chapter. Finally, it would be of interest to try various other kinds of stochastic stability (Turnovsky and Weintraub, 1971) in the context of diagonal dominance, and provide a less conservative stability criterion. This could open up a real possibility of including price expectations (Arrow and Nerlove, 1958; Arrow and Hurwicz, 1962; Turnovsky and Weintraub, 1971) in an essential way in our nonlinear and nonstationary competitive-equilibrium models.

4.7. DISCRETE MODELS

In constructing a dynamic model from static equations describing an equilibrium situation instead of differential equations one can use difference equations (Solow, 1952; Arrow and Hahn, 1971). In order to give a brief account of how discrete models arise in competitive equilibrium analysis, let us consider the algebraic equations

$$x_i = \sum_{j=1}^{n} a_{ij} x_j + b_i, \qquad i = 1, 2, \ldots, n. \tag{4.120}$$

If x_i is the national income of the ith country (or region) of an n-country system, a_{ij} the marginal propensity of the jth country to import from the ith country, a_{ii} the marginal propensity of the ith country to consume domestic commodities, and b_i the autonomous expenditure in the ith country, then the equations (4.120) represent the international trade system of Metzler (1950). On the other hand, if x_i is interpreted as the output of the ith industry, a_{ij} as the input of the ith commodity per unit output of the jth commodity, and b_i as the amount of the ith commodity in the "bill of commodities", then (4.120) is a Leontief (1936, 1948) open-end input-output system. Finally, when x_i is the price of the ith commodity, (4.120) can be used to describe the competitive-equilibrium condition discussed in Section 4.1.

The linear equation system (4.120) is easily seen to be the static solution of the linear difference-equation system

$$x(t + 1) = Ax(t) + b, \tag{4.121}$$

which is the dynamic model of Solow (1952) written in vector notation with time t taking integer values. The equilibrium solution x^* of (4.121) for which $x^*(t + 1) = x^*(t)$ for all t is obtained from (4.120) as

$$x^* = (I - A)^{-1}b. \tag{4.122}$$

For x^* to be economically meaningful, we must have $x \geqslant 0$, that is, all components x_i of the vector x should be nonnegative. Furthermore, in order to have a workable system (4.121), we require that the equilibrium be asymptotically stable. It turns out that stability of x^* implies existence and nonnegativity of $(I - A)^{-1}$. Thus, stability of x^* and nonnegativity of b (that is, $b \geqslant 0$), imply $x^* \geqslant 0$.

It was shown by Metzler (1950) that if $a_{ij} \geqslant 0$, then the necessary and sufficient condition for stability of x^* in (4.121) is that $A - I$ is a Hicks matrix, that is, $I - A$ has all its principal minors positive. It is a well-known fact (e.g. Hahn, 1967) that stability of x^* in (4.121) is equivalent to the condition that the eigenvalues of A are less than one in absolute value, that is, $|\lambda_i(A)| < 1$, $i = 1, 2, \ldots, n$. Now, Metzler's result is an application of the classical Perron-Frobenius theorem (Gantmacher, 1960; Seneta, 1973): If A is a nonnegative matrix (that is, $a_{ij} \geqslant 0$ for all i, j), then there exists an eigenvalue $\lambda_P(A)$ of A, the "Perron root of A", such that $\lambda_P(A) \geqslant 0$ and $|\lambda_i(A)| \leqslant \lambda_P(A)$ for all $i = 1, 2, \ldots, n$. We see that $I - A$ has all off-diagonal elements nonpositive (that is, $a_{ij} \leqslant 0$, $i \neq j$), and from the Appendix we recall that positivity of principal minors of $I - A$ is equivalent to the existence of a positive eigenvalue $\lambda_m(I - A)$ of $I - A$ such that the real part of any eigenvalue of $I - A$ is at least $\lambda_m(I - A)$. Now, $\lambda_P(A) = 1$

$-\lambda_m(I - A)$ and $|\lambda_i(A)| \leqslant 1$. This established Metzler's result.

We can go a step farther and derive a quasidominant diagonal stability criterion for the discrete system (4.121) as we did for the continuous system (4.17). We need only to recall from the Appendix that for a matrix $I - A$ with nonnegative off-diagonal elements, saying that all principal minors of $I - A$ are positive is equivalent to saying that there exist positive numbers d_i, $i = 1, 2, \ldots, n$, such that

$$d_i|1 - a_{jj}| > \sum_{\substack{j=1 \\ j \neq i}}^{n} d_j|a_{ij}|, \qquad i = 1, 2, \ldots, n, \tag{4.123}$$

with $1 - a_{jj} > 0$, $j = 1, 2, \ldots, n$. When all d_i's are chosen as $d_i = 1$, (4.123) is reduced to the dominant diagonal condition of Solow (1952), which is only sufficient for stability of (4.121).

More details to aid in the above derivation of Metzler's result and quasidominant diagonal condition for stability of the discrete systems can be found in the Appendix.

As far as competitive equilibrium is concerned, discrete dynamic models are closer to the reality of multiple markets. *Tâtonnement* as proposed by Walras is a discrete procedure which is used to calculate iteratively the equilibrium prices for an economy. The discrete price adjustments were extensively analyzed by Marshall (1948), whose intervals were Day I, Day II, etc., and later by Hicks (1939), whose "units of time" were "weeks". Dynamic versions of discrete market models were introduced by Samuelson (1947), who also discussed their stability properties. Nonlinear discrete market models were considered recently by Arrow and Hahn (1971), who suggested that under fairly restrictive conditions (such as gross substitutability and specific rules that *tâtonnement* has to follow), it can be shown that the diagonal-dominance condition ensures global stability of the equilibrium price. In the rest of this section, we shall show that a great deal of the pessimism expressed about the discrete models by Arrow and Hahn can be removed at the expense of a more elaborate analysis. We shall actually show that a complete discrete counterpart to the analysis of continuous models of Section 4.4 is possible and that a "discrete" version of quasidiagonal dominance (4.123) can be used to establish the global stability of a general class of nonlinear nonstationary *tâtonnement* processes.

Let us consider a discrete matrix equation

$$x(t_{k+1}) = A[t_k, x(t_k)]x(t_k), \tag{4.124}$$

where $x(t_k) \in \mathcal{R}^n$ is the state of the system represented by Equation (4.124), which can stand for the vector of prices on a market, the income vector of an international trade system, or the output vector of a dynamic input-output model. In (4.124), $t_k \in \mathcal{T}$ is the discrete time, k being an integer; $\mathcal{T}$ is the time interval $(\tau, +\infty)$, where τ is a number or a symbol $-\infty$; and $t_k \to +\infty$ when $k \to +\infty$. The matrix function $A: \mathcal{T} \times \mathcal{R}^n \to \mathcal{R}^{n^2}$ is such that $A[t_k, x(t_k)]x(t_k)$ is bounded, continuous, and one-to-one for any fixed $t_k \in \mathcal{T}$, so that the solution $x(t_k; t_0, x_0)$ which starts at $(t_0, x_0) \in \mathcal{T} \times \mathcal{R}^n$ has the property

$$x(t_b; t_0, x_0) = x[t_b; t_a, x(t_a; t_0, x_0)] \qquad \forall t_0 \in \mathcal{T}, \quad \forall t_a,\, t_b \in \mathcal{T}_0 \qquad (4.125)$$

with $t_a \leqslant t_b$ and $\mathcal{T}_0 = [t_0, +\infty)$. As before, the coefficients of the matrix function $A[t_k, x(t_k)]$ are denoted by $a_{ij}[t_k, x(t_k)]$.

We state

Definition 4.7. *The equilibrium $x^* = 0$ of the system (4.124) is exponentially stable in the large if there exist numbers $\Pi \geqslant 1$, $\pi > 0$, which do not depend on the initial conditions (t_0, x_0), such that*

$$\|x(t_k; t_0, x_0)\| \leqslant \Pi \|x_0\| \exp[-\pi(t_k - t_0)] \qquad \forall t_k \in \mathcal{T}_0 \qquad (4.126)$$

for all $(t_0, x_0) \in \mathcal{T} \times \mathcal{R}^n$.

This kind of stability for the system (4.124) can be established by the following Grujić-Šiljak (1973) discrete version of the classical Krassovskii (1959) theorem established for continuous systems:

Theorem 4.11. *The equilibrium $x^* = 0$ of the system (4.124) is exponentially stable in the large if and only if there exists a continuous function $v: \mathcal{T} \times \mathcal{R}^n \to \mathcal{R}_+$ and numbers $\eta_1 > 0$, $\eta_2 > 0$, $0 < \eta_3 < 1$ such that*

$$\begin{aligned} \eta_1 \|x\|^2 \leqslant v(t_k, x) &\leqslant \eta_2 \|x\|^2, \\ v[t_{k+1}, x(t_{k+1}; t_0, x_0)] &\leqslant \eta_3 v[t_k, x(t_k; t_0, x_0)] \end{aligned} \qquad \forall (t_k, x) \in \mathcal{T}_0 \times \mathcal{R}^n \quad (4.127)$$

for all $(t_0, x_0) \in \mathcal{T} \times \mathcal{R}^n$.

Proof. To prove the "only if" part of the theorem, we need

$$v(t_k, x) = \sum_{i=k}^{k+m-1} x^T(t_i; t_k, x) H x(t_i; t_k, x), \qquad (4.128)$$

where H is a constant, symmetric, and positive definite $n \times n$ matrix. In (4.128), the integer $m > 0$ is chosen so that

$$\sup_{t_k \in \mathcal{T}_0} (t_{k+m} - t_k) \leqslant \pi^{-1} \ln(\mu\Pi), \qquad (4.129)$$

where $\mu > [\lambda_M(H)\lambda_m^{-1}(H)]^{1/2}$, and $\lambda_M(H)$ and $\lambda_m(H)$ are the maximum and the minimum eigenvalue of H, respectively.

Using (4.126), from (4.128), we get

$$\nu(t_k, x) \geqslant x^T(t_k; t_k, x) H x(t_k; t_k, x) \geqslant \lambda_m(H)\|x\|^2 \tag{4.130}$$

and

$$\nu(t_k, x) \leqslant \sum_{i=k}^{k+m-1} \lambda_M(H)\|x(t_i; t_k, x)\|^2 \leqslant m\Pi^2\lambda_M(H)\|x\|^2. \tag{4.131}$$

The choice

$$\eta_1 = \lambda_m(H), \qquad \eta_2 = m\Pi^2\lambda_M(H) \tag{4.132}$$

implies the first two inequalities of (4.127).

Making use of the solution property (4.125), we derive

$$\begin{aligned} \nu[t_{k+1}, x(t_{k+1}; t_0, x_0)] &= \nu\{t_{k+1}, x[t_{k+1}; t_k, x(t_k; t_0, x_0)]\} \\ &\leqslant \sum_{i=k+1}^{k+m} x^T[t_i; t_k, x(t_k; t_0, x_0)] H x[t_i; t_k, x(t_k; t_0, x_0)]. \end{aligned} \tag{4.133}$$

Adding and subtracting $\nu(t_k, x)$ on the right side of (4.133), and applying the first two inequalities of (4.126), (4.127), and (4.132), we arrive at

$$\nu[t_{k+1}, x(t_{k+1}; t_0, x_0)] \leqslant \eta_1^{-1}\lambda_M(H)\Pi^2\nu[t_k, x(t_k; t_0, x_0)]\exp[-2\pi(t_{k+m} - t_k)]. \tag{4.134}$$

Finally, from (4.129) and (4.134), we get

$$\nu[t_{k+1}, x(t_{k+1}; t_0, x_0)] \leqslant \eta_3 \nu[t_k, x(t_k; t_0, x_0)], \tag{4.135}$$

where

$$\eta_3 = \lambda_m^{-1}(H)\lambda_M(H)\mu^{-2} < 1, \tag{4.136}$$

and thus we have the "only if" part of the theorem.

The "if" part of the theorem is almost automatic. From (4.127), it follows that

$$\begin{aligned} \nu[t_{k+1}, x(t_{k+1}; t_0, x_0)] &\leqslant \Pi_1 \nu[t_k, x(t_k; t_0, x_0)]\exp(-2\pi T_M) \\ &\leqslant \Pi_1 \nu[t_k, x(t_k; t_0, x_0)]\exp[-2\pi(t_{k+1} - t_k)], \end{aligned} \tag{4.137}$$

where $\eta_3 = \Pi_1 \exp(-2\pi T_M)$, $T_M = \sup_k(t_{k+1} - t_k)$. From (4.127) and (4.137), we get (4.126) with $\Pi = (\Pi_1\eta_1^{-1}\eta_2)^{1/2}$. The proof of Theorem 4.11 is completed.

With Theorem 4.11 at our disposal, we can prove the desired result formulated as follows:

Theorem 4.12. *The equilibrium* $x^* = 0$ *of the system* (4.124) *is exponentially stable in the large if there exist numbers* $d_i > 0$, $i = 1, 2, \ldots, n$, *and* $0 < \pi < 1$, *such that*

$$d_i^{-1} \sum_{j=1}^{n} d_j |a_{ij}[t_k, x(t_k)]| \leqslant \pi, \qquad i = 1, 2, \ldots, n, \tag{4.138}$$

holds for all $(t_k, x) \in \mathcal{T} \times \mathcal{R}^n$.

Proof. Consider a function ν: $\mathcal{R}^n \to \mathcal{R}_+$ defined by

$$\nu(x) = \max_i \{d_i^{-1}|x_i|\}, \qquad i = 1, 2, \ldots, n. \tag{4.139}$$

With $\eta_1 = n^{-1}d_M^{-1}$ and $\eta_2 = d_m^{-1}$, $d_M = \max_i d_i$, $d_m = \min_i d_i$, the function $\nu(x)$ satisfies the first two inequalities in (4.127). To show the third, we compute $\nu[x(t_{k+1}; t_0, x_0)]$ using (4.124) and (4.138) as follows:

$$\begin{aligned} \nu[x(t_{k+1}; t_0, x_0)] &= \max_i \{d_i^{-1}|x_i(t_{k+1}; t_0, x_0)|\} \\ &= \max_i \left\{ d_i^{-1} \left| \sum_{j=1}^{n} a_{ij} x_j(t_k; t_0, x_0) \right| \right\} \\ &\leqslant \max_i \left\{ \sum_{j=1}^{n} d_j d_i^{-1} |a_{ij}| d_j^{-1} |x_j(t_k; t_0, x_0)| \right\} \\ &\leqslant \max_i \left\{ d_i^{-1} \sum_{j=1}^{n} d_j |a_{ij}| \right\} \max_j \{d_j^{-1}|x_j(t_k; t_0, x_0)|\} \\ &\leqslant \pi\nu[x(t_k; t_0, x_0)] \qquad \forall (t_k, x) \in \mathcal{T} \times \mathcal{R}^n. \end{aligned} \tag{4.140}$$

By applying Theorem 4.11 to (4.140), we get (4.126), and the proof of Theorem 4.12 is complete.

We should note here that a similar proof of Theorem 4.12 can be constructed (Grujić and Šiljak, 1974) using the absolute-value norm (4.44) as a Liapunov function instead of the maximum-value norm (4.139). These results are straightforward generalizations of those reported by Kalman and Bertram (1960), and Newman (1961).

Having in mind that $\pi < 1$, we can replace a_{ij} by $a_{ij}[t_k, x(t_k)]$ in (4.123) and use this condition instead of (4.138) to prove the stability of $x^* = 0$ in (4.124). Furthermore, it is obvious that we can redefine the coefficients a_{ij} of A to include the elements of interconnection matrices and show connective exponential stability of the discrete market model. Therefore, by Theorem 4.12, we established a discrete version of the stability result which has the same generality as its continuous counterpart presented in Section 4.4. Using Theorem 4.12, it would be possible to obtain discrete variations of the results obtained in Section 4.5 for models with forcing terms. Stochastic discrete versions of the results of Section 4.6 can be attempted using the results of Green and Majubdar (1975), Barta and Varaiya (1976), and Diamond (1975).

4.8. COMPOSITE COMMODITIES: A HIERARCHIC MODEL

The stability analysis presented so far is effective when the number of economic variables is not very large. For systems of high dimension, we have to use aggregates, either to gain in conceptual clarity of the analysis or to make computations feasible and reliable when numerical results are desired. Leontief (1936) and Hicks (1939) independently proposed the notion of composite commodities in order to extend the validity of partial equilibrium analysis to markets with many commodities. The idea is as follows: If the relative prices of some set of commodities remain constant, then for analytical and practical purposes the commodities can be combined into one composite commodity. The price of the composite commodity can be taken to be proportional to that of any commodity of the set, so as to have the price times the quantity of the composite commodity equal to the sum of the prices times the quantities of the commodities included in the set. This motivated Lange (1945) to formulate the *law of composition of goods* (the terms "goods" and "commodities" are used interchangeably): "Any goods the prices of which always vary in the same proportion can be combined into one composite good; and, conversely, any good can be split up into an arbitrary number of separate goods with prices varying always in the same proportion". Simon and Ando (1961) successfully applied this law to nearly decomposable systems by considering them as composite systems, constructed by the superposition of terms representing interactions of the variables within each subsystem and terms representing interactions among the subsystems, thus proposing a hierarchic model of the economy.

As pointed out by Arrow and Hahn (1971), the strict constancy of relative prices may hold approximately in many cases of practical interest, but it cannot be taken to hold in the general case. By imitating the spirit of the composition of goods but using entirely different techniques, a general decomposition-aggregation approach to the stability of large economic systems was proposed by Šiljak (1976a), which does not require "strict constancy" or any other specific relation among prices. The approach is based on the concept of vector Liapunov functions introduced independently by Matrosov (1962) and Bellman (1962) (see Section 2.4) and the strategy proposed by Simon and Ando (1961). The approach was described in Section 1.8, where the following scheme was adapted:

(1) Decompose the economic system by classifying the economic variables into a number of subeconomies.

(2) Decouple the subeconomies and represent stability properties for each individual subeconomy by an appropriate scalar Liapunov function (stability analysis on the first hierarchic level).

(3) Formulate constraints on the interactions among the subeconomies.

(4) Construct an aggregate model involving the vector Liapunov function and interaction constraints, and deduce the stability of the entire economy from the stability of the aggregate model (stability analysis on the second hierarchic level).

In the above scheme, there is a great deal of freedom in decomposing and aggregating the economic equations and variables. We do not have to require "that q prices must be kept rigid in order to secure stability" as proposed by Lange (1945). Furthermore, no linearity assumption is necessary either for the subsystems or for their interactions.

To illustrate the application of the proposed decomposition-aggregation scheme of Šiljak (1976a), let us consider an economic system described by an equation of the type (4.40),

$$\dot{x} = A(t,x)x, \tag{4.141}$$

where $x(t) \in \mathcal{R}^n$ is the state of the system, and the $n \times n$ matrix function $A(t,x)$ is defined, bounded, and continuous on the domain $\mathcal{T} \times \mathcal{R}^n$. The system (4.141) is decomposed into s interconnected subsystems

$$\dot{x}_i = A_i(t,x_i)x_i + \sum_{j=1}^{s} e_{ij}(t)A_{ij}(t,x)x_j, \qquad i = 1, 2, \ldots, s, \tag{4.142}$$

where $x_i(t) \in \mathcal{R}^{n_i}$ is the state of the subsystem, so that

$$\mathcal{R}^n = \mathcal{R}^{n_1} \times \mathcal{R}^{n_2} \times \cdots \times \mathcal{R}^{n_s}, \tag{4.143}$$

and $A_i(t,x_i)$ and $A_{ij}(t,x)$ are $n_i \times n_i$ and $n_i \times n_j$ matrix functions, respectively.

With each decoupled subsystem

$$x_i = A_i(t,x_i)x_i, \qquad i = 1, 2, \ldots, s, \tag{4.144}$$

we associate a scalar Liapunov function $\nu_i\colon \mathcal{R}^{n_i} \to \mathcal{R}_+$ as provided by Theorem 4.5:

$$\nu_i(x_i) = \sum_{k=1}^{n_i} d_{ik} |x_{ik}|, \tag{4.145}$$

with the numbers $d_{i1}, d_{i2}, \ldots, d_{in_i}$ positive, and with the estimates

$$\begin{aligned} d_{im} \|x_i\| \leqslant \nu_i(x_i) \leqslant n_i^{1/2} d_{iM} \|x_i\|, \\ D^+ \nu_i(x_i)_{(4.144)} \leqslant -\pi_i \nu_i(x_i) \end{aligned} \qquad \forall (t, x_i) \in \mathcal{T} \times \mathcal{R}^{n_i}. \tag{4.146}$$

These functions $\nu_i(x_i)$ serve as indices of composite commodities as far as stability of the partial equilibrium of (4.144) is concerned. In fact, the functions can be viewed as derived prices of composite commodities, whose nonnegativity is guaranteed by (4.146).

As for the interconnections among the subeconomies, we assume that they are bounded as

$$\|A_{ij}(t, x) x_j\| \leqslant \xi_{ij} \|x_j\| \qquad \forall (t, x) \in \mathcal{T} \times \mathcal{R}^n, \tag{4.147}$$

where ξ_{ij} are nonnegative numbers.

To construct the aggregate model and conclude stability of the entire economic system (4.141), we use Theorem 2.14. For this purpose, we compute the Lipschitz constant κ_i for $\nu_i(x_i)$ as

$$\begin{aligned} \|\nu_i(x_i) - \nu_i(y_i)\| &= \left\| \left(\sum_{k=1}^{n_i} d_{ik} |x_{ik}| - \sum_{k=1}^{n_i} d_{ik} |y_{ik}| \right) \right\| \\ &\leqslant \left\| \left(\sum_{k=1}^{n_i} d_{ik} |x_{ik} - y_{ik}| \right) \right\| \\ &\leqslant n_i^{1/2} d_{iM} \|x_i - y_i\|, \end{aligned} \tag{4.148}$$

and get $\kappa_i = n_i^{1/2} d_{iM}$.

Finally, we compute, as in Theorem 2.14,

$$D^+ \nu_i(x_i)_{(4.142)} \leqslant D^+ \nu_i(x_i)_{(4.144)} + \kappa_i \left\| \sum_{j=1}^{s} e_{ij}(t) A_{ij}(t, x) x_j \right\|, \qquad i = 1, 2, \ldots, s. \tag{4.149}$$

By using (4.147) and (4.148), and defining the vector Liapunov function $v\colon \mathcal{R}^n \to \mathcal{R}_+^s$ as

$$v = [\nu_1(x_1), \nu_2(x_2), \ldots, \nu_s(x_s)]^T, \tag{4.150}$$

the inequalities (4.149) can be rewritten as a vector differential inequality

$$\dot{v} \leqslant \overline{W} v, \tag{4.151}$$

which is an aggregate model of the economy (4.141). In (4.151), the $s \times s$

aggregate matrix $\overline{W} = (\overline{w}_{ij})$ is defined as

$$\overline{w}_{ij} = \begin{cases} -\pi_i, & i = j, \\ \bar{e}_{ij} \xi_{ij} n_i^{1/2} d_{iM} d_{im}^{-1}, & i \neq j. \end{cases} \tag{4.152}$$

By applying Theorem 2.14 to the aggregate system (4.151), we conclude that the equilibrium $x^* = 0$ of the original system (4.141) is connectively exponentially stable in the large if the $s \times s$ Metzler matrix $\overline{W} = (\overline{w}_{ij})$ defined by (4.152) is a Hicks matrix, that is, if the matrix $\overline{W}$ is a quasidominant diagonal matrix. In other words, if each subeconomy is diagonally dominant and the aggregate model is also diagonally dominant, then the entire economy is connectively and exponentially stable, the connectivity aspect being related to structural perturbations among the subeconomies.

As for the reduction of dimensionality of the stability problem, we see that by the proposed decomposition-aggregation method one determines stability of an nth-order system (4.141) by demonstrating the stability of s subsystems (4.144) of the n_ith order ($\sum_{i=1}^{s} n_i = n$) and one aggregate model (4.151) of the sth order. If the dimension n of the system (4.141) is high, the procedure becomes increasingly important despite its conservativeness due to approximations involved in aggregating the economic variables into the model (4.151). An important byproduct of the decomposition-aggregation process is the automatic tolerance of nonlinearities in the coupling among the subeconomies and insensitivity to any sign pattern of coefficients in the original subsystem matrices. Furthermore, one also obtains the important connective property of stability, which represents valuable information about markets of composite goods. This is true in particular when the decomposition-aggregation scheme is not used solely to reduce the dimensionality of the stability problems, but to give insight into special structural characteristics of a given economy. It is, however, left largely unclear how decomposition and aggregation of the equilibrium models should be performed to balance the gains in simplifications against errors resulting from the approximations involved in the decomposition-aggregation process.

REFERENCES

Aoki, M. (1976), *Optimal Control and System Theory in Dynamic Economic Analysis*, North-Holland, New York.

Arrow, K. J. (1966), "Price-Quantity Adjustments in Multiple Markets with Rising Demands", *Proceedings of a Symposium on Mathematical Methods in the Social Sciences*, K. J. Arrow, S. Karlin, and P. Suppes (eds.), Stanford University Press, Stanford, California, 3–15.

Arrow, K. J., Block, H. D., and Hurwicz, L. (1959), "On the Stability of Competitive Equilibrium", II, *Econometrica*, 27, 82–109.

Arrow, K. J., and Hahn, F. H. (1971), *General Competitive Analysis*, Holden-Day, San Francisco, California.

Arrow, K. J., and Hurwicz, L. (1958), "On the Stability of Competitive Equilibrium", I, *Econometrica*, 26, 522–552.

Arrow, K. J., and Hurwicz, L. (1962), "Competitive Stability Under Weak Gross Substitutability: Nonlinear Price Adjustment and Adaptive Expectations", *International Economic Review*, 3, 233–255.

Arrow, K. J., and Hurwicz, L. (1963), "Decentralization and Computation in Resource Allocation", *Essays in Economics and Econometrics*, R. W. Pfouts (ed.), The University of North Carolina Press, Chapel Hill, North Carolina, 34–104.

Arrow, K. J., and Nerlove, M. (1958), "A Note on Expectations and Stability", *Econometrica*, 26, 297–305.

Bailey, F. N. (1966), "The Application of Lyapunov's Second Method to Interconnected Systems", *SIAM Journal of Control*, 3, 443–462.

Barta, S., and Varaiya, P. (1976), "Stochastic Models of Price Adjustment", *Annals of Economic and Social Measurement*, 5, 267–281.

Bassett, L., Habibagahi, H., and Quirk, J. (1967), "Qualitative Economic and Morishima Matrices", *Econometrica*, 35, 221–233.

Bellman, R. (1962), "Vector Lyapunov Functions", *SIAM Journal of Control*, 1, 32–34.

Bellman, R. (1970), "Topics in Pharmocokinetics I: Concentration-Dependent Rates", *Mathematical Biosciences*, 6, 13–17.

Debreu, G. (1976), "The Application to Economics of Differential Topology and Global Analysis", *American Economic Review*, 66, 280–287.

Demidovich, B. P. (1967), *Lectures on Mathematical Theory of Stability* (in Russian), Nauka, Moscow.

Diamond, P. (1975), "Stochastic Exponential Stability Concepts and Large-Scale Discrete Systems", *International Journal of Control*, 22, 141–145.

Fiedler, M., and Pták, V. (1962), "On Matrices with Non-Positive Off-Diagonal Elements and Positive Principal Minors", *Czechoslovakian Mathematical Journal*, 12, 382–400.

Fisher, F. M. (1972), "On Price Adjustment Without an Auctioneer", *Review of Economic Studies*, 39, 1–15.

Gantmacher, F. N. (1959), "*The Theory of Matrices*", Vols. I and II, Chelsea, New York.

Gikhman, I. I., and Skorodkhod, A. V. (1969), *Introduction to the Theory of the Random Processes*, Saunders, Philadelphia, Pennsylvania.

Green, J. R., and Majumdar, M. (1975), "The Nature of Stochastic Equilibria", *Econometrica*, 43, 647–660.

Grujić, Lj. T., and Šiljak, D. D. (1973), "On Stability of Discrete Composite Systems", *IEEE Transactions*, AC-18, 522–524.

Grujić, Lj. T., and Šiljak, D. D. (1974), "Exponential Stability of Large-Scale Discrete Systems", *International Journal of Control*, 19, 481–491.

Habibagahi, H., and Quirk, J. (1973), "Hicksian Stability and Walras' Law", *The Review of Economic Studies*, 40, 249–258.

Hahn, W. (1967), *Stability of Motion*, Springer, New York.

Hale, J. K. (1969), *Ordinary Differential Equations*, Wiley, New York.

Hawkins, D., and Simon, H. (1949), "Some Conditions of Macroeconomic Stability", *Econometrica*, 17, 245–248.

Herstein, I. N. (1953), "Some Mathematical Methods and Techniques in Economics", *Quarterly of Applied Mathematics*, 11, 249–262.

Hicks, J. R. (1939), *Value and Capital*, Oxford University Press, Oxford, (2nd Ed., 1946).

Intriligator, M. D. (1971), *Mathematical Optimization and Economic Theory*, Prentice-Hall, Englewood Cliffs, New Jersey.

Kalman, R. E., and Bertram, J. E. (1960), "Control System Analysis and Design via the 'Second Method' of Lyapunov", *Transactions ASME*, 82; Part I: 371–393; Part II: 394–400.

Kotelyanskii, I. N. (1952), "On Some Properties of Matrices with Positive Elements" (in Russian), *Matematicheskii Sbornik*, 31, 497–506.

Krassovskii, N. N. (1959), "*Certain Problems of the Theory of Stability of Motion*" (in Russian), Fimatgiz, Moscow. (English translation: *Stability of Motion*, Stanford University Press, Stanford, California, 1963.)

Ladde, G. S. (1975), "Systems of Differential Inequalities and Stochastic Differential Equations II", *Journal of Mathematical Physics*, 16, 894–900.

Ladde, G. S. (1976a), "Stability of Large-Scale Hereditary Systems Under Structural Perturbations", *Proceedings of the IFAC Symposium on Large Scale Systems Theory and Applications*, G. Guardabassi and A. Locatelli (eds.), Instrument Society of America, Pittsburgh, Pennsylvania, 215–226.

Ladde, G. S. (1976b), "Cellular Systems - I. Stability of Chemical Systems", *Mathematical Biosciences*, 29, 309–330.

Ladde, G. S. (1976c), "Cellular Systems - II. Stability of Compartmental Systems", *Mathematical Biosciences*, 30, 1–21.

Ladde, G. S., Lakshmikantham, V., and Liu, P. T. (1973), "Differential Inequalities and Itô Type Stochastic Differential Equations", *Proceedings of the International Conference on Nonlinear Differential and Functional Equations*, Bruxelles et Louvain, Belgium, Herman, Paris, 611–640.

Lakshmikantham, V., and Leela, S. (1969), "*Differential and Integral Inequalities*", Vols. I and II, Academic, New York.

Lange, O. (1945), "The Stability of Economic Equilibrium", Appendix to *Price Flexibility and Employment*, Cowles Commision Monograph, 8, Bloomington, Indiana, 91–108. Also in: *Readings in Mathematical Economics*, Part I, Editor: P. Newman, The John Hopkins Press, Baltimore, 1968, 178–196.

Lange, O. (1964), "On the Economic Theory of Socialism", *On the Economic Theory of Socialism*, B. E. Lippincott (ed.), McGraw-Hill, New York.

Leontief, W. W. (1936), "Composite Commodities and the Problem of Index Numbers", *Econometrica*, 4, 39–59.

Leontief, W. W. (1948), "Computational Problems Arising in Connection with Economic Analysis of Interindustrial Relationships", *Proceedings of a Symposium on Large-Scale Digital Calculating Machinery*, Harvard University Press, Cambridge, Massachusetts, 169–175.

Leontief, W. W. (1966), *Input-Output Economics*, Oxford University Press, Oxford, England.

Marshall, A. (1948), *Principles of Economics*, Eighth Edition, Macmillan, New York.

Matrosov, V. M. (1962), "On the Theory of Stability of Motion" (in Russian), *Prikladnaya Mathematika i Mekhanika*, 26, 992–1002.

McKenzie, L. (1966), "Matrices with Dominant Diagonal and Economic Theory", *Proceedings of a Symposium on Mathematical Methods in the Social Sciences*, K. J. Arrow, S. Karlin, and P. Suppes (eds.), Stanford University Press, Stanford, California, 47–62.

Metzler, L. (1945), "Stability of Multiple Markets: The Hicks Conditions", *Econometrica*, 13, 277–292.

Metzler, L. A. (1950), "A Multiple-Region Theory of Income and Trade", *Econometrica*, 18, 329–354.

Mill, J. S. (1848), *Principles of Political Economy*, Parker, London. Also in: "*Collected Works*", Vols. I and II, University of Toronto Press, Toronto, Canada, 1965.

Morgenstern, O. (1972), "Thirteen Critical Points in Contemporary Economic Theory: An Interpretation", *The Journal of Economic Literature*, 10, 1163–1189.

Morishima, M. (1952), "On the Laws of Change of the Price-System in an Economy Which Contains Complementary Commodities", *Osaka Economic Papers*, 1, 101–113.

Morishima, M. (1964), "*Equilibrium, Stability, and Growth*", Claredon, Oxford, England.

Mosak, J. L. (1944), *General Equilibrium Theory in International Trade*, Cowles Commission Monograph No. 7, Principia, Bloomington, Indiana.

Negishi, T. (1962), "The Stability of a Competitive Economy: A Survey Article", *Econometrica*, 30, 635–669.

Newman, P. K. (1959), "Some Notes on Stability Conditions", *The Review of Economic Studies*, 72, 1–9.

Newman, P. (1961), "Approaches to Stability Analysis", *Economica*, 28, 12–29.

Nikaido, H. (1969), *Convex Structures and Economic Theory*, Academic, New York.

Nikaido, H., and Uzawa, H. (1960), "Stability and Non-Negativity in a Walrasian Tatônnement Process", *International Economic Review*, 1, 50–59.

Pareto, V. (1927), *Manuel d'Economic Politique*, Deuxieme Edition, Marcel Giard, Paris.

Patinkin, D. (1956), *Money, Interest, and Prices*, Row, Peterson and Co., Evanston, Illinois.

Persidskii, S. K. (1969), "Problems of Absolute Stability" (in Russian), *Automatika i Telemekhanika*, 29, 5–11.

Pliss, B. A. (1964), *Nonlocal Problems in Theory of Oscillations* (in Russian), Nauka, Moscow.

Quirk, J., and Saposnik (1968), *Introduction to General Equilibrium Theory and Welfare Economics*, McGraw-Hill, New York.

Radner, R. (1968), "Competitive Equilibrium Under Uncertainty", *Econometrica*, 36, 31–58.

Rapoport, A. (1960), "*Flights, Games and Debates*", The University of Michigan Press, Ann Arbor, Michigan.

Richardson, L. F. (1960), "*Arms and Insecurity*", Boxwood, Chicago.

Rosenbrock, H. H. (1963), "A Lyapunov Function for Some Naturally Occuring Linear Homogeneous Time-Dependent Equations", *Automatica*, 1, 97–109.

Samuelson, P. A. (1941), "The Stability of Equilibrium: Comparative Statics and Dynamics", *Econometrica*, 9, 97–120.

Samuelson, P. A. (1947), *Foundations of Economic Analysis*, Harvard University Press, Cambridge, Massachusetts.

Samuelson, P. A. (1974), "Complementarity: An Essay on the 40th Anniversary of the Hicks-Allen Revolution in Demand Theory", *The Journal of Economic Literature*, 12, 1255–1289.

Sandberg, I. W. (1969), "Some Theorems on the Dynamic Response of Nonlinear Transistor Networks", *The Bell Technical Journal*, 48, 35–54.

Sandberg, I. W. (1974), "On the Mathematical Theory of Interactions in Social Groups", *IEEE Transactions*, SMC-4, 432–445.

Seneta, E. (1973), "*Non-Negative Matrices: An Introduction to Theory and Applications*", Wiley, New York.

Sevastyanov, B. A. (1951), "Theory of Branching Stochastic Processes" (in Russian), *Uspekhi Matematicheskih Nauk*, 6, 47–99.

Šiljak, D. D. (1969), *Nonlinear Systems*, Wiley, New York.

Šiljak, D. D. (1972), "Stability of Large-Scale Systems Under Structural Perturbations", *IEEE Transactions*, SMC-2, 657–663.

Šiljak, D. D. (1973), "On Stability of Large-Scale Systems Under Structural Perturbations", *IEEE Transactions*, SMC-3, 415–417.

Šiljak, D. D. (1974), "Connective Stability of Complex Ecosystems", *Nature*, 249, 280.

Šiljak, D. D. (1975a), "When is a Complex Ecosystem Stable?", *Mathematical Biosciences*, 25, 25–50.

Šiljak, D. D. (1975b), "Connective Stability of Competitive Equilibrium", *Automatica*, 11, 389–400.

Šiljak, D. D. (1975c), "On Total Stability of Competitive Equilibrium", *International Journal of Systems Science*, 10, 951–964.

Šiljak, D. D. (1976a), "Competitive Economic Systems: Stability, Decomposition, and Aggregation", *IEEE Transactions*, AC-21, 149–160. (See also: *Proceedings of the 1973 IEEE Conference on Decision and Control*, San Diego, California, December 5–7, 1973, 265–275.)

Šiljak, D. D. (1976b), "Competitive Analysis of the Arms Race", *Annals of Economic and Social Measurement*, 5, 283–295.

Šiljak, D. D. (1977a), "On Stochastic Stability of Competitive Equilibrium", *Annals of Economic and Social Measurement*, 6, 315–323.

Šiljak, D. D. (1977b), "On Stability of the Arms Race", *Proceedings of the NSF Conference on Control Theory in International Relations Research*, J. V. Gillespie and D. A. Zinnes (eds.), Praeger, New York, 264–304.

Simon, H. A. (1957), *Models of Man*, Wiley, New York.

Simon, H. A., and Ando, A. (1961), "Aggregation of Variables in Dynamic Systems", *Econometrica*, 29, 111–138.

Smale, S. (1976), "Dynamics in General Equilibrium Theory", *American Economic Reviews*, 66, 288–294.

Smith, A. (1776), "*An Inquiry into the Nature and Cause of the Wealth of Nations*", Vols. I and II, Strachan and Cadell, London.

Solow, R. (1952), "On the Structure of Linear Models", *Econometrica*, 20, 29–47.

Thom, R. (1975), *Structural Stability and Morphogenesis*, Benjamin, Reading, Massachusetts.

Turnovsky, S. J., and Weintraub, E. R. (1971), "Stochastic Stability of a General Equilibrium System Under Adaptive Expectations", *International Economic Review*, 12, 71–86.

Uzawa, H. (1961), "The Stability of Dynamic Processes", *Econometrica*, 29, 617–631.

von Neuman, J., and Morgenstern, O. (1947), *Theory of Games and Economic Behavior*, Second Edition, Princeton University Press, Princeton, New Jersey.

Walras, L. (1874), "*Elements d'economic politique pure*", L. Corbaz, Lausanne, (English translation by W. Jaffe: *Elements of Pure Economics*, George Allen & Unwin, London.)

Weissenberger, S. (1973), "Stability Regions of Large-Scale Systems", *Automatica*, 9, 653–663.

Yoshizawa, T. (1963), "Stability of Sets and Perturbed Systems", *Funkcialaj Ekvacioj*, 5, 31–69.

Yoshizawa, T. (1966), *The Stability Theory by Liapunov's Second Method*, Mathematical Society of Japan, Tokyo.

Zubov, V. I. (1957), *Methods of A. M. Liapunov and Their Applications* (in Russian), LGU Press, Leningrad, USSR. (English translation: Noordhoff, Groningen, The Nederlands, 1964).

5

ECOLOGY

Multispecies Communities

A central question in population ecology is: Does increased trophic web complexity lead to a higher degree of community stability, or is it the other way around? There is a good deal of experimental evidence collected by many people, as surveyed recently by May (1973a) in his comprehensive monograph, which supports each alternative. The inconclusive empirical evidence calls for attempts to answer the above question by theoretical stability studies, and in this chapter we will provide a partial answer to the complexity-vs.-stability problem using Liapunov's stability concept. A general solution to this difficult problem is not available at present, nor is such a result to be expected in the near future. Only by recasting the problem in the framework of connective stability, and thus focusing our attention on the structural vulnerability of interacting populations, will we be able to expose the effects of complexity on the stability properties of general ecomodels. By arguing that it is not sufficient to determine that a given community is merely stable, but that it should also be stable under a wide range of perturbations in its interconnection structure, we shall come up with both qualitative and quantitative results regarding the mechanisms of complexity and stability in multispecies communities.

In stability investigations of a linear model of n competing species, MacArthur (1970) formulated the complexity-vs.-stability problem in an important way as the effect on stability of increasing the number of species by one, that is, to $n + 1$. By making a symmetry assumption in the model, he was able to use a well-known property of symmetric matrices and conclude that the $(n + 1)$-species community is never more stable than the n-species community. The symmetry assumption, however, is overrestrictive and can hardly be used in stability studies of communities with more than one trophic level. In the spirit of MacArthur's investigations, but using computer experiments, Gardner and Ashby (1970) observed that increasing the levels of interactions in a community of density-dependent species leads to instability regardless of the sign pattern in the interactions among species. These experiments were confirmed by May (1972) for a large number of species, and a functional relationship was studied between the number of variables in the system and the average connectedness and interaction magnitude among various variables. Although the experiments were useful in shedding more light on the complexity-vs.-stability problem, they suffer from the lack of mathematical generality and cannot be used to predict stability in new situations not actually tested.

By exposing general ecomodels to a wide range of both structural (internal) and environmental (external) disturbances, we shall be able to determine the level of complexity under which community stability is invulnerable. By the detailed mathematical arguments, we will show not only that MacArthur's conclusion can be valid for asymmetrical linear constant models, but that it can be proved to hold in *communities with nonlinear time-varying interactions*. The stability condition is the *diagonal dominance* of the community matrices, which proves to be an ideal mechanism for establishing the trade-off between complexity and stability, thus providing a natural mathematical setting for interpretations of the Gardner-Ashby and May experimental investigations. It is this condition that reflects in mathematical terms May's (1973a) observation, made on the basis of computer studies, about the average interaction strength s and connectedness C of a community: "Roughly speaking, this suggests that, within a web, species which interact with many others (large C) should do so weakly (small s), and conversely those which interact strongly should do so with but a few species". That this indeed takes place in natural ecosystems was concluded from empirical evidence by Margalef (1968). It is interesting to note that the diagonal-dominance condition, which was borrowed from economic studies of competitive equilibrium (see Chapter 4), appears to be a good tool for balancing out complexity and stability of multispecies communities under structural and environmental perturbations.

Since most of the models in multispecies ecology are not globally stable, we will also consider *stability regions* in the population space. Estimates of the regions are obtained using Liapunov functions. How this is carried out in a particular situation will be illustrated using the Lotka-Volterra model of competing species.

The diagonal-dominance property of community matrices is for large classes of ecomodels merely a sufficient condition for stability. For this reason, we will investigate several causes for *community instability*, and the resulting sufficiency conditions for instability will provide a necessity part in our stability requirements.

Although the superiority of *stochastic models* over their deterministic counterparts in studying multispecies communities has long been recognized and well documented (see May, 1973a), they have remained restricted almost invariably to linear interactions among species. It was only recently that a general approach to stochastic stability of ecomodels was proposed by Ladde and Šiljak (1976a), which makes it possible to study large variations of population size in communities exposed to a randomly varying environment. Itô differential equations were used to describe stochastic ecomodels and recast the stability analysis in the framework of the Liapunov stability theory. For our analysis of ecosystems which can include nonlinear phenomena such as predator switching, food limitations, saturation of predator attack capacities, etc., the Itô-Liapunov approach to community stability is superior to the use of the Fokker-Planck-Kolmogorov equation (May, 1973b), which is restricted to linear constant models. We will show again that the diagonal dominance is an appropriate tool for establishing the degree of tolerance of stable communities to a wide variety of nonlinear nonstationary interaction effects as well as the level of random fluctuations that can be absorbed by stabilized populations.

In large multispecies communities, it was observed (May, 1973a) that the interactions among species are concentrated in small blocks. This fact made *hierarchic models* a suitable framework for stability studies of communities with a large number of species (Šiljak, 1975a). By raising the level of abstraction we will use the decomposition-aggregation scheme in the context of vector Liapunov functions, which was outlined in Chapter 2, to show that the diagonal-dominance property of the aggregate community matrix is again a sufficient condition for stability. This result makes it possible to carry most of our conclusions drawn from the direct analysis over to the aggregate model, where each block of the species is considered as an individual part of the overall community model. The hierarchic analysis effects a considerable reduction in the dimensionality of stability problems, and at the same time provides an insight into essential structural properties of large multispecies communities.

5.1. LINEAR CONSTANT MODELS

To answer some of the basic questions raised above and to set the stage for more elaborate analysis of nonlinear and time-varying effects in

$$\begin{bmatrix} + & - \\ - & + \end{bmatrix} \qquad \begin{bmatrix} - & + \\ + & - \end{bmatrix}$$

Competition   Symbiosis

$$\begin{bmatrix} + & - \\ + & - \end{bmatrix} \qquad \begin{bmatrix} + & 0 \\ + & - \end{bmatrix}$$

Predation   Saprophytism

FIGURE 5.1. Basic types of interactions.

complex ecosystems, let us consider the linear constant model of n interacting species which are close to an equilibrium situation,

$$\dot{x} = Ax, \tag{5.1}$$

where $x = x(t)$ is an n-vector $x = \{x_1, x_2, \ldots, x_n\}$ the components of which represent the populations. The $n \times n$ constant *community matrix* $A = (a_{ij})$ has elements a_{ij} which characterize the influence of the population x_j of the jth species on the population x_i of the ith species in the community.

We assume at the start that all species are density-dependent, so that the matrix A has a negative diagonal, that is, $a_{ii} < 0$ for all $i = 1, 2, \ldots, n$. Such an assumption reflects the resource limitation in the community, and as pointed out by Tanner (1966) is unlikely to be seriously violated by any of the commonly studied populations but that of man. With regard to the off-diagonal elements a_{ij}, we do not specify their sign, thus allowing for a "mixed" competitive-predator-symbiotic-saprophytic interactions among species.

Since the characterizations of interactions among the species, such as "competition", are different from those used in the economic considerations of the previous chapter, we list in Figure 5.1 the sign patterns in the community matrix of two species. Of course, for the moment we assume that all diagonal signs on Figure 5.1 are minus, to uphold the density-dependence assumption. We will waive this assumption later when we consider hierarchic models of multispecies communities.

In order to study the effect of the strength of interactions among populations on the stability of the equilibrium $x^* = 0$ of the community model (5.1), let us assume that the elements a_{ij} of the community matrix A have the form

$$a_{ij} = \begin{cases} -\alpha_i + e_{ii}\alpha_{ii}, & i = j, \\ e_{ij}\alpha_{ij}, & i \neq j. \end{cases} \tag{5.2}$$

The numbers e_{ij} are the elements of the $n \times n$ constant *interconnection matrix* $E = (e_{ij})$, and can take any value between zero and one, that is,

$$0 \leqslant e_{ij} \leqslant 1 \tag{5.3}$$

for all $i, j = 1, 2, \ldots, n$. Therefore, by e_{ij} we can represent the influence of the population x_j on the population x_i, ranging from a disconnection ($e_{ij} = 0$) to a full connection ($e_{ij} = 1$) measured by the number α_{ij}. For the diagonal elements a_{ii}, to uphold again the density-dependence condition, we assume that α_i and α_{ii} are related as $\alpha_i > \alpha_{ii} \geqslant 0$.

An important role in the following analysis of community stability is played by the $n \times n$ *fundamental interconnection matrix* $\bar{E} = (\bar{e}_{ij})$. The elements $\bar{e}_{ij}$ of $\bar{E}$ have binary values: 1 if the jth population x_j influences the ith population x_i directly, and 0 if x_j has no direct influence on x_i regardless of the changes in the community environment or the community itself. The matrix $\bar{E}$ is a binary matrix which reflects the basic structure of the community. Therefore, any interconnection matrix E can be generated from $\bar{E}$ by replacing the unit elements of $\bar{E}$ with the corresponding elements e_{ij} of E. The fact that E is generated by a given $\bar{E}$ is denoted by $E \in \bar{E}$.

Since the underlying characteristics of ecosystems are hardly ever known precisely and never fixed in time, it is necessary to assume possible perturbations in the interactions among the species, in both magnitude and direction. The influence of such perturbations on the stability properties of the system (5.1) was studied experimentally by Gardner and Ashby (1970). On the basis of a computer study, they concluded that if the strength of interconnections among the populations increases beyond a certain critical value, the system becomes unstable. This fact was confirmed by May (1972), who used it to draw conclusions about the relationship between the complexity of large systems, expressed by connectedness among the variables, and its stability properties.

The interconnection matrix represents a suitable means for describing the perturbations of interconnections among the species and formulating the connective-stability concept for multispecies communities:

A community is connectively stable if it is stable for all admissible interconnection matrices.

By "admissible" interconnection matrices E we mean such as are generated by the fundamental interconnection matrix $\bar{E}$ of the community. In the context of connective stability, for example, one can consider structural perturbations whereby a certain species (or group of species) is disconnected from and again connected with the rest of the community. Such was the case with a community of marine invertebrates reported by Paine (1966), where removal of a species caused the community to become unstable and shrink from a fifteen to an eight species community. The structural perturbations of this kind can be readily described by interconnection matrices. Removal (or destruction) of the kth species from the community (5.1) is represented by

$$\begin{aligned} e_{ik} &= 0, \qquad i = 1, 2, \ldots, n, \\ e_{kj} &= 0, \qquad j = 1, 2, \ldots, n. \end{aligned} \tag{5.4}$$

If we show that the community (5.1) is connectively stable, then the structural perturbation (5.4) cannot produce instability of the equilibrium. Structural perturbations that change the number of species were analytically considered by MacArthur (1970) assuming a special symmetric form of the community matrix. Not only do we not require this special symmetric form of A, but also we will waive the assumption of constancy of the system (5.1), thus allowing for time-varying structural perturbations which take place at disequilibrium populations and give rise to a time-dependent community matrix.

Let us now present the conditions for the connective stability property of the community (5.1). We recall that an $n \times n$ matrix $A = (a_{ij})$ was called quasidominant diagonal by McKenzie (1966) if there exists an n-vector $d = \{d_1, d_2, \ldots, d_n\}$ with positive components ($d > 0$) such that

$$d_j |a_{jj}| > \sum_{\substack{i=1 \\ i \neq j}}^{n} d_i |a_{ij}|, \qquad j = 1, 2, \ldots, n. \tag{5.5}$$

As shown by McKenzie, for a negative diagonal matrix A, the inequalities (5.5) are sufficient for stability (all eigenvalues of A have negative real parts; Theorem A.11). That is, if the community matrix A is negative and quasidominant diagonal, the community (5.1) is globally asymptotically stable.

The condition (5.5) includes the usual diagonal dominance

$$|a_{jj}| > \sum_{\substack{i=1 \\ i \neq j}}^{n} |a_{ij}|, \qquad j = 1, 2, \ldots, n, \tag{5.6}$$

as a special case when $d = \{1, 1, \ldots, 1\}$. Although the condition (5.6) is easier to interpret in the context of the system (5.1), it is more restrictive than that of (5.5). It should be noted, however, that for large classes of nonlinear time-varying models we shall be able to replace the condition (5.5) with readily interpretable inequalities involving only a_{ij}'s.

Now, if we establish (5.5) for a given $\bar{E}$, then it will be satisfied for any E generated by that $\bar{E}$. To see this, we use (5.2) to rewrite (5.5) as

$$d_j|-\alpha_j + e_{jj}\alpha_{jj}| > \sum_{\substack{i=1 \\ i\neq j}}^{n} d_i|e_{ij}\alpha_{ij}|, \qquad j = 1, 2, \ldots, n. \tag{5.7}$$

Obviously, if (5.7) holds for $\bar{E}$, it will hold for any E which is obtained from $\bar{E}$ by replacing unit elements of $\bar{E}$ by the numbers e_{ij} which lie between zero and one as defined in (5.3). Specifically, (5.7) will hold in case of the structural perturbation (5.4), and removal of the kth species from a stable community described by (5.1) will not cause instability.

From the above analysis, it is possible to draw conclusions about the tolerance of stability for increasing complexity of the system: So long as the strength of interactions does not violate the inequalities (5.5), the stability of the community is assured. This conclusion is compatible with the empirical evidence noticed by Margalef (1968): "It seems that species that interact feebly with others do so with a great number of other species. Conversely, species with strong interactions are often part of a system with a small number of species . . . ", and it does not contradict the statement of Levins (1968) that "there is often a limit to the complexity of systems".

Since the conditions (5.5) are only sufficient for stability, they cannot be used to test for instability and verify entirely the Gardner-Ashby and May experiments. The conditions (5.5), however, are both necessary and sufficient for stability of (5.1) if the interactions obey the rules "friends of friends are friends, enemies of enemies are friends, and friends of enemies and enemies of friends are enemies". This situation is important in the analysis of substitute-complement multiple markets of commodities or services and is known as the Morishima case (see Šiljak, 1975a, 1977a).In the context of the community model (5.1), it means that the community is composed of species in either competitive or symbiotic mutual relationships according to the rules given above. A special case of the Morishima matrix (Definition A.13) is that of the Metzler matrix (Definition A.10) used in Chapter 4 to represent the case when all commodities are substitutes. A community model (5.1) described by a Metzler matrix is of little significance in population dynamics, since it represents a rare "gross mutualism" among species. It should be noted, however, that in the case of the Metzler matrix A, the conditions (5.5) are necessary and sufficient for stability (Theorem A.10), and if they were violated, the community would suddenly become unstable. This conclusion certainly agrees with the May's (1973a) speculation that "predator-prey bonds are more common than symbiotic ones", and via the Morishima matrix the conclusion can be extended to his statement that "competition or mutualism between two species is less conducive to overall web stability than is a predator-prey relationship".

Finally, in the context of the model (5.1) we should mention that if the community matrix A is negative diagonal, then the conditions (5.5) imply that it is D-stable, that is, the product DA is stable for any positive diagonal matrix $D = \text{diag}\{d_{11}, d_{22}, \ldots, d_{nn}\}$ in which $d_{ii} > 0$ for all i and $d_{ij} = 0$ for all $i \neq j$. This fact demonstrates that stability is highly reliable in communities described by negative quasidominant diagonal matrices, and that they are stable under arbitrary "speeds of adjustments" of populations in the community. Unfortunately, neither D-stability nor the closely related qualitative stability which was applied to population dynamics by May (1973a) can be used for other than linear constant models. This is not true, however, for the connective-stability concept outlined in Chapter 2, which can be used to establish the stability of nonlinear time-varying model ecosystems at the expense of more refined analysis (Šiljak, 1975a).

5.2. NONLINEAR MODELS: LINEARIZATION

Interaction in multispecies communities is a highly nonlinear and nonstationary affair. For this reason, it is not surprising that the first population model of Lotka (1925) and Volterra (1926) was a system of nonlinear differential equations. In Section 1.6, where the model was considered, it was shown that there are two population equilibria. This fact is good enough to discourage any attempt to approximate large fluctuations of populations by means of a linear model which is limited to a single equilibrium state. Nevertheless, local behavior of the community in the neighborhood of the equilibrium populations, can be successfully predicted by standard linearization techniques (Allen, 1975; May, 1973a). A simple demonstration of this fact in the case of the Lotka-Volterra model, was given in Section 1.6.

To describe a general linearization mechanism, let us assume that a community is described by a nonautonomous differential equation

$$\dot{x} = f(t, x), \tag{5.8}$$

where $x = \{x_1, x_2, \ldots, x_n\}$ is again the population vector and the function $f(t, x)$ has continuous first derivatives. Then we set

$$x(t) = x^*(t) + y(t), \tag{5.9}$$

where $x^*(t)$ is a solution of Equation (5.8), whose neighborhood is of our concern. As shown at the end of Section 2.2, we can assume without loss of generality that $x^*(t)$ is an equilibrium, that is, a constant solution of Equation (5.8) such that

$$f(t, x^*) = 0 \qquad \text{for all } t. \tag{5.10}$$

Expanding $f(t, x)$ about $x = x^*$ in a Taylor series expansion and using the transformation (5.9), we get

$$\dot{y} = A(t)y + g(t, y), \tag{5.11}$$

where

$$A(t) = \left.\frac{\partial f(t, x)}{\partial x}\right|_{x=x^*} \tag{5.12}$$

is the well-known $n \times n$ Jacobian matrix, $g(t, y)$ represents the higher-order terms in the Taylor series expansion, and $g(t, 0) = 0$, $\partial g(t, 0)/\partial y = 0$. Now, for sufficiently small initial values $x_0 = x(t_0)$, the solutions $x(t; t_0, x_0)$ of the nonlinear equation (5.8) can be studied by considering the corresponding solutions of the linear equation

$$\dot{x} = A(t)x, \tag{5.13}$$

which is obtained from (5.11) by removing the higher-order terms and replacing y by x.

When we have an autonomous equation

$$\dot{x} = f(x), \tag{5.14}$$

the above linearization process yields the linear constant system

$$\dot{x} = Ax, \tag{5.15}$$

where A is an $n \times n$ constant Jacobian matrix defined by surpressing t in (5.12).

To illustrate the power and limitation of the linearization, let us consider the well-known Verhulst-Pearl logistic equation (Pielou, 1969),

$$\dot{x} = (\alpha - \beta x)x, \tag{5.16}$$

which describes the growth of a population $x(t)$ in a restricted environment. We assume that α, $\beta > 0$, and set

$$(\alpha - \beta x)x = 0 \tag{5.17}$$

to determine the two equilibrium states $x_1^* = 0$ and $x_2^* = \alpha/\beta$. Since at least one member of the community is needed for it to grow, the growth rate is zero at $x_1^* = 0$. For a small initial population $x_0 = x(0)$, we linearize Equation (5.16) about $x_1^* = 0$ and get the linear equation

$$\dot{x} = \alpha x, \tag{5.18}$$

which has the solution

$$x(t) = x_0 e^{\alpha t}. \tag{5.19}$$

Therefore, for a small population, there is no interference among its members, and the population grows in a Malthusian manner at the exponential rate α. Since the environment is restricted, the population growth is limited by a shortage of resources, and it reaches the saturation level $x_2^* = \alpha/\beta$ set by the carrying capacity of the environment. If the initial population x_0 is larger than the saturation level $x_2^* = 0$, the population will decrease, approaching the equilibrium x_2^* asymptotically from above. To see this, we use the transformation

$$x = x_2^* + y \tag{5.20}$$

and rewrite (5.16) as

$$\dot{y} = (-\alpha - \beta y)y. \tag{5.21}$$

The linearized version of (5.21) around $y^* = 0$ is

$$\dot{y} = -\alpha y, \tag{5.22}$$

which has the solution

$$y(t) = y_0 e^{-\alpha t}. \tag{5.23}$$

Both predictions made by linearization—that the equilibrium $x_1^* = 0$ is unstable, and that the equilibrium $x_2^* = \alpha/\beta$ is stable—can be confirmed by solving the original logistic equation to get

$$x(t) = \frac{\alpha}{\beta}\left[1 + \left(\frac{\alpha/\beta - x_0}{x_0}\right)e^{-\alpha t}\right]^{-1}. \tag{5.24}$$

Two typical solutions (5.24) are shown in Figure 5.2.

Although the linearization predicted the stability properties of the two equilibria, in each case the predictions were correct only locally, in the immediate neighborhood of the equilibrium states. For large deviations from either of the equilibrium states, which exceed the distance α/β between them, the predictions are clearly inconsistent. Therefore, for large perturbations of populations and global analysis of community stability, nonlinear analysis is necessary.

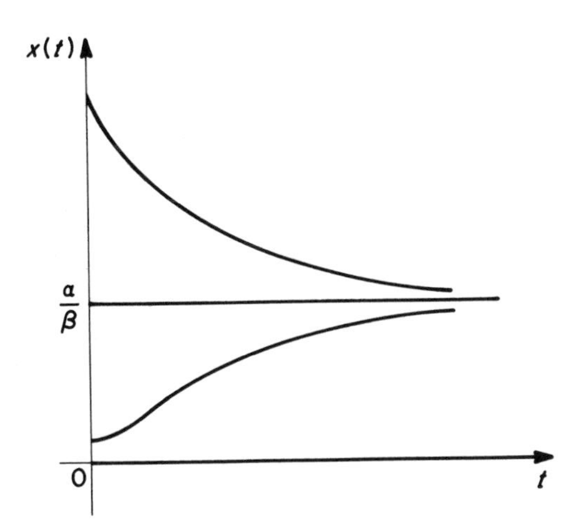

FIGURE 5.2. Logistic curves.

5.3. NONLINEAR MATRIX MODELS

There are several ways to bring the stability analysis of the preceeding section closer to reality, and a step in that direction is to consider the nonlinear matrix model

$$\dot{x} = A(t, x)x. \tag{5.25}$$

The model (5.25) is an obvious generalization of the linear constant linear model

$$\dot{x} = \bar{A}x, \tag{5.26}$$

in which the constant elements $\bar{a}_{ij}$ of the $n \times n$ matrix $\bar{A} = (\bar{a}_{ij})$ are replaced by nonlinear time-dependent functions

$$a_{ij} = a_{ij}(t, x). \tag{5.27}$$

Equation (5.26) is the same as (5.1), and the bar on $\bar{A}$ is used to distinguish it from $A(t, x)$, since we will often omit the arguments (t, x) of A for simplicity.

Stability analysis cannot be carried out in algebraic terms for the nonlinear matrix model (5.25) as it can for the linear model (5.26), and we need to use the Liapunov direct method outlined in Chapter 2. It is, however, of fundamental importance to note that we shall be able to deduce various stability properties of the nonlinear time-varying model (5.25) from stability properties of the linear constant model (5.26) established by simple algebraic criteria involving only the elements $\bar{a}_{ij}$ of the matrix $\bar{A}$ in (5.26). It is fair to say, at the same time, that this powerful result can be accomplished only at the expense of a more refined analysis.

In (5.25), $x(t) \in \mathcal{R}^n$ is again the population vector, and the $n \times n$ matrix function $A: \mathcal{T} \times \mathcal{R}^n \rightarrow \mathcal{R}^{n^2}$ is defined, continuous and bounded on $\mathcal{T} \times \mathcal{R}^n$, so that the solutions $x(t; t_0, x_0)$ of (5.25) exist for all initial conditions $(t_0, x_0) \in \mathcal{T} \times \mathcal{R}^n$ and $t \in \mathcal{T}_0$. The symbol $\mathcal{T}$ represents the time interval $(\tau, +\infty)$, where τ is a number or the symbol $-\infty$, and $\mathcal{T}_0$ is the semi-infinite time interval $[t_0, +\infty)$.

In the following analysis, we will consider stability of the equilibrium population $x^* = 0$ of the model (5.25). If $A(t, x^*)x^* = 0$ for all $t \in \mathcal{T}$, and $x^* \neq 0$ is of interest, then we can define the nonlinear matrix function $B(t, y)y \equiv A(t, x^* + y)(x^* + y)$ and consider the equation $\dot{y} = B(t, y)y$ instead of (5.25), where $y^* = 0$ represents the equilibrium x^* under investigation. Although it is always possible in principle to use the transformation $x = x^* + y$ introduced in (5.9) and place the equilibrium at the origin of $\mathcal{R}^n$, the new transformed equation is usually more cumbersome to handle.

In order to include the connective property of stability in the analysis, we write the elements $a_{ij}(t, x)$ of $A(t, x)$,

$$a_{ij}(t, x) = \begin{cases} -\varphi_i(t, x) + e_{ii}(t)\varphi_{ii}(t, x), & i = j, \\ e_{ij}(t)\varphi_{ij}(t, x), & i \neq j, \end{cases} \tag{5.28}$$

where the functions $\varphi_i(t, x), \varphi_{ij}(t, x) \in C(\mathcal{T} \times \mathcal{R}^n)$.

In (5.28), we consider the elements $e_{ij} = e_{ij}(t)$ of the $n \times n$ interconnection matrix $E = (e_{ij})$ as functions of time such that $e_{ij}(t) \in C(\mathcal{T})$ and restricted only by the condition $e_{ij}(t) \in [0, 1]$. This represents a major improvement over the use of E in Section 5.1, since structural perturbations can take place at the disequilibrium populations without any prescribed pattern—the elements $e_{ij}(t)$ may be arbitrary functions of time without any statistical description. This is remarkable in that we allow the interactions among the species in a community to vary arbitrarily in strength during the population adjustment process without any *a priori* specification of such variations except that they are bounded.

To derive necessary and sufficient conditions for connective stability of the equilibrium $x^* = 0$ of (5.25), we need further specifications of the elements $a_{ij}(t, x)$ of the matrix A. We assume that the functions $\varphi_i(t, x)$, $\varphi_{ij}(t, x)$ in (5.28) are bounded functions on $\mathcal{T} \times \mathcal{R}^n$ and that there exist positive numbers $\alpha_i > 0$, $\alpha_{ij} \geqslant 0$ such that

$$\alpha_i = \inf_{\mathcal{T} \times \mathcal{R}^n} \varphi_i(t, x), \qquad \alpha_{ij} = \sup_{\mathcal{T} \times \mathcal{R}^n} |\varphi_{ij}(t, x)|, \qquad i, j = 1, 2, \ldots, n, \tag{5.29}$$

and $\alpha_i > \alpha_{ii}$.

Since we want the connective stability of the system (5.25) to be *absolute* (Definition 2.8), we need to establish the connective stability of $x^* = 0$ in (5.25) for any set of functions $\varphi_i(t,x)$, $\varphi_{ij}(t,x)$ satisfying the conditions (5.29). For this purpose, we define the following classes of functions:

$$\begin{aligned}\Phi_i &= \{\varphi_i(t,x)\colon \varphi_i(t,x) \geqslant \alpha_i\},\\ \Phi_{ij} &= \{\varphi_{ij}(t,x)\colon |\varphi_{ij}(t,x)| \leqslant \alpha_{ij}\},\end{aligned} \tag{5.30}$$

where $\alpha_{ij} \geqslant 0$, $\alpha_i > \alpha_{ii} \geqslant 0$. As in the case of the elements of the interconnection matrix, we do not have to specify the precise shape of the nonlinear interactions among the species in the community so long as the interactions are bounded. This is a realistic assumption, since most of the familiar models of ecosystems surveyed by Pielou (1969), Goel, Maitra, and Montroll (1971), and May (1973) have this property. In particular, the restrictions (5.30) include the cases of saturation in predator attack capability, changes in predator searching behavior, and its "switching" mechanism. Furthermore, the conditions (5.30) include any sign pattern of the interactions, thus allowing for the "mixed" (competitive-predator-symbiotic-saprophitic) communities, as well as for changes of a predator to a prey of another species over a time interval. The constraints in (5.30) are illustrated in Figure 5.3.

Besides concluding merely that a community is stable, we would also like to estimate the rate of convergence of populations to their equilibrium values. The stability conditions to be derived in this section imply *exponential* stability, that is, the populations approach the equilibrium faster than an exponential (Definition 2.7).

Combining the two aspects of connective stability discussed above, we come up with the following:

Definition 5.1. *The equilibrium* $x^* = 0$ *of the system* (5.25) *is absolutely and exponentially connectively stable if there exist two positive numbers* Π *and* π *independent of the initial conditions* (t_0, x_0) *such that*

$$\|x(t; t_0, x_0)\| \leqslant \Pi \|x_0\| \exp[-\pi(t - t_0)] \qquad \forall t \in \mathcal{T}_0 \tag{5.31}$$

for all $(t_0, x_0) \in \mathcal{T} \times \mathcal{R}^n$, *all* $\varphi_i \in \Phi_i$, $\varphi_{ij} \in \Phi_{ij}$, *and all interconnection matrices* $E \in \overline{E}$.

To establish the kind of stability specified by Definition 5.1, we denote by $\overline{A} = (\overline{a}_{ij})$ the constant $n \times n$ matrix with elements

$$\overline{a}_{ij} = \begin{cases} -\alpha_i + \overline{e}_{ii}\alpha_{ii}, & i = j,\\ \overline{e}_{ij}\alpha_{ij}, & i \neq j,\end{cases} \tag{5.32}$$

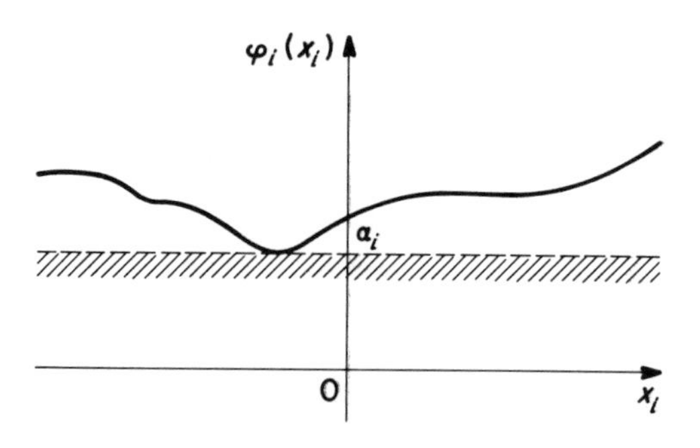

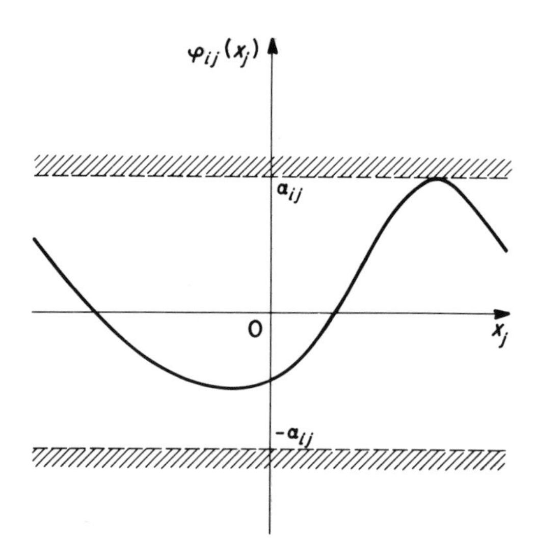

FIGURE 5.3. Nonlinear constraints.

where the $\bar{e}_{ij}$'s take on values 1 or 0 according to the fundamental interconnection matrix $\bar{E}$. Since the elements $\bar{a}_{ij}$ of $\bar{A}$ are such that

$$\bar{a}_{ij} \begin{cases} < 0, & i = j, \\ \geqslant 0, & i \neq j, \end{cases} \tag{5.33}$$

$\bar{A}$ is a Metzler matrix regardless of the sign of interactions in the original nonlinear matrix function $A(t, x)$. The Metzler matrix was introduced in mathematical economics and was used considerably in Chapter 4. We recall that a Metzler matrix $\bar{A}$ is stable if and only if it is quasidominant diagonal (Theorem A.10), that is, satisfies the conditions (5.5). This is equivalent to saying that $\bar{A}$ is a Hicks matrix, that is, the sign of its kth-order principal minor is $(-1)^k$. This is also a well-known result in economics (see Chapter 4). The Hicksian property of a Metzler matrix $\bar{A}$, and thus stability, is equivalent to the Sevastyanov-Kotelyanskii inequalities (Theorem A.9)

$$(-1)^k \begin{vmatrix} \bar{a}_{11} & \bar{a}_{12} & \cdots & a_{1k} \\ \bar{a}_{21} & \bar{a}_{22} & \cdots & \bar{a}_{2k} \\ \cdots & \cdots & \cdots & \cdots \\ \bar{a}_{k1} & \bar{a}_{k2} & \cdots & a_{kk} \end{vmatrix} > 0, \qquad k = 1, 2, \ldots, n, \tag{5.34}$$

whereby only the signs of the leading principal minors of $\bar{A}$ need be tested for stability. For these properties of Metzler matrices see the Appendix.

We prove the following:

Theorem 5.1. *The equilibrium population* $x^* = 0$ *of the community* (5.25) *is absolutely, exponentially, and connectively stable if and only if the* $n \times n$ *constant matrix* $\bar{A}$ *defined by* (5.29) *and* (5.32) *is a quasidominant diagonal matrix.*

Proof. To prove the "if" part of Theorem 5.1, let us consider a decrescent, positive definite, and radially unbounded function $v: \mathfrak{R}^n \to \mathfrak{R}_+$,

$$v(x) = \sum_{i=1}^{n} d_i |x_i|, \tag{5.35}$$

which was used by Rosenbrock (1963), as a candidate for Liapunov's function. Here $d = (d_1, d_2, \ldots, d_n)^T$ is a constant, as yet unspecified n-vector with positive components ($d > 0$).

We calculate the derivative $D^+ v[x(t)]$ of the function $v(x)$ with respect to (5.25). Since the derivative of $|x_i(t)|$ need not exist at a point where $x_i(t) = 0$, it is necessary to use the right-hand derivative $D^+|x_i(t)|$. For this purpose, a functional σ_i is defined as

$$\sigma_i = \begin{cases} 1 & \text{if } x_i(t) > 0, \quad \text{or if } x_i(t) = 0 \text{ and } \dot{x}_i(t) > 0, \\ 0 & \text{if } x_i(t) = 0 \quad \text{and } \dot{x}_i = 0, \\ -1 & \text{if } x_i(t) < 0, \quad \text{or if } x_i(t) = 0 \text{ and } \dot{x}_i(t) < 0, \end{cases} \tag{5.36}$$

where $x_i(t) \in C^1(\mathfrak{T})$. Then

$$\begin{aligned} D^+\nu(x) &= \sum_{i=1}^{n} d_i\sigma_i\dot{x}_i \\ &= \sum_{i=1}^{n} d_i\sigma_i \sum_{j=1}^{n} a_{ij}x_j \\ &= \sum_{j=1}^{n} d_j\sigma_j x_j a_{jj} + \sum_{j=1}^{n} x_j \sum_{\substack{i=1\\ i\neq j}}^{n} d_i\sigma_i a_{ij} \\ &\leqslant -\sum_{j=1}^{n} d_j|x_j||\bar{a}_{jj}| + \sum_{j=1}^{n}|x_j| \sum_{\substack{i=1\\ i\neq j}}^{n} d_i|\bar{a}_{ij}|. \end{aligned} \tag{5.37}$$

Since $\bar{A}$ is quasidominant diagonal, there exist a vector $d > 0$ and a number $\pi > 0$ such that

$$|\bar{a}_{jj}| - d_j^{-1}\sum_{\substack{i=1\\ i\neq j}}^{n} d_i|\bar{a}_{ij}| \geqslant \pi, \qquad j = 1, 2, \ldots, n. \tag{5.38}$$

From (5.37) and (5.38), we get the differential inequality

$$D^+\nu \leqslant -\pi\nu \qquad \forall t \in \mathscr{T}, \quad \forall \nu \in \mathscr{R}_+, \tag{5.39}$$

valid $\forall\varphi_i(t,x) \in \Phi_i$, $\forall\varphi_{ij}(t,x) \in \Phi_{ij}$, $\forall E \in \bar{E}$. By integrating (5.39) we get

$$\nu[x(t)] \leqslant \nu(x_0)\exp[-\pi(t-t_0)] \qquad \forall t \in \mathscr{T}_0, \quad \forall(t_0,x_0) \in \mathscr{T}\times\mathscr{R}^n. \tag{5.40}$$

Using the well-known relationship between the Euclidean and absolute-value norms, $\|x\| \leqslant |x| \leqslant n^{1/2}\|x\|$, we can rewrite (5.40) as

$$\begin{aligned} &\|x(t;t_0,x_0)\| \leqslant \Pi\|x_0\|\exp[-\pi(t-t_0)] \qquad \forall t \in \mathscr{T}_0, \quad \forall(t_0,x_0) \in \mathscr{T}\times\mathscr{R}^n, \\ &\qquad \forall\varphi_i \in \Phi_i, \quad \forall\varphi_{ij} \in \Phi_{ij}, \quad \forall E \in E, \end{aligned} \tag{5.41}$$

with

$$\Pi = n^{1/2}d_M d_m^{-1}, \tag{5.42}$$

where $d_M = \max_i d_i$, $d_m = \min_i d_i$. This proves the "if" part of Theorem 5.1.

For the "only if" part, we select the particular system (5.25) specified by

$$\varphi_i(t,x) = \alpha_i, \quad \varphi_{ij}(t,x) = \alpha_{ij}, \qquad i,j = 1, 2, \ldots, n, \tag{5.43}$$

and the fundamental interconnection matrix $\bar{E}$; that is, the matrix $A(t,x)$ in (5.25) is the constant Metzler matrix $\bar{A}$. If the matrix $\bar{A}$ is not quasidominant diagonal, then the chosen system is unstable for $E = \bar{E}$, and the system (5.25) is not connectively stable for $\varphi_i(t,x) = \alpha_i \in \Phi_i$, $\varphi_{ij}(t,x) = \alpha_{ij} \in \Phi_{ij}$. This proves Theorem 5.1.

The term "resilience" was used by Holling (1973) to designate the ability of an ecosystem to withstand a wide range of general types of shocks while moving from one regime of evolution to another. He points out that this notion of resilience is qualitative and is not amenable to analytical considerations. Our use of the term "reliability" is in the similar spirit, but it is suitable for mathematical analysis.

The result of Theorem 5.1 is compatible with experiments by Gardner and Ashby (1970) and by May (1972): In varying the values of the off-diagonal elements of the community matrix, one has to violate the conditions (5.34)—or, which is the same, the quasidominant conditions (5.38)—in order to produce instability. Furthermore, according to Theorem 5.1, it is irrelevant how one changes the interactions in sign and shape. Stability will prevail as long as the magnitude constraints (5.29) and inequalities (5.5) are satisfied. This fact establishes a considerable robustness of the ecological models that can be represented by Equation (5.25).

5.4. STABILITY REGIONS

In all predator-prey models ranging from the original one of Lotka and Volterra to the models recently examined by Ayala, Gilpin, and Ehrenfeld (1973) and May (1973a), there exists more than one equilibrium. Even in the simplest case of a logistic model for the population growth of a single species, we had two equilibrium populations. Multiple equilibria rule out global stability, and stability (if present) is restricted to a finite region of the population space. In this section, we show how the theory of the preceeding section can be applied to estimate regions of connective stability of the nonlinear and time-varying matrix models of multispecies communities. For this purpose, we need the results of Section 2.8 obtained by Weissenberger (1973), but modified to consider each species of the community as a subsystem.

Once stability is limited to a finite region, the constraints (5.29) need not hold globally for all $x \in \mathfrak{R}^n$, but can be required to hold only on a bounded region $\tilde{\mathfrak{X}}_0 \subset \mathfrak{R}^n$. That is, we set

$$\alpha_i = \inf_{\mathfrak{T}\times\tilde{\mathfrak{X}}_0} \varphi_i(t, x), \qquad \alpha_{ij} = \sup_{\mathfrak{T}\times\tilde{\mathfrak{X}}_0} |\varphi_{ij}(t, x)| \tag{5.44}$$

and appropriately define the modified versions Φ_i^0, Φ_{ij}^0 of the classes of functions (5.30) which correspond to a region

$$\tilde{\mathfrak{X}}_0 = \{x \in \mathfrak{R}^n \colon |x_i| < v_i, \quad i = 1, 2, \ldots, n\}, \tag{5.45}$$

where v_i are positive numbers. Now we can state an "absolute" version of Definition 2.27 as follows:

Definition 5.2. *The region of absolute, exponential, and connective stability of the equilibrium* $x^* = 0$ *of the system* (5.25) *is the set* $\mathfrak{X}$ *of all points* x_0 *for which the inequality* (5.31) *is satisfied for all* $t_0 \in \mathfrak{T}$, *all* $\varphi_i \in \Phi_i^0$, $\varphi_{ij} \in \Phi_{ij}^0$, *and all* $E \in \bar{E}$.

By using the Liapunov functions defined in Section 2.8 as

$$\nu_1(x) = \sum_{i=1}^{n} d_i |x_i|, \tag{5.46}$$

$$\nu_2(x) = \max_{1 \leqslant i \leqslant n} \{d_i^{-1} |x_i|\}, \tag{5.47}$$

$$\nu_3(x) = x^T H x, \tag{5.48}$$

and shown in Figure 2.9, we can provide *estimates*

$$\tilde{\mathfrak{X}}^k = \{x \in \mathfrak{R}^n \colon \nu_k(x) < \gamma_k\}, \qquad k = 1, 2, 3, \tag{5.49}$$

of the region $\mathfrak{X}$ such that $\tilde{\mathfrak{X}}^k \subset \mathfrak{X}$ and $x^* \in \tilde{\mathfrak{X}}^k$ for $k = 1, 2, 3$.

The results obtained in Section 2.8 enable us to prove the following:

Theorem 5.2. *The regions* $\tilde{\mathfrak{X}}^k \subset \tilde{\mathfrak{X}}_0$,, $k = 1, 2, 3$, *defined by* 5.49 *and*

$$\gamma_1 = \min_{1 \leqslant i \leqslant n} \{d_i v_i\}, \tag{5.50}$$

$$\gamma_2 = \min_{1 \leqslant i \leqslant n} \{d_i^{-1} v_i\}, \tag{5.51}$$

$$\gamma_3 = (\det H) \min_{1 \leqslant i \leqslant n} \{\mu_i^{-1} v_i^2\}, \tag{5.52}$$

where μ_i *is the ith leading principle minor of the* $n \times n$ *matrix* H, *are estimates of the region* $\mathfrak{X}$ *if the* $n \times n$ *constant matrix* $\bar{A}$ *defined by* (5.32) *and* (5.44) *is a quasidominant diagonal matrix.*

Proof. We establish Theorem 5.2 for the region $\tilde{\mathfrak{X}}_1$. From (5.39) and (5.46), we get

$$\begin{rcases} d_m \|x\| \leqslant \nu_1(x) \leqslant d_M n^{1/2} \|x\| \\ D^+ \nu_1(x) \leqslant -\pi \nu_1(x) \end{rcases} \quad \forall x \in \tilde{\mathfrak{X}}_0. \tag{5.53}$$

Since $\tilde{\mathfrak{X}}_1 \subset \tilde{\mathfrak{X}}_0$, the inequalities (4.53) imply that the solutions $x(t; t_0, x_0)$ of Equation (5.25) which start in $\tilde{\mathfrak{X}}_1$ (that is, $x_0 \in \tilde{\mathfrak{X}}_1$) satisfy the exponential inequality (5.31) so long as they stay in $\tilde{\mathfrak{X}}_1$. But once $x(t; t_0, x_0)$ starts in $\tilde{\mathfrak{X}}_1$, it cannot leave $\tilde{\mathfrak{X}}_1$, since that would require crossing of the boundary of $\tilde{\mathfrak{X}}_1$ from the inside out, which is impossible by the definition of $\tilde{\mathfrak{X}}_1$ and the inequalities (4.53). The boundary of $\tilde{\mathfrak{X}}_1$ is a Liapunov-function surface, and crossing it from the inside out would mean that the Liapunov function $\nu_1(x)$ increases along a solution $x(t; t_0, x_0)$ of (5.25). This is clearly in contradiction with the second inequality in (5.53), which states that $\nu_1(x)$ decreases throughout the region $\tilde{\mathfrak{X}}_1$ including its boundary, since $\tilde{\mathfrak{X}}_1 \subset \tilde{\mathfrak{X}}_0$. Similar arguments can be advanced for the two remaining regions $\tilde{\mathfrak{X}}_2$ and $\tilde{\mathfrak{X}}_3$ corresponding to the functions $\nu_2(x)$ and $\nu_3(x)$ defined in (5.46)–(5.48). This proves Theorem 5.2.

An application of Theorem 5.2 is presented in the next section, where we consider regions of stability for the classical Lotka-Volterra models of interacting species. The exposition uses the results obtained by Šiljak (1977d) in the context of a Lotka-Volterra model for the arms race.

5.5. LOTKA-VOLTERRA MODEL

Let us consider now the Lotka-Volterra model (Section 1.6) generalized to n interacting species, which is described by the equations

$$\dot{x}_i = x_i\left(c_i + \sum_{j=1}^{n} b_{ij}x_j\right), \qquad i = 1, 2, \ldots, n. \tag{5.54}$$

These equations are different from the original Lotka-Volterra equations which can be obtained from (5.54) by setting all b_{ii}'s equal to zero. The inclusion of nonzero numbers b_{ii} follows from the requirement that each species has a logistic population growth when all are decoupled from each other, that is, when $b_{ij} = 0$, $i \neq j$. This requirement was justified in Section 5.2, by invoking the limitation of resources in the environment and the effect of *crowding*, as reflected in the frequency of collisions between pairs of individuals in the competition for food and space. Therefore, we again assume that $c_i, b_{ii} > 0$ for all $i = 1, 2, \ldots, n$. No assumptions are made concerning the sign of the interaction coefficients b_{ij}, $i \neq j$.

Equilibrium populations are determined by

$$x_i\left(c_i + \sum_{j=1}^{n} b_{ij}x_j\right) = 0. \tag{5.55}$$

From (5.55), we immediately conclude that the origin $x = 0$ of the population space is an equilibrium for the community (5.54). This equilibrium is not interesting, so we assume that $x \neq 0$. In that case, (5.55) becomes a matrix equation

$$c + Bx = 0, \tag{5.56}$$

where $B = (b_{ij})$ is an $n \times n$ constant matrix and $c = \{c_1, c_2, \ldots, c_n\}$ is an n-vector.

We assume that there exists an equilibrium population $x^* > 0$ as a positive solution

$$x^* = -B^{-1}c \tag{5.57}$$

of Equation (5.56). This assumption is consistent with consideration of community stability, but it is not simple to express the conditions for its validity in terms of the B. In the special case when the matrix B has all off-diagonal elements nonnegative (that is, it is a Metzler matrix), then as shown in the preceeding chapter, stability of B implies (and is implied by) the fact that $x^* > 0$ exists. In other words, it is possible to show (see Appendix) that for a Metzler matrix B, the diagonal-dominance condition (5.5) is equivalent to saying that $-B^{-1}$ is nonnegative element by element (Theorem A.2 and 4). Since B^{-1} cannot have a row of zeros, positivity of c implies positivity of x^*. Nonnegativity of the off-diagonal elements of B implies a "gross mutualism" among the species in the community, which is of limited interest in population ecology. Considerations of population equilibria in the context of general interactions among species can be found in the book by Goel, Maitra, and Montroll (1971). One can certainly benefit from the similar considerations in the framework of the general economic equilibrium theory discussed in Chapter 4.

To investigate the stability of $x^* > 0$, we use the standard transformation $x = x^* + y$ of (5.9) and rewrite the equations (4.54) as

$$\dot{y}_i = (y_i + x_i^*) \sum_{j=1}^{n} b_{ij} y_j, \qquad i = 1, 2, \ldots, n. \tag{5.58}$$

The equilibrium $y^* = 0$ of (5.58) corresponds now to the equilibrium $x^* > 0$ of (5.54), and we can readily identify the model (5.58) as our general matrix model (5.25), where the $n \times n$ community matrix $A(y)$ has the coefficients $a_{ij}(y)$ specified as

$$a_{ij}(y) = \begin{cases} -b_{ii}(y_i + x_i^*), & i = j, \\ e_{ij} b_{ij}(y_i + x_i^*), & i \neq j. \end{cases} \tag{5.59}$$

Therefore, the nonlinear functions of (5.28) are

$$\begin{gathered} \varphi_i(y) = b_{ii}(y_i + x_i^*), \qquad \varphi_{ij}(y) = b_{ij}(y_i + x_i^*), \\ i, j = 1, 2, \ldots, n, \quad i \neq j. \end{gathered} \tag{5.60}$$

We should note here that the functions $\varphi_i(y)$, $\varphi_{ij}(y)$ are precisely defined, and the "absolute" aspect of Theorem 5.2 should be dropped.

From (5.60) and Figure 5.3, we see that the constraints

$$\varphi_i(y) \geqslant \alpha_i, \quad |\varphi_{ij}(y)| \leqslant \alpha_{ij}, \qquad i, j = 1, 2, \ldots, n, \tag{5.61}$$

with $\alpha_i > 0$, $\alpha_{ij} \geqslant 0$, are satisfied on the open region

$$\tilde{Y}_0 = \{y \in \mathcal{R}^n : |y_i| < \beta_i x_i^*, i = 1, 2, \ldots, n\} \tag{5.62}$$

defined by (5.45) when $\tilde{X}_0$ and x are replaced by $\tilde{Y}_0$ and y. In (5.62), β_i are numbers that measure the size of the region $\tilde{Y}_0$, and are such that

$$0 < \beta_i < 1, \qquad i = 1, 2, \ldots, n. \tag{5.63}$$

From the diagram of the function $y_i + x_i^*$ given in Figure 5.4, we calculate the numbers α_i, α_{ij} as

$$\alpha_i = (1 - \beta_i)|b_{ii}|x_i^*, \qquad \alpha_{ij} = (1 + \beta_i)|b_{ij}|x_i^*, \qquad i, j = 1, 2, \ldots, n. \tag{5.64}$$

Now an estimate $\tilde{Y}$ for a stability region Y of the ecomodel (5.58) can be obtained by imbedding a Liapunov function

$$v_1(y) = \sum_{i=1}^{n} d_i|y_i| \tag{5.65}$$

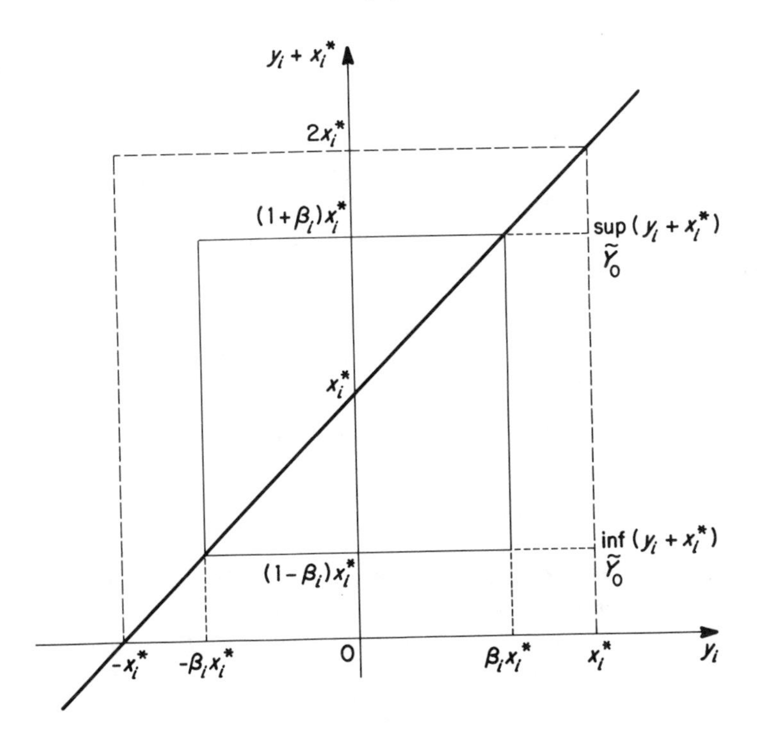

FIGURE 5.4. The Function $y_i + x_i^*$.

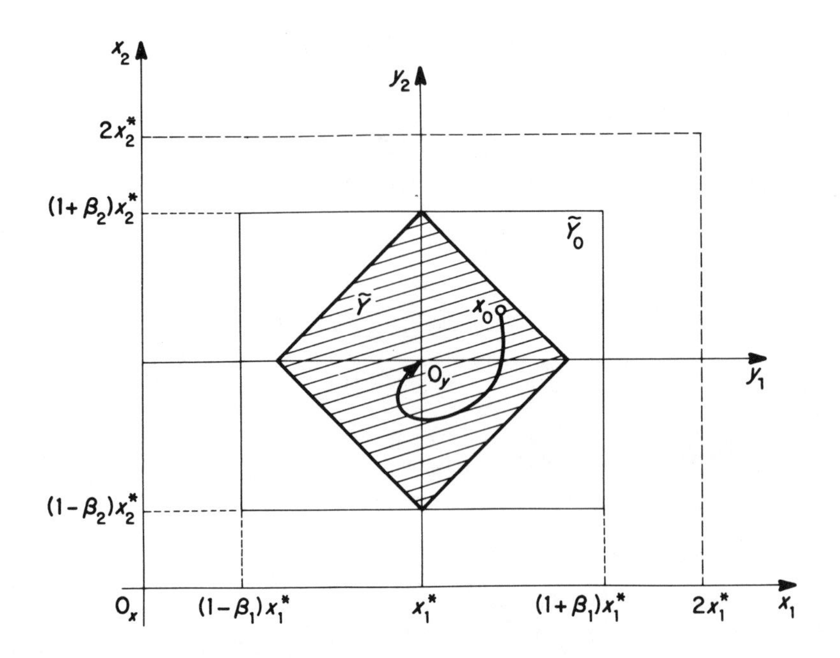

FIGURE 5.5. Stability region.

in the region $\tilde{Y}_0$ as shown in Figure 5.5, which corresponds to the case $n = 2$. Although we can establish conditions under which $\nu_1(y) > 0$, $D^+\nu_1(y) < 0$ throughout the region $\tilde{Y}_0$, we need to imbed a Liapunov surface in $\tilde{Y}_0$. Only then we can be sure that all solutions $y(t; t_0, y_0)$ of (5.58) that start inside $\tilde{Y}$ stay there for all future times and approach the equilibrium $y^* = 0$ as time progresses.

If we identify the model (5.58) as

$$\dot{y} = A(y)y, \tag{5.66}$$

then the $n \times n$ constant matrix $\bar{A} = (\bar{a}_{ij})$ defined by (5.32) is given by

$$\bar{a}_{ij} = \begin{cases} -\alpha_i, & i = j, \\ \bar{e}_{ij}\alpha_{ij}, & i \neq j, \end{cases} \tag{5.67}$$

where $\bar{e}_{ij}$ are binary elements of the $n \times n$ fundamental interconnection matrix $\bar{E}$. Furthermore, if we define a diagonal matrix $D = \text{diag}\{x_1^*, x_2^*, \ldots, x_n^*\}$, then from (5.64) and (5.67) we get $\bar{A} = D\bar{B}$, where the $n \times n$ matrix $\bar{B} = (\bar{b}_{ij})$ is defined as

$$\bar{b}_{ij} = \begin{cases} -(1 - \beta_i)|b_{ii}|, & i = j, \\ (1 + \beta_i)|b_{ij}|, & i \neq j. \end{cases} \tag{5.68}$$

It is a well-known fact (Theorem A.7) that for a pair $(\overline{A}, \overline{B})$ of Metzler matrices, the quasidominant diagonal property of $\overline{B}$ implies and is implied by the same property of $\overline{A} = D\overline{B}$ for any positive diagonal matrix D. The condition

$$|\overline{b}_{ii}| - d_i^{-1} \sum_{\substack{j=1 \\ j \neq i}}^{n} d_j |\overline{b}_{ij}| \geqslant \pi_b, \qquad i = 1, 2, \ldots, n, \tag{5.69}$$

is equivalent to

$$|x_i^* \overline{b}_{ii}| - d_i^{-1} \sum_{\substack{j=1 \\ j \neq i}}^{n} d_j |x_j^* \overline{b}_{ij}| \geqslant x_i^* \pi_b, \qquad i = 1, 2, \ldots, n, \tag{5.70}$$

which is the condition

$$|\overline{a}_{ii}| - d_i^{-1} \sum_{\substack{j=1 \\ j \neq i}}^{n} d_j |\overline{a}_{ij}| \geqslant \pi_a, \qquad i = 1, 2, \ldots, n, \tag{5.71}$$

where $\pi_a = x_m^* \pi_b$ and $x_m^* = \min_i x_i^*$. The condition (5.71) is equivalent to the original quasidominant diagonal condition (5.38) or (5.5). This is because for a Metzler matrix $\overline{A}$, (5.38) is equivalent to the determinantal conditions (5.34). These conditions, however, always hold for $\overline{A}^T$ whenever they hold for $\overline{A}$. That is, the original quasidominant diagonal property (5.38) is equivalent to the conditions

$$(-1)^k \begin{vmatrix} \overline{a}_{11} & \overline{a}_{21} & \cdots & \overline{a}_{k1} \\ \overline{a}_{12} & \overline{a}_{22} & \cdots & \overline{a}_{k2} \\ \cdots & \cdots & \cdots & \cdots \\ \overline{a}_{1k} & \overline{a}_{2k} & \cdots & \overline{a}_{kk} \end{vmatrix} > 0, \qquad k = 1, 2, \ldots, n. \tag{5.72}$$

Since $\overline{A}^T$ is also a Metzler matrix, these conditions are equivalent to (5.71) and thus to (5.69). Therefore, $\overline{A}$ is a quasidominant diagonal matrix if and only if $\overline{B}$ is. The quasidominant diagonal property of $\overline{B}$ is equivalent to the conditions

$$(-1)^k \begin{vmatrix} \overline{b}_{11} & \overline{b}_{12} & \cdots & \overline{b}_{1k} \\ \overline{b}_{21} & \overline{b}_{22} & \cdots & \overline{b}_{2k} \\ \cdots & \cdots & \cdots & \cdots \\ \overline{b}_{k1} & \overline{b}_{k2} & \cdots & \overline{b}_{kk} \end{vmatrix} > 0, \qquad k = 1, 2, \ldots, n, \tag{5.73}$$

which can be easily verified.

Now we are ready to prove the following:

Theorem 5.3. *Assume that for a given ecomodel* (5.58) *a region* $\tilde{Y}_0$ *is chosen as in* (5.62), *and the matrix* $\overline{B}$ *is shown to be quasidominant diagonal. Then the open region*

$$\tilde{Y} = \{y \in \mathfrak{R}^n : \nu_1(y) < \gamma_1\} \tag{5.74}$$

with

$$\gamma_1 = \min_{1 \leqslant i \leqslant n} \{d_i \beta_i x_i\} \tag{5.75}$$

is the largest estimate of the region Y of exponential connective stability of the equilibrium population $y^* = 0$ *in the ecomodel* (5.58), *which can be obtained by the Liapunov function* $\nu_1(y)$ *defined in* (5.65).

Proof. If $\overline{B}$ is quasidominant diagonal, so is $\overline{A}$, as shown above. Then by computing $D^+\nu_1(y)$ of $\nu_1(y)$ with respect to the equation (5.66), we get

$$D^+\nu_1(y) \leqslant -\pi\nu_1(y) \qquad \forall y \in \tilde{Y}, \quad \forall E \in \overline{E}, \tag{5.76}$$

as in the proof of Theorem 5.1. Now, we imbed the largest diamond $\tilde{Y}$ inside $\tilde{Y}_0$, as shown in Figure 5.5, which is specified by γ_1 of (5.75). If $y_0 \in \tilde{Y}$, then $y(t; t_0, y_0) \subset \tilde{Y}$ for all $t \in \mathfrak{T}_0$, and $\tilde{Y}$ is an invariant region. This is simply because $y(t; t_0, y_0)$ cannot pierce the surface $\nu_1(y) = \gamma_1$ from the inside, since that would violate the inequality (5.76) and the fact that $\nu_1(y) > 0$ for all $y \in \mathfrak{R}^n_+$ except $y = 0$. Therefore, from (5.76), we derive

$$\|y(t; t_0, y_0\| \leqslant \Pi\|y_0\| \exp[\pi(t - t_0)] \qquad \forall t \in \mathfrak{T}_0, \tag{5.77}$$

which is valid for all $(t_0, y_0) \in \mathfrak{T} \times \tilde{Y}$ and all $E \in \overline{E}$. The region $\tilde{Y}$ is the largest estimate of the region Y of exponential connective stability of $y^* = 0$ in (5.58), which corresponds to the equilibrium $x^* > 0$ of the original Lotka-Volterra model (5.54). This proves Theorem 5.3.

It is of interest to explain the role of the numbers β_i which determine the size of the region $\tilde{Y}_0$ (Figure 5.4). As the β_i's increase, so does the region $\tilde{Y}_0$, and so accordingly does the stability-region estimate $\tilde{Y}$. At the same time, however, it becomes more difficult to establish the quasidominant-diagonal property of the matrix $\overline{B}$ defined in (5.69), since an increase in the β_i's means a decrease in the diagonal elements of $\overline{B}$ and an increase in the off-diagonal elements of $\overline{B}$. This also produces a smaller π_b, and ultimately a smaller degree of exponential stability π. Therefore, the numbers β_i establish a trade-off between the estimates of the stability region $\tilde{Y}$ and the stability margin represented by π.

We should also note that computation of the d_i's in (5.75) is similar to that explained in Section 2.8, except that the aggregation matrix W should be replaced by the matrix $\bar{A}$ defined in (5.67).

We should also mention a possibility (Šiljak, 1978) of improving the estimates $\tilde{Y}$ by using the Liapunov function $\nu_2(y)$ defined by (5.47). An improvement is possible because the estimates $\tilde{Y}$ and the regions $\tilde{Y}_0$ have the same form. Finally, we should point out that such results, as well as those obtained above, can easily incorporate the generalized Lotka-Volterra models (Ayala, Gilpin, and Ehrenfeld, 1973). When the coefficients b_{ij} in (5.54) are nonlinear and time-varying functions $b_{ij}(t, x)$, the community model $\dot{x} = A(t, x)x$ is ideally suitable for the analysis of interaction effects on the stability of the population equilibria.

An immediate question is: Can we show that the entire first quadrant $\mathcal{R}_+^n$ of the population space $\mathcal{R}^n$ is a stable region? A positive answer to this question was proposed by MacArthur (1970), who showed that $\mathcal{R}_+^n$ is a stable region provided the $n \times n$ matrix $B = (b_{ij})$ of the model (5.54) is a symmetric positive definite matrix. The stringent symmetry condition was removed by Goh (1976, 1977) using Volterra's function as a Liapunov function. For the Lotka-Volterra model

$$\dot{x}_i = x_i\left(c_i + \sum_{j=1}^{n} b_{ij} x_j\right), \qquad i = 1, 2, \ldots, n \tag{5.54}$$

with the nontrivial equilibrium at $x^* > 0$, let us use Volterra's function

$$\nu(x) = \sum_{i=1}^{n} d_i[(x_i - x_i^* - x_i \ln(x_i/x_i^*)], \tag{5.78}$$

where d_i's are all positive (yet unspecified) numbers, and compute

$$\begin{aligned} \dot{\nu}(x) &= \sum_{i=1}^{n} d_i(x_i - x_i^*)(\dot{x}_i/x_i) \\ &= \sum_{i=1}^{n} d_i(x_i - x_i^*)\left(c_i + \sum_{j=1}^{n} b_{ij} x_j\right) \\ &= \sum_{i=1}^{n} d_i(x_i - x_i^*) \sum_{j=1}^{n} b_{ij}(x_j - x_j^*) \\ &= \tfrac{1}{2}(x - x^*)^T(B^T D + DB)(x - x^*) \end{aligned} \tag{5.79}$$

where $D = \text{diag}\{d_1, d_2, \ldots, d_n\}$. Obviously, we have $\nu(x)$, $-\dot{\nu}(x) > 0$ for all $x > 0$ if the matrix G defined as

$$G = B^T D + DB, \tag{5.80}$$

is negative definite. Asymptotic stability of x^* in the region

$$\mathcal{X} = \{x \in \mathcal{R}_+^n : x > 0\}, \tag{5.81}$$

follows from $\nu(x)$, $-\dot{\nu}(x) > 0$, and the fact that $\nu(x) \to +\infty$ as $x_i \to 0^+$ and $x_i \to +\infty$, $i = 1, 2, \ldots, n$ (Goh, 1978). A question now arises: Under what conditions on B there exists a D so that G is negative definite? If B is a Metzler matrix, then from Section 2.6, equation (2.245), we conclude that there exists a D with positive elements such that G is negative definite if and only if B is a quasidominant diagonal matrix. In the context of the Lotka-Volterra model, the Metzler structure of B is too restrictive. Therefore, when comparing Goh's approach with the one described previously, we conclude that there is a trade–off between the extent of the stability region that can be established for a given model, and simplicity of the corresponding stability conditions.

Toward the end of his paper, MacArthur considers structural perturbations of the Lotka-Volterra model when a species is added to (or removed from) the community. The negative deffinitness of the matrix B makes it possible to study such perturbations, as can be seen in Section 2.6, where they were called the "principal structural perturbations". The simple well-known fact is the if a matrix B is positive definite, so are all its principal submatrices. In the context of the Lotka-Volterra model, the principal submatrices represent situations when a number of species are removed from the community. Such situations include the case when a number of species perish due to changes in the environment of the community. By using the analysis of Section 2.6, MacArthur's results can be further generalized to include nonlinear and time-varying coefficients b_{ij} of more complex Lotka-Volterra models. In this case, however, one should be aware of the fact that in presence of structural perturbations, the equilibrium is "moving" (see Section 4.5), and the connective property of stability should be introduced with care (see also Goh, 1978).

5.6. INSTABILITY

Instability analysis of population dynamics appears in connection with the problem of stability at one trophic level versus web stability, and so far is on a conjecture-experiment level, as reported by May (1973a). Two basic questions are: (1) if one particular trophic level is unstable, can the total community be stable? (Paine's conjecture, 1966); and (2) if one level in isolation is stable, can the total system be unstable? (Watt's conjecture, 1968). As May pointed out, by judicious choice of interactions, both situations are possible. In this section, we will not try to answer either of the two questions, but rather determine conditions under which the instability of species when isolated implies community instability. Such a result, however, provides a partial answer to both of the above questions, and forms a rigorous basis for further exploration of these questions in the multilevel stability analysis presented at the end of this chapter.

In proving Theorem 5.1, we implicitly used the notion of connective instability. The system (5.25) was said to be connectively unstable if it was not stable in the sense of Liapunov for at least one interconnection matrix. This is a natural instability counterpart to Definition 5.1 and can be resolved using the results of Grujić and Šiljak (1973b). Since Theorem 5.1 provides necessary and sufficient conditions for the stability of (5.25), they can be used to test for instability in the context of structural perturbations. Therefore, we are not interested in pursuing such investigations, but rather show that the mechanism used to prove connective stability can also be used to prove connective instability for all interconnection matrices—this is *complete connective instability*. In the ecological context, it means that no matter how the direction and strength of interactions may vary, the community remains unstable. More precisely, we state the following:

Definition 5.3. *The equilibrium* $x^* = 0$ *of the system* (5.25) *is completely and exponentially connectively unstable in the large if there exist two positive numbers* Π *and* π *independent of the initial conditions* (t_0, x_0) *such that*

$$\|x(t; t_0, x_0)\| \geqslant \Pi \|x_0\| \exp[\pi(t - t_0)] \qquad \forall t \in \mathcal{T}_0 \tag{5.82}$$

for all $(t_0, x_0) \in \mathcal{T} \times \mathcal{R}^n$ *and all interconnection matrices* $E \in \bar{E}$.

Let us assume that the coefficients $a_{ij}(t, x)$ of the system matrix $A(t, x)$ in (5.25) are defined by

$$a_{ij}(t, x) = \begin{cases} \varphi_i(t, x) - e_{ii}(t)\varphi_{ii}(t, x), & i = j, \\ -e_{ij}(t)\varphi_{ij}(t, x), & i \neq j, \end{cases} \tag{5.83}$$

and that there exist numbers $\alpha_i > 0$, $\alpha_{ij} \geqslant 0$ such that $\alpha_i > \alpha_{ii}$ and the constraints (5.29) are satisfied by the functions $\varphi_i(t, x)$, $\varphi_{ij}(t, x)$. Then we specify the elements $\bar{a}_{ij}$ of the matrix $\bar{A} = (\bar{a}_{ij})$ by the corresponding matrix $\bar{E}$.

We prove the following:

Theorem 5.4. *The equilibrium population* $x^* = 0$ *of the community* (5.25) *is completely and exponentially connectively unstable if the* $n \times n$ *constant matrix* $-\bar{A} = (-\bar{a}_{ij})$ *defined by* (5.29) *and* (5.32) *is a quasidominant diagonal matrix.*

Proof. Let us consider again the function $\nu(x)$ of (5.35), and let us compute the function $D^+\nu(x)$ with respect to (5.25) as

$$\begin{aligned} D^+\nu(x) &= \sum_{i=1}^{n} d_i \sigma_i \dot{x}_i \\ &= \sum_{i=1}^{n} d_i \sigma_i \sum_{j=1}^{n} a_{ij} x_j \\ &= \sum_{j=1}^{n} d_j \sigma_j x_j a_{jj} + \sum_{j=1}^{n} x_j \sum_{\substack{i=1 \\ i \neq j}}^{n} d_i \sigma_i a_{ij} \\ &\geqslant \sum_{j=1}^{n} d_j |x_j| \bar{a}_{jj} - \sum_{j=1}^{n} |x_j| \sum_{\substack{i=1 \\ i \neq j}}^{n} d_i |\bar{a}_{ij}|. \end{aligned} \tag{5.84}$$

Since $-\bar{A}$ is a Metzler matrix and has a quasidominant diagonal, there exist a vector $d > 0$ and a number $\pi > 0$ such that

$$\bar{a}_{jj} - d_j^{-1} \sum_{\substack{i=1 \\ i \neq j}}^{n} d_i |\bar{a}_{ij}| \geqslant \pi, \qquad j = 1, 2, \ldots, n. \tag{5.85}$$

Therefore, from (5.84) and (5.85) we get

$$D^+\nu \geqslant \pi\nu \qquad \forall t \in \mathfrak{T}_0, \quad \forall \nu \in \mathfrak{R}_+, \quad \forall E \in \bar{E} \tag{5.86}$$

Integrating (5.86) and manipulating the result as we did (5.39), we get the inequality (5.82) with

$$\Pi = n^{-1/2} d_m d_M^{-1}, \tag{5.87}$$

where $d_m = \min_i d_i$, $d_M = \max_i d_i$. This completes the proof of Theorem 5.4.

It is of interest to note that the quasidominant diagonal property of the matrix $-\bar{A}$ can be a necessary and sufficient condition if the instability in Theorem 5.4 is required to be absolute as in Theorem 5.1. If instability is going to hold for all $\varphi_i(t, x) \in \Phi_i$, $\varphi_{ij}(t, x) \in \Phi_{ij}$, it must hold also for $\varphi_i(t, x) = \alpha_i$, $\varphi_{ij}(t, x) = -\alpha_{ij}$, which makes the system (5.25) with $A(t, x) = \bar{A}$ unstable, since $-\bar{A}$ is a quasidominant diagonal matrix.

Theorem 5.4 has little application to population models, since it implies that none of the species in the community is density-dependent. That is, the diagonal elements $a_{ii}(t, x)$ of the matrix $A(t, x)$ are all positive.

There is a possibility of relaxing the positivity conditions placed on all a_{ii}'s and requiring only that one of the populations be not density-dependent. Then, however, we require positivity of all off-diagonal coefficients a_{ij} for that population. More precisely, we have the following:

Theorem 5.5. *The equilibrium population* $x^* = 0$ *of the community* (5.25) *is completely connectively unstable if for some* $i \in \{1, 2, \ldots, n\}$ *there exist numbers* $\alpha_i > 0$, $\alpha_{ij} \geqslant 0$ *such that the coefficients* $a_{ij}(t, x)$ *of the* $n \times n$ *matrix* $A(t, x)$ *defined by* (5.84) *are such that the conditions*

$$\varphi_i(t,x)x_i \geqslant \alpha_i x_i, \quad \varphi_{ij}(t,x)x_j \leqslant -\alpha_{ij}|x_j| \qquad \forall t \in \mathfrak{T}, \quad \forall x \in \Delta \tag{5.88}$$

are satisfied, where $\Delta = \{x \in \mathfrak{R}^n\colon \|x\| < \rho\}$ *and* $\rho > 0$.

Proof. The proof of Theorem 5.5 is almost automatic. Using (5.88), the ith state equation can be rewritten as

$$\begin{aligned} \dot{x}_i &= \varphi_i(t,x)x_i - \sum_{j=1}^{n} e_{ij}(t)\varphi_{ij}(t,x)x_j \\ &\geqslant \alpha_i x_i + \sum_{j=1}^{n} e_{ij}(t)\alpha_{ij}|x_j| \\ &\geqslant \alpha_i x_i \qquad \forall (t,x) \in \mathfrak{T} \times \Delta, \quad \forall E \in \overline{E}. \end{aligned} \tag{5.89}$$

By integrating the last inequality, we get

$$\begin{aligned} |x_i(t; t_0, x_0)| \geqslant x_{i0} \exp[\alpha_i(t - t_0)] \qquad \forall t \in \mathfrak{T}_0, \quad \forall (t_0, x_0) \in \mathfrak{T} \times \Delta, \\ \forall E \in \overline{E}, \end{aligned} \tag{5.90}$$

where $x_{i0} = x_i(t_0) > 0$. The inequality (5.90) implies complete connective instability of the equilibrium $x^* = 0$ in (5.25), and the proof of Theorem 5.5 is completed.

5.7. ENVIRONMENTAL FLUCTUATIONS

The ecomodels considered so far have described only the internal interactions among the species in a community which is "closed". That is, there were no terms in the models that would explicitly reflect influences of the environment on the affairs that go on in the community. Among other things, such terms may represent changes in the climate conditions, pollution effects, application of pesticides, etc. A problem of interest in this context is to find out how much of the environmental fluctuations can be tolerated by stable communities.

Again, a good place to start is a linear model

$$\dot{x} = Ax + b(t, x), \tag{5.91}$$

where $A = (a_{ij})$ is an $n \times n$ constant matrix and $b(t, x)$ is continuous for small $\|x\|$ and $t \in \mathfrak{T}_0$, and $b(t, 0) = 0$ for all $t \in \mathfrak{T}$ so that $x^* = 0$ is the equilibrium of (5.91). As expected, the nature of solutions of Equation (5.91) depends to a great extent on the properties of the linear model

$$\dot{x} = Ax, \tag{5.1}$$

especially when the perturbation function $b(t, x)$ is dominated by the function Ax. There are numerous results (e.g. Coddington and Levinson, 1955; Hale, 1969) which relate the stability of the system (5.1) to that of (5.91), and they can be traced back to Liapunov's work. The basic result is that if the matrix A has all eigenvalues with negative real parts and

$$\|b(t, x)\| = o(\|x\|) \tag{5.92}$$

as $\|x\| \to 0$ uniformly in $t \in \mathfrak{T}_0$ (that is, $\|b(t, x)\|/\|x\| \to 0$ uniformly in t with $\|x\| \to 0$), then the equilibrium $x^* = 0$ of (5.91) is exponentially stable. This result can be established by using Liapunov's direct method. We recall from Section 1.8 that when the linear system (5.1) is asymptotically stable, there is a Liapunov function

$$\nu(x) = x^T H x, \tag{5.93}$$

with the estimates

$$\begin{aligned} \eta_1 \|x\| &\leqslant \nu(x) \leqslant \eta_2 \|x\|, \\ \dot{\nu}(x)_{(5.1)} &\leqslant -\eta_3 \|x\|, \\ \|\text{grad } \nu(x)\| &\leqslant \eta_4, \end{aligned} \tag{5.94}$$

where the η_i's are positive numbers calculated using the symmetric positive definite matrices H and G of the Liapunov matrix equation (Section 1.8)

$$A^T H + HA = -G. \tag{5.95}$$

Taking the total time derivative $\dot{\nu}(x)_{(5.91)}$ of $\nu(x)$ along the solutions of the perturbed Equation (5.91), we get

$$\begin{aligned} \dot{\nu}(x)_{(5.91)} &= \dot{\nu}(x)_{(5.1)} + [\text{grad } \nu(x)]^T b(t, x) \\ &\leqslant -\eta_3 \|x\| + \eta_4 \|b(t, x)\| \\ &\leqslant -\eta_3 \|x\| + \eta_4 o(\|x\|). \end{aligned} \tag{5.96}$$

Therefore, for sufficiently small $\|x\|$, we can find a number $\eta > 0$ such that from (5.96) we obtain

$$\dot{\nu}(x)_{(5.91)} \leqslant -\eta \|x\|, \tag{5.97}$$

and asymptotic stability of $x^* = 0$ in (5.91) follows. When we assume that

$\|b(t,x)\| \leqslant \eta_5 \|x\|^2$, then the extent of the neighborhood of $x^* = 0$ in which (5.97) is valid, is given by $\|x\| \leqslant \eta_3(\eta_4\eta_5 + \eta)^{-1}$.

The above result verifies the linearization procedure of Section 5.2 in the following way: If $Ax = [\partial f(t,0)/\partial x]x$ and $b(t,x) = f(t,x) - Ax$, then $b(t,0) = 0$ and $\partial b(t,0)/\partial x = 0$. The function Ax obviously represents the linear approximation of the function $f(t,x)$ on the right-hand side of the equation $\dot{x} = f(t,x)$.

There are several objections that can be raised concerning the applicability of the above results to ecomodels. The most important one comes from the fact that the external perturbations of an ecological community are not of a form that permits the estimate (5.92). All we may know about the function $b(t,x)$ is that it is bounded, and the condition $b(t,0) = 0$ is not met. In this case, the role of the population equilibrium is taken by a compact region which is reached by all population processes in finite time, and once in the region, they stay there for all future times. Such a property will be established for a nonlinear time-varying model (Šiljak, 1975b),

$$\dot{x} = A(t,x)x + b(t,x), \tag{5.98}$$

where the function b: $\mathcal{T} \times \mathcal{R}^n \to \mathcal{R}^n$ is added on the right-hand side of (5.25). Under the condition that $b(t,x)$ is bounded, we will provide an upper estimate of the mentioned region by the same Liapunov function used to determine global stability properties of the model (5.25). The estimate is directly proportional to the size of the function $b(t,x)$, and it is invariant under structural perturbations. Furthermore, we will show that all population processes $x(t; t_0, x_0)$ of (5.98) reach the estimated region faster than an exponential despite structural changes in the model.

Let us assume that $b(t,x)$ has components of the form

$$b_i(t,x) = l_i(t)\psi_i(t,x), \tag{5.99}$$

where $l_i(t)$ are components of the interconnection vector $l(t) = [l_1(t), l_2(t), \ldots, l_n(t)]^T$, such that $l_i(t) \in [0,1]$ for all $t \in \mathcal{T}$. Similarly, as in case of the matrix $\bar{E}$, we define the binary vector $\bar{l} \in \mathcal{R}_+^n$ as

$$\bar{l}_i = \begin{cases} 1, & b_i(t,x) \not\equiv 0, \\ 0, & b_i(t,x) \equiv 0, \end{cases} \tag{5.100}$$

which means that $\bar{l}_i = 1$ when the ith species is perturbed by $b(t,x)$, and $\bar{l} = 0$ when the ith species is not perturbed by $b(t,x)$.

In (5.99), the functions $\psi_i(t,x) \in C(\mathcal{T} \times \mathcal{R}^n)$ satisfy the conditions

$$|\psi_i(t,x)| \leqslant \beta_i, \qquad i = 1, 2, \ldots, n, \tag{5.101}$$

where the β_i's are nonnegative numbers. The inequalities (5.101) mean that

$b(t, x)$ is bounded. Special cases of perturbation constraints (5.101) are illustrated in Figure 5.6.

We also define a constant vector $\bar{b} \in \mathcal{R}_+^n$ as

$$\bar{b}_i = \bar{l}_i \beta_i. \tag{5.102}$$

With the model (5.98) we associate a compact region

$$\mathfrak{X} = \{x \in \mathcal{R}^n : \|x\| \leqslant \rho\}, \tag{5.103}$$

where ρ is a nonnegative number. In the absence of perturbations, $b(t, x) \equiv 0$ and the region $\mathfrak{X}$ is reduced to the equilibrium $x^* = 0$, that is,

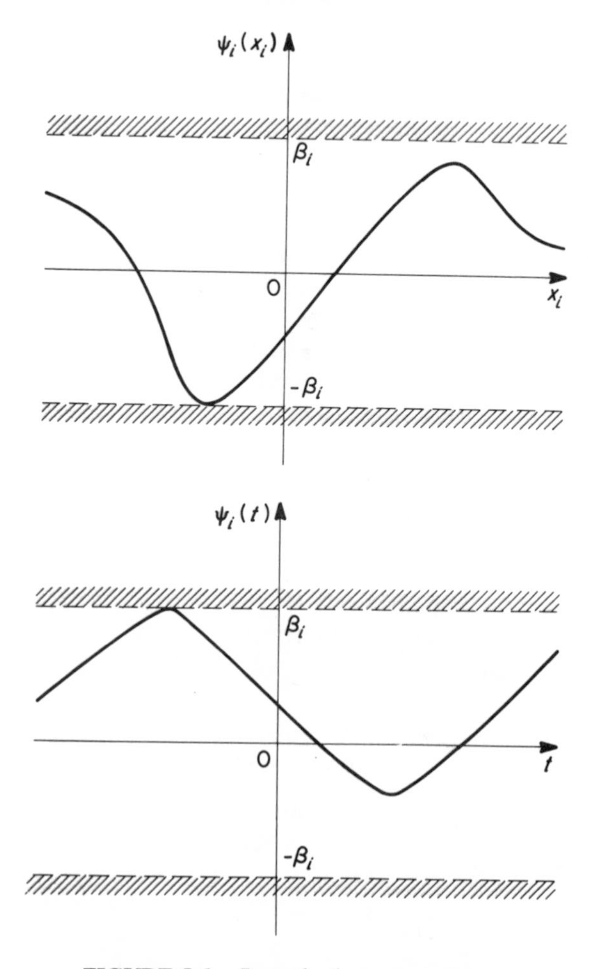

FIGURE 5.6. Perturbation constraints.

$\rho = 0$. By $\mathfrak{X}^c$ we denote the complement of the region $\mathfrak{X}$.

Now we state the following:

Definition 5.4. *The population process* $x(t; t_0, x_0)$ *of the model* (5.91) *is exponentially and ultimately connectively bounded in the large with respect to a region* $\mathfrak{X}$ *defined by* (5.103) *if there exist positive numbers* μ, Π, *and* π *independent of the initial conditions* (t_0, x_0) *such that*

$$\|x(t; t_0, x_0)\| \leqslant \mu + \Pi\|x_0\| \exp[-\pi(t - t_0)] \qquad \forall t \in \mathfrak{T}_0 \qquad (5.104)$$

for all $t_0 \in \mathfrak{T}$, $x_0 \in \mathfrak{X}^c$, $E \in \bar{E}$, *and* $l \in \bar{l}$.

We assume that the community (5.91) is stable in the absence of perturbations, and prove the following:

Theorem 5.6. *The population process* $x(t; t_0, x_0)$ *of the model* (5.91) *is exponentially and ultimately connectively bounded with respect to the region* $\mathfrak{X}$ *for which*

$$\rho = \eta^{-1/2} d_M^{-1} \pi^{-1}(\beta + \varepsilon), \qquad (5.105)$$

where $d_M = \max_i d_i$, $\beta = \sum_{i=1}^n d_i \bar{b}_i$, *and* $\varepsilon > 0$ *is an arbitrarily small number, if the* $n \times n$ *constant matrix* $\bar{A}$ *defined by* (5.32) *and* (5.44) *is a quasidominant diagonal matrix.*

Proof. We use again the function $\nu(x)$ defined in (5.35), and calculate $D^+\nu(x)$ with respect to the system (5.91) to get

$$\begin{aligned} D^+\nu(x) &= \sum_{i=1}^n d_i \sigma_i \left(\sum_{j=1}^n a_{ij} x_j + b_i \right) \\ &\leqslant -\sum_{j=1}^n d_j |x_j| \, |\bar{a}_{jj}| + \sum_{j=1}^n |x_j| \sum_{\substack{i=1 \\ i \neq j}}^n d_i |\bar{a}_{ij}| + \sum_{i=1}^n d_i \bar{b}_i . \end{aligned} \qquad (5.106)$$

Using the fact that $\bar{A}$ is quasidominant diagonal, we can rewrite (5.106) as

$$D^+\nu \leqslant -\pi\nu + \beta \qquad \forall t \in \mathfrak{T}, \quad \forall \nu \in \mathfrak{R}_+, \quad \forall E \in \bar{E}, \quad \forall l \in \bar{l}. \quad (5.107)$$

From (5.107), we can conclude that the region $\tilde{\mathfrak{X}}$ is reached in finite time by every solution $x(t; t_0, x_0)$ which starts in its complement $\tilde{\mathfrak{X}}^c = \{x \in \mathfrak{R}^n : \nu(x) \geqslant \pi^{-1}(\beta + \varepsilon)\}$, where $\varepsilon > 0$ can be chosen arbitrarily small. This is because from (5.107) we get

$$\nu[x(t)] \leqslant \pi^{-1}\beta + \nu(x_0)\exp[-\pi(t - t_0)] \qquad \forall t \in \mathfrak{T}_0, \qquad (5.108)$$

which is valid for all $t_0 \in \mathfrak{T}$, $x_0 \in \tilde{\mathfrak{X}}^c$, and all $E \in \bar{E}$, $l \in \bar{l}$. Imitating the development of (5.40) in the proof of Theorem 5.1, we obtain the inequality (5.104) with

$$\mu = d_m^{-1} \pi^{-1} \beta, \qquad \Pi = n^{1/2} d_M d_m^{-1}, \qquad (5.109)$$

and π given by the quasidominant diagonal condition on $\bar{A}$,

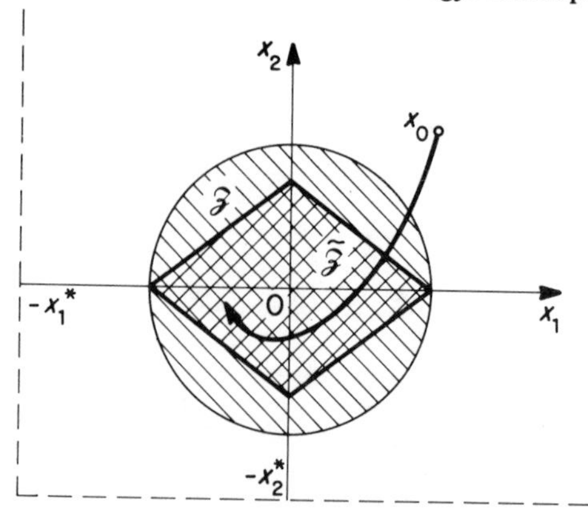

FIGURE 5.7. Ultimate boundedness region.

$$|\bar{a}_{jj}| - d_j^{-1} \sum_{\substack{i=1 \\ i \neq j}}^{n} d_i |\bar{a}_{ij}| \geqslant \pi, \qquad j = 1, 2, \ldots, n. \tag{5.38}$$

The region $\mathfrak{Z} \supset \tilde{\mathfrak{Z}}$ is now given by Equation (5.105). Since (5.108) is valid for all $E \in \bar{E}$, $l \in \bar{l}$, so are Π, π, μ, ρ, and thus the region $\mathfrak{Z}$. This proves Theorem 5.6.

It is now simple to compute from (5.107) an upper estimate

$$t_1 = t_0 + \pi^{-1} \ln[\varepsilon^{-1}(\pi\nu_0 - \beta)] \tag{5.110}$$

of the time necessary for the population process to enter the computed region $\tilde{\mathfrak{Z}}$.

For the case $n = 2$, the two regions $\mathfrak{Z}$ and $\tilde{\mathfrak{Z}}$ are shown in Figure 5.7.

The case of a linear system

$$\dot{x} = Ax + b(t) \tag{5.111}$$

is of special interest. We assume that A is a constant negative diagonal matrix, and that $b(t)$ has the property that for some $t_1 \in \mathfrak{T}_0$

$$b(t) = c \qquad \forall t \geqslant t_1, \tag{5.112}$$

where $c \in \mathfrak{R}_+^n$ is a constant vector. Then if the matrix A is a quasidominant diagonal matrix, we have

$$\lim_{t \to +\infty} x(t; t_0, x_0) = -A^{-1}c \qquad \forall (t_0, x_0) \in \mathfrak{T} \times \mathfrak{R}^n, \tag{5.113}$$

and the convergence is exponential. That is, the population processes track the equilibrium levels determined by the environmental fluctuations.

5.8. STOCHASTIC MODELS: STABILITY

The effects of unpredictable changes in weather, resources, etc., on multispecies communities cannot be satisfactorily estimated from the deterministic models. This fact was recognized long ago by several authors

(Erlih and Birch, 1967; Levins, 1969; Lewontin and Cohen, 1969), who proposed stochastic models as better descriptions of the real community environment. Stability as a central aspect of stochastic multispecies models was considered by May (1973a, b). In order to contrast the two kinds of models, May showed how various notions of stability are related in deterministic and stochastic environments. The mathematical apparatus that May applies in his studies of random disturbances in otherwise linear constant multispecies models is the Fokker-Planck-Kolmogorov equation, also called simply the diffusion equation. Since our analysis in this section goes beyond linear constant models, the intractability of the diffusion equation makes the Liapunov method via Itô's calculus a much more attractive framework for investigating the nonlinear phenomena in stochastic environments. This other approach was developed by Ladde and Šiljak (1976a, b), and the results are the subject of this section.

Our stability criterion is again the diagonal-dominance property of the community matrix. It will be shown that this property guarantees stability under both structural and random perturbations. Moreover, it is an ideal mechanism for establishing the trade-off between the degree of community stability and the size of environmental stochastic fluctuations that can be absorbed by stable communities.

Before we start our analysis of community models in the framework of the Itô differential equations and Liapunov direct method, it is of interest to list the appropriate references. Besides the book by Kushner (1967) on this subject, there is a survey paper on stochastic stability by Kozin (1969). These references should be supplemented by the original articles written by Bertram and Sarachik (1959), Kats and Krasovskii (1960), Bunke (1963), Khas'minskii (1962), and Kats (1964). Introductory tutorial papers on Itô differential equations as models of physical systems have been written by Mortensen (1969) and Papanicolaou (1973). The results presented in this section are based on the work of Ladde, Lakshmikantham, and Liu (1973) and of Ladde (1974, 1975), devoted to the development of the stochastic comparison principle.

Let us start our stochastic stability analysis with a linear constant equation of the Itô type,

$$dx = Ax\,dt + Bx\,dz, \tag{5.114}$$

where $x = x(t)$ is an n-vector $x = \{x_1, x_2, \ldots, x_n\}$ the components of which represent the populations. The $n \times n$ matrix $A = (a_{ij})$ is a constant community matrix that reflects the interactions among the populations in the community. Here $z = z(t)$ is a scalar function, representing the random environmental fluctuations, which is a normalized Wiener process with

$$\mathcal{E}\{[z(t_1) - z(t_2)]^2\} = |t_1 - t_2|, \tag{5.115}$$

where $\mathcal{E}$ denotes expectation, that is, averaging across the statistical ensemble. The constant $n \times n$ matrix $B = (b_{ij})$ is the diffusion community matrix which specifies how the random variable $z(t)$ influences the community.

Stochastic stability of the equilibrium $x^* = 0$ of the community model (5.114) means convergence to equilibrium of the solution process $x(t; t_0, x_0)$ starting at time t_0 and the initial population vector $x_0 = x(t_0)$. The convergence is measured in terms of "stochastic closeness" (e.g., in the mean, almost sure, in probability, etc.), which, in turn, generates various notion of stochastic stability. In case of the linear model under consideration, we are interested in establishing conditions for *global asymptotic stability in the mean* (Ladde and Šiljak, 1976c)—that is, conditions under which the expected value of the distance between the solution process and the equilibrium $\mathcal{E}\{\|x(t; t_0, x_0)\|\}$ tends to zero as $t \to \infty$ for all initial data (t_0, x_0).

We assume that each population in the community (5.114) is density-dependent, which is a realistic assumption (Tanner, 1966). This means that all diagonal elements a_{ii} of the community matrix A are negative. We make no assumption on the off-diagonal elements a_{ij} of A, thus allowing for "mixed" (competitive-predator-symbiotic-saprophitic) interactions among species in the community. The coefficients d_{ij} of the community diffusion matrix B can have arbitrary signs, which allows a good deal of freedom in the stochastic interactions among species and their environment.

To establish stochastic stability of the equilibrium of the chosen model, we will use the Liapunov direct method and the comparison principle. We propose the function

$$\nu(x) = \sum_{i=1}^{n} x_i^2 \tag{5.116}$$

as a candidate for Liapunov's function for the system (5.114). Using Itô's calculus we examine the expression

$$\mathfrak{L}\nu(x) = \frac{\partial \nu(x)}{\partial x} Ax + \frac{1}{2} \sum_{i,j=1}^{n} \frac{\partial^2 \nu(x)}{\partial x_i \partial x_j} s_{ij}(x), \tag{5.117}$$

where $\partial \nu / \partial x = (\partial \nu / \partial x_1, \partial \nu / \partial x_2, \ldots, \partial \nu / \partial x_n)$ is the gradient of $\nu(x)$, $\partial^2 \nu / \partial x_i \partial x_j$ is the (i,j)th element of the Hessian matrix related to $\nu(x)$, and the s_{ij}'s are the elements of the $n \times n$ matrix $S = Bxx^T B^T$. To establish the stability of the equilibrium, we observe that $\nu(x)$ is a positive definite function, and demonstrate that $\mathfrak{L}\nu(x)$ is negative definite.

Let us calculate $\mathfrak{L}\nu(x)$ as

$$\begin{aligned} \mathfrak{L}\nu(x) &= \sum_{i=1}^{n} 2x_i \Big(\sum_{j=1}^{n} a_{ij} x_j \Big) + \sum_{i=1}^{n} \Big(\sum_{j=1}^{n} b_{ij} x_j \Big)^2 \\ &= \sum_{j=1}^{n} 2a_{jj} x_j^2 + \sum_{j=1}^{n} x_j \sum_{\substack{i=1 \\ i \neq j}}^{n} 2a_{ij} x_i + \sum_{i=1}^{n} \Big(\sum_{j=1}^{n} b_{ij} x_j \Big)^2. \end{aligned} \tag{5.118}$$

Let us define the elements of the matrices $\overline{A} = (\bar{a}_{ij})$, $\overline{B} = (\bar{b}_{ij})$ as

$$\bar{a}_{ij} = \begin{cases} -|a_{ii}|, & i = j, \\ |a_{ij}|, & i \neq j, \end{cases} \qquad \bar{b}_{ij} = |b_{ij}| \sum_{k=1}^{n} |b_{ik}| \tag{5.119}$$

and use the inequality $2|x_i x_j| \leqslant x_i^2 + x_j^2$ to rewrite (5.118) as

$$\mathfrak{L}v(x) \leqslant \sum_{j=1}^{n} \left[\left(\bar{a}_{jj} + \sum_{\substack{i=1\\ i\neq j}}^{n} \bar{a}_{ij} \right) + \left(\bar{a}_{jj} + \sum_{\substack{i=1\\ i\neq j}}^{n} \bar{a}_{ji} \right) + \sum_{i=1}^{n} \bar{b}_{ij} \right] x_j^2. \tag{5.120}$$

Our central interest is to estimate how much of random perturbations can be absorbed by the deterministic stable version of the model (5.114). Therefore, we assume that the community matrix A is stable. Since A is negative diagonal, in view of relations (5.119) and (5.120), we assume such property of A by the diagonal-dominance conditions

$$\bar{a}_{jj} + \sum_{\substack{i=1\\ i\neq j}}^{n} \bar{a}_{ij} \leqslant -\pi_c, \qquad \bar{a}_{jj} + \sum_{\substack{i=1\\ i\neq j}}^{n} \bar{a}_{ji} \leqslant -\pi_r, \qquad i, j = 1, 2, \ldots, n, \tag{5.121}$$

where π_c and π_r are positive numbers.

If the diffusion matrix B is zero [which implies that the model (5.114) ignores the random disturbances] and all $\bar{b}_{ij}$'s are zero due to (5.119), then

$$\dot{v}(x) \leqslant -(\pi_c + \pi_r)v(x). \tag{5.122}$$

Integrating the inequality (5.122) and taking into account the definition (5.116) of the function $v(x)$, we arrive at the inequality

$$\|x(t; t_0, x_0\| \leqslant \|x_0\| \exp[-\tfrac{1}{2}(\pi_c + \pi_r)(t - t_0)], \qquad t \geqslant t_0, \tag{5.123}$$

which establishes the global exponential stability of the equilibrium.

We would like to obtain conditions on the matrices A and B which would guarantee stability in the stochastic sense (in the mean), and thus establish the tolerance of random fluctuations by stable community models. The conditions are expressed in terms of the matrix $\overline{A} + \overline{A}^T + \overline{B}$. We first require that this matrix be negative diagonal, which amounts to

$$2\bar{a}_{jj} + \bar{b}_{jj} < 0, \qquad j = 1, 2, \ldots, n. \tag{5.124}$$

Then stability of the model (5.114) is established by the diagonal-dominance property

$$2\bar{a}_{jj} + \bar{b}_{jj} + \sum_{\substack{i=1\\ i\neq j}}^{n} (\bar{a}_{ij} + \bar{a}_{ji} + \bar{b}_{ij}) \leqslant -\pi, \qquad j = 1, 2, \ldots, n, \tag{5.125}$$

of the matrix $\overline{A} + \overline{A}^T + \overline{B}$. We observe that the stability conditions (5.124) and (5.125) are expressed explicitly in terms of the elements a_{ij}, b_{ij} of the model matrices A, B, since they are simply given in terms of their absolute values.

From (5.120) and (5.125), we get

$$\mathfrak{L}v(x) \leqslant -\pi v(x). \tag{5.126}$$

Imitating in stochastic terms the development that led from (5.122) to (5.123), we obtain

$$\mathcal{E}\{\|x(t; t_0, x_0)\|\} \leqslant \|x_0\| \exp[-\tfrac{1}{2}\pi(t - t_0)], \qquad t \geqslant t_0, \tag{5.127}$$

which is a stochastic version of (5.123). The inequality (5.127) says that the expected value of the distance between the equilibrium $x = 0$ and the solution process $x(t; t_0, x_0)$ decreases faster than an exponential. The exponential decrement π can be determined directly from the community matrices A and B as specified in the inequalities (5.125).

It is important to note that the algebraic conditions (5.124) and (5.125) imply stability for a range of values of $\bar{a}_{ij}$ and $\bar{b}_{ij}$. The off-diagonal elements $\bar{a}_{ij}(i \neq j)$ and all the elements $\bar{b}_{ij}$ can have any values (including zero) smaller than those for which the conditions (5.124) and (5.125) hold. In particular, (5.124) and (5.125) imply stability when all $\bar{b}_{ij}$'s are zero (that is, $\bar{B} = B = 0$), in which case the model ignores the random disturbances of the environment. This confirms our earlier assumption that the deterministic community matrix A satisfies the conditions (5.121) and that the deterministic part of the model is stable.

From the above analysis, we conclude that if we ignore the nature of the interactions among the species in a community, then the smaller the magnitude of the interactions, the better the chances for an increase of community stability.

To consider large variations of population size and broaden the type of interactions among species to include phenomena such as predator switching, resource limitations, saturation of predator attack capacity, and the like, it is imperative to widen the scope of stochastic stability analysis and incorporate nonlinear time-varying community models. The fact that our stability conditions obtained for linear constant models are insensitive to magnitude variations in both deterministic and stochastic interactions, and that the chosen Liapunov function tolerates such variations, makes it possible for such variations to be time- and state-dependent. This leads to the following nonlinear time-varying stochastic model:

$$dx = A(t, x)x\,dt + B(t, x)x\,dz. \tag{5.128}$$

Again, as in (5.114), $x(t) \in \mathfrak{R}^n$ is the population vector, and $z(t) \in \mathfrak{R}$ is a random variable. The community matrices $A(t, x)$, $B(t, x)$ are now $n \times n$ matrix functions A, B: $\mathfrak{T} \times \mathfrak{R}^n \to \mathfrak{R}^{n^2}$ which are smooth enough that the solution process $x(t; t_0, x_0)$ of (5.128) exists for all initial conditions $(t_0, x_0) \in \mathfrak{T} \times \mathfrak{R}^n$ and all $t \in \mathfrak{T}_0$. The symbol $\mathfrak{T}$ represents the time interval $(\tau, +\infty)$, where τ is a number or $-\infty$, and $\mathcal{T}_0$ is the semi-infinite time interval $[t_0, +\infty)$.

In the following analysis, we will consider stochastic stability of the equilibrium population $x^* = 0$ of the model (5.128). If $A(t, x^*)x^* = 0$, $B(t, x^*)x^* = 0 \; \forall t \in \mathcal{T}$, and $x^* \neq 0$ is of interest, then we can define the nonlinear matrix functions $\hat{A}(t, y)y \equiv A(t, y + x^*)(y + x^*)$, $\hat{B}(t, y)y = B(t, y + x^*)(y + x^*)$ and consider the equation $dy = \hat{A}(t, y)y\,dt + B(t, y)y\,dz$ instead of (5.128), where $y^* = 0$ represents the equilibrium x^* under investigation.

In order to include the connective property of stochastic stability, we write the elements

$$a_{ij} = a_{ij}(t, x), \qquad b_{ij} = b_{ij}(t, x) \tag{5.129}$$

of the matrices $A(t, x)$, $B(t, x)$ as

$$a_{ij}(t, x) = \begin{cases} -\varphi_i(t, x) + e_{ii}(t)\varphi_{ii}(t, x), & i = j, \\ e_{ij}(t)\varphi_{ij}(t, x), & i \neq j, \end{cases} \tag{5.130}$$
$$b_{ij}(t, x) = l_{ij}(t)\psi_{ij}(t, x),$$

where the functions $\varphi_i, \varphi_{ij}, \psi_{ij} \in C(\mathcal{T} \times \mathcal{R}^n)$.

In (5.130), $e_{ij} = e_{ij}(t)$ and $l_{ij} = l_{ij}(t)$ are elements of the $n \times n$ interconnection matrices $E = (e_{ij})$ and $L = (l_{ij})$ which are defined and continuous on the time interval $\mathcal{T}$ with values in [0, 1]. The interconnection matrices reflect structural changes in both the deterministic and the stochastic interactions among species in the community. In particular, a disconnection of a trophic link between ith and jth species in the community is represented by $e_{ij} = e_{ji} = 0$ for all i, j. Such structural perturbations may occur independently in the stochastic interconnections involving the elements l_{ij}. Therefore, a wide variety of situations can take place as an interplay among deterministic and stochastic interconnections, and they can be conveniently described by various forms of interconnection matrices E and L.

It is important to note that community stability will be established for arbitrary forms of the functions $e_{ij}(t)$, $l_{ij}(t) \in [0, 1]$. This fact implies a high degree of reliability of the stability properties of communities for which our stability conditions hold.

Now we introduce the notion of stochastic connective stability as follows:

Definition 5.5. *The equilibrium* $x^* = 0$ *of the system* (5.128) *is stochastically connectively stable in the mean if and only if it is stable in the mean for all interconnection matrices* $E(t) \in \overline{E}$, $L(t) \in \overline{L}$.

To derive sufficient conditions for stochastic connective stability, we need to impose certain bounds on the coefficients of the matrices $A(t,x)$ and $B(t,x)$. We assume that the functions in (5.130) satisfy the constraints

$$\inf_{\mathcal{T}\times\mathcal{R}^n} \varphi_i(t,x) = \alpha_i, \qquad \sup_{\mathcal{T}\times\mathcal{R}^n} |\varphi_{ij}(t,x)| = \alpha_{ij}, \qquad \sup_{\mathcal{T}\times\mathcal{R}^n} |\psi_{ij}(t,x)| = \beta_{ij}, \qquad i,j = 1, 2, \dots, n, \tag{5.131}$$

for some numbers $\alpha_{ij} \geqslant 0$, $\alpha_i > \alpha_{ii}$, $\beta_{ij} \geqslant 0$. The constraints (5.131) specify the following classes of functions:

$$\begin{aligned} \Phi_i &= \{\varphi_i(t,x) \colon \varphi_i(t,x) \geqslant \alpha_i\}, \\ \Phi_{ij} &= \{\varphi_{ij}(t,x) \colon |\varphi_{ij}(t,x)| \leqslant \alpha_{ij}\}, \\ \Psi_{ij} &= \{\psi_{ij}(t,x) \colon |\psi_{ij}(t,x)| \leqslant \beta_{ij}\}. \end{aligned} \tag{5.132}$$

The classes of function Φ_i, Φ_{ij} include nonlinear interactions among species such as saturation of predator attack capacities and death among predators and prey, as well as predator switching, nonlinearity in the food supply, etc. Moreover, they include the changing of a species from predator to prey of another species over a finite time interval. The class of functions Ψ_{ij} allows for a possibility that the interactions of the community with the random environment are not known precisely, but are specified only by their magnitude.

It should be noted that our stability conditions will assure that the solution process approaches the equilibrium faster than an exponential. This additional property of stability provides an estimate of transient process in the community.

More precisely, we are going to establish stability as defined by the following:

Definition 5.6. *The equilibrium $x^* = 0$ of the system (5.128) is stochastically, exponentially, and connectively stable in the mean if and only if there exists two positive numbers Π and π independent of initial conditions (t_0, x_0) such that*

$$\mathcal{E}\{\|x(t; t_0, x_0)\|\} \leqslant \Pi \|x_0\| \exp[-\pi(t - t_0)] \qquad \forall t \in \mathcal{T}_0 \tag{5.133}$$

for all $(t_0, x_0) \in \mathcal{T} \times \mathcal{R}^n$ and all interconnection matrices $E(t) \in \overline{E}$, $L(t) \in \overline{L}$.

Actually, the stability conditions to be derived will assure the validity of the inequality (5.133) for all $\varphi_i \in \Phi_i$, $\varphi_{ij} \in \Phi_{ij}$, $\psi_{ij} \in \Psi_{ij}$, and thus add the "absolute" aspect to Definition 5.6 as in Definition 5.1.

In order to establish stability as given in Definition 5.6, we denote $\overline{A} = (\overline{a}_{ij})$ and $\overline{B} = (\overline{b}_{ij})$ the constant $n \times n$ matrices with elements

$$\bar{a}_{ij} = \begin{cases} -\alpha_i + \bar{e}_{ii}\alpha_{ii}, & i = j, \\ \bar{e}_{ij}\alpha_{ij}, & i \neq j, \end{cases} \qquad \bar{b}_{ij} = \bar{l}_{ij}\beta_{ij} \sum_{k=1}^{n} \bar{l}_{ik}\beta_{ik}, \tag{5.134}$$

where α_i, α_{ij}, β_{ij} are as in (5.131). Here $\bar{e}_{ij}$ and $\bar{l}_{ij}$ are elements of the $n \times n$ fundamental interconnection matrices $\bar{E}$ and $\bar{L}$. The matrices $\bar{E}$ and $\bar{L}$ are binary matrices in which each element is equal to 1 if there is an interaction between the corresponding species, or 0 if there is none.

As in the linear constant case, we assume that each species is in the "stabilized" form ($\bar{a}_{jj} < 0$), and moreover we have

$$2\bar{a}_{jj} + \bar{b}_{jj} < 0, \qquad j = 1, 2, \ldots, n. \tag{5.135}$$

In other words, when we take $E = 0$, $L = I$, where 0 and I are the zero and the identity $n \times n$ matrices, the system (5.128) reduces to

$$dx_j = -\varphi_j(t, x)x_j\,dt + \psi_{jj}(t, x)x_j\,dz, \qquad j = 1, 2, \ldots, n. \tag{5.136}$$

To examine stability of the "disconnected" community model (5.136), we can use the Liapunov function

$$v_j(x_j) = x_j^2. \tag{5.137}$$

Computing $\mathfrak{L}v_j(x_j)$ with respect to (5.136), we get

$$\begin{aligned} \mathfrak{L}v_j(x_j) &= -2\varphi_j(t, x)x_j^2 + \psi_{jj}^2(t, x)x_j^2 \\ &\leqslant (-2\alpha_j + \beta_{jj}^2)v_j(x_j). \end{aligned} \tag{5.138}$$

Applying the comparison principle to (5.138) and using (5.137), we conclude that

$$\mathcal{E}\{v_j(x_j)\} \leqslant v_{j0}\exp[-(2\alpha_j + \beta_{jj}^2)(t - t_0)] \qquad \forall t \in \mathcal{T}_0 \tag{5.139}$$

and

$$\mathcal{E}\{|x_j(t; t_0, x_0)|\} \leqslant |x_{j0}|\exp[-(\alpha_j + \tfrac{1}{2}\beta_{jj}^2)(t - t_0)] \qquad \forall t \in \mathcal{T}_0. \tag{5.140}$$

From this we conclude that the "disconnected" system is stable in the mean whenever

$$-\alpha_j + \tfrac{1}{2}\beta_{jj}^2 < 0, \tag{5.141}$$

which is equivalent to the inequality (5.135).

Now, we are interested in finding out under what conditions stability of the interconnected community (5.128) is implied by stability of the disconnected community (5.136). That is, we would like to estimate how much deterministic and stochastic interaction can be tolerated in a community of stable density-dependent species.

Let us prove the following:

Theorem 5.7. *The equilibrium* $x^* = 0$ *of the system* (5.128) *is stochastically, exponentially, and connectively stable if the* $n \times n$ *constant matrix* $\bar{A} + \bar{A}^T + \bar{B}$ *is dominant diagonal.*

Proof. Let us consider a decrescent, positive definite, and radially unbounded function $\nu: \mathcal{R}^n \to \mathcal{R}_+$,

$$\nu(x) = \sum_{i=1}^{n} x_i^2, \tag{5.142}$$

as a candidate for Liapunov's function for the system (5.128). By following (5.118) and (5.120), and using the constraints (5.131), we obtain

$$\begin{aligned} \mathcal{L}\nu(x) &= \sum_{i=1}^{n} 2x_i \left[\sum_{j=1}^{n} a_{ij}(t,x)x_j \right] + \sum_{i=1}^{n} \left[\sum_{j=1}^{n} b_{ij}(t,x)x_j \right]^2 \\ &= \sum_{j=1}^{n} 2a_{jj}(t,x)x_j^2 + \sum_{j=1}^{n} x_j \sum_{\substack{i=1 \\ i\neq j}}^{n} 2a_{ij}(t,x)x_i + \sum_{i=1}^{n} \left[\sum_{j=1}^{n} b_{ij}(t,x)x_j \right]^2 \\ &\leqslant \sum_{j=1}^{n} \left[\left(\bar{a}_{jj} + \sum_{\substack{i=1 \\ i\neq j}}^{n} \bar{a}_{ij} \right) + \left(\bar{a}_{jj} + \sum_{\substack{i=1 \\ i\neq j}}^{n} \bar{a}_{ji} \right) + \sum_{i=1}^{n} \bar{b}_{ij} \right] x_j^2. \end{aligned} \tag{5.143}$$

Since $\bar{A} + \bar{A}^T + \bar{B}$ is a negative diagonal matrix as stated by (5.135), and it is a dominant diagonal matrix, that is,

$$2\bar{a}_{jj} + \bar{b}_{jj} + \sum_{\substack{i=1 \\ i\neq j}}^{n} (\bar{a}_{ij} + \bar{a}_{ji} + \bar{b}_{ij}) \leqslant -\pi, \qquad j = 1, 2, \ldots, n, \tag{5.144}$$

the inequality (5.143) can be reduced to

$$\mathcal{L}\nu(x) \leqslant -\pi\nu(x) \qquad \forall t \in \mathcal{T}_0, \quad \forall x \in \mathcal{R}^n, \tag{5.145}$$

which is valid for all interconnection matrices $E(t) \in \bar{E}$, $L(t) \in \bar{L}$.

By applying Ladde's (1974) stochastic comparison principle to (5.145), we get

$$\mathcal{E}\{\nu[x(t; t_0, x_0)]\} \leqslant \nu(x_0)\exp[-\pi(t - t_0)] \qquad \forall t \in \mathcal{T}_0, \quad \forall (t_0, x_0) \in \mathcal{T} \times \mathcal{R}^n. \tag{5.146}$$

From (5.142) and (5.146), we obtain

$$\begin{aligned} \mathcal{E}\{\|x(t; t_0, x_0)\|\} \leqslant \|x_0\| \exp[-\tfrac{1}{2}\pi(t - t_0)] \qquad &\forall t \in \mathcal{T}_0, \quad \forall (t_0, x_0) \in \mathcal{T} \times \mathcal{R}^n, \\ \forall E(t) \in \bar{E}, \quad \forall L(t) \in \bar{L}, \end{aligned} \tag{5.147}$$

which proves Theorem 5.7.

We should immediately notice that from our proof of Theorem 5.7, it follows that the inequality (5.147) holds for all nonlinear interactions $\varphi_i(t, x)$, $\varphi_{ij}(t, x)$, $\psi_{ij}(t, x)$ which belong to the classes of functions Φ_i, Φ_{ij}, Ψ_{ij} defined in (5.132). This simply means that we do not have to know the actual shape of these interactions to establish stability so long as they are bounded as in (5.131). Imitating the consideration of deterministic models outlined in Section 5.3, such stability can be termed *absolute stochastic stability*.

Theorem 5.7 provides an estimate of the size of random disturbances that can be tolerated by stable communities. The trade-off obtained between the size of random disturbances and the degree of stability is established by the diagonal-dominance conditions applied to the community matrices. Furthermore, the conditions can serve as a measure of the magnitudes of interactions which do not disturb community stability, thus providing a suitable mechanism for resolving the complexity-vs.-stability problem in stochastic community models.

The stability conditions provided by Theorem 5.7 can be immediately extended to cases when there is a vector stochastic perturbation $z = \{z_1, z_2, \ldots, z_m\}$,

$$dx = A(t, x)x\,dt + \sum_{k=1}^{m} B_k(t, x)x\,dz_k, \tag{5.148}$$

where $z(t) \in \mathfrak{R}^m$ is a normalized Wiener process with

$$\mathcal{E}\{[z(t_1) - z(t_2)][z(t_1) - z(t_2)]^T\} = |t_1 - t_2|I, \tag{5.149}$$

where I is the $m \times m$ identity matrix. The elements $a_{ij}(t, x)$ and $b_{ij}^k(t, x)$ of the community matrices $A(t, x)$, $B_k(t, x)$ satisfy the same conditions as in (5.131). Specifically, the conditions on $b_{ij}^k(t, x)$ imply that there exist positive numbers $\bar{b}_{ij}$ such that

$$\sum_{k=1}^{m} \left[\sum_{j=1}^{n} b_{ij}^k(t, x)x_j \right]^2 \leqslant \sum_{j=1}^{n} \bar{b}_{ij} x_j^2 . \tag{5.150}$$

By using the same Liapunov function $\nu(x)$ defined in (5.142) and following the proof of Theorem 5.7, we arrive at the inequality (5.143). Then diagonal dominance of the matrix $\bar{A} + \bar{A}^T + \bar{B}$, where $\bar{B} = (\bar{b}_{ij})$ is an $n \times n$ matrix with elements defined in (5.150), implies stochastic exponential stability of the equilibrium.

Finally, it should be mentioned that by relaxing the nonlinear constraints (5.131) using the general comparison functions instead of the absolute values, we can establish the weaker asymptotic property of stability (Ladde and Šiljak, 1976a). The condition for stochastic and connective asymptotic stability in the mean is the quasidominant diagonal property of community matrices, which is a less restrictive condition than the ordinary diagonal dominance used in this section.

5.9. STOCHASTIC MODELS: INSTABILITY

Since our conditions for stochastic stability are only sufficient, it is of interest to consider instability in our nonlinear community models. Such considerations will provide sufficient conditions for stochastic instability which, in turn, can be used as a necessity part of our stability conditions.

The instability conditions we are about to derive are established for the fundamental interconnection matrices $\bar{E}$, $\bar{L}$, but are valid for all interconnection matrices $E(t) \in \bar{E}$, $L(t) \in \bar{L}$. Therefore, we introduce the following:

Definition 5.7. *The equilibrium $x^* = 0$ of the system* (5.128) *is stochastically and completely connectively unstable in the mean if and only if it is unstable in the mean for all interconnection matrices $E(t) \in \bar{E}$, $L(t) \in \bar{L}$.*

Let us assume that the coefficients $a_{ij}(t,x)$ of the community matrix $A(t,x)$ in (5.128) are redefined as

$$a_{ij}(t,x) = \begin{cases} \varphi_i(t,x) - e_{ii}(t)\varphi_{ii}(t,x), & i = j, \\ -e_{ij}(t)\varphi_{ij}(t,x), & i \neq j. \end{cases} \tag{5.151}$$

The elements $b_{ij}(t,x) = l_{ij}(t)\psi_{ij}(t,x)$ of the matrix $B(t,x)$ remain as in (5.130). We also assume that the functions φ_i, φ_{ij}, ψ_{ij} are smooth enough, and that the functions φ_i, φ_{ij} satisfy the constraints (5.131) for some numbers $\alpha_{ij} \geqslant 0$, $\alpha_i > \alpha_{ii}$.

We prove the following:

Theorem 5.8. *The equilibrium $x^* = 0$ of the system* (5.128) *is stochastically and completely connectively unstable in the mean if the $n \times n$ constant matrix $-(\bar{A} + \bar{A}^T)$ is dominant diagonal.*

Proof. We consider again the function $\nu(x)$ of (5.142) and compute

$$\begin{aligned} \mathfrak{L}\nu(x) &= \sum_{i=1}^{n} 2x_i \left[\sum_{j=1}^{n} a_{ij}(t,x)x_j \right] + \sum_{i=1}^{n} \left[\sum_{j=1}^{n} b_{ij}(t,x)x_j \right]^2 \\ &\geqslant \sum_{j=1}^{n} \left[\left(\bar{a}_{jj} - \sum_{\substack{i=1 \\ i\neq j}}^{n} \bar{a}_{ij} \right) + \left(\bar{a}_{jj} - \sum_{\substack{i=1 \\ i\neq j}}^{n} \bar{a}_{ij} \right) \right] x_j^2 . \end{aligned} \tag{5.152}$$

Since $-(\bar{A} + \bar{A}^T)$ is dominant diagonal, we can use the conditions (5.121) to rewrite (5.152) as

$$\mathfrak{L}\nu(x) \geqslant (\pi_c + \pi_r)\nu(x) \qquad \forall t \in \mathfrak{T}_0, \quad \forall (t_0, x_0) \in \mathfrak{T} \times \mathfrak{R}^n. \quad (5.153)$$

The inequality (5.153) is valid for all interconnection matrices $E(t) \in \bar{E}$, $L(t) \in \bar{L}$.

By applying again Ladde's (1974) stochastic comparison principle, we get from (5.153)

$$\mathcal{E}\{\|x(t; t_0, x_0)\|\} \geqslant \|x_0\| \exp[\tfrac{1}{2}(\pi_c + \pi_r)(t - t_0)] \qquad \forall t \in \mathfrak{T}_0, \quad (5.154)$$

which establishes Theorem 5.8.

From (5.154), it follows that instability under the conditions of Theorem 5.8 is exponential. Furthermore, it is easy to show that the inequality (5.155) is valid for all nonlinearities φ_i, φ_{ij}, ψ_{ij} that belong to classes of functions defined in (5.132), and that the instability is also absolute.

Theorem 5.7 is over restrictive, in the sense that the negative-diagonal property of $-(\bar{A} + \bar{A}^T)$ specified by $\alpha_i > \alpha_{ii} \geqslant 0$ implies that all species are unstable if disconnected from the community. We can relax this restriction and ask that only one species be unstable when disconnected, at the expense of specifying the sign of interactions between the species and the community.

Theorem 5.9. *The equilibrium $x^* = 0$ of the system* (5.128) *is stochastically and completely connectively unstable in the mean if for some $i = 1, 2, \ldots, n$, there exist numbers $\alpha_{ij} \geqslant 0$, $\alpha_i > 0$, such that the coefficients $a_{ij}(t, x)$ of the $n \times n$ matrix $A(t, x)$ defined by* (5.151) *satisfy the conditions*

$$\varphi_i(t, x) \geqslant \alpha_i, \qquad \varphi_{ij}(t, x)x_i x_j \leqslant -\alpha_{ij} x_j^2 \qquad \forall t \in \mathfrak{T}, \quad \forall x \in \mathfrak{R}^n. \quad (5.155)$$

Proof. Let us consider the Liapunov function of (5.137) as

$$v_i(x_i) = x_i^2. \quad (5.156)$$

Computing $\mathfrak{L}v_i(x_i)$ with respect to (5.128), we get

$$\begin{aligned} \mathfrak{L}v_i(x_i) &= 2\varphi_i(t, x)x_i^2 - 2 \sum_{j=1}^n e_{ij}(t)\varphi_{ij}(t, x)x_i x_j + \left[\sum_{j=1}^n b_{ij}(t, x)x_j\right]^2 \\ &\geqslant 2\alpha_i v_i(x_i) + 2 \sum_{j=1}^n e_{ij}(t)\alpha_{ij} v_j(x_j) \\ &\geqslant 2\alpha_i v_i(x_i) \qquad \forall t \in \mathfrak{T}_0, \quad \forall x \in \mathfrak{R}^n \end{aligned} \quad (5.157)$$

for any interconnection matrices $E(t) \in \bar{E}$, $L(t) \in \bar{L}$. By following the same argument as in Theorem 5.8, we obtain from (5.157)

$$\mathcal{E}\{|x_i(t; t_0, x_0)|\} \geqslant |x_{i0}| \exp[\alpha_i(t - t_0)] \qquad \forall t \in \mathcal{T}_0, \tag{5.158}$$

where $x_{i0} = x_i(t_0) \neq 0$. This proves Theorem 5.9.

If we dispose of completely connective instability and ask simply that the system (5.128) be unstable for fundamental interconnection matrices, then we can relax the conditions of Theorem 5.9. Let us assume that for some $i = 1, 2, \ldots, n$ the functions in (5.130) satisfy the following constraints:

$$\begin{aligned} &\varphi_i(t, x) \leqslant \alpha_i, \\ &2 \sum_{j=1}^{n} \varphi_{ij}(t, x) x_i x_j \geqslant \sum_{j=1}^{n} \alpha_{ij}(x_i^2 + x_j^2), \qquad \forall (t, x) \in \mathcal{T} \times \mathcal{R}^n, \\ &\left[\sum_{j=1}^{n} \psi_{ij}(t, x) x_j\right]^2 \geqslant \sum_{j=1}^{n} \beta_{ij} x_j^2 \end{aligned} \tag{5.159}$$

where $\alpha_{ij} \geqslant 0$, $\alpha_i > \alpha_{ii}$, $\beta_{ij} \geqslant 0$, and

$$\beta_{ii} > 2(\alpha_i - \bar{e}_{ii}\alpha_{ii}). \tag{5.160}$$

In Theorem 5.9, we required that at least one of the species be not density-dependent, so that it would be unstable if disconnected regardless of the presence of random disturbances. The condition (5.160) is weaker in that it does not require the instability of any species when isolated. Instability is introduced solely by the random perturbation. This shows that under the constraint (5.159), for stability of the community under stochastic disturbances we have to have a certain finite degree of stability of the deterministic part as well as a limitation on the random fluctuations.

We prove the following:

Theorem 5.10. *The equilibrium $x^* = 0$ of the system* (5.128) *is stochastically unstable in the mean if for some $i = 1, 2, \ldots, n$ there exist numbers $\alpha_{ij} \geqslant 0$, $\alpha_i > \alpha_{ii}$, $\beta_{ij} \geqslant 0$, such that the coefficients $a_{ij}(t, x)$, $b_{ij}(t, x)$ of the $n \times n$ matrices $A(t, x)$, $B(t, x)$ defined by* (5.130) *satisfy the conditions* (5.159) *and* (5.160) *for the fundamental interconnection matrices $\bar{E}$ and $\bar{L}$, where $\bar{l}_{ii} = 1$.*

Proof. We start again with the Liapunov function of (5.156) and compute

$$\begin{aligned} \mathcal{L}v_i(x_i) &= -2\varphi_i(t, x)x_i^2 + 2\sum_{j=1}^{n} \bar{e}_{ij}\varphi_{ij}(t, x)x_i x_j + \left[\sum_{j=1}^{n} \bar{l}_{ij}\psi_{ij}(t, x)x_j\right]^2 \\ &\geqslant [-2(\alpha_i - \bar{e}_{ii}\alpha_{ii}) + \bar{l}_{ii}\beta_{ii}]v_i(x_i) \qquad \forall t \in \mathcal{T}_0, \quad \forall x \in \mathcal{R}^n \end{aligned} \tag{5.161}$$

for the fundamental interconnection matrices $\bar{E}$, $\bar{L}$. Using the same argument as in the proof of Theorem 5.9, we obtain the inequality (5.158) where α_i is replaced by $-(\alpha_i - \bar{e}_{ii}\alpha_{ii}) + \frac{1}{2}\bar{l}_{ii}\beta_{ii}$. In view of the condition (5.160) and the fact that $\bar{l}_{in} = 1$, the proof of Theorem 5.10 is completed.

In the framework of the Itô calculus and the Liapunov direct method, we have derived sufficient conditions under which communities become unstable. The analysis exposed two major destabilizing factors: density independence of species and random fluctuations in the community environment. The latter factor was identified also by May (1973a) as the source of instability in linear constant community models, by using the Fokker-Planck-Kolmogorov equation.

5.10. HIERARCHIC MODELS

On the basis of Gardner and Ashby's (1970) computer studies of large complex systems, May (1973a) made an important observation: "A second feature of the models may be illustrated by using Gardner's and Ashby's computations (which are for a particular interaction strength) to see, for example, that 12-species communities with 15% connectance have a probability essentially zero of being stable, whereas if the interactions be organized into three separate 4×4 blocks of 4-species communities, each with a consequent 45% connectance, the 'organized' 12-species models will be stable with probability 35%". From this observation, May goes on to conclude that "Such examples suggest that our model multispecies communities, with given average interaction strength and web connectance, may do better if the interactions tend to be concentrated in small blocks, rather than distributed uniformly throughout the web—again a feature observed in many natural ecosystems". This is strikingly similar to Simon's (1962) intuitive arguments about the hierarchic structure of evolutionary systems mentioned in Section 1.10. Motivated by such observations, a stability study of hierarchic ecosystems was initiated by Šiljak (1975a) on deterministic models. The study was later extended by Ladde and Šiljak (1976a) to include ecosystems in a randomly varying environment. It is this stochastic analysis that will be presented in this section.

The hierarchical nature of a complex ecosystem is a matter of its structure. Therefore, any analysis that recognizes the hierarchic organization of ecomodels as composite systems composed of interconnected subsystems should start with a consideration of the *food* or *trophic web structure* of the community. Since matter and energy flows within community are fundamental ecological processes, the feeding relationships represented by the web reflect the basic structure of the community. Once this fact is recognized as an important preliminary step in the stability studies of ecosystems, it is only natural to use directed graphs as abstract descriptions of the trophic relations in the community. An excellent account of structural properties of food webs in the framework of graph theory was given by Gallopin (1972). Various uses of the digraphs to study ecosystem structure were presented also by Goel, Maitra, and Montroll (1971), Levins (1975), and Šiljak (1978).

We recall from Section 3.1 that a directed graph $\mathscr{D} = (V, R)$ consists of a set V of points and a family R of lines connecting the points of $\mathscr{D}$. An obvious way to associate a digraph $\mathscr{D}$ with a food web in a community of species is to let each vertex represent a species and lines represent the feeding relations between them. However, it is often difficult to sort organisms of a community into species. For example, the larvae of a species may have entirely different trophic links from those of the adults. "Thus", as Gallopin concludes, "even when taxonomic units reflect approximately the trophic characteristics of a population of organisms, they should be partitioned into more meaningful subunits whenever there are subsets of the same species that perform different trophic roles in a community". On the other hand, assigning separate vertices to larvae and adults of the same species may be misleading if we do not bear in mind that apart from the feeding patterns, the two organisms are strongly related. Any change in one of the organisms immediately affects the other.

Carrying the above reasoning a step further, Gallopin suggests that "a number of species' subgroups may be lumped into a single food web vertex when they have similar trophic requirements or functions". An example for such an approach was given by Paine (1966), who lumped the herbivores of intertidal trophic subwebs into such groups of species as chitons, barnacles, limpets, etc. This grouping Paine based upon the feeding patterns of predators. A convenient way to study structural properties of such complex communities is to use condensations $\mathscr{D}^* = (V^*, R^*)$, where each point in V^* stands for a group of species and R^* describes the influences among the groups rather than the individual species of a given community. The concept of input and output reachability outlined in Section 3.1 and the theory of directed graphs become a convenient framework for resolving structural problems of complex ecosystems.

A structural analysis of model ecosystems is not merely a desirable preliminary consideration prior to stability studies involving various configurations and condensations; it alone can only provide hints of how community complexity relates to stability. The purely structural information offered by digraphs of food webs ignores the magnitudes of interactions, which are crucial in establishing the stability of the communities they represent. Therefore, in partitioning the organisms of a community and using condensation to represent such grouping, we should be guided by purely structural properties of the food web, but also by the size of interactions among the organisms or species in that community. In the following development, we will show how the size of interactions as a measure of community complexity can be studied explicitly as a major factor in establishing community stability.

In this section, we study stability properties of stochastic hierarchic models of multispecies communities. By grouping species of a large community placed in the random environment into a relatively small number of blocks, the community is decomposed into subcommunities with both stochastic and deterministic interactions. Stability of the entire community is established in terms of stability of the subcommunities and stability of the aggregate community model formed by the decomposition-aggregation method.

An important feature of the decomposition-aggregation method is that it does not require that each species in the community be density-dependent. Species within the subcommunities need not be in the "stabilized form", but instead satisfy a weaker condition: that each group of species constitutes a stable community.

Let us consider a mathematical model represented by the stochastic differential equation

$$dx = f(t,x)\,dt + F(t,x)\,dz, \tag{5.162}$$

where $x(t) \in \mathcal{R}^n$ is the population vector, and $z(t) \in \mathcal{R}^m$ is a random variable which is a normalized Wiener process with

$$\mathcal{E}\{[z(t_1) - z(t_2)][z(t_1) - z(t_2)]^T = |t_2 - t_1|I. \tag{5.149}$$

Here $\mathcal{E}$ denotes expectation, and I is the $m \times m$ identity matrix. In (5.162), the n-vector function f: $\mathcal{T} \times \mathcal{R}^n \to \mathcal{R}^n$ and the $n \times m$-matrix function F: $\mathcal{T} \times \mathcal{R}^n \to \mathcal{R}^{n\times m}$ are smooth enough so that the soultion process $x(t; t_0, x_0)$ exists for all initial values $(t_0, x_0) \in \mathcal{T} \times \mathcal{R}^n$ and $t \in \mathcal{T}_0$. We recall that $\mathcal{T}$ represents the time interval $(\tau, +\infty)$, where τ is a number or the symbol $-\infty$, and $\mathcal{T}_0$ is the semi-infinite time interval $[t_0, +\infty)$. We also assume again that $f(t, 0) = 0$, $F(t, 0) = 0$ for all $t \in \mathcal{T}$, and $x^* = 0$ is the unique equilibrium of (5.162).

Now, we decompose Equation (5.162) into s interconnected subcommunities described by the equations

$$dx_i = g_i(t, x_i)\,dt + h_i(t,x)\,dt + F_i(t,x)\,dz, \qquad i = 1, 2, \ldots, s, \tag{5.163}$$

where $x_i(t) \in \mathcal{R}^{n_i}$ is the population vector of the ith subcommunity and represents the ith component of the population vector

$$x(t) = [x_1^T(t), x_2^T(t), \ldots, x_s^T(t)]^T \tag{5.164}$$

of the community. In (5.163), the function g_i: $\mathcal{T} \times \mathcal{R}^{n_i} \to \mathcal{R}^{n_i}$ represents the interactions among species within the ith subcommunity, the function $h_i: \mathcal{T} \times \mathcal{R}^n \to \mathcal{R}^{n_i}$ describes the deterministic interaction between the subcommunities, and $F_i: \mathcal{T} \times \mathcal{R}^n \to \mathcal{R}^{n_i \times m}$ is the matrix function which specifies how the random disturbance $z(t)$ affects the ith subcommunity.

We assume that deterministic interconnections depend on the $s \times s$ interconnection matrix $E = (e_{ij})$ as

$$h_i(t, x) \equiv h_i(t, e_{i1} x_1, e_{i2} x_2, \ldots, e_{is} x_s), \qquad i = 1, 2, \ldots, s. \quad (5.165)$$

Similarly, we express the stochastic interconnections as a function of the $n \times m$ interconnection matrix $L = (l_{ij})$,

$$F_i(t, x) = [l^i_{pq} f^i_{pq}(t, x)], \qquad i = 1, 2, \ldots, s, \quad (5.166)$$

where $l^i_{pq} f^i_{pq}(t, x)$ is the (p, q)th element of the $n_i \times m$ matrix $F_i(t, x)$.

In (5.165), $e_{ij}: \mathcal{T} \to [0, 1]$ are coupling functions which are the elements of the $s \times s$ interconnection matrix $E(t)$ associated with the deterministic interactions. Similarly, $l^i_{pq}: \mathcal{T} \to [0, 1]$ are coupling functions which are elements of the ith partition $L_i(t)$ of the $n \times m$ interconnection matrix $L(t) = [L_1^T(t), L_2^T(t), \ldots, L_s^T(t)]^T$.

When $E(t) \equiv 0$, $L(t) \equiv 0$, from (5.163) we get the free subcommunities described by

$$\dot{x}_i = g_i(t, x_i), \qquad i = 1, 2, \ldots, s, \quad (5.167)$$

where $g_i(t, 0) = 0$ for all $t \in \mathcal{T}$, and $x_i = 0$ is the unique equilibrium of the ith subcommunity.

For each subcommunity, we assume that there exists a scalar function $v_i: \mathcal{T} \times \mathcal{R}^{n_i} \to \mathcal{R}_+$ such that $v_i(t, x_i) \in C^{(1,2)}(\mathcal{T} \times \mathcal{R}^{n_i})$ which satisfies the following inequalities:

$$\begin{aligned} \phi_{i1}(\|x_i\|) \leqslant v_i(t, x_i) \leqslant \phi_{i2}(\|x_i\|), \\ \dot{v}_i(t, x_i) \leqslant -\alpha_i \phi_{i3}(\|x_i\|) \end{aligned} \qquad \forall (t, x_i) \in \mathcal{T} \times \mathcal{R}^{n_i}, \quad (5.168)$$

where α_i is a positive number, the functions ϕ_{i1}, ϕ_{i2}, ϕ_{i3} are comparison functions (see Definition 2.11), and $\phi_{i1}(\rho)$ is convex and $\phi_{i3}(\rho) \to +\infty$ as $\rho \to +\infty$ for all $i = 1, 2, \ldots, s$. In (5.168)

$$\dot{v}_i(t, x_i) = \frac{\partial}{\partial t} v_i(t, x_i) + [\text{grad } v_i(t, x_i)]^T g_i(t, x_i). \quad (5.169)$$

Under the conditions (5.168), each free subcommunity (5.167) has globally asymptotically stable equilibrium.

For deterministic interactions $h_i(t, x)$, we assume that

$$[\text{grad } v_i(t, x_i)]^T h_i(t, x) \leqslant \sum_{j=1}^{s} \bar{e}_{ij} \alpha_{ij} \phi_{j3}(\|x_j\|), \qquad i = 1, 2, \ldots, s, \tag{5.170}$$
$$\forall (t, x) \in \mathcal{T} \times \mathcal{R}^n,$$

where $\bar{e}_{ij}$ are elements of the $s \times s$ fundamental interconnection matrix $\bar{E}$, and α_{ij} are nonnegative numbers.

For stochastic interactions $F_i(t, x)$, we assume that

$$\frac{\partial^2 v_i(t, x_i)}{\partial x_{ip} \partial x_{iq}} l_{pk}^i l_{kq}^i f_{pk}^i(t, x) f_{kq}^i(t, x) \leqslant \sum_{j=1}^{s} \bar{l}_{pk}^i \bar{l}_{kq}^i \beta_{ij} \phi_{j3}(\|x_j\|), \qquad i = 1, 2, \ldots, s,$$
$$\forall (t, x) \in \mathcal{T} \times \mathcal{R}^n, \tag{5.171}$$

where $\bar{l}_{ij}^i$ are elements of the $n_i \times m$ fundamental interconnection submatrix $\bar{L}_i$, and β_{ij} are nonnegative numbers.

Under the constraints (5.168)–(5.171), we can establish the asymptotic stability properties of the system (5.162) as follows:

Definition 5.8. *The equilibrium* $x^* = 0$ *of the system* (5.162) *is globally asymptotically and connectively stable in the mean if and only if it is asymptotically stable in the mean for all interconnection matrices* $E(t) \in \bar{E}$, $L(t) \in \bar{L}$.

By $\bar{A} = (\bar{a}_{ij})$, $b = (\bar{b}_{ij})$, we denote the constant $s \times s$ matrices with the coefficients

$$\bar{a}_{ij} = \begin{cases} -\alpha_i + \bar{e}_{ii} \alpha_{ii}, & i = j, \\ \bar{e}_{ij} \alpha_{ij}, & i \neq j, \end{cases} \qquad \bar{b}_{ij} = \beta_{ij} \sum_{p,q=1}^{n_i} \left[\sum_{k=1}^{m} \bar{l}_{pk}^i \bar{l}_{kq}^i \right], \tag{5.172}$$

where α_i, α_{ij}, and β_{ij} are as defined in (5.168), (5.170), and (5.171). We also assume that the $s \times s$ matrix

$$\bar{C} = \bar{A} + \bar{B} \tag{5.173}$$

has negative diagonal elements $\bar{c}_{jj}$, that is,

$$\bar{a}_{jj} + \bar{b}_{jj} < 0, \qquad j = 1, 2, \ldots, s. \tag{5.174}$$

We prove the following result obtained by Ladde and Šiljak (1976a):

Theorem 5.11. *The equilibrium* $x^* = 0$ *of the system* (5.162) *is globally asymptotically and connectively stable in the mean if the constant* $s \times s$ *matrix* $\bar{C}$ *defined in* (5.173) *is a quasidominant diagonal matrix.*

Proof. Let us consider the function ν: $\mathcal{T} \times \mathcal{R}^n \rightarrow \mathcal{R}_+$,

$$\nu(t, x) = \sum_{i=1}^{s} d_i v_i(t, x_i). \tag{5.175}$$

In view of the inequalities (5.168), the function $v(t, x)$ satisfies the inequalities

$$\phi_{\mathrm{I}}(\|x\|) \leqslant v(t,x) \leqslant \phi_{\mathrm{II}}(\|x\|), \tag{5.176}$$

where $\phi_{\mathrm{I}}(\|x\|) = d_m \sum_{i=1}^{s} \phi_{i1}(\|x_i\|)$, $\phi_{\mathrm{II}}(\|x\|) = d_M \sum_{i=1}^{s} \phi_{i2}(\|x_i\|)$, $d_m = \min_i d_i$, $d_M = \max_i d_i$, $v(t, x)$ is a positive definite, decrescent, and radially unbounded function, and $\phi_{\mathrm{I}}(\|x\|)$ is a convex function.

Using (5.169) and imitating (5.117), we compute

$$\begin{aligned}\mathfrak{L}v(t,x) = \sum_{i=1}^{s} d_i \Bigg\{ &\dot{v}_i(t,x) + [\operatorname{grad} v_i(t,x_i)]^T h_i(t,x) \\ &+ \frac{1}{2} \sum_{p,q=1}^{n_i} g_{pq}^{i}(t,x) \frac{\partial^2 v_i(t,x_i)}{\partial x_{ip}\,\partial x_{iq}} \Bigg\},\end{aligned} \tag{5.177}$$

where $g_{pq}^{i}(t,x) = \sum_{k=1}^{m} l_{pk}^{i} l_{kq}^{i} f_{pk}^{i}(t,x) f_{kq}^{i}(t,x)$ is the (p,q)th element of the $n_i \times n_i$ matrix $G_i(t,x) = F_i(t,x) F_i^T(t,x)$.

By using the assumptions (5.168) and (5.170)–(5.174), we arrive at

$$\begin{aligned}\mathfrak{L}v(t,x) &\leqslant \sum_{i=1}^{s} d_i \sum_{j=1}^{s} (\bar{a}_{ij} + \bar{b}_{ij}) \phi_{j3}(\|x_j\|) \\ &\leqslant \sum_{j=1}^{s} d_j \Bigg[-|\bar{a}_{jj} + \bar{b}_{jj}| + d_j^{-1} \sum_{\substack{i=1 \\ i \neq j}}^{s} d_i (\bar{a}_{ij} + \bar{b}_{ij}) \Bigg] \phi_{j3}(\|x_j\|) \\ &\leqslant -\sum_{j=1}^{s} d_j \Bigg(|\bar{c}_{jj}| - d_j^{-1} \sum_{\substack{i=1 \\ i \neq j}}^{s} d_i \bar{c}_{ij} \Bigg) \phi_{j3}(\|x_j\|).\end{aligned} \tag{5.178}$$

Since the matrix $\bar{C}$ is a quasidominant diagonal matrix, that is, there exist positive numbers d_i, $i = 1, 2, \ldots, s$, and $\bar{\pi}$ such that

$$|\bar{c}_{jj}| - d_j^{-1} \sum_{\substack{i=1 \\ i \neq j}}^{n} d_i \bar{c}_{ij} \geqslant \bar{\pi}, \qquad j = 1, 2, \ldots, s, \tag{5.179}$$

we can rewrite (5.178) as

$$\mathfrak{L}v(t,x) \leqslant -\phi_{\mathrm{III}}(\|x\|) \qquad \forall (t,x) \in \mathfrak{T} \times \mathfrak{R}^n, \tag{5.180}$$

where $\phi_{\mathrm{III}}(\|x\|) = \bar{\pi} \sum_{j=1}^{s} d_j \phi_{j3}(\|x_j\|)$. This establishes the global asymptotic stability in the mean of $x^* = 0$ in (5.162).

To appreciate the constraints (5.170) and (5.171) imposed on the deterministic and stochastic interconnection functions, we will interpret them by choosing the specific form $\phi_{j3}(\|x_j\|) = \|x_j\|$ for the comparison function. Then these constraints become

$$[\operatorname{grad} v_i(t,x_i)]^T h_i(t,x) \leqslant \sum_{j=1}^{s} \bar{e}_{ij} \alpha_{ij} \|x_j\| \tag{5.181}$$

and

$$\frac{\partial^2 v_i(t, x_i)}{\partial x_{ip}\,\partial x_{iq}} l_{pk}^i l_{kq}^i f_{pk}^i(t, x) f_{kq}^i(t, x) \leqslant \sum_{j=1}^{s} \bar{l}_{pk}^i \bar{l}_{kq}^i \beta_{ij} \|x_j\|, \qquad i = 1, 2, \ldots, s, \qquad \forall (t, x) \in \mathcal{T} \times \mathcal{R}^n. \tag{5.182}$$

Furthermore, we assume that

$$\phi_{i1}(\|x_i\|) = \eta_{i1}\|x_i\|, \quad \phi_{i2}(\|x_i\|) = \eta_{i2}\|x_i\|, \qquad i = 1, 2, \ldots, s, \tag{5.183}$$

where η_{i1}, η_{i2} are positive numbers.

Under these specialized conditions, we can use Theorem 5.11 and establish the following:

Theorem 5.12. *The equilibrium* $x^* = 0$ *of the system* (5.162) *is globally exponentially and connectively stable in the mean if the constant* $s \times s$ *matrix* $\bar{C}$ *defined in* (5.173) *is a quasidominant diagonal matrix.*

Proof. By substituting $\phi_{j3}(\|x_j\|) = \|x_j\|$ in (5.178) and using (5.180), we get

$$\mathcal{L}v(t, x) \leqslant -\bar{\pi}\eta_{M2}^{-1} v(t, x) \qquad \forall (t, x) \in \mathcal{T} \times \mathcal{R}^n, \tag{5.184}$$

which is valid for all interconnection matrices $E(t)$, $L(t)$, and $\eta_{M2} = \max_i \eta_{i2}$. By applying the comparison principle proposed by Ladde, Lakshmikantham, and Liu (1973) to (5.184), we obtain

$$\mathcal{E}\{v[t, x(t; t_0, x_0)]\} \leqslant v(t_0, x_0)\exp[-\pi(t - t_0)] \qquad \forall t \in \mathcal{T}_0, \qquad \forall (t_0, x_0) \in \mathcal{T} \times \mathcal{R}^n. \tag{5.185}$$

Using (5.168) and (5.183), the inequality (5.185) yields (5.133), where $\Pi = n_{m1}^{-1}\eta_{M2}$, $\pi = \bar{\pi}\eta_{M2}^{-1}$, with $\eta_{m1} = \min_i \eta_{i1}$. The proof of Theorem 5.12 is complete.

It has been demonstrated that the comparison principle and vector Liapunov functions in the context of Itô's differential equations are powerful tools in stability investigations of multispecies communities in a stochastic environment. The diagonal dominance of the aggregate community matrix is shown to be a sufficient condition for community stability in the mean. Besides stability, the diagonal-dominance property assures the tolerance by the community models of a wide class of nonlinearities and time-varying structural perturbations in the interactions among species. This fact makes the diagonal dominance a suitable measure of the community complexity and therefore an ideal mechanism for dealing with the central problem of complexity vs. stability in the context of the robustness of model ecosystems.

REFERENCES

Allen, J. C. (1975), "Mathematical Models of Species Interactions in Time and Space", *The American Naturalist*, 109, 319–341.

Ayala, F. J., Gilpin, M. E., and Ehrenfeld, J. G. (1973), "Competition Between Species: Theoretical Models and Experimental Tests", *Theoretical Population Biology*, 4, 331–356.

Bellman, R. (1970), "Topics in Pharmacokinetics-I: Concentration Dependent Rates", *Mathematical Biosciences*, 6, 13–17.

Bertram, J. E., Sarachik, P. E. (1959), "Stability of Circuits with Randomly Time-Varying Parameters", *IRE Transactions*, CT-6, 260–270.

Bunke, H. (1963), "Stabilität bei stochastischen Differentialgleichungssystemen", *Zeitschrift für Angewandte Mathematik und Mechanik*, 43, 63–70.

Coddington, E. A., and Levinson, N. (1955), *Theory of Ordinary Differential Equations*, McGraw-Hill, New York.

Ehrlich, P. R., and Birch, L. C. (1967), "The 'Balance of Nature' and 'Population Control'", *The American Naturalist*, 101, 97–107.

Fiedler, M., and Pták, V. (1962), "On Matrices with Non-Positive Off-Diagonal Elements and Positive Principal Minors", *Czechoslovakian Mathematical Journal*, 12, 382–400.

Gallopin, G. C. (1972), "Structural Properties of Food Webs", *Systems Analysis and Simulation in Ecology*, B. C. Patten (ed.), Vol. II, Academic Press, New York, 241–282.

Gantmacher, F. R. (1960), *The Theory of Matrices*, Chelsea, New York.

Gardner, M. R., and Ashby, W. R. (1970), "Connectedness of Large Dynamical (Cybernetic) Systems: Critical Values for Stability", *Nature*, 228, 784.

Gato, M., and Rinaldi, S. (1977), "Stability Analysis of Predator-Prey Models Via the Liapunov Method", *Bulletin of Mathematical Biology*, 39, 339–347.

Gikhman, I. I., and Skorokhod, A. V. (1969), *Introduction to the Theory of Random Processes*, Saunders, Philadelphia, Pennsylvania.

Goel, N. S., Maitra, S. C., and Montroll, E. W. (1971), "*Nonlinear Models of Interacting Populations*", Academic Press, New York.

Goh, B. S. (1975), "Stability, Vulnerability and Persistence of Complex Ecosystems", *Ecological Modeling*, 1, 105–116.

Goh, B. S. (1976), "Nonvulnerability of Ecosystems in Unpredictable Environments", *Theoretical Population Biology*, 10, 83–95.

Goh, B. S. (1977), "Global Stability in Many–Species Systems", *The American Naturalist*, 111, 135–143.

Goh, B. S. (1978), "Robust Stability Concepts", *Mathematical Systems Ecology*, E. Halfon (ed.), Academic Press, New York, to appear.

Goh, B. S., and Jennings, L. S. (1977), "Feasibility and Stability in Randomly Assembled Lotka-Volterra Models", *Ecological Modeling*, 3, 63–71.

Goh, B. S., Leitmann, G., and Vincent, T. L. (1974), "Optimal Control of a Prey-Predator Systems", *Mathematical Biosciences*, 19, 263–286.

Grujić, Lj. T., and Šiljak, D. D. (1973a), "On Stability of Discrete Composite Systems", *IEEE Transactions*, AC-18, 522–524.

Grujić, Lj. T., and Šiljak, D. D. (1973b), "Asympotic Stability and Instability of Large-Scale Systems", *IEEE Transactions*, AC-18, 636–645.

Grujić, Lj. T., and Šiljak, D. D. (1974), "Exponential Stability of Large-Scale Discrete Systems", *International Journal of Control*, 19, 481–491.

Hale, J. K. (1969), "*Ordinary Differential Equations*", Wiley, New York.

Halfon, E. (1976), "Relative Stability of Eco-System Linear Models", *Ecological Modeling*, 2, 279–286.

Halfon, E. (ed.) (1978), *Mathematical Systems Ecology*, Academic, New York.

Harrary, F. (1969), *Graph Theory*, Addison-Wesley, Reading, Massachusetts.

Harrary, F., Norman, R. Z., and Cartwright, D. (1965), *Structural Models: An Introduction to the Theory of Directed Graphs*, Wiley, New York.

Holling, C. S. (1973), "Resilience and Stability of Ecological Systems", *Annual Review of Ecology and Systematics*, 4, 1–23.

Kats, I. Ia. (1964), "On the Stability of Stochastic Systems in the Large" (in Russian), *Prikladnaia Matematika i Mekhanika*, 28, 366–372.

Kats, I. Ia., and Krassovskii, N. N. (1960), "On the Stability of Systems with Random Parameters" (in Russian), *Prikladnaia Matematika i Mekhanika*, 24, 809–823.

Keyfitz, N. (1968), *Introduction to the Mathematics of Populations*, Addison-Wesley, Reading, Massachusetts.

Khas'minskii, R. Z. (1962), "On the Stability of the Trajectory of Markov Processes", *Prikladnaia Matematika i Mekhanika*, 26, 1554–1565.

Kloeden, P. E. (1975), "Aggregation-Decomposition and Equi-Ultimate Boundedness", *The Journal of the Australian Mathematical Society*, 19, 249–258.

Kozin, F. (1969), "A Survey of Stability of Stochastic Systems", *Automatica*, 5, 95–112.

Kushner, H. (1967), *Stochastic Stability and Control*, Academic, New York.

Ladde, G. S. (1974), "Differential Inequalities and Stochastic Functional Differential Equations", *Journal of Mathematical Physics*, 15, 738–743.

Ladde, G. S. (1975), "Systems of Differential Inequalities and Stochastic Differential Equations II", *Journal of Mathematical Physics*, 16, 894–900.

Ladde, G. S. (1976a), "Cellular Systems-I. Stability of Chemical Systems", *Mathematical Biosciences*, 29, 309–330.

Ladde, G. S. (1976b), "Stability of Large-Scale Hereditary Systems Under Structural Perturbations", *Proceedings of the IFAC Symposium on Large Scale Systems Theory and Applications*, Udine, Italy, 215–226.

Ladde, G. S., Lakshmikantham, V., and Liu, P. T. (1973), "Differential Inequalities and Itô Type Stochastic Differential Equations", *Proceedings of the International Conference on Nonlinear Differential and Functional Equations*, Bruxelles et Louvain, Belgium, Herman, Paris, 611–640.

Ladde, G. S., and Šiljak, D. D. (1976a), "Stability of Multispecies Communities in Randomly Varying Environment", *Journal of Mathematical Biology*, 2, 165–178.

Ladde, G. S., and Šiljak, D. D. (1976b), "Stochastic Stability and Instability of Model Ecosystems", *Proceedings of the Sixth IFAC World Congress*, Boston, Massachusetts, 55.4: 1–7.

Ladde, G. S., and Šiljak, D. D. (1976c), "Connective Stability of Large-Scale Stochastic Systems", *International Journal of Systems Science*, 6, 713–721.

LaSalle, J., and Lefschetz, S. (1961), *Stability by Liapunov's Direct Method With Applications*, Academic, New York.

Levin, S. A. (Editor) (1975), *Ecosystem Analysis and Prediction*, SIAM, Philadelphia, Pennsylvania.

Levin, S. H., Scudo, F. M., and Plunkett, D. J. (1977), "Persistence and Convergence of Ecosystems: An Analysis of Some Second Order Difference Equations", *Journal of Mathematical Biology*, 4, 171–182.

Levins, R. (1968), "Ecological Engineering: Theory and Practice", *Quarterly Review of Biology*, 43, 301–305.

Levins, R. (1969), "The Effect of Random Variations of Different Types on Population Growth", *Proceedings of the National Academy of Sciences*, USA, 62, 1061–1065.

Levins, R. (1975), "Problems of Signed Digraphs in Ecological Theory", *Ecosystem Analysis and Prediction*, S. A. Levin (ed.), SIAM, Philadelphia, Pennsylvania, 264–277.

Lewontin, R. C., and Cohen, D. (1969), "On Population Growth in a Randomly Varying Environment", *Proceedings of the National Academy of Sciences*, USA, 62, 1056–1060.

Lotka, A. J. (1925), *Elements of Physical Biology*, Williams and Wilkins, Baltimore, Maryland. (Reissued as *Elements of Mathematical Biology*, Dover, New York, 1956).

MacArthur, R. H. (1970), "Species Packing and Competitive Equilibrium for Many Species", *Theoretical Population Biology*, 1, 1–11.

Margalef, R. (1968), *Perspectives in Ecology Theory*, Chicago University Press, Chicago, Illinois.

May, R. M. (1972), "Will a Large Complex System be Stable?", *Nature*, 238, 413–414.

May, R. M. (1973a), *Stability and Complexity in Model Ecosystems*, Princeton University Press, Princeton, New Jersey.

May, R. M. (1973b), "Stability in Randomly Fluctuating Versus Deterministic Environments", *The American Naturalist*, 107, 621–650.

McKenzie, L. (1966), "Matrices with Dominant Diagonals and Economic Theory", *Proceedings of the Symposium on Mathematical Methods in the Social Sciences*, K. J. Arrow, S. Karlin, and P. Suppes (eds.), Stanford University Press, Stanford, California, 47–62.

Mortensen, R. E. (1969), "Mathematical Problems of Modeling Stochastic Nonlinear Dynamic Systems", *Journal of Statistical Physics*, 1, 271–296.

Murdoch, W. W. (1969), "Switching in General Predators: Experiments on Predator Specificity and Stability of Prey Populations", *Ecological Monographs*, 39, 335–354.

Newman, P. K. (1959), "Some Notes on Stability Conditions", *Review of Economic Studies*, 72, 1–9.

Paine, R. T. (1966), "Food Web Complexity and Species Diversity", *American Naturalist*, 100, 65–75.

Papanicolauo, G. C. (1973), "Stochastic Equations and Their Applications", *The American Mathematical Monthly*, 80, 526–545.

Persidskii, S. K. (1968), "Investigation of Stability of Solutions of Some Nonlinear Systems of Differential Equations" (in Russian), *Prikladnaia Matematika i Mekhanika*, 32, 1122–1125.

Persidskii, S. K. (1969), "Problems of Absolute Stability" (in Russian), *Avtomatika i Telemekhanika*, 29, 5–11.
Peridskii, S. K. (1970), "Investigating the Stability of Solutions of Systems of Differential Equations" (in Russian), *Prikladnaia Matematika i Mekhanika*, 34, 219–226.
Pielou, E. C. (1969), *An Introduction to Mathematical Ecology*, Wiley, New York.
Quirk, J., and Saposnik (1968), *Introduction to General Equilibrium Theory and Welfare Economics*, McGraw-Hill, New York.
Rosenbrock, H. H. (1963), "A Lyapunov Function for Some Naturally Occurring Linear Homogeneous Time-Dependent Equations", *Automatica*, 1, 97–109.
Sandberg, I. W. (1969), "Some Theorems on the Dynamic Response of Nonlinear Transistor Networks", *Bell Systems Technical Journal*, 48, 35–54.
Simon, H. A. (1962), "The Architecture of Complexity", *Proceedings of the American Philosophical Society*, 106, 467–482.
Šiljak, D. D. (1969), *Nonlinear Systems*, Wiley, New York.
Šiljak, D. D. (1972), "Stability of Large-Scale Systems Under Structural Perturbations", *IEEE Transactions*, SMC-2, 657–663.
Šiljak, D. D. (1974), "Connective Stability of Complex Ecosystems", *Nature*, 249, 280.
Šiljak, D. D. (1975a), "When is a Complex Ecosystem Stable?", *Mathematical Biosciences*, 25, 25–50.
Šiljak, D. D. (1975b), "Connective Stability of Competitive Equilibrium", *Automatica*, 11, 389–400.
Šiljak, D. D. (1977a), "On Reachability of Dynamic Systems", *International Journal of Systems Science*, 8, 321–338.
Šiljak, D. D. (1977b), "On Pure Structure of Dynamic Systems", *Nonlinear Analysis, Theory, Methods and Applications*, 1, 397–413.
Šiljak, D. D. (1977c), "Vulnerability of Dynamic Systems", *Proceedings of the IFAC Workshop on Control and Management of Integrated Industrial Complexes*, Toulouse, France, 133–144.
Šiljak, D. D. (1977d), "On Stability of the Arms Race", *Proceedings of the NSF Conference on Systems Theory and International Relations Research*, D. Zinnes and J. Gillespie, (eds.), Praeger, New York, 264–304.
Šiljak, D. D. (1978), "Structure and Stability of Model Ecosystems", *Mathematical Systems Ecology*, E. Halfon (ed.), Academic, New York, to appear.
Smith, J. M. (1974), *Models in Ecology*, Cambridge University Press, Cambridge, England.
Tanner, J. T. (1966), "Effects of Population Density on Growth Rates of Animal Populations", *Ecology*, 47, 733–745.
Volterra, V. (1926), "Variazioni e fluttuazioni del numero d'individui in specie animali conviventi", *Memorie della R. Academia Natzionale dei Lincei*, 2, 31–113. (Translated in *Animal Ecology*, R. N. Chapman, McGraw-Hill, New York, 1931).
Watt, K. E. F. (1968), *Ecology and Resource Management*, McGraw-Hill, New York.
Weissenberger, S. (1973), "Stability Regions of Large-Scale Systems", *Automatica*, 9, 653–663.

6

ENGINEERING

Spacecraft Control Systems

The purpose of this chapter is to apply the multilevel decomposition-aggregation stability analysis to the design of control systems for the two NASA spacecraft, the Large Space Telescope (LST) and the Skylab. A common feature in the two applications is the fact that the equations of motion are decomposed so that individual motions (roll, pitch, and yaw for the LST, and spin and wobble for the Skylab) are chosen as subsystems. Then a feedback control is used to stabilize the two systems, stability being established on the basis of the corresponding aggregate models. A major distinction between the two designs is in the nature and use of feedback. In the case of the LST, we assume that all the states are available for control either by physical measurements or by use of the state estimators. To control the Skylab, we apply an output feedback control scheme.

The LST model is nonlinear but composed of linear subsystems with bilinear interactions. Local state-variable feedback is used as in Section 3.3 to stabilize each subsystem separately. Since the subsystems are linear, stability is global. The introduction of bilinear interconnections among the subsystems limits stability considerations by aggregate models to a finite region of the state space. Estimates of the region are provided readily by using the methods of Section 2.8, and their size can be controlled by the gains of the local controllers.

The structure of the LST model is suitable for the suboptimal design presented in Section 3.5, since it can easily handle nonlinearities in the interconnections among the subsystems. It is important and somewhat surprising that the use of global controllers in this context can result in an optimal design because the bilinear interactions can be completely neutralized.

There is still another advantage of the decomposition-aggregation method: one can determine explicitly the effect on system stability of structural parameters which appear in the interactions among the subsystems. The problem can be formulated in the mathematical-programming format, and the effect can be calculated by standard numerical techniques. As a byproduct of the solution process, we will construct an aggregate model which is optimal in the sense that it yields the *best* estimates of the interaction parameters. The construction is based upon a certain form of Liapunov function which provides an *exact* estimate of the degree of exponential stability in linear constant systems.

An 11th-order model for the Skylab is used to illustrate the study of the effect of structural parameters on overall system stability. The model includes both passive stabilization by extendable booms and active stabilization by control torques about the body-fixed axes. The model is decomposed into two sets of equations describing the wobble motion and the spin motion. The two sets of equations are treated as subsystems, which are interconnected by a coupling structural parameter representing the asymmetry in the boom setting. A linear second-order model is formed by the decomposition-aggregation method, which is used to determine the stability of the overall system and, at the same time, provide an estimate of the stability region for the structural parameter.

6.1. LARGE SPACE TELESCOPE: A MODEL

The Large Space Telescope (LST), presently being developed by NASA (National Aeronautics and Space Administration), is to be a long-lived, versatile space observatory for optical stellar astronomy (O'Dell, 1973). The LST is expected to aid significantly in discoveries of the origin and structure of our galaxy as well as the other galaxies and the universe. Its high-precision imaging of less than 0.005 arc sec rms should open up a plethora of new research areas for studies of the energy processes in galactic nuclei, of the structure of gaseous and planetary nebulae and protostars, and of such highly evolved phenomena as supernovae and white dwarfs.

One of the principal tasks of the LST system analysis is the image-motion stabilization to better than 0.005 arc sec rms. A number of studies have been conducted to explore the capabilities of body pointing by using single-gimbaled and double-gimbaled control moment gyros and reaction wheels. All these studies show that the required pointing accuracy can be met, but that it can be severely jeopardized by the gravity-gradient and sensor noise, by vibrations of the solar panels, gyros, and reaction wheels, and by the effect of various bending modes and nonlinearities. In a report by Schiehlen (1973), a fine guidance system is proposed to body-point the spacecraft, which is modeled as a rigid body having reaction-wheel actuators and subject to gravitational and magnetic disturbance torques. The proposed control has an optimal performance close to 0.0001 arc sec rms, which is well within the pointing-accuracy limit.

The model considered by Schielen (1973) is linear, so that it ignores the nonlinear coupling phenomena and thus may not be an entirely satisfactory description of the spacecraft. The principal objective of this section is to develop a nonlinear model of the LST as in (Šiljak, Sundareshan, and Vukčević, 1975), and then use the model for the design of the control system by the multilevel stabilization and optimization schemes outlined in Chapter 3. The description of the LST for purposes of modeling follows that of Schielen (1973).

The LST spacecraft configuration is shown in Figure 6.1. It consists of the optical telescope assembly, the scientific instruments, and the support systems module. The attached solar panels are a source of electrical power for the spacecraft. The LST is assumed to be a rigid body. This assumption is justified for the control-system analysis, since the structural frequencies are considerably higher than those of the control system.

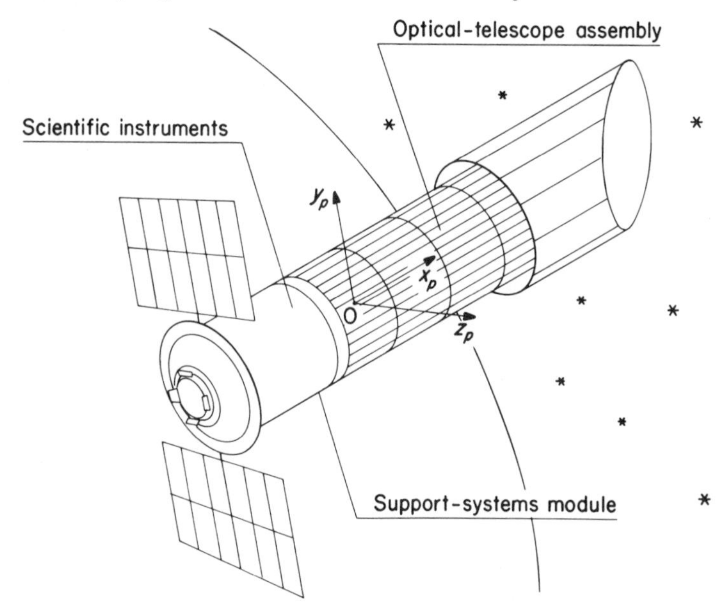

FIGURE 6.1. LST configurations.

To develop the equations of motion for the LST spacecraft, we consider the principal-axis frame P with its origin at the center of mass O, indicated in Figure 6.1. The x_p-axis is along the telescope axis, the y_p-axis is along the solar-panel axis, and the z_p-axis augments the x_p- and y_p-axes to form the orthogonal coordinate system. The task of the fine guidance system is to maintain the frame P as close as possible to a given inertial frame I, the transformation matrix between the two frames being

$$\begin{bmatrix} 1 & -\psi & \theta \\ \psi & 1 & -\phi \\ -\theta & \phi & 1 \end{bmatrix}, \tag{6.1}$$

where ϕ is the roll angle, θ the pitch angle, and ψ the yaw angle.

The spacecraft equation of motion can be written down from the Euler equations (Schiehlen, 1973) as

$$I \cdot \dot{\omega} + \omega \times I \cdot \omega + \sum_{i=1}^{3} \{\omega \times \omega_i \operatorname{tr} I_i + 2\omega_i \times I_i \cdot \omega + I_i \cdot \dot{\omega}_i + \omega_i \times I_i \cdot \omega_i\} = M \tag{6.2}$$

and

$$I_i \cdot \dot{\omega}_i + \omega \times I_i \cdot \omega + \omega \times \omega_i \operatorname{tr} I_i + 2\omega_i \times I_i \cdot \omega + I_i \cdot \dot{\omega}_i + \omega_i \times I_i \cdot \omega_i = M_i, \quad i = 1, 2, 3, \tag{6.3}$$

where

I is the inertia tensor of the LST, given by

$$I = \begin{bmatrix} I_x & 0 & 0 \\ 0 & I_y & 0 \\ 0 & 0 & I_z \end{bmatrix}, \tag{6.4}$$

I_x, I_y, I_z denoting the components along the three axes constituting an inertial reference frame I_{rf};

I_i, $i = 1, 2, 3$, are the inertia tensors of the three reaction wheels, which are mounted orthogonally and parallel to the axes constituting the standard body-fixed reference frame B_{rf}; they can be therefore expressed as

$$I_1 = \begin{bmatrix} I_{1x} & 0 & 0 \\ 0 & I_{1y} & 0 \\ 0 & 0 & I_{1y} \end{bmatrix}, \quad I_2 = \begin{bmatrix} I_{2z} & 0 & 0 \\ 0 & I_{2y} & 0 \\ 0 & 0 & I_{2z} \end{bmatrix}, \quad I_3 = \begin{bmatrix} I_{3x} & 0 & 0 \\ 0 & I_{3x} & 0 \\ 0 & 0 & I_{3z} \end{bmatrix}; \tag{6.5}$$

ω is the angular-velocity vector of the LST relative to the frame I_{rf};
ω_i, $i = 1, 2, 3$, are the angular-velocity vectors of the reaction wheels relative to the frame I_{rf};
M is the total external torque acting on the LST; and
M_i, $i = 1, 2, 3$, are the internal torques on the reaction wheels.

The angular velocity ω can be expressed in terms of the rates of angular deviations along the three axes of I_{rf} as

$$\omega = \begin{bmatrix} \dot{\phi} \\ \dot{\theta} \\ \dot{\psi} \end{bmatrix}, \tag{6.6}$$

where ϕ is the roll angle, θ is the pitch angle, and ψ is the yaw angle. Similarly the angular velocities ω_i of the reaction wheels can be expressed in terms of the components as

$$\omega_1 = \begin{bmatrix} \nu_1 \\ 0 \\ 0 \end{bmatrix}, \qquad \omega_2 = \begin{bmatrix} 0 \\ \nu_2 \\ 0 \end{bmatrix}, \qquad \omega_3 = \begin{bmatrix} 0 \\ 0 \\ \nu_3 \end{bmatrix}. \tag{6.7}$$

Equations (6.2) and (6.3) can now be simplified into the following four sets of scalar equations:

(1) Equations governing the motion of the LST body:

$$\begin{aligned} I_x\ddot{\phi} + \dot{\theta}\dot{\psi}(I_z - I_y) + I_{3z}\nu_3\dot{\theta} - I_{2y}\nu_2\dot{\psi} + I_{1x}\dot{\nu}_1 &= M_x, \\ I_y\ddot{\theta} + \dot{\phi}\dot{\psi}(I_x - I_z) + I_{1x}\nu_1\dot{\psi} - I_{3z}\nu_3\dot{\phi} + I_{2y}\dot{\nu}_2 &= M_y, \\ I_z\ddot{\psi} + \dot{\phi}\dot{\theta}(I_y - I_x) + I_{2y}\nu_2\dot{\phi} - I_{1x}\nu_1\dot{\theta} + I_{3z}\dot{\nu}_3 &= M_z. \end{aligned} \tag{6.8}$$

(2) Equations for the reaction wheel mounted parallel to the x-axis:

$$\begin{aligned} I_{1x}\ddot{\phi} + I_{1x}\dot{\nu}_1 &= M_{1x}, \\ I_{1y}\ddot{\theta} + (I_{1x} - I_{1y})\dot{\theta}\dot{\psi} + I_{1x}\nu_1\dot{\psi} &= M_{1y}, \\ I_{1y}\ddot{\psi} + (I_{1y} - I_{1x})\dot{\phi}\dot{\theta} - I_{1x}\nu_1\dot{\theta} &= M_{1z}. \end{aligned} \tag{6.9}$$

(3) Equations for the reaction wheel mounted parallel to the y-axis:

$$\begin{aligned} I_{2z}\ddot{\phi} + (I_{2y} - I_{2z})\dot{\phi}\dot{\psi} + I_{2y}\nu_2\dot{\phi} &= M_{2x}, \\ I_{2y}\ddot{\theta} + I_{2y}\dot{\nu}_2 &= M_{2y}, \\ I_{2z}\ddot{\psi} + (I_{2z} - I_{2y})\dot{\theta}\dot{\psi} - I_{2y}\nu_2\dot{\psi} &= M_{2z}. \end{aligned} \tag{6.10}$$

(4) Equations for the reaction wheel mounted parallel to the z-axis:

$$
\begin{aligned}
I_{3x}\ddot{\phi} + (I_{3x} - I_{3z})\dot{\phi}\dot{\theta} + I_{3z}\nu_3\dot{\theta} &= M_{3x}, \\
I_{3x}\ddot{\theta} + (I_{3z} - I_{3x})\dot{\phi}\dot{\psi} + I_{3z}\nu_3\dot{\phi} &= M_{3y}, \\
I_{3z}\ddot{\psi} + I_{3z}\dot{\nu}_3 &= M_{3z}.
\end{aligned} \tag{6.11}
$$

For further simplification, we shall assume that the reaction wheels are small, so that $I_{1x} \ll I_x$, $I_{2y} \ll I_y$, $I_{3z} \ll I_z$, and they have one degree of freedom only. Then Equations (6.8)–(6.11) can be simplified to

$$
\begin{aligned}
I_x\ddot{\phi} + \dot{\theta}\dot{\psi}(I_z - I_y) + I_{1x}\dot{\nu}_1 &= M_x, \\
I_y\ddot{\theta} + \dot{\phi}\dot{\psi}(I_x - I_z) + I_{2y}\dot{\nu}_2 &= M_y, \\
I_z\ddot{\psi} + \dot{\phi}\dot{\theta}(I_y - I_x) + I_{3z}\dot{\nu}_3 &= M_z
\end{aligned} \tag{6.12}
$$

and

$$
\begin{aligned}
I_{1x}\ddot{\phi} + I_{1x}\dot{\nu}_1 &= M_{1x}, \\
I_{2y}\ddot{\theta} + I_{2y}\dot{\nu}_2 &= M_{2y}, \\
I_{3z}\ddot{\psi} + I_{3z}\dot{\nu}_3 &= M_{3z}.
\end{aligned} \tag{6.13}
$$

Substitution of (6.13) into (6.12) will result in the following three equations describing the motions along the individual axes and their interconnections:

$$
\begin{aligned}
I_x\ddot{\phi} + (I_z - I_y)\dot{\theta}\dot{\psi} &= (M_x - M_{1x}), \\
I_y\ddot{\theta} + (I_x - I_z)\dot{\phi}\dot{\psi} &= (M_y - M_{2y}), \\
I_z\ddot{\psi} + (I_y - I_x)\dot{\theta}\dot{\phi} &= (M_z - M_{3z}).
\end{aligned} \tag{6.14}
$$

It is now necessary to evaluate the various torques. Since the internal torques on the reaction wheels are small, it may be assumed that these are proportional to the control signals actuating the wheels. Hence,

$$
\begin{aligned}
M_{1x} &= -K_1 u_1, \\
M_{2y} &= -K_2 u_2, \\
M_{3z} &= -K_3 u_3,
\end{aligned} \tag{6.15}
$$

where K_1, K_2, and K_3 are the drive-motor constants [the negative signs in (6.15) merely indicate the directions of these torques].

The external torques acting on the body of the LST are mainly due to environmental disturbance forces and are composed of gravity-gradient, magnetic, aerodynamic, and solar-pressure torques. The latter two will be negligibly small compared to the others and will usually be accounted for in control-system designs by treating them as equivalent zero-mean stationary white-noise processes. The gravity-gradient and magnetic torques can be represented as purely deterministic signals involving a constant term and a sinusoidal function of time with twice the orbital frequency. Hence, following the analysis of Schiehlen (1973), the external torques can be obtained as

$$\begin{aligned} M_x &= \{\gamma_{11} + \gamma_{12}\cos(\omega t + x) + s_1\} I_x, \\ M_y &= \{\gamma_{21} + \gamma_{22}\cos(\omega t + x) + s_2\} I_y, \\ M_z &= \{\gamma_{31} + \gamma_{32}\cos(\omega t + x) + s_3\} I_z, \end{aligned} \tag{6.16}$$

where γ_{ij}, $i = 1, 2, 3$, are constants that can be determined from the inertia components I_x, I_y, I_z, the magnitude of the LST dipole moment, and the earth's magnetic field intensity; and s_i, $i = 1, 2, 3$, are white-noise processes characterizing the aerodynamic and solar-pressure torques.

Substitution of (6.15) and (6.16) in (6.14) and further simplification results in the following system of equations:

$$\begin{aligned} \ddot{\phi} + \alpha_1 \dot{\theta}\dot{\psi} &= \beta_1 u_1 + M_x, \\ \ddot{\theta} + \alpha_2 \dot{\phi}\dot{\psi} &= \beta_2 u_2 + M_y, \\ \ddot{\psi} + \alpha_3 \dot{\phi}\dot{\theta} &= \beta_3 u_3 + M_z, \end{aligned} \tag{6.17}$$

where $\alpha_1 = (I_z - I_y)/I_x$, $\alpha_2 = (I_x - I_z)/I_y$, $\alpha_3 = (I_y - I_x)/I_z$, $\beta_1 = K_1/I_x$, $\beta_2 = K_2/I_y$, $\beta_3 = K_3/I_z$, and M_x, M_y, M_z are the external disturbance torques given by (6.16).

It is now simple to obtain a state-space representation of the LST by choosing the state vector

$$x = (\phi, \dot{\phi}, \theta, \dot{\theta}, \psi, \dot{\psi})^T, \tag{6.18}$$

which results in the time-invariant model

$$\dot{x} = Ax + h(x) + Bu + FM, \tag{6.19}$$

where

$$A = \begin{bmatrix} 0 & 1 & 0 & 0 & 0 & 0 \\ 0 & 0 & 0 & 0 & 0 & 0 \\ 0 & 0 & 0 & 1 & 0 & 0 \\ 0 & 0 & 0 & 0 & 0 & 0 \\ 0 & 0 & 0 & 0 & 0 & 1 \\ 0 & 0 & 0 & 0 & 0 & 0 \end{bmatrix}, \qquad h(x) = \begin{bmatrix} 0 \\ -\alpha_1 \dot{\theta}\dot{\psi} \\ 0 \\ -\alpha_2 \dot{\phi}\dot{\psi} \\ 0 \\ -\alpha_3 \dot{\phi}\dot{\theta} \end{bmatrix},$$

$$B = \begin{bmatrix} 0 & 0 & 0 \\ \beta_1 & 0 & 0 \\ 0 & 0 & 0 \\ 0 & \beta_2 & 0 \\ 0 & 0 & 0 \\ 0 & 0 & \beta_3 \end{bmatrix}, \qquad F = \begin{bmatrix} 0 & 0 & 0 \\ 1 & 0 & 0 \\ 0 & 0 & 0 \\ 0 & 1 & 0 \\ 0 & 0 & 0 \\ 0 & 0 & 1 \end{bmatrix}. \tag{6.20}$$

The diagonal structure of the matrices A, B, and F permits us to partition the state vector as

$$x = (x_1^T, x_2^T, x_3^T)^T, \tag{6.21}$$

where

$$x_1 = \begin{bmatrix} x_{11} \\ x_{12} \end{bmatrix} = \begin{bmatrix} \phi \\ \dot{\phi} \end{bmatrix}, \quad x_2 = \begin{bmatrix} x_{21} \\ x_{22} \end{bmatrix} = \begin{bmatrix} \theta \\ \dot{\theta} \end{bmatrix}, \quad x_3 = \begin{bmatrix} x_{31} \\ x_{32} \end{bmatrix} = \begin{bmatrix} \psi \\ \dot{\psi} \end{bmatrix}. \tag{6.22}$$

With this, (2.18) can be described as a set of interconnected subsystems,

$$\dot{x}_i = A_i x_i + h_i(x) + b_i u_i + f_i d_i, \qquad i = 1, 2, 3, \tag{6.23}$$

where

$$A_i = \begin{bmatrix} 0 & 1 \\ 0 & 0 \end{bmatrix}, \quad b_i = \begin{bmatrix} 0 \\ \beta_i \end{bmatrix}, \quad f_i = \begin{bmatrix} 0 \\ 1 \end{bmatrix}, \qquad i = 1, 2, 3,$$
$$h_1(x) = \begin{bmatrix} 0 \\ -\alpha_1 x_{22} x_{32} \end{bmatrix}, \quad h_2(x) = \begin{bmatrix} 0 \\ -\alpha_2 x_{32} x_{12} \end{bmatrix}, \tag{6.24}$$
$$h_3(x) = \begin{bmatrix} 0 \\ -\alpha_3 x_{12} x_{22} \end{bmatrix},$$

with $d_1 = M_x$, $d_2 = M_y$, and $d_3 = M_z$ the external disturbances.

It may be observed that when $h_i(x) \equiv 0$, $i = 1, 2, 3$, (6.23) represents three decoupled subsystems which describe the motions of the spacecraft along the three axes. However, $h_i(x)$ are not zero and constitute the interconnections among the subsystems, thus making an analysis based on the smaller-dimensional decoupled subsystems alone inaccurate.

The system represented by (6.23) is driven by the disturbance forces d_i in addition to the control signals u_i. However, these external disturbances can be completely canceled by constructing a disturbance-accommodating controller as described by Schiehlen (1973). This involves the determination of a suitable differential-equation model for the disturbances and, after augmenting the disturbance variables with the state variables of the system, designing a feedback controller that counteracts the disturbance forces by feeding back the estimated disturbance variables. Although this analysis is conducted for a single-axis model of the LST (only for the pitch motion control) by Schiehlen (1973), a straightforward extension that uses three separate disturbance-accommodating controllers can be obtained for the three-axis model here considered. For this reason, we will ignore the disturbance terms in our model and conduct all further analysis on the system

$$\dot{x}_i = A_i x_i + h_i(x) + b_i u_i, \qquad i = 1, 2, 3, \tag{6.25}$$

obtained from (6.23) with the substitution $d_i \equiv 0$, $i = 1, 2, 3$.

6.2. BILINEAR SYSTEMS: LOCAL STABILIZATION

The LST model developed in the last section is a nonlinear model of a special type: The nonlinearity is a product of the states. The situation is similar to that of multispecies communities modeled by the Lotka-Volterra equations, but not quite. The nonlinearities in the LST model are *bilinear forms* in the state variables that exclude the variable of the corresponding state equation, which is not the case with Lotka-Volterra equations. It is, however, true that the two models have similar structures, and quite a few results obtained in this section can be carried over to the ecomodels considered in the preceding chapter.

Stabilization of the bilinear interconnected systems will be achieved by the local linear controllers using the decomposition-aggregation method of Chapter 3. In this respect, the decentralized stabilization scheme is similar to that for linear large-scale systems considered in Section 3.3. The difference is, however, in the fact that stability of bilinear interconnected systems can be established in a finite region of the state space (Šiljak and Vukčević, 1976, 1977). Therefore, the results of Section 2.8 are used extensively in the stabilization scheme proposed in this section.

We observe that the LST model (6.23) belongs to the general class of bilinear systems described by the equations

$$\begin{aligned} \dot{x}_i &= A_i x_i + a_i x_l^T \sum_{\substack{j=1 \\ j \neq i,l}}^{s} A_{ij} x_j + b_i u_i, \qquad i = 1, 2, \ldots s, \\ l &= \begin{cases} 1, & i = s, \\ i + 1, & i \neq s, \end{cases} \end{aligned} \tag{6.26}$$

where $x_i(t) \in \mathfrak{R}^{n_i}$ is the state of each interconnected subsystem $\mathcal{S}_i$, A_i are constant $n_i \times n_i$ matrices, A_{ij} are constant $n_l \times n_j$ matrices, and a_i and b_i are constant n_i-vectors.

To stabilize the system (6.26), we choose the local control

$$u_i = -k_i^T x_i, \qquad i = 1, 2, \ldots, s, \tag{6.27}$$

so that each uncoupled subsystem

$$\dot{x}_i = (A_i - b_i k_i^T)x_i, \qquad i = 1, 2, \ldots, s, \tag{6.28}$$

has a prescribed set of distinct eigenvalues

$$\mathfrak{L}_i = \{-\sigma_1^i \pm j\omega_1^i, \ldots, \sigma_p^i \pm j\omega_p^i, -\sigma_{p+1}^i, \ldots, -\sigma_{n_i-p}^i\}, \tag{6.29}$$

where $\sigma_q^i > 0$, $q = 1, 2, \ldots, n_i - p$, and the integer p is such that $0 \leqslant p \leqslant [n_i/2]$.

By using the transformation

$$x_i = T_i \tilde{x}_i, \tag{6.30}$$

introduced in Section 3.3, we get the system (6.26) in the form

$$\begin{aligned} \dot{\tilde{x}}_i &= \Lambda_i \tilde{x}_i + \tilde{a}_i \tilde{x}_l^T \sum_{\substack{j=1 \\ j\neq i,l}}^{s} \tilde{A}_{ij}\tilde{x}_j, \qquad i = 1, 2, \ldots, s, \\ l &= \begin{cases} 1, & i = s, \\ i+1, & i \neq s, \end{cases} \end{aligned} \tag{6.31}$$

where $\Lambda_i = T_i^{-1}(A_i - b_i k_i^T)T_i$ has the quasidiagonal form

$$\Lambda_i = \operatorname{diag}\left\{\begin{bmatrix} -\sigma_1^i & \omega_1^i \\ -\omega_1^i & -\sigma_1^i \end{bmatrix}, \ldots, \begin{bmatrix} -\sigma_p^i & \omega_p^i \\ -\omega_p^i & -\sigma_p^i \end{bmatrix}, -\sigma_{p+1}^i, \ldots, -\sigma_{n_i-p}^i\right\}, \tag{6.32}$$

and

$$\tilde{A}_{ij} = T_l^T A_{ij} T_j, \qquad \tilde{a}_i = T_i^{-1} a_i. \tag{6.33}$$

We define the interaction function $h_i\colon \mathfrak{T} \times \mathfrak{R}^n \to \mathfrak{R}^{n_i}$ among the subsystems of (6.31) as

$$h_i(\tilde{x}) = \tilde{a}_i \tilde{x}_l^T \sum_{\substack{j=1 \\ j\neq i,l}}^{s} \tilde{A}_{ij}\tilde{x}_j. \tag{6.34}$$

The interactions $h_i(\tilde{x})$ can be bounded as

$$\|h_i(\tilde{x})\| \leqslant \upsilon_l \sum_{\substack{j=1 \\ j\neq i,l}}^{s} \xi_{ij}\|\tilde{x}_j\| \qquad \forall \tilde{x} \in \tilde{\mathfrak{X}}_0 \tag{6.35}$$

on the region

$$\tilde{\mathfrak{X}}_0 = \{\tilde{x} \in \mathfrak{R}^n\colon \|\tilde{x}_i\| < \upsilon_i, i = 1, 2, \ldots, s\}, \tag{6.36}$$

where υ_i are positive constants which are as yet unspecified, and $\tilde{\xi}_{ij} = (\tilde{a}_i^T \tilde{a}_i)^{1/2}\lambda_M^{1/2}(\tilde{A}_{ij}^T \tilde{A}_{ij})$.

The aggregate $s \times s$ matrix $\tilde{W} = (\tilde{w}_{ij})$ which corresponds to the system (6.31) and constraints (6.35) is obtained by using the results of Weissenberger (1973) as shown in Section 2.8:

$$\tilde{W} = D\overline{W}, \tag{6.37}$$

where

$$D = \text{diag}\{v_2, v_3, \ldots, v_s, v_1\}, \tag{6.38}$$

and the $s \times s$ matrix $\overline{W} = (\overline{w}_{ij})$ is defined by

$$\overline{w}_{ij} = -\delta_{ij} v_l^{-1} \pi_i + (1 - \delta_{ij})\tilde{\xi}_{ij} \tag{6.39}$$

with π_i defined in Section 3.3 as

$$\pi_i \doteq \min_q \sigma_q^i, \qquad q = 1, 2, \ldots, n_i - p. \tag{6.40}$$

It is a known fact (see Appendix) that the matrix $\tilde{W}$ in (6.37) satisfies the Sevastyanov-Kotelyanskii inequalities

$$(-1)^k \begin{vmatrix} -\pi_1 & \tilde{\xi}_{12} & \cdots & \tilde{\xi}_{1k} \\ \tilde{\xi}_{21} & -\pi_2 & \cdots & \tilde{\xi}_{2k} \\ \cdots & \cdots & \cdots & \cdots \\ \tilde{\xi}_{k1} & \tilde{\xi}_{k2} & \cdots & -\pi_k \end{vmatrix} > 0, \qquad k = 1, 2, \ldots, s, \tag{6.41}$$

if and only if the matrix $\overline{W}$ does. These inequalities applied to $\overline{W}$ determine the constants $v_1, v_2, \ldots, v_s$ in (6.35). It is possible to calculate these constants recursively. To see this, we note that the kth leading principal $k \times k$ submatrix $\overline{W}_k$ can be expressed as

$$\begin{aligned} \overline{W}_k &= \begin{bmatrix} \overline{W}_{k-1} & f_k \\ g_k^T & \overline{w}_{kk} \end{bmatrix} \\ &= \begin{bmatrix} I & 0 \\ g_k^T \overline{W}_{k-1}^{-1} & 1 \end{bmatrix} \begin{bmatrix} \overline{W}_{k-1} & 0 \\ 0 & \overline{w}_{kk} - g_k^T \overline{W}_{k-1}^{-1} f_k \end{bmatrix} \begin{bmatrix} I & \overline{W}_{k-1}^{-1} f_k \\ 0 & 1 \end{bmatrix}. \end{aligned} \tag{6.42}$$

Therefore, the kth leading principle minor of $\overline{W}$ is

$$\det \overline{W}_k = (\det \overline{W}_{k-1})(\overline{w}_{kk} - g_k^T \overline{W}_{k-1}^{-1} f_k). \tag{6.43}$$

For the inequalities (6.43) to be satisfied by $\overline{W}$, it is necessary and sufficient that

$$-\overline{w}_{kk} + g_k^T \overline{W}_{k-1}^{-1} f_k > 0, \qquad k = 1, 2, \ldots, s. \tag{6.44}$$

From (6.39), we have

$$f_k^T = (\tilde{\xi}_{1k}, \tilde{\xi}_{2k}, \ldots, \tilde{\xi}_{sk}), \qquad g_k^T = (\tilde{\xi}_{k1}, \tilde{\xi}_{k2}, \ldots, \tilde{\xi}_{ks}), \tag{6.45}$$

and from (6.39) and (6.44), we get the constants v_l as

$$v_l < -\pi_k (g_k^T \overline{W}_{k-1}^{-1} f_k)^{-1}, \qquad l = \begin{cases} 1, & k = s, \\ k+1, & k \neq s. \end{cases} \tag{6.46}$$

Once the constants v_l are calculated by (6.46), the region $\tilde{\mathscr{X}}_0$ of (6.36) is determined. Now it remains to imbed a Liapunov function ν: $\mathscr{R}^n \to \mathscr{R}_+$ inside the region $\tilde{\mathscr{X}}_0$ and determine an estimate of the stability region as

$$\tilde{\mathscr{X}} = \{\tilde{x} \in \mathscr{R}^n: \nu(\tilde{x}) < \gamma\}. \tag{6.47}$$

Here we choose

$$\nu(\tilde{x}) = \sum_{i=1}^{s} d_i |v_i(\tilde{x}_i)|, \tag{6.48}$$

where d_i are positive numbers, and $v_i(\tilde{x}_i) = \|\tilde{x}_i\|$. Following Section 2.9, we calculate the positive constant γ in (6.47) using (6.46) and

$$\gamma = \min_i d_i v_i, \qquad i = 1, 2, \ldots, s, \tag{6.49}$$

where the positive vector $d^T = (d_1, d_2, \ldots, d_s)$ is computed by

$$d^T = -c^T \tilde{W}^{-1}, \tag{6.50}$$

c being any positive s-vector ($c > 0$).

Since $\tilde{x}_i = T_i^{-1} x_i$ and $\|\tilde{x}_i\| \leqslant \|T_i^{-1}\| \, \|x_i\|$, from (6.48) and (6.49) we get finally the region estimate $\hat{\mathscr{X}}$ in the original state space, which is

$$\hat{\mathscr{X}} = \left\{x \in \mathscr{R}^n: \sum_{i=1}^{s} d_i \|T_i^{-1}\| \, \|x_i\| < \gamma\right\}. \tag{6.51}$$

6.3. STABILIZATION OF LST

Now we consider the nonlinear model of the LST given in Section 6.2, which belongs to the class of systems described by (6.26) with

$$A_i = \begin{bmatrix} 0 & 1 \\ 0 & 0 \end{bmatrix}, \quad A_{ij} = \begin{bmatrix} 0 & 0 \\ 0 & 1 \end{bmatrix}, \quad a_i = \begin{bmatrix} 0 \\ -\alpha_i \end{bmatrix}, \quad b_i = \begin{bmatrix} 0 \\ \beta_i \end{bmatrix}, \tag{6.52}$$

$i = 1, 2, 3.$

Applying the control law

$$u_i = -k_i^T x_i, \qquad i = 1, 2, 3, \tag{6.53}$$

where

$$k_i^T = \beta_i^{-1} \bar{k}_i^T, \qquad i = 1, 2, 3, \tag{6.54}$$

and $\bar{k}_i^T = (\bar{k}_{i1}, \bar{k}_{i2})$, we obtain the closed-loop uncoupled subsystems (6.28) with

$$A_i - b_i k_i^T = \begin{bmatrix} 0 & 1 \\ -\bar{k}_{i1} & -\bar{k}_{i2} \end{bmatrix}, \qquad i = 1, 2, 3. \tag{6.55}$$

The gains $\bar{k}_i$ are chosen so that each subsystem has a set of eigenvalues

$$\mathfrak{L}_i = \{-\sigma_1^i, -\sigma_2^i\}, \qquad i = 1, 2, 3. \tag{6.56}$$

To get the transformed system corresponding to (6.31), we use the transformation matrix

$$T_i = \begin{bmatrix} 1 & 1 \\ -\sigma_1^i & -\sigma_2^i \end{bmatrix}, \qquad i = 1, 2, 3, \tag{6.57}$$

and get

$$\Lambda_i = \begin{bmatrix} -\sigma_1^i & 0 \\ 0 & -\sigma_2^i \end{bmatrix}, \qquad h_i(\tilde{x}_i) = \tilde{a}_i \tilde{x}_l^T \tilde{A}_{ij} \tilde{x}_j, \quad i, j = 1, 2, 3, \quad i \neq j;$$
$$l = \begin{cases} i + 1, & i \neq 3, \\ 1, & i = 3. \end{cases} \tag{6.58}$$

To compute $\tilde{\xi}_{ij}$, we choose $\sigma_1^i = \sigma_1$, $\sigma_2^i = \sigma_2$, $i = 1, 2, 3$, and calculate $\|\tilde{A}_{ij}\| = (\sigma_1)^2 + (\sigma_2)^2$, $(\tilde{a}_i^T \tilde{a}_i)^{1/2} = \sqrt{2}\,|\alpha_i|(|\sigma_1 - \sigma_2|)^{-1}$. We can minimize the numbers $\tilde{\xi}_{ij}$ with respect to the distance $\rho = \sigma_2 - \sigma_1$ between the two subsystem eigenvalues. This yields

$$\tilde{\xi}_{ij} = \sqrt{2}\,|\alpha_i|\rho^{-1}[(\sigma_1)^2 + (\sigma_1 + \rho)^2], \tag{6.59}$$

and we get the minimal values $\tilde{\xi}_{ij}^m$ for $\tilde{\xi}_{ij}$ as

$$\tilde{\xi}_{ij}^m = (4 + 2\sqrt{2})|\alpha_i|\sigma_1, \tag{6.60}$$

which is obtained for $\rho = \sqrt{2}\,\sigma_1$.

The corresponding matrix $\overline{W}$ in (6.37) is

$$\overline{W} = \begin{bmatrix} -\sigma_1/\upsilon_2 & 0 & \tilde{\xi}_{13} \\ \tilde{\xi}_{21} & -\sigma_1/\upsilon_3 & 0 \\ 0 & \tilde{\xi}_{32} & -\sigma_1/\upsilon_1 \end{bmatrix}. \tag{6.61}$$

From (6.46) and (6.61), we get

$$\upsilon_1 \upsilon_2 \upsilon_3 < \frac{\sigma_1^3}{\tilde{\xi}_{13} \tilde{\xi}_{21} \tilde{\xi}_{32}}. \tag{6.62}$$

Choosing $\upsilon_1 = \upsilon_2 = \upsilon_3 = \upsilon$, and using (6.60) and (6.62), we compute $\upsilon < 0.584$. Selecting $\upsilon = 0.574$, $\sigma_1 = 10$, and $c = (1, 1, 1)^T$, we further compute from (6.50) the vector

$$d = (4.8, 13.7614, 4.2963)^T. \tag{6.63}$$

From (6.49), we calculate

$$\gamma = 2.4663, \tag{6.64}$$

and the region $\tilde{\mathfrak{X}}$ in the transformed state space as

$$\tilde{\mathfrak{X}} = \{\tilde{x} \in \mathfrak{R}^n\colon 4.8\|\tilde{x}_1\| + 13.7614\|\tilde{x}_2\| + 4.2963\|\tilde{x}_3\| < 2.4663\}. \tag{6.65}$$

In the original space, the stability region $\hat{\mathfrak{X}}$ is finally obtained as

$$\hat{\mathfrak{X}} = \{x \in \mathfrak{R}^n\colon 4.8\|x_1\| + 13.7614\|x_2\| + 4.2963\|x_3\| < 1.3331\}, \tag{6.66}$$

where we have used $\|T_i^{-1}\| = 1.8500$, $i = 1, 2, 3$.

The feedback gains that yield the region $\hat{\mathfrak{X}}$ are computed from (6.54) and

$$\bar{k}_i^T = (\sigma_1 \sigma_2, \sigma_1 + \sigma_2)^T = (241.4213, 34.1421)^T, \qquad i = 1, 2, 3, \tag{6.67}$$

as

$$\begin{aligned} k_1^T &= (2.8196, 0.3988)^T, \\ k_2^T &= (17.6348, 2.4939)^T, \\ k_3^T &= (18.2756, 2.5846)^T. \end{aligned} \tag{6.68}$$

This completes the design of the LST control system.

6.4. OPTIMAL CONTROL OF LST

The LST model developed and stabilized so far is also suitable for applying the multilevel optimization scheme outlined in Section 3.5. We only need to note that the model is composed of three linear subsystems interconnected by nonlinear functions. This fact enables us to use the well-known solution of the linear-quadratic regulator problem for the subsystems, and then handle the nonlinearities as disturbance terms to come up with a suboptimal design. As it turns out, we will be able to do quite a bit more than the multilevel scheme suggests. In the particular model of the LST, we will be able to construct a "decoupling" global control and achieve an optimal design. The price for this unexpected success is that the global control is nonlinear, which may be a problem in implementation of the real control system. The presentation in this section follows that of Sundareshan given in the report by Šiljak, Sundareshan, and Vukčević (1975).

We recall the LST model given by the equations

$$\dot{x}_i = A_i x_i + b_i u_i + h_i(x), \qquad i = 1, 2, 3, \tag{6.23}$$

where $x_i = (x_{i1}, x_{i2})^T$, $i = 1, 2, 3$, and

$$A_i = \begin{bmatrix} 0 & 1 \\ 0 & 0 \end{bmatrix}, \qquad b_i = \begin{bmatrix} 0 \\ \beta_i \end{bmatrix},$$

$$h_1(x) = \begin{bmatrix} 0 \\ -\alpha_2 x_{32} x_{12} \end{bmatrix}, \quad h_2(x) = \begin{bmatrix} 0 \\ -\alpha_2 x_{32} x_{12} \end{bmatrix}, \quad h_3(x) = \begin{bmatrix} 0 \\ -\alpha_3 x_{12} x_{22} \end{bmatrix}. \tag{6.69}$$

Following our multilevel control policy, we split each of the control functions u_i into a local component u_i^l and a global component u_i^g and optimize the decoupled subsystems

$$\dot{x}_i = A_i x_i + b_i u_i^l, \qquad i = 1, 2, 3, \tag{6.70}$$

with respect to the performance indices

$$J_i = \int_{t_0}^{\infty} \{\|x_i\|^2 + \|u_i^l\|^2\} dt, \qquad i = 1, 2, 3, \tag{6.71}$$

obtained with the choice $Q_i = I_{2\times 2}$ and $R_i = 1$, $i = 1, 2, 3$. The solution of this linear-quadratic optimal-control problem is simple and involves the solution of the associated Riccati equations,

$$A_i^T P_i + P_i A_i - P_i b_i b_i^T P_i + Q_i = 0, \qquad i = 1, 2, 3. \tag{6.72}$$

With the specified structure of A_i and b_i, the solution of (6.72) can be obtained as

$$P_i = \begin{bmatrix} \left(1 + \dfrac{2}{\beta_i}\right)^{1/2} & \dfrac{1}{\beta_i} \\ \dfrac{1}{\beta_i} & \dfrac{1}{\beta_i}\left(1 + \dfrac{2}{\beta_i}\right)^{1/2} \end{bmatrix}, \qquad i = 1, 2, 3, \tag{6.73}$$

and the local optimal controls are

$$u_i^l = -b_i^T P_i x_i = -\left[1, \quad \left(1 + \frac{2}{\beta_i}\right)^{1/2}\right] x_i, \qquad i = 1, 2, 3. \quad (6.74)$$

In the absence of global control functions u_i^g, (6.74) will be suboptimal for the composite system (6.23) with the index of suboptimality ε. With global control, however, we have the effective interconnections,

$$\hat{h}_i(x, u_i^g) = h_i(x) + b_i u_i^g, \qquad i = 1, 2, 3. \quad (6.75)$$

The equations (6.75) can be simplified to yield

$$\begin{aligned} \hat{h}_1(x, u_1^g) &= \begin{bmatrix} 0 \\ -\alpha_1 x_{22} x_{32} + \beta_1 u_1^g \end{bmatrix}, \\ \hat{h}_2(x, u_2^g) &= \begin{bmatrix} 0 \\ -\alpha_2 x_{12} x_{32} + \beta_2 u_2^g \end{bmatrix}, \\ \hat{h}_3(x, u_3^g) &= \begin{bmatrix} 0 \\ -\alpha_3 x_{12} x_{22} + \beta_3 u_3^g \end{bmatrix}. \end{aligned} \quad (6.76)$$

It is now simple to observe that the choice of the global control

$$\begin{aligned} u_1^g(x) &= \frac{\alpha_1}{\beta_1} x_{22} x_{32}, \\ u_2^g(x) &= \frac{\alpha_2}{\beta_2} x_{12} x_{32}, \\ u_3^g(x) &= \frac{\alpha_3}{\beta_3} x_{12} x_{22} \end{aligned} \quad (6.77)$$

achieves the required optimality with $\varepsilon = 0$.

It is of interest to evaluate the control functions for a set of the LST parameters. For the values of the inertia components, we choose $I_x = 14,656$ kg m^2, $I_y = 91,772$ kg m^2, $I_z = 95,027$ kg m^2; and for the reaction-wheel constants we assume the typical values $K_1 = K_2 = K_3 = 12.57 \times 10^5$ N m/rad. These values correspond to the on-orbit configuration of the LST with extended light shield and solar wings, with the corresponding mass of the body totalling 9380 kg, as outlined in the report by Schiehlen (1973). The values of α_i, β_i, $i = 1, 2, 3$, can now be calculated as

$$\begin{aligned} \alpha_1 &= 0.2221, & \beta_1 &= 85.62, \\ \alpha_2 &= -0.08754, & \beta_2 &= 13.69, \\ \alpha_3 &= 0.8112, & \beta_3 &= 13.21. \end{aligned} \quad (6.78)$$

Hence, the control components can be evaluated from (6.74) and (6.77) as

$$\begin{aligned} u_1^l &= -[1 \quad 1.012]\begin{bmatrix} x_{11} \\ x_{12} \end{bmatrix}, \\ u_2^l &= -[1 \quad 1.061]\begin{bmatrix} x_{21} \\ x_{22} \end{bmatrix}, \\ u_3^l &= -[1 \quad 1.070]\begin{bmatrix} x_{31} \\ x_{32} \end{bmatrix}, \end{aligned} \tag{6.79}$$

and

$$\begin{aligned} u_1^g &= 0.0026 x_{22} x_{32}, \\ u_2^g &= -0.064 x_{12} x_{32}, \\ u_3^g &= 0.0613 x_{12} x_{22}. \end{aligned} \tag{6.80}$$

This completes the multilevel optimization of the LST control system.

6.5. MAXIMIZATION OF STRUCTURAL PARAMETERS

In this section, we will show still another aspect of the decomposition-aggregation stability analysis and decentralized control scheme promoted in Chapters 2 and 3: One can determine explicitly the effect on the overall system stability of the interconnection parameters that couple the subsystems. Such explicit formulation of the interconnection effects is suitable for parameter maximization in the mathematical-programming format. This aspect of the method turns out to be useful in the stability analysis of the Skylab control system (Šiljak, 1975), which is the subject of the following section.

Let us now consider a linear system $\mathcal{S}$ described by a linear time-constant differential equation

$$\dot{x} = Ax, \tag{6.81}$$

where $x \in \mathcal{R}^n$ is the state of the system and A is a constant $n \times n$ matrix. For stability studies, the system $\mathcal{S}$ is decomposed into s interconnected subsystems $\mathcal{S}_i$ represented by the equations

$$\dot{x}_i = A_i x_i + \sum_{j=1}^{s} e_{ij} A_{ij} x_j, \qquad i = 1, 2, \ldots, s, \tag{6.82}$$

where $x_i \in \mathcal{R}^{n_i}$ is the state of the subsystem $\mathcal{S}_i$, so that $x = (x_1^T, x_2^T, \ldots, x_s^T)^T$; and A_i and A_{ij} are constant $n_i \times n_i$ and $n_i \times n_j$ matrices.

In contrast with the structural analysis conducted so far, we consider the elements e_{ij} of the constant $s \times s$ interconnection matrix $E = (e_{ij})$ as structural parameters which should be maximized, with the stability of the system $\mathbb{S}$ as a constraint. To show how this can be done, we will briefly review the decomposition-aggregation analysis of Chapters 1 and 2.

For $E = 0$, all subsystems are decoupled, and we have the free subsystems $\mathbb{S}_i$ described by the equations

$$\dot{x}_i = A_i x_i, \qquad i = 1, 2, \ldots, s. \tag{6.83}$$

As required by the decomposition-aggregation stability analysis outlined in Chapter 2, we choose a Liapunov function v_i: $\mathcal{R}^{n_i} \to \mathcal{R}_+$ for each $\mathbb{S}_i$ of (6.83) as

$$v_i(x_i) = (x_i^T H_i x_i)^{1/2}, \tag{6.84}$$

and compute the total time derivative $\dot{v}_{(6.83)}$ using (6.83) to get

$$\dot{v}_i(x_i)_{(6.83)} = -\tfrac{1}{2} v_i^{-1}(x_i) x_i^T G_i x_i, \tag{6.85}$$

where

$$A_i^T H_i + H_i A_i = -G_i, \tag{6.86}$$

and H_i, G_i are symmetric positive definite $n_i \times n_i$ matrices ($H_i^T = H_i > 0$, $G_i^T = G_i > 0$). That is, each function $v_i(x_i)$ satisfies the following estimates:

$$\begin{aligned} \eta_{i1} \|x_i\| \leqslant v_i(x_i) &\leqslant \eta_{i2} \|x_i\|, \\ \dot{v}_i(x_i)_{(6.83)} &\leqslant -\eta_{i3} v_i(x_i), \qquad \forall x_i \in \mathcal{R}^{n_i}, \\ \|\text{grad } v_i(x_i)\| &\leqslant \eta_{i4} \end{aligned} \tag{6.87}$$

where

$$\begin{aligned} \eta_{i1} &= \lambda_m^{1/2}(H_i), \qquad & \eta_{i2} &= \lambda_M^{1/2}(H_i), \\ \eta_{i3} &= \frac{1}{2} \frac{\lambda_m(G_i)}{\lambda_M(H_i)}, \qquad & \eta_{i4} &= \frac{\lambda_M(H_i)}{\lambda_m^{1/2}(H_i)}. \end{aligned} \tag{6.88}$$

Here λ_m and λ_M are the minimum and maximum eigenvalues of the indicated matrices.

We recall from Section 2.5, that the interactions among the subsystems $\mathbb{S}_i$ in (6.82) can be estimated by the inequality

$$\left\| \sum_{j=1}^{s} e_{ij} A_{ij} x_j \right\| \leqslant \sum_{j=1}^{s} e_{ij} \xi_{ij} \|x_j\| \qquad \forall x \in \mathcal{R}^n, \tag{6.89}$$

where $\xi_{ij} = \lambda_M^{1/2}(A_{ij}^T A_{ij})$ are nonnegative numbers.

The aggregate model of the system $\mathcal{S}$ can now be formed using the vector Liapunov function v: $\mathcal{R}^n \to \mathcal{R}_+^s$,

$$v = (v_1, v_2, \ldots, v_s)^T. \tag{6.90}$$

Taking the total time derivative $\dot{v}_i(x_i)_{(6.82)}$ of $v_i(x_i)$ along the solutions of the interconnected subsystems (6.82), we get

$$\dot{v}_{i(6.82)} \leqslant \dot{v}_{i(6.83)} + (\text{grad } v_i)^T \sum_{j=1}^{s} e_{ij} A_{ij} x_j, \qquad i = 1, 2, \ldots, s. \tag{6.91}$$

By applying the estimates (6.87) and (6.88), we can rewrite (6.91) as

$$\dot{v}_{i(6.82)} \leqslant -\eta_{i2}^{-1} \eta_{i3} v_i + \eta_{i4} \sum_{j=1}^{s} e_{ij} \xi_{ij} \eta_{j1}^{-1} v_j, \qquad i = 1, 2, \ldots, s. \tag{6.92}$$

With the notation (6.90), these scalar inequalities can be finally rewritten as a vector differential inequality

$$\dot{v} \leqslant Wv, \tag{6.93}$$

where the $s \times s$ constant aggregate matrix $W = (w_{ij})$ has the elements defined by

$$w_{ij} = \begin{cases} -\eta_{i2}^{-1} \eta_{i3} + e_{ii} \xi_{ii} \eta_{i1}^{-1} \eta_{i4}, & i = j, \\ e_{ij} \xi_{ij} \eta_{j1}^{-1} \eta_{i4}, & i \neq j. \end{cases} \tag{6.94}$$

Applying Theorem 2.16, we conclude that if the matrix W satisfies Sevastyanov-Kotelyanskii inequalities

$$(-1)^k \begin{vmatrix} w_{11} & w_{12} & \cdots & w_{1k} \\ w_{21} & w_{22} & \cdots & w_{2k} \\ \cdots & \cdots & \cdots & \cdots \\ w_{k1} & w_{k2} & \cdots & w_{kk} \end{vmatrix} > 0, \qquad k = 1, 2, \ldots, s, \tag{6.95}$$

then the equilibrium $x^* = 0$ of the linear system $\mathcal{S}$ described by the equations (6.82) is globally exponentially stable, that is, there exist two positive numbers Π and π independent of the initial conditions (t_0, x_0) such that the solution $x(t; t_0, x_0)$ of (6.82) satisfies the inequality

$$\|x(t; t_0, x_0)\| \leqslant \Pi \|x_0\| \exp[-\pi(t - t_0)] \qquad \forall t \in \mathcal{T}_0 \tag{6.96}$$

for all $(t_0, x_0) \in \mathcal{T} \times \mathcal{R}^n$, where $\mathcal{T} = (-\infty, +\infty)$, $\mathcal{T}_0 = [t_0, +\infty)$. At present, we are not interested in the connectivity aspects of the stability property (6.96).

We also recall from Section 2.5 (see also Appendix), that the inequalities (6.95) are equivalent to saying that there exist numbers $d_i > 0 (i = 1, 2, \ldots, s)$ and $\pi > 0$ such that

$$|w_{jj}| - d_j^{-1} \sum_{\substack{i=1 \\ i \neq j}}^{s} d_i |w_{ij}| \geqslant \pi, \qquad j = 1, 2, \ldots, s, \tag{6.97}$$

that is, W is a quasidominant diagonal matrix.

In order to study the effect of structural parameters e_{ij} on the overall system stability, it is more convenient to use the diagonal-dominance conditions (6.97) than the Sevastyanov-Kotelyanskii determinantal inequalities (6.95). Instead of the quasidominant-diagonal property (6.97), we will apply the weaker but simpler dominant-diagonal conditions

$$|w_{jj}| > \sum_{\substack{i=1 \\ i \neq j}}^{s} w_{ij}, \qquad j = 1, 2, \ldots, s, \tag{6.98}$$

which are obtained from (6.97) when the d_i's are all set equal to one. By using the definition (6.94) of the w_{ij}'s and (6.88), we can rewrite (6.98) in terms of the matrices H_i, G_i as

$$\frac{1}{2} \frac{\lambda_m(G_j)}{\lambda_M(H_j)} > \frac{1}{\lambda_m^{1/2}(H_j)} \sum_{i=1}^{s} e_{ij} \xi_{ij} \frac{\lambda_M(H_i)}{\lambda_m^{1/2}(H_i)}, \qquad j = 1, 2, \ldots, s. \tag{6.99}$$

From (6.99), it is obvious that our ability to determine stability by the decomposition-aggregation method depends crucially on the choice of the matrix $G_i > 0$ and the corresponding solution $H_i > 0$ of the Liapunov matrix equation (6.86) for each free subsystem (6.83). It is also clear from (6.98) and (6.99) that the best estimates of the values of the structural parameters e_{ij} are provided by the optimal aggregate matrix W^0 which is the solution of the following:

Problem 6.1.

Find: $\quad W^0$

subject to: $\quad W^0 - W \leqslant 0 \quad \forall W$

$$A_i^T H_i + H_i A_i = -G_i, \qquad i = 1, 2, \ldots, s.$$

Here the matrix inequality is taken element by element.

Furthermore, if the decomposition (6.82) is performed with "the least violence" done to the system $\mathbb{S}$, as proposed by Steward (1965), the interaction matrices A_{ij} are sparse, and their respective norms ξ_{ij} are small. From (6.99), we conclude that the largest estimates for the parameters e_{ij} are obtained for the pair of matrices (G_i, H_i) with the largest ratio $\lambda_m(G_i)/\lambda_M(H_i)$. Therefore, we replace Problem 6.1 with the following:

Problem 6.2.

$$\textit{Find}: \qquad \max_{G_i}\left\{\frac{\lambda_m(G_i)}{\lambda_M(H_i)}\right\}$$

$$\textit{subject to}: \qquad A_i^T H_i + H_i A_i = -G_i,$$

which should be solved for each of the s free subsystems $\mathbb{S}_i$ of (6.83) separately. As a byproduct of the solution, we obtain the Liapunov function (6.84) which provides the exact value of the degree of exponential stability for each isolated subsystem $\mathbb{S}_i$.

To solve Problem 6.2, we assume that all eigenvalues of the matrix A_i are known numerically and all of them are distinct. This would be a prohibitive assumption for the overall system $\mathbb{S}$, but since the decomposition-aggregation method (Section 2.5) requires such assumptions for the low-order subsystems, it is quite realistic for them. Furthermore, in the proposed multilevel stabilization scheme (Section 3.3), such an assumption can readily be realized using local linear feedback controls and any of the conventional techniques such as pole assignment, the root-locus method, the parameter plane method, etc.

To solve Problem 6.2, we use again the transformation introduced in Section 3.3,

$$x_i = T_i \tilde{x}_i, \tag{6.30}$$

where T_i is a nonsingular constant $n \times n$ matrix. This produces the free subsystem $\mathbb{S}_i$ in the form

$$\dot{\tilde{x}}_i = \Lambda_i \tilde{x}, \tag{6.100}$$

with $\Lambda_i = T_i^{-1} A_i T_i$ having the following canonical quasidiagonal form:

$$\Lambda_i = \text{diag}\left\{\begin{bmatrix} -\sigma_1^i & \omega_1^i \\ -\omega_1^i & -\sigma_1^i \end{bmatrix}, \ldots, \begin{bmatrix} -\sigma_p^i & \omega_p^i \\ -\omega_p^i & -\sigma_p^i \end{bmatrix}, -\sigma_{p+1}^i, \ldots, -\sigma_{n_i-p}^i\right\}, \tag{6.32}$$

where $-\sigma_p^i$ and $-\sigma_p^i \pm j\omega_p^i$ $(\sigma_q^i > 0\ q = 1, 2, \ldots, n_i - p$, and $0 \leqslant p \leqslant [n_i/2])$ are real and complex eigenvalues of the matrix Λ_i and, thus, A_i. The Liapunov matrix equation (6.86) becomes

$$\Lambda_i^T \tilde{H}_i + \tilde{H}_i \Lambda_i = -\tilde{G}_i, \tag{6.101}$$

where $\tilde{H}_i = T_i^T H_i T_i$ and $\tilde{G}_i = T_i^T G_i T_i$. Now we replace Problem 6.2 by the following:

Problem 6.3.

$$\textit{Find}: \qquad \max_{\tilde{G}_i} \left\{ \frac{\lambda_m(\tilde{G}_i)}{\lambda_M(\tilde{H}_i)} \right\}$$

$$\textit{subject to}: \qquad \tilde{P}_i^T \tilde{H}_i + \tilde{H}_i \tilde{P}_i = -\tilde{G}_i,$$

and prove the following:

Theorem 6.1. *The solution to Problem 6.3 is provided by the choice*

$$\tilde{H}_i^0 = \theta_i I_i, \quad \tilde{G}_i^0 = 2\theta_i \operatorname{diag}\{\sigma_1^i, \sigma_1^i, \ldots, \sigma_p^i, \sigma_p^i, \sigma_{p+1}^i, \ldots \sigma_{n_i-p}^i\}, \tag{6.102}$$

where $\theta_i > 0$ is an arbitrary number and I_i is the $n_i \times n_i$ identity matrix.

Proof. The proof is almost automatic if we recognize that the ratio

$$\pi_i^0 = \frac{1}{2} \frac{\lambda_m(\tilde{G}_i^0)}{\lambda_M(\tilde{H}_i^0)} = \min_q \{\sigma_q^i\} = \sigma_M^i \tag{6.103}$$

is the largest estimate of the degree of exponential stability for the ith subsystem (6.100). That is, by integrating the second inequality in (6.87) we get

$$v_i(\tilde{x}_i) \leqslant v_i(\tilde{x}_{i0}) \exp[-\eta_{i3}(t - t_0)] \qquad \forall t \in \mathcal{T}_0. \tag{6.104}$$

By using (6.87), from (6.104) we get

$$\|\tilde{x}_i(t; t_0, \tilde{x}_{i0})\| \leqslant \eta_{i1}^{-1} \eta_{i2} \|\tilde{x}_{i0}\| \exp[-\eta_{i3}(t - t_0)] \qquad \forall t \in \mathcal{T}_0, \tag{6.105}$$

which is globally valid for all $(t_0, \tilde{x}_{i0}) \in \mathcal{T} \times \mathcal{R}^{n_i}$, and where $\eta_{i3} = \pi_i = \frac{1}{2}\lambda_m(\tilde{G}_i)/\lambda_M(\tilde{H}_i)$. If we assume that the largest eigenvalue of the subsystem matrix Λ_i in (6.100) is a real root $-\sigma_M^i$, then it is obvious that the ratio $\pi_i^0 = \sigma_M^i$ is the largest possible. Otherwise, $\eta_{i3} > \sigma_M^i$, and by solving (6.100) with $\tilde{x}_{i0} = \tilde{x}_M^i$, where $\tilde{x}_M^i$ is the eigenvector corresponding to $-\sigma_M^i$ and Λ_i, we get $\tilde{x}_i(t; t_0, \tilde{x}_M^i) = \tilde{x}_M^i \exp[-\sigma_M^i(t - t_0)]$, which contradicts (6.105). Similar reasoning leads to a contradiction in the case when the largest eigenvalue of Λ_i is a complex number $-\sigma_M^i \pm j\omega_M^i$. This proves Theorem 6.1.

As a byproduct of the above proof, we get the Liapunov function

$$v_i(\tilde{x}_i) = \|\tilde{x}_i\|, \tag{6.106}$$

which provides the exact estimate of the degree of exponential stability, $\pi_i^0 = \sigma_M^i$. Furthermore, the solution $\tilde{H}_i^0 = \theta I_i$ yields the lowest possible value for the ratio $\lambda_M(\tilde{H}_i)/\lambda_m^{1/2}(\tilde{H}_i)\lambda_m^{1/2}(\tilde{H}_j)$ which appears on the right-hand side of the inequalities (6.99). Therefore, the solution $(\tilde{G}_i^0, \tilde{H}_i^0)$ of Problem

6.3 in the transformed space provides the best aggregate matrix $\tilde{W}^0$ required by Problem 6.1. Consequently, the best estimates for the structural parameters e_{ij} are obtained by using the inequalities (6.99) in the transformed space,

$$\sigma_M^j > \sum_{i=1}^{s} e_{ij} \tilde{\xi}_{ij}, \qquad j = 1, 2, \ldots, s, \tag{6.107}$$

provided $\tilde{\xi}_{ij} \leqslant \xi_{ij}$ $(i,j = 1,2,\ldots,s)$ and $\tilde{\xi}_{ij} = \lambda_M^{1/2}(\tilde{A}_{ij}^T \tilde{A}_{ij})$, $\tilde{A}_{ij} = T_i^{-1} A_{ij} T_j$.

Maximization of the interconnection parameters e_{ij} can be now performed in the mathematical-programming format (Kuhn and Tucker, 1951) for each subsystem separately. Let us define the s-vectors $e_j = (e_{1j}, e_{2j}, \ldots, e_{sj})^T$, $b_j = (\xi_{1j}, \xi_{2j}, \ldots, \xi_{sj})^T$, and the positive number $\sigma_j = \sigma_M^j + \varepsilon$, where $\varepsilon > 0$ is an arbitrarily small number. Then, we can state the following *vector maximization problem* for the jth subsystem:

Problem 6.4.

$$\begin{aligned} &\textit{Maximize:} && e_j \\ &\textit{subject to:} && \sigma_j - b_j^T e_j \geqslant 0, \quad e_j \geqslant 0. \end{aligned}$$

That is, we are interested in finding an s-vector e_j^0 constrained by $\sigma_j - b_j^T e_j \geqslant 0$, $e_j \geqslant 0$ such that $e_j \geqslant e_j^0$ for no e_j satisfying the constraints (the vector inequalities taken element by element). A Pareto-optimal solution to Problem 6.4 can be obtained using the results of Kuhn and Tucker (1951), and DaCunha and Polak (1967).

If weights can be assigned to the components of the interconnection vector e_j by choosing an s-vector $c_j = (c_{1j}, c_{2j}, \ldots, c_{sj})^T$ such that $c_{ij} \geqslant 0$ $(i = 1,2,\ldots,s)$ and $\sum_{i=1}^{s} c_{ij} = 1$, then Problem 6.4 can be reformulated as a linear-programming problem:

Problem 6.5.

$$\begin{aligned} &\textit{Maximize:} && c_j^T e_j \\ &\textit{subject to:} && \sigma_j - b_j^T e_j \geqslant 0, \quad e_j \geqslant 0, \end{aligned}$$

which can be solved by known techniques (e.g. Dantzig, 1963; Zukhovitskii and Avdeyeva, 1966).

To establish stability of the large-scale system $\mathbb{S}$ by the method outlined, we rely entirely on the stability properties of the free subsystems $\mathbb{S}_i$. If active feedback elements are available, we can use them in a multilevel stabilization scheme as proposed in Section 3.3, and stabilize unstable large-scale systems. Furthermore, the scheme can be used to increase the values of the interconnection parameters e_{ij}. The local controllers for the decoupled subsystems can be used to raise the level of σ_M^i, while the global controllers can be applied to minimize some (or all) numbers $\tilde{\xi}_{ij}$ in the inequalities (6.95). This control strategy generally leads to the satisfaction of the stability conditions (6.95) with higher values of the interconnection parameters e_{ij} without a decrease in the degree π of exponential stability of the overall system.

So far, it has been shown how the multilevel stability analysis can be directed towards obtaining the maximum estimates of the interconnection parameters. These estimates can be further improved by introducing either output or state feedback.

The output-feedback scheme (Šiljak, 1975) is a straightforward application of the decomposition-aggregation method. We consider a linear system

$$\dot{x} = Ax + Bu, \qquad y = Cx, \tag{6.108}$$

where $x \in \mathcal{R}^n$ is the state of the system; $u \in \mathcal{R}^m$ is the control; $y \in \mathcal{R}^p$ is the output; and A, B, and C are constant $n \times n$, $n \times m$, and $p \times n$ matrices. The linear feedback

$$u(y) = -Ky \tag{6.109}$$

is introduced directly into the system (6.108), where K is a constant $m \times p$ matrix, and we get

$$\dot{x} = A^c x, \tag{6.110}$$

where the closed-loop system matrix is $A^c = A - BKC$. Then the system (6.110) is decomposed into s dynamic subsystems (6.82), and the stabilization proceeds in pretty much the same way as the stability analysis outlined in this section. The only difference is in the freedom provided by the gain matrix K of the linear control law (6.109). The elements of K can be chosen on the subsystem level to produce an optimal aggregate matrix and the largest estimates of the structural system parameters. The application of this approach is presented in the following section.

The above stabilization scheme is basically a trial-and-error procedure, since there is no systematic way of choosing the elements of the gain matrix K. The only advantage that the decomposition-aggregation scheme has over the conventional techniques is that the stabilization is performed "piece by piece" and the structural parameters appear explicitly in the aggregate model. It is possible to retain this advantage and improve the search for appropriate feedback gains a great deal, if one uses the multilevel control scheme based upon the state feedback, as outlined in Section 3.3

Let us consider a linear constant dynamic system $\mathbb{S}$ described by

$$\dot{x} = Ax + Bu, \tag{6.111}$$

where again $x \in \mathcal{R}^n$ is the state of the system, $u \in \mathcal{R}^m$ is the control, and A and B are constant $n \times n$ and $n \times m$ matrices. In order to stabilize the system, it is decomposed into s dynamic subsystems $\mathbb{S}_i$ represented by the equations

$$\dot{x}_i = A_i x_i + \sum_{j=1}^{s} e_{ij} A_{ij} x_j + B_i u_i, \qquad i = 1, 2, \ldots, s, \tag{6.112}$$

where $x_i \in \mathcal{R}^{n_i}$ is the state of $\mathbb{S}_i$, so that

$$\mathcal{R}^n = \mathcal{R}^{n_1} \times \mathcal{R}^{n_2} \times \cdots \times \mathcal{R}^{n_s}; \tag{6.113}$$

$u_i \in \mathcal{R}^{m_i}$ is the decentralized control, so that

$$\mathcal{R}^m = \mathcal{R}^{m_1} \times \mathcal{R}^{m_2} \times \cdots \times \mathcal{R}^{m_s}; \tag{6.114}$$

A_i, A_{ij}, B_i are constant matrices of appropriate dimension; and the pairs (A_i, B_i) are controllable for all $i = 1, 2, \ldots, s$.

The control functions $u_i\colon \mathcal{R}^n \to \mathcal{R}^{m_i}$ are chosen as

$$u_i(x) = u_i^l(x_i) + u_i^g(x), \tag{6.115}$$

where the local control $u_i^l(x_i)$ and the global control $u_i^g(x)$ are linear functions of the states x_i and x,

$$u_i^l(x_i) = -K_i x_i, \qquad u_i^g(x) = -\sum_{j=1}^{s} e_{ij} K_{ij} x_j. \tag{6.116}$$

By substituting (6.116) into (6.112) and using the transformation (6.30), we get the closed-loop system as

$$\dot{\tilde{x}}_i = \Lambda_i^c x_i + \sum_{j=1}^{s} e_{ij} \tilde{A}_{ij}^c \tilde{x}_j, \qquad i = 1, 2, \ldots, s, \tag{6.117}$$

where the quasidiagonal matrix Λ_i^c and the matrix $\tilde{A}_{ij}^c$ are given as

$$\Lambda_i^c = T_i^{-1}(A_i - B_i K_i) T_i, \qquad \tilde{A}_{ij}^c = T_i^{-1}(A_{ij} - B_j K_{ij}) T_j, \qquad i, j = 1, 2, \ldots, s. \tag{6.118}$$

The gain matrices K_i can be chosen to fix the eigenvalues of the Λ_i^c's and get a sufficiently high degree of stability π_i of the subsystems. As suggested in Section 3.3, this can be accomplished by the pole-shifting technique, or by solution of the optimal-linear-regulator problem as in Section 3.5.

Once the local controllers are selected, one is interested in choosing the global controllers so as to maximize the estimates of the interconnection parameters. From (6.107), it is clear that the smaller $\tilde{\xi}_{ij}$ is, the larger e_{ij} can be chosen without violating stability. That is, we are interested in a solution of the following:

Problem 6.6.

$$\textit{Find:} \quad \min_{\tilde{K}_{ij}} \tilde{\xi}_{ij}.$$

A solution to this problem is offered in Sections 3.3 and 3.5, where the gain matrix $\tilde{K}_{ij}$ is chosen as

$$\tilde{K}_{ij}^0 = \tilde{B}_j^{\dagger} \tilde{A}_{ij}, \tag{6.119}$$

in which $\tilde{B}_j^{\dagger}$ is the Moore-Penrose generalized inverse of the matrix $\tilde{B}_j = T_i^{-1} B_j$. This choice of $\tilde{K}_{ij}$ is made to minimize the numbers $\tilde{\xi}_{ij} = \lambda_M^{1/2}[(\tilde{A}_{ij}^c)^T \tilde{A}_{ij}^c]$ and obtain the optimal aggregate matrix $\tilde{W}^0$ as formulated in Problem 6.1. If rank $\tilde{B}_j = m_j$, then $\tilde{B}_j^{\dagger} = (\tilde{B}_j^T \tilde{B}_j)^{-1} \tilde{B}_j$, and $\tilde{K}_{ij}$ is simply calculated as

$$\tilde{K}_{ij}^0 = (\tilde{B}_j^T \tilde{B}_j)^{-1} \tilde{B}_j^T \tilde{A}_{ij}. \tag{6.120}$$

The design process can be entirely and effectively computerized as shown by Šiljak, Sundareshan, and Vukčević (1975).

6.6. STABILIZATION OF THE SKYLAB

To provide an artificial-gravity environment, NASA initiated and conducted a study to determine the feasibility of spinning the Skylab (Seltzer, Justice, Patel, and Schweitzer, 1972; Seltzer, Patel, and Schweitzer, 1973). In spinning the spacecraft, it is necessary to point the solar panels toward the sun, which requires the vehicle to spin about a principal axis of intermediate moment of inertia. Since such spin cannot be achieved without stabilization, it was proposed to establish passive stability by deploying masses either on cables or extendable booms attached to the Skylab as shown on Figure 6.2. Such configuration has the principal axis of maximum moment of inertia pointing (in the same direction as the solar panels) to the sun. In order to inertially fix the axis in presence of disturbance torques, attitude-control torques must be applied to the vehicle, which depend on error signals that are proportional to the angle between the principal 3-axis and the solar vector. Sun sensors and rate gyros on the present Skylab can

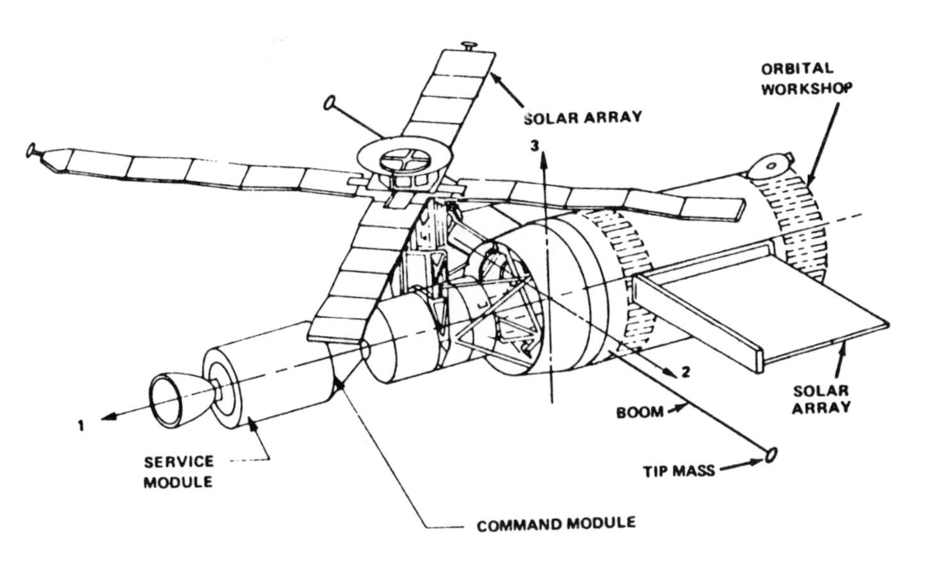

FIGURE 6.2. Skylab.

readily provide the control signals ϕ_1, ϕ_2, w_1, and w_2 shown on the simplified model of the spinning Skylab, which consists of a core mass with two tip masses connected to it by flexible massless beams lying in two different planes as shown in Figure 6.3.

The angular-velocity vector of the vehicle may be written in body-fixed coordinates 1, 2, 3 as $[w_1, w_2, w_3 + \Omega]^T$, where $|w_i| \ll 1$ $(i = 1,2,3)$ represent small perturbations about the steady-state velocity Ω. Small displacements of the two tip masses m from the steady state are denoted by u_i^k $(i = 1,2,3;\ k = \mathrm{I},\mathrm{II})$. The rotational dynamics of the Skylab may be represented by a set of nine differential equations written in terms of w_i and u_i^k. It is possible to reduce the set of nine equations to six by using the substitution $u_i = u_i^{\mathrm{I}} - u_i^{\mathrm{II}}$, where u_i now represents the skew-symmetric mode of the elastic deformation and hence causes angular motion about the vehicle's steady-state attitude. Since the stability of rotational motion will be of interest, only the skew-symmetric mode is considered.

The linearized equations of motion are

wobble motion:

$$
\begin{aligned}
I_1\dot{w}_1 + (I_3 - I_2)\Omega w_2 + m\Gamma_2(\ddot{u}_3 + \Omega^2 u_3) - m\Gamma_3(2\Omega\dot{u}_1 + \ddot{u}_2 - \Omega^2 u_2) &= T_1,\\
(I_1 - I_3)\Omega w_1 + I_2\dot{w}_2 + m\Gamma_3(\ddot{u}_1 - \Omega^2 u_1 - 2\Omega\dot{u}_2) &= T_2,\\
2m\Gamma_2(\dot{w}_1 + \Omega w_2) + m\ddot{u}_3 + d_3\dot{u}_3 + (k_3 + m\Omega^2)u_3 &= 0;
\end{aligned}
\tag{6.121}
$$

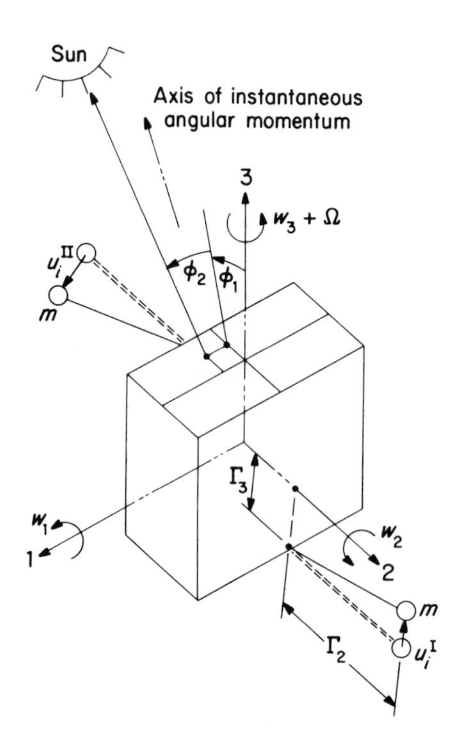

FIGURE 6.3. Simplified model.

spin motion:

$$I_3 \dot{w}_3 - m\Gamma_2(\ddot{u}_1 - 2\Omega\dot{u}_2) = T_3,$$
$$2m\Gamma_3(\Omega w_1 + \dot{w}_2) - 2m\Gamma_2 \dot{w}_3 + m\ddot{u}_1 + d_1 \dot{u}_1 + k_1 u_1 - 2m\Omega\dot{u}_2 = 0,$$
$$2m\Gamma_3(-\dot{w}_1 + \Omega w_2) - 4m\Gamma_2 \Omega w_3 + 2m\Omega\dot{u}_1 + m\ddot{u}_2 + d_2 \dot{u}_2 + (k_2 - m\Omega^2)u_2 = 0. \quad (6.122)$$

The spin velocity and its perturbation w_3 are controlled separately and are not considered here. Consequently, we assume that $w_3 \equiv \dot{w}_3 \equiv 0$ (Seltzer, Patel, Schweitzer, 1973).

The linear control is postulated (Seltzer, Justice, Patel, Schweitzer, 1972) as

$$T = \alpha\phi + \beta\omega, \quad (6.123)$$

where $T = (T_1, T_2, T_3)^T$ is the vector of control torques; $\phi = (\phi_1, \phi_2, \phi_3)^T$ is the vector of angular rotations; $\omega = (w_1, w_2, w_3 + \Omega)^T$ is the vector of angular velocities; α, β are 3×3 matrices

$$\alpha = \begin{bmatrix} \alpha_{11} & \alpha_{12} & 0 \\ \alpha_{21} & \alpha_{22} & 0 \\ 0 & 0 & 0 \end{bmatrix}, \qquad \beta = \begin{bmatrix} \beta_{11} & \beta_{12} & 0 \\ \beta_{21} & \beta_{22} & 0 \\ 0 & 0 & \beta_{33} \end{bmatrix}; \tag{6.124}$$

and the kinematic relations are

$$\omega = \begin{bmatrix} -1 & 0 & 0 \\ 0 & -1 & 0 \\ 0 & 0 & -1 \end{bmatrix} \dot{\phi} + \begin{bmatrix} 0 & \Omega & 0 \\ -\Omega & 0 & 0 \\ 0 & 0 & 0 \end{bmatrix} \phi. \tag{6.125}$$

The control law in this study is chosen as

$$\begin{aligned} \alpha_{12} &= I_1 \Omega^2; & \text{all other } \alpha_{ij} &= 0; \\ \beta_{11} &= I_1 \Omega \delta, \quad \beta_{33} = -I_1 \Omega \rho; & \text{all other } \beta_{ij} &= 0; \end{aligned} \tag{6.126}$$

so that the normalized control torques $v = [v_1, v_2, v_3]^T = [T_1/I_1\Omega^2, T_2/I_1\Omega^2, T_3/I_1\Omega^2]^T$ are

$$v_1 = (\varepsilon + \delta)\phi_2 - \delta\phi_1', \qquad v_2 = 0, \qquad v_3 = \rho\phi_3'. \tag{6.127}$$

Referring to Equations (6.121) and (6.122), the control torque T_1 is used to stabilize the wobble motion, and the torque T_3 is used to stabilize the spin motion. In (6.127), ε, δ, ρ are control parameters to be selected in the stabilization process.

Upon introducing these transformations in addition to the dimensionless variables and constants as defined in Table 6.1, the linearized equations of motion become

wobble motion:

$$\begin{aligned} \phi_1'' - (1 + K_1)\phi_2' - K_1\phi_1 - \gamma(\mu_3'' + \mu_3) + \xi\gamma(2\mu_1' + \mu_2'' - \mu_2) + (\varepsilon + \delta)\phi_2 & \\ - \delta\phi_1' &= 0, \\ (1 + K_1)\phi_1' + K_2\alpha\phi_2 + \alpha\phi_2'' - \xi\gamma(\mu_1'' - \mu_1 - 2\mu_2') &= 0, \\ -\phi_1'' - \phi_1 + \mu_3'' + \Delta_3\mu_3' + (\sigma_3^2 + 1)\mu_3 &= 0; \end{aligned} \tag{6.128}$$

spin motion:

$$\begin{aligned} \beta\phi_3'' + \gamma(\mu_1'' - 2\mu_2') + \rho\phi_3' &= 0, \\ -2\xi\phi_1' - \xi\phi_2'' + \phi_3'' + \mu_1'' + \Delta_1\mu_1' + \sigma_1^2\mu_1 - 2\mu_2' + \xi\phi_2 &= 0, \\ \xi\phi_1'' - 2\xi\phi_2' - \xi\phi_1 + 2\phi_3' + 2\mu_1' + \mu_2'' + \Delta_2\mu_2' + (\sigma_2^2 - 1)\mu_2 &= 0. \end{aligned} \tag{6.129}$$

An important feature of these equations is that when $\xi = 0$ (that is, $\Gamma_3 = 0$), they become uncoupled into two sets of equations: the wobble motion described by (6.128) and the spin motion described by (6.129). This suggests that the influence of the asymmetry in the arrangements of the booms ($\Gamma_3 \neq 0$ or $\xi \neq 0$) can be treated as the structural parameter between the two motions, so that $|\xi| = e_{12} = e_{21}$ and $\xi^2 = e_{11} = e_{22}$. In the decomposition-aggregation analysis each motion represents a subsystem.

The state-space representation of the overall system (6.128)–(6.129) is obtained as

$$x'(\tau) = Ax(\tau), \tag{6.130}$$

where the state 11-vector $x(\tau)$ is chosen as

$$x(\tau) = (\phi_1, \phi_2, \mu_3, \phi_1', \phi_2', \mu_3', \phi_3', \mu_1, \mu_1', \mu_2, \mu_2')^T, \tag{6.131}$$

and the 11×11 matrix A^c is given in (6.132)—see pages 358 and 359.

The system of equations (6.130) can be decomposed into two interconnected subsystems described by
wobble motion:

$$x_1'(\tau) = A_1 x_1(\tau) + \xi^2 A_{11}(\xi) x_1(\tau) + \xi A_{12}(\xi) x_2(\tau), \tag{6.133}$$

spin motion:

$$x_2'(\tau) = A_2 x_2(\tau) + \xi A_{21}(\xi) x_1(\tau) + \xi^2 A_{22}(\xi) x_2(\tau), \tag{6.134}$$

where the state vectors $x(\tau)$, $x_1(\tau)$, $x_2(\tau)$ of the system (6.130) and the two subsystems are

$$x(\tau) = \begin{bmatrix} x_1(\tau) \\ x_2(\tau) \end{bmatrix}; \quad \begin{cases} x_1(\tau) = (\phi_1, \phi_2, \mu_3, \phi_1', \phi_2', \mu_3')^T, \\ x_2(\tau) = (\phi_3', \mu_1, \mu_1', \mu_2, \mu_2')^T. \end{cases} \tag{6.135}$$

The following identity relationships were used in order to get the subsystem matrices A_1 and A_2 independent of the coupling parameter ξ:

$$\begin{aligned} \frac{1}{\alpha(1-\gamma_3) - \xi^2\gamma} &\equiv \frac{1}{\alpha(1-\gamma_3)} + \frac{\xi^2\gamma}{\alpha(1-\gamma_3)[\alpha(1-\gamma_3) - \xi^2\gamma]}, \\ \frac{1}{1-\gamma-\xi^2} &\equiv \frac{1}{1-\gamma} + \frac{\xi^2\gamma}{(1-\gamma)(1-\gamma-\xi^2\gamma)}. \end{aligned} \tag{6.136}$$

The structural configuration of the system (6.130), as composed of the two subsystems (6.133) and (6.134) and the interconnections between them through the coupling parameter ξ, is represented by the direct graph in Figure 6.4(a). It is obvious that the digraph of Figure 6.4(a) becomes that

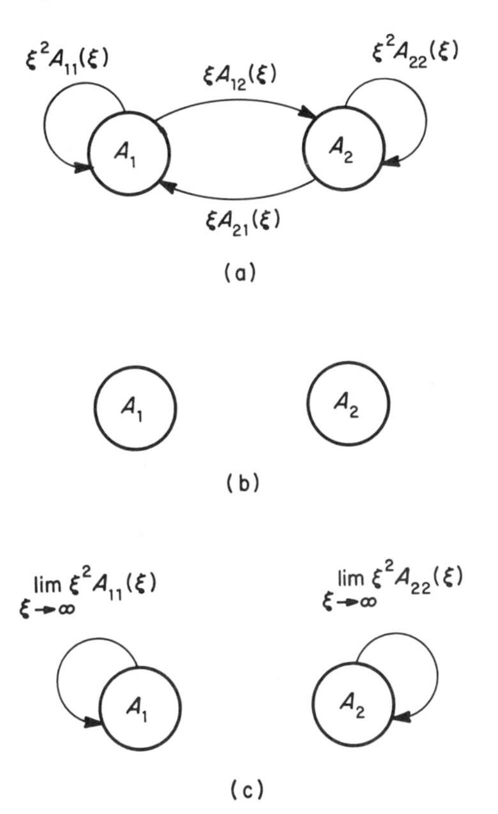

FIGURE 6.4. Structural decomposition.

of Figure 6.4(b) when $\xi = 0$. When $\xi \to \infty$, the digraph of Figure 6.4(a) is again decomposed into the two separate digraphs shown in Fig. 6.4(c), because the interconnection matrices $\xi A_{12}(\xi)$, $\xi A_{21}(\xi)$ and $\xi^2 A_{11}(\xi)$, $\xi^2 A_{22}(\xi)$ become zero and constant matrices, respectively, as seen in (6.132) or (6.133)–(6.134).

The subsequent stability analysis shows that the free subsystems (6.133) and (6.134) ($\xi = 0$) are stable. It is easy to check, however, that the decoupled subsystems in Figure 6.4(c) are unstable. Therefore, our main objective is to determine the best estimate of the maximum allowable value of ξ which lies between the two extremes $\xi = 0$ and $\xi = \infty$, and for which the overall system of Figure 6.4(a) is stable.

On the basis of the Skylab physical characteristics given in Table 6.2, $A_{ij}(\xi)$ ($i = 1, 2$) of (6.132) can be made independent of ξ and denoted by A_{ij}. This is achieved by neglecting the term $\xi^2 = 8.5 \times 10^{-4}$ in comparison with the terms $1 - \gamma = 0.803$ and $\alpha(1 - \gamma_3) = 5.33$.

After this simplification, the numbers ξ_{12} and ξ_{21} are the norms of the coupling matrices A_{12} and A_{21}, which are computed in (6.89). The subsystem

$$
P^c = \left[\begin{array}{cccccc}
0 & 0 & 0 & 1 & 0 & 0 \\
0 & 0 & 0 & 0 & 1 & 0 \\
0 & 0 & 0 & 0 & 0 & 1 \\
\dfrac{K_1 + \gamma - \xi^2\gamma}{1 - \gamma - \xi^2\gamma} & -\dfrac{\epsilon + \delta}{1 - \gamma - \xi^2\gamma} & -\dfrac{\gamma\sigma_3^2}{1 - \gamma - \xi^2\gamma} & \dfrac{\delta}{1 - \gamma - \xi^2\gamma} & \dfrac{1 + K_1 - 2\xi^2\gamma}{1 - \gamma - \xi^2\gamma} & -\dfrac{\gamma\Delta_3}{1 - \gamma - \xi^2\gamma} \\
0 & -\dfrac{(1 - \gamma_3)K_2\alpha + \xi^2\gamma}{\alpha(1 - \gamma_3) - \xi^2\gamma} & 0 & -\dfrac{(1 - \gamma_3)(1 + K_1) + 2\xi^2\gamma}{\alpha(1 - \gamma_3) - \xi^2\gamma} & 0 & 0 \\
1 + \dfrac{K_1 + \gamma - \xi^2\gamma}{1 - \gamma - \xi^2\gamma} & -\dfrac{\epsilon + \delta}{1 - \gamma - \xi^2\gamma} & -(\sigma_3^2 + 1) - \dfrac{\gamma\sigma_3^2}{1 - \gamma - \xi^2\gamma} & \dfrac{\delta}{1 - \gamma - \xi^2\gamma} & \dfrac{1 + K_1 - 2\xi^2\gamma}{1 - \gamma - \xi^2\gamma} & -\dfrac{\gamma\Delta_3}{1 - \gamma - \xi^2\gamma} - \Delta_3 \\
\hline
0 & \dfrac{\xi\gamma_3\alpha(K_2 + 1)}{\alpha(1 - \gamma_3) - \xi^2\gamma} & 0 & -\dfrac{\xi\gamma_3(2\alpha - 1 - K_1)}{\alpha(1 - \gamma_3) - \xi^2\gamma} & 0 & 0 \\
0 & 0 & 0 & 0 & 0 & 0 \\
0 & -\dfrac{\xi\alpha(K_2 + 1)}{\alpha(1 - \gamma_3) - \xi^2\gamma} & 0 & \dfrac{\xi(2\alpha - 1 - K_1)}{\alpha(1 - \gamma_3) - \xi^2\gamma} & 0 & 0 \\
0 & 0 & 0 & 0 & 0 & 0 \\
\dfrac{\xi(1 - K_1 - 2\gamma)}{1 - \gamma - \xi^2\gamma} & \dfrac{\xi(\epsilon + \delta)}{1 - \gamma - \xi^2\gamma} & \dfrac{\xi\gamma\sigma_3^2}{1 - \gamma - \xi^2\gamma} & -\dfrac{\xi\delta}{1 - \gamma - \xi^2\gamma} & \dfrac{\xi(1 - K_1 - 2\gamma)}{1 - \gamma - \xi^2\gamma} & \dfrac{\xi\gamma\Delta_3}{1 - \gamma - \xi^2\gamma}
\end{array}\right.
$$

$$
\left.\begin{array}{ccccc}
0 & 0 & 0 & 0 & 0 \\
0 & 0 & 0 & 0 & 0 \\
0 & 0 & 0 & 0 & 0 \\
\dfrac{2\xi\gamma}{1-\gamma-\xi^2\gamma} & 0 & 0 & \dfrac{\xi\gamma\sigma_2^2}{1-\gamma-\xi^2\gamma} & \dfrac{\xi\gamma\Delta_2}{1-\gamma-\xi^2\gamma} \\
\dfrac{\xi\gamma\rho}{\beta[\alpha(1-\gamma_3)-\xi^2\gamma]} & -\dfrac{\xi\gamma(1-\gamma_3+\sigma_1^2)}{\alpha(1-\gamma_3)-\xi^2\gamma} & -\dfrac{\xi\gamma\Delta_1}{\alpha(1-\gamma_3)-\xi^2\gamma} & 0 & 0 \\
\dfrac{2\xi\gamma}{1-\gamma-\xi^2\gamma} & 0 & 0 & \dfrac{\xi\gamma\sigma_2^2}{1-\gamma-\xi^2\gamma} & \dfrac{\xi\gamma\Delta_2}{1-\gamma-\xi^2\gamma} \\
\hline
-\dfrac{\rho}{\beta}-\dfrac{\alpha\rho\gamma_3}{\beta[\alpha(1-\gamma_3)-\xi^2\gamma]} & \dfrac{\gamma_3(\sigma_1^2\alpha+\xi^2\gamma)}{\alpha(1-\gamma_3)-\xi^2\gamma} & \dfrac{\alpha\gamma_3\Delta_1}{\alpha(1-\gamma_3)-\xi^2\gamma} & 0 & 0 \\
0 & 0 & 1 & 0 & 0 \\
\dfrac{\alpha\rho}{\beta[\alpha(1-\gamma_3)-\xi^2\gamma]} & -\dfrac{\sigma_1^2\alpha+\xi^2\gamma}{\alpha(1-\gamma_3)-\xi^2\gamma} & -\dfrac{\alpha\Delta_1}{\alpha(1-\gamma_3)-\xi^2\gamma} & 0 & 2 \\
0 & 0 & 0 & 0 & 1 \\
-\dfrac{2(1-\gamma)}{1-\gamma-\xi^2\gamma} & 0 & -2 & -\dfrac{(1-\gamma)(\sigma_2^2-1)+\xi^2\gamma}{1-\gamma-\xi^2\gamma} & -\dfrac{(1-\gamma)\Delta_2}{1-\gamma-\xi^2\gamma}
\end{array}\right]
\tag{6.132}
$$

Table 6.1. Nomenclature for the Skylab

I_1, I_2, I_3	Principal moment of inertia of body, ith coordinate: $I_1^* + 2m\Gamma_2^2$, I_2^*, $I_3^* + 2m\Gamma_2^2$, respectively
I_i^*	Principal moment of inertia of rigid core body about ith body-fixed coordinate
k_i	Stiffness coefficient characterizing nonrotating-boom stiffness
m	Tip mass
T_i	Applied torque about ith coordinate
t	Time
$(\cdot) = d/dt$	Differentiation with respect to real time t
$u_i = u_i^{\mathrm{I}} - u_i^{\mathrm{II}}$	Skew-symmetric mode of elastic deformations
u_i^k	Displacement of kth tip mass from spinning steady state in ith direction (k = I, II)
w_i	Perturbation (about spinning steady state) of angular velocity about ith coordinate
ϕ_i	Angular rotation about ith coordinate
Γ_2	Steady-state boom dimension in 2-axis direction from center of mass to tip mass
Γ_3	Asymmetry in setting of booms
K_1, K_2	Ratios of inertia $(I_2 - I_3)/I_1$ and $(I_3 - I_1)/I_2$, respectively
$\alpha = \dfrac{1 + K_1}{1 - K_2} = \dfrac{I_2}{I_1}$	Ratio of inertia $\dfrac{I_2}{I_1}$
$\gamma = 2\Gamma m_2^2/I_1$	Dimensionless inertia ratio
$\Delta_i = d_i/m\Omega$	Dimensionless damping ratio
$\mu_i = u_i/2\Gamma_2$	General skew-symmetric coordinate
$\sigma_i^2 = k_i/m\Omega^2$	Dimensionless natural frequency coefficient of boom
$\xi = \Gamma_3/\Gamma_2$	Dimensionless length ratio
$\tau = \Omega t$	Dimensionless time
Ω	Steady-state spin rate about 3-axis
$\nu_i = w_i/\Omega$	Dimensionless wobble ratio ($i = 1, 2, 3$)
$(') = d/d\tau$	Differentiation with respect to τ
$\gamma_3 = 2m\Gamma_2^2/I_3$	Dimensionless inertia ratio
$\beta = I_3/I_1$	Dimensionless inertia ratio

Liapunov functions v_1 and v_2 are chosen as in (6.106), and the aggregate comparison system is constructed, where the aggregate matrix $\tilde{W}$ is given by

$$\tilde{W} = \begin{bmatrix} -\dfrac{1}{2}\dfrac{\lambda_m(\tilde{G}_1)}{\lambda_M(\tilde{H}_1)} + \xi^2\dfrac{\lambda_M(\tilde{H}_1)}{\lambda_m(\tilde{H}_1)} & \dfrac{\tilde{\xi}_{12}|\xi|\lambda_M(\tilde{H}_1)}{\lambda_m^{1/2}(\tilde{H}_1)\lambda_m^{1/2}(\tilde{H}_2)} \\ \dfrac{\tilde{\xi}_{21}|\xi|\lambda_M(\tilde{H}_2)}{\lambda_m^{1/2}(\tilde{H}_1)\lambda_m^{1/2}(\tilde{H}_2)} & -\dfrac{1}{2}\dfrac{\lambda_m(\tilde{G}_2)}{\lambda_M(\tilde{H}_2)} + \xi^2\dfrac{\lambda_M(\tilde{H}_2)}{\lambda_m(\tilde{H}_2)} \end{bmatrix}. \tag{6.137}$$

In (6.137), λ_m and λ_M denote the minimum and maximum eigenvalues of the indicated matrices, respectively.

From (6.95), we have that

$$\tilde{w}_{11} < 0, \qquad \det \tilde{W} > 0 \tag{6.138}$$

Table 6.2. Physical Characteristics of the Skylab

$I_1 = 1.25 \times 10^6\,\text{kg m}^2$	$m = 227\,\text{kg}$
$I_2 = 6.90 \times 10^6\,\text{kg m}^2$	$k_1 = k_3 = 146\,\text{N/m}$
$I_3 = 7.10 \times 10^6\,\text{kg m}^2$	$k_2 = 7.4 \times 10^4\,\text{N/m}$
$\Gamma_1 = 0$	$d_1 = d_3 = 0.04(k_3 m)^{1/2}$
$\Gamma_2 = 23.3\,\text{m}$	$d_2 = 0.04(k_2 m)^{1/2}$
	$\Omega = 0.06\,\text{s}^{-1}$

are necessary and sufficient for stability of $\tilde{W}$ in (6.137), and sufficient for stability of the overall system.

After several trial-and-error steps, the control parameters are chosen as

$$\varepsilon = 25, \qquad \delta = -14, \qquad \rho = 0.5, \tag{6.139}$$

and the aggregate matrix is obtained as

$$\tilde{W} = \begin{bmatrix} -7.279 \times 10^{-4} + 42.3371\xi^2 & 2.2625|\xi| \\ 1.3944|\xi| & -19.321 \times 10^{-4} + 11.5150\xi^2 \end{bmatrix}. \tag{6.140}$$

The conditions (6.138) yield the best estimate of the parameter ξ as

$$|\xi| \leqslant 6.584 \times 10^{-4}. \tag{6.141}$$

This concludes the design of the Skylab control system.

REFERENCES

DaCunha, N. O., and Polak, E. (1967), "Constrained Minimization Under Vector-Valued Criteria in Finite-Dimensional Spaces", *Journal of Mathematical Analysis and Applications*, 19, 103–124.

Dantzig, G. B. (1963), "*Linear Programming and Extensions*", Princeton University Press, Princeton, New Jersey.

Kuhn, H. W., and Tucker, A. W. (1951), "Nonlinear Programming", *Proceedings of the Second Berkeley Symposium on Mathematical Statistics and Probability*, J. Neyman (ed.), University of California Press, Berkeley, California, 481–492.

O'Dell, C. R. (1973), "Optical Space Astronomy and Goals of the Large Scale Telescope", *Astronautics and Aeronautics*, 11, 22–27.

Schiehlen, W. O. (1973), "A Fine Pointing System for the Large Space Telescope", *NASA Report, No. NASA TN D-7500,* National Aeronautics and Space Administration, Washington, D.C.

Seltzer, S. M., Justice, D. W., Patel, J. S., and Schweitzer, G. (1972), "Stabilizing a Spinning Skylab", *Proceedings of the Fifth IFAC World Congress*, Paris, 17.2:1–7.

Seltzer, S. M., Patel, J. S., and Schweitzer, G. (1973), "Attitude Control of a Spinning Flexible Spacecraft", *Computers and Electrical Engineering*, 1, 323–339.

Seltzer, S. M., and Šiljak, D. D. (1972), "Absolute Stability Analysis of Attitude Control Systems for Large Boosters", *Journal of Spacecraft and Rockets*, 9, 506–510.

Šiljak, D. D. (1969), "*Nonlinear Systems*", Wiley, New York.

Šiljak, D. D. (1975), "Stabilization of Large-Scale Systems: A Spinning Flexible Spacecraft", *Proceedings of the Sixth IFAC World Congress*, Boston, Massachusetts, 35.1:1-10. (See also: *Automatica*, 12, 1976, 309–320).

Šiljak, D. D., Sundareshan, M. K., and Vukčević, M., B. (1975), "A Multilevel Control System for the Large Space Telescope", *NASA Contract Report*, No. NAS 8-27799, University of Santa Clara, Santa Clara, California.

Šiljak, D. D., and Vukčević, M. B. (1976), "Multilevel Control of Large-Scale Systems: Decentralization, Stabilization, Estimation, and Reliability", *Large-Scale Dynamical Systems*, R. Saeks (ed.), Point Lobos Press, Los Angeles, California, 34–57.

Šiljak, D. D., and Vukčević, M. B. (1977), "Decentrally Stabilizable Linear and Bilinear Large-Scale Systems", *International Journal of Control*, 26, 289–305.

Steward, D. V. (1965), "Partitioning and Tearing Systems of Equations", *SIAM Journal of Numerical Analysis*, 2, 345–365.

Weissenberger, S. (1973), "Stability Regions of Large-Scale Systems", *Automatica*, 9, 653–663.

Zukhovitskii, S. I., and Avdeyeva, L. I. (1966), "*Linear and Convex Programming*", Saunders, Philadelphia, Pennsylvania.

7

ENGINEERING

Power Systems

In this chapter, we consider two distinct problems in power systems: transient stability and automatic generation control. Both problems are formulated, analyzed, and resolved by partitioning appropriate power-system models into interconnected subsystems. In the case of transient stability, a multimachine system is decomposed into two-machine subsystems, whereas an automatic generation control system is of decentralized type, with each subsystem representing an area and all its tie lines originating from that area. In both situations, such decomposition is only possible if overlapping is allowed among the subsystems. In the transient-stability partition, all subsystems overlap the reference machine. In automatic generation control system, tie lines are common parts of the subsystems. Overlapping of subsystems in both cases is a necessity dictated by physical characteristics of the power-system models. It should be pointed out, however, that the fact that overlapping is permitted in our methods of analysis shows a considerable flexibility of our decomposition and decentralization approach.

Although we plan to describe both problems and the corresponding power system models in some detail, it may not be enough for a reader who is not familiar with the subject. For further reading—advanced as well as tutorial—the following references are recommended. For transient-stability analysis, the book by Anderson and Fouad (1977), as well as the papers by Willems (1971) and Fouad (1975) are a good place to start. For automatic generation control, the books by Kirchmayer (1959) and Cohn (1966), the

ERDA conference proceedings edited by Fink and Carlsen (1975), as well as the papers by Elgerd and Fosha (1970), Ćalović (1972), Ewart (1975), and Reddoch (1975) are recommended.

7.1. TRANSIENT STABILITY

Before we engage in the problem of model specifications for multimachine power systems, let us briefly review the physical problem of transient stability. Assume that a power system is in its steady-state operation and the mechanical input power to the generators is equal to the electrical power delivered to the network. When a large disturbance occurs (such as a short circuit, sudden loss of a large load, etc.), the steady-state operation is perturbed. Unless the fault is cleared before a certain maximum time (the critical clearing time), a loss of synchronism may take place, so that the system is not capable of recovering steady-state operation. The transient-stability problem consists in finding out whether or not the system motion converges to a steady state after clearing the fault. If at the time the fault is cleared the state of the system is within the stability region of the post-fault equilibrium (the post-fault steady state), convergence to steady-state operation takes place. Therefore, solutions of the transient-stability problem have two distinct phases. The first phase consists in keeping track of the system motion during the disturbance, whereas in the second phase one determines the stability of the post-fault equilibrium state.

Both solution phases can be carried out by standard numerical techniques for integration of the system equations of motion. These techniques become increasingly unattractive as the size of multimachine power systems becomes large, because of the excessive computer time and memory required. It is for this reason that in the second phase of the solution, numerical techniques can be advantageously replaced by the direct method of Liapunov.

Initial applications of Liapunov's method to transient-stability analysis were made in the 1960s by Gless (1966), and by El-Abiad and Nagapan (1966), and later developed by a large number of authors, as surveyed by Willems (1971), Ribbens-Pavella (1971), and Fouad (1975). Since power systems are never asymptotically stable in the large, Liapunov's method is used to estimate the region of attraction of the post-fault equilibrium state. Once a multimachine system was recognized as a multinonlinear Lur'e-Postnikov system by Willems (1971), the method of Popov (1973) and the Kalman-Yakubovich lemma in matrix form (Kalman, 1963; Popov, 1973) were available for systematic construction of Liapunov functions (Willems,

1970, 1971; Henner, 1974). These functions were used to estimate stability regions by extending the procedures of Walker and McClamroch (1967) and Weissenberger (1968), which were originally applied to Lur'e-Postnikov systems with a single nonlinearity. As the size of the system increases, these direct methods lose much of their appeal, mainly because of the conceptual and numerical difficulties involved in considering a large number of unstable equilibrium states of the system.

In 1975, Pai and Narayana (1975) made an attempt to apply the Bellman-Matrosov concept of the vector Liapunov function, as proposed in the decomposition-aggregation method developed by Šiljak (1972) and Grujić and Šiljak (1973), to the transient-stability analysis of multimachine systems. After the method was applied to construct an aggregate model involving the vector Liapunov function, a single Liapunov function was used to estimate the stability region by the procedure proposed by Weissenberger (1973). The main drawback of the Pai-Narayana approach is in that an n-machine system with uniform damping is decomposed into $n(n-1)/2$ second-order subsystems, so that the most significant characteristic of the decomposition-aggregation method, that of reducing the dimensionality of stability problems, is lost. In addition to this main drawback, there were some unresolved details of varied significance that prevented computation of valid estimates of the stability regions.

Nevertheless, the new approach of Pai and Narayana is quite promising, and in the first part of this chapter we will present the works of Jocić, Ribbens-Pavella, and Šiljak (1977), and Jocić and Šiljak (1977), which is in the same direction but eliminates all major difficulties encountered in the Pai-Narayana approach. In particular, it reduces the order of the aggregate model of an n-machine system to $n - 1$.

As compared to the existing direct methods (Willems, 1971; Fouad, 1975) for estimating stability regions of multimachine power systems, the following are the advantages of the method proposed by Jocić, Ribbens-Pavella, and Šiljak (1977), and Jocić and Šiljak (1977):

(1) The method is rigorous and can be carried out systematically with parametrically simple and explicit estimates of the stability region obtained as the end result.

(2) In the course of transient-stability analysis, decomposition can be used to take advantage of (or obtain information about) the special structural features of the power system.

(3) The method opens up a real possibility for more refined models of the subsystems to be included in the analysis of large power systems.

(4) In particular, the transfer conductances can readily be incorporated in the analysis, a feature which is missing in almost all previous considerations.

The plan of our exposition is as follows: In the next section, we will formulate a model of an n-machine power system, and then perform a pairwise decomposition of the model into $n - 1$ interconnected second-order subsystems. Since each subsystem is of the Lur'e-Postnikov type with one nonlinearity, a simple analysis in Section 7.3 produces estimates of the stability regions for each free (disconnected) subsystem. Finally, in Section 7.4, we construct an aggregate model involving a vector Liapunov function, and use the subsystem-region estimates to determine an estimate for the stability region of the overall system.

7.2. A MODEL AND DECOMPOSITION

We consider an n-machine power system in which the absolute motion of the ith machine is described by the equation

$$M_i \ddot{\delta}_i + D_i \dot{\delta}_i = P_{mi} - P_{ei}, \qquad i = 1, 2, \ldots, n, \tag{7.1}$$

where $P_{ei} = \sum_{j=1}^{n} E_i E_j Y_{ij} \cos(\delta_i - \delta_j - \theta_{ij})$ and

δ_i is the absolute rotor angle,
M_i is the inertia coefficient,
D_i is the damping coefficient,
P_{mi} is the mechanical power delivered to the ith machine,
P_{ei} is the electrical power delivered by the ith machine,
E_i is the internal voltage,
Y_{ij} is the modulus of the transfer admittance between the ith and jth machines,
θ_{ij} is the phase angle of transfer admittance between the ith and the jth machines.

In Equation (7.1), it is assumed that M_i, P_{mi}, and E_i are constant for all machines. In addition, we assume uniform damping, that is,

$$\frac{D_i}{M_i} = \lambda, \qquad i = 1, 2, \ldots, n. \tag{7.2}$$

With the above assumptions, we must refer the system motion to the motion of one of the machines (say the nth) taken arbitrarily as the comparison machine and choose the state vector $x \in \mathcal{R}^{2(n-1)}$ as

$$x = (\omega_{1n}, \omega_{2n}, \ldots, \omega_{n-1,n}, \delta_{1n} - \delta_{1n}^0, \ldots, \delta_{n-1,n} - \delta_{n-1,n}^0)^T, \tag{7.3}$$

where $\delta_{in} = \delta_{in} - \delta_n$, $\omega_{in} = \omega_i - \omega_n$, $\omega_i = \dot{\delta}_i$, and the δ_{ij}^0's are obtained by solving the equilibrium equations

$$P_{ei}(\delta_{ij}^0) = P_{mi}, \qquad i = 1, 2, \ldots, n-1. \tag{7.4}$$

With the choice of the state vector x in (7.3), the state model is derived from the Equations (7.1) as a multinonlinear Lur'e-Postnikov system

$$\begin{aligned} \dot{x} &= Ax + Bf(y), \\ y &= C^T x, \end{aligned} \tag{7.5}$$

where $x \in \mathcal{R}^{2(n-1)}$, $y \in \mathcal{R}^{n-1}$, and the function f: $\mathcal{R}^{n-1} \to \mathcal{R}^{2(n-1)}$ is defined by

$$f = \left[\sum_{\substack{j=1\\ j\neq 1}}^{n} f_{1j}, \sum_{\substack{j=1\\ j\neq 2}}^{n} f_{2j}, \ldots, \sum_{\substack{j=1\\ j\neq n}}^{n} f_{nj} \right]^T \tag{7.6}$$

and

$$f_{ij} = A_{ij}[\cos(\delta_{ij} - \theta_{ij}) - \cos(\delta_{ij}^0 - \theta_{ij})]. \tag{7.7}$$

In (7.5), the matrices A, B, C are defined as

$$A = \left[\begin{array}{c|c} -\lambda I_{n-1} & 0_{n-1} \\ \hline I_{n-1} & 0_{n-1} \end{array}\right], \qquad B = \left[\begin{array}{cccc|c} -M_1^{-1} & & & 0 & M_n^{-1} \\ & -M_2^{-1} & & & M_n^{-1} \\ 0 & & \ddots & & \vdots \\ & & & -M_{n-1}^{-1} & M_n^{-1} \\ \hline & & 0_{n-1} & & 0_{n-1,1} \end{array}\right],$$

$$C^T = \left[0_{n-1} \mid I_{n-1} \right]. \tag{7.8}$$

In order to decompose the system (7.5) into $n-1$ interconnected Lur'e-Postnikov-type subsystems with a single nonlinearity, we choose the vector

$$x_i = (\omega_{in}, \delta_{in} - \delta_{in}^0)^T \tag{7.9}$$

as a state vector for the ith subsystem. Then, by using the simple trigonometric relation $\cos(\delta_{ij} - \theta_{ij}) = \cos\delta_{ij}\cos\theta_{ij} + \sin\delta_{ij}\sin\theta_{ij}$, Equation (7.5) can be rewritten as

$$\begin{aligned} \dot{x}_i &= \begin{bmatrix} -\lambda & 0 \\ 1 & 0 \end{bmatrix} x_i + \begin{bmatrix} -1 \\ 0 \end{bmatrix} \varphi_i(y_i) + h_i(x), \qquad i = 1, 2, \ldots, n-1, \\ y_i &= [0 \quad 1] x_i, \end{aligned} \tag{7.10}$$

where the nonlinear function $\varphi_i(y_i)$ is

$$\varphi_i(y_i) = (M_i^{-1} + M_n^{-1})A_{in}\sin\theta_{in}[\sin(y_i + \delta_{in}^0) - \sin\delta_{in}^0] \qquad (7.11)$$

and the interconnection function $h_i(x)$ is defined as

$$h(x) = \begin{bmatrix} (M_n^{-1} - M_i^{-1})A_{in}\cos\theta_{in}(\cos\delta_{in} - \cos\delta_{in}^0) + \sum_{\substack{j=1 \\ j\neq i}}^{n-1}(M_n^{-1}f_{nj} - M_i^{-1}f_{ij}) \\ 0 \end{bmatrix}. \qquad (7.12)$$

The system (7.10) has the form of interconnected Lur'e-Postnikov subsystems, similar to that of (2.186) studied in Section 2.5. The material of Section 2.5 related to the system (2.186) will be used in the next section to analyze the system (7.10).

A graphical representation of the physical pairwise decomposition of a three-machine system is shown in Figure 7.1, where the overlapping of the two subsystems $\mathbb{S}_1$ and $\mathbb{S}_2$ is displayed. After the states of the subsystems are selected as in (7.9), a digraph describing the structure of the system has the form shown on Figure 7.2(a). The corresponding condensation digraph, with x_1 and x_2 as state vectors of the subsystems, is given in Figure 7.2(b). Due to the choice (7.9) of subsystem states, the "physical" overlapping of Figure 7.1 does not take place in the system digraph of Figure 7.2(a). If we allow for nonuniform damping, then as shown by Jocić and Šiljak (1977), overlapping appears in the system digraph. This fact, however, necessitates only a reformulation of the system condensation digraph, but leaves virtually unaffected the principal features of the stability-region computations for the uniform-damping case which is presented here.

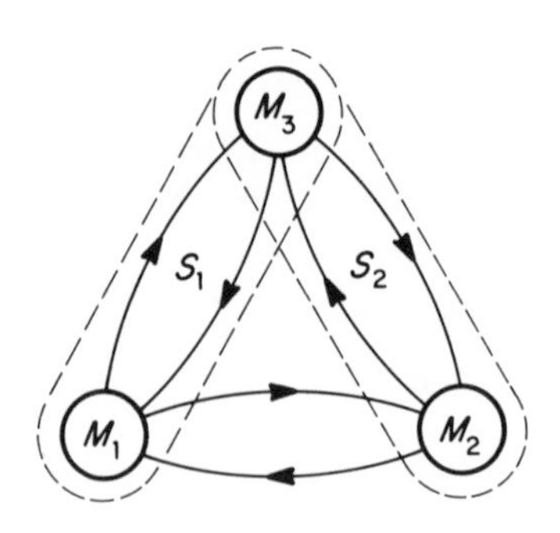

FIGURE 7.1. Pairwise decomposition.

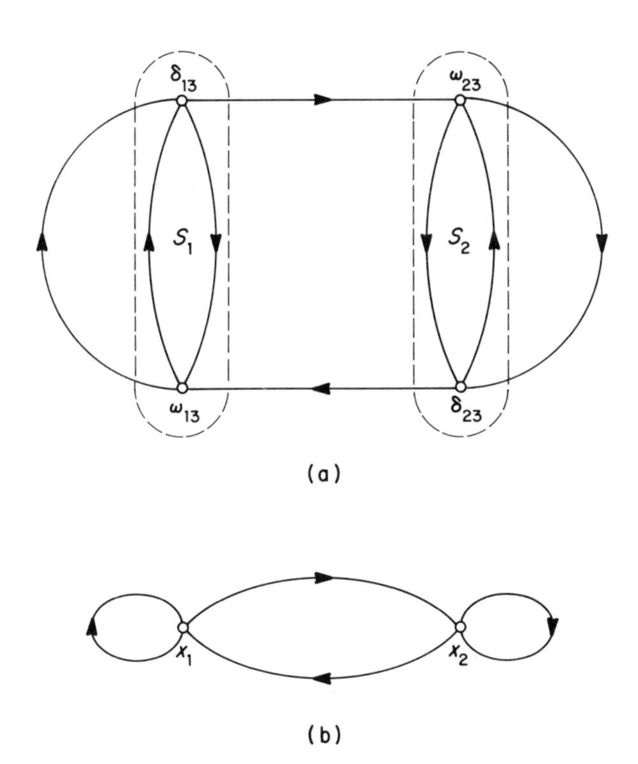

FIGURE 7.2. Three-machine system: (a) Decomposition. (b) Condensation.

7.3. SUBSYSTEM ANALYSIS

Now, we focus our attention on each free subsystem

$$\dot{x}_i = \begin{bmatrix} -\lambda & 0 \\ 1 & 0 \end{bmatrix} x_i + \begin{bmatrix} -1 \\ 0 \end{bmatrix} \varphi_i(y_i),$$
$$y_i = [0 \quad 1] x_i, \tag{7.13}$$

where $x_i = (x_{i1}, x_{i2})^T$ is the state vector defined in (7.9) with $x_{i1} = \omega_{in}$, $x_{i2} = \delta_{in} - \delta_{in}^0$, and $\varphi_i(y_i)$ is specified in (7.11). The main objective of this section is to construct a Lur'e-Postnikov type of Liapunov function and compute an estimate for the stability region of the equilibrium $x_i^* = 0$ of the subsystem (7.13).

Since we use Kalman's (1963) construction to get an appropriate Liapunov function for the system (7.13), we have to transform the system so that it has an asymptotically stable linear part (Seltzer and Šiljak, 1972).

Therefore, we add to the right side of the first equation in (7.13) the zero term $\alpha\mu_i y_i - \alpha\mu_i[0 \;\; 1]x_i$, where

$$\mu_i = (M_i^{-1} + M_n^{-1})A_{1n}\sin\theta_{in}, \tag{7.14}$$

and α is a positive number. After a simple regrouping of the terms, we get the transformed subsystem as

$$\begin{aligned} \dot{x}_i &= A_i x_i + b_i\varphi_i(y_i), \\ y_i &= c_i^T x_i, \end{aligned} \tag{7.15}$$

where

$$A_i = \begin{bmatrix} -\lambda & -\alpha\mu_i \\ 1 & 0 \end{bmatrix}, \qquad b_i = \begin{bmatrix} -1 \\ 0 \end{bmatrix}, \qquad c^T = [0 \;\; 1] \tag{7.16}$$

and the matrix A is stable (it has all eigenvalues with negative real parts).

From the definition (7.11) of the function $\varphi_i(y_i)$, we see that for $y_{i1} = -\pi - 2\delta_{in}^0$, $y_{i2} = \pi - 2\delta_{in}^0$ we have $\varphi_i(y_{i1}) = \varphi_i(y_{i2}) = 0$, and we cannot establish global asymptotic stability of $x_i^* = 0$ in (7.15). The nonlinearity $\varphi_i(y_i)$ belongs to the class

$$\Phi_i = \{\varphi_i(y_i)\colon 0 \leqslant y_i\varphi_i(y_i) \leqslant \kappa_i y_i^2 \;\forall y_i \in [y_{i1}, y_{i2}];\; \varphi_i(0) = 0\}, \tag{7.17}$$

and the stability analysis is restricted to a finite region of the state space, as considered by Walker and McClamroch (1967) and Weissenberger (1968). Following their analysis, we start with the Lur'e-Postnikov type of Liapunov function (2.178), defined here by

$$V_i(x_i) = x_i^T H_i x_i + \zeta_i \int_0^{c_i^T x_i} \varphi_i(y_i)\, dy_i, \tag{7.18}$$

where H_i is a constant matrix and ζ_i is a scalar. As is well known (e.g. Šiljak, 1969), the Popov condition

$$\kappa_i^{-1} + \mathrm{Re}\{(1 + j\zeta_i\omega)c_i^T(A_i - j\omega I)^{-1}b_i\} \geqslant 0 \qquad \forall\omega \geqslant 0 \tag{7.19}$$

is necessary and sufficient for the existence of a function $V_i(x_i)$ such that its time derivative with respect to (7.15) is

$$\dot{V}_i(x_i)_{(7.15)} = [\gamma_i^{1/2}\varphi_i(y_i) + g_i^T x_i]^2 + \Gamma_i, \tag{7.20}$$

where

$$
\begin{aligned}
-g_i g_i^T &= A_i^T H_i + H_i A_i, \\
-\gamma_i^{1/2} g_i &= H_i b_i + \tfrac{1}{2}(\zeta_i A_i^T + I)c_i, \\
-\gamma_i &= \zeta_i c_i^T b_i - \kappa_i^{-1}, \\
-\Gamma_i &= [\kappa_i^{-1}\varphi_i(y_i) - y_i]\varphi_i(y_i);
\end{aligned} \tag{7.21}
$$

H_i is a constant, symmetric, and positive definite matrix; I is the identity matrix; g is a constant vector; and γ_i, Γ_i are positive scalars. Therefore, we have $V_i(x_i)$, $-\dot{V}_i(x_i)_{(7.15)} > 0$ for $y_i \in (y_{i1}, y_{i2})$.

Now, to estimate the stability region corresponding to $x_i^* = 0$, we follow the procedure of Walker and McClamroch (1967). We first conclude that the Popov condition (7.19) for the system (7.15) and $\kappa_i = +\infty$,

$$
\frac{\alpha\mu_i + (\lambda\zeta_i - 1)\omega^2}{(\alpha\mu_i - \omega^2)^2 + \lambda^2\omega^2} > 0 \qquad \forall\omega \geqslant 0, \tag{7.22}
$$

is satisfied for $\lambda\zeta_i \geqslant 1$. Taking $\lambda\zeta_i > 1$, we can factor the left side of the inequality (7.22) as

$$
\frac{\alpha\mu_i + (\lambda\zeta_i - 1)\omega^2}{(\alpha\mu_i - \omega^2)^2 + \lambda\omega^2} = \frac{(\alpha\mu_i)^{1/2} + j\omega(\lambda\zeta_i - 1)^{1/2}}{-\alpha\mu_i + j\omega(-\lambda + j\omega)} \, \frac{(\alpha\mu_i)^{1/2} - j\omega(\lambda\zeta_i - 1)^{1/2}}{-\alpha\mu_i - j\omega(-\lambda - j\omega)}. \tag{7.23}
$$

Then the vector g_i in (7.21) can be computed from the identity

$$
g_i^T(A_i - j\omega I)^{-1} b_i \equiv \frac{(\alpha\mu_i)^{1/2} + j\omega(\lambda\zeta_i - 1)^{1/2}}{-\alpha\mu_i + j\omega(-\lambda + j\omega)} \tag{7.24}
$$

as

$$
g_i = [-(\lambda\zeta_i - 1)^{1/2}, \qquad -(\alpha\mu_i)^{1/2}]^T. \tag{7.25}
$$

By solving the first equation in (7.21) with g_i of (7.25), we get the matrix H_i as

$$
H_i = \frac{1}{2}\begin{bmatrix} \zeta_i & 1 \\ 1 & \lambda + \alpha\zeta_i\mu_i - 2[\alpha\mu_i(\lambda\zeta_i - 1)]^{1/2} \end{bmatrix}. \tag{7.26}
$$

We can choose now $\alpha = 0$ and obtain from (7.18) the Liapunov function $V_i(x_i)$ in the form

$$
V_i(x_i) = \tfrac{1}{2}(\zeta_i x_{i1}^2 + 2x_{i1}x_{i2} + \lambda x_{i2}) + \zeta_i\mu_i \int_0^{x_{i2}} [\sin(y_i + \delta_{in}^0) - \sin\delta_{in}^0]\, dy_i. \tag{7.27}
$$

Computing the time derivative of $V_i(x_i)$ with respect to (7.15) and noting that $y_i = x_{i2}$, we get

$$\dot{V}_i(x_i)_{(7.15)} = -(\lambda\zeta_i - 1)x_{i1}^2 - x_{i2}\varphi_i(x_{i2}), \tag{7.28}$$

where $\varphi_i(x_{i2})$ is defined by (7.11). Since $\dot{V}_i(x_i)_{(7.15)}$ is zero at the limits of the interval $[y_{i1}, y_{i2}]$ in (7.17), in order to have $\dot{V}_i(x_i)_{(7.15)} < 0$ we must reduce the interval to $[y'_{i1}, y'_{i2}]$. The choice of the reduced interval is considered next.

We can always select a positive number ε_i so that the inequality

$$y_i\varphi_i(y_i) \geqslant \varepsilon_i \mu_i y_i^2 \tag{7.29}$$

is satisfied on an interval $[y'_{i1}, y'_{i2}]$ contained in $[y_{i1}, y_{i2}]$. For a chosen ε_i, the limits y'_{i1} and y'_{i2} are the nonzero solution of the equation

$$\varepsilon_i y'_{ik} = \sin(y'_{ik} + \delta_{in}^0) - \sin\delta_{in}^0, \qquad k = 1, 2, \tag{7.30}$$

within the interval $(-\pi - 2\delta_{in}^0, \pi - 2\delta_{in}^0)$.

Now, using (7.29), we get from (7.28) the inequality

$$\dot{V}_i(x_i)_{(7.15)} \leqslant -x_i^T G_i x_i, \tag{7.31}$$

where the positive definite matrix G_i has the form

$$G_i = \begin{bmatrix} \lambda\zeta_i - 1 & 0 \\ 0 & \varepsilon_i\mu_i \end{bmatrix}. \tag{7.32}$$

Since $V_i(x_i)$, $-\dot{V}_i(x_i)_{(7.15)} > 0$ is assured for all $y_i \in [y'_{i1}, y'_{i2}]$, we can proceed to calculate an estimate for the stability region for each subsystem (7.15) using the function $V_i(x_i)$ in (7.27). We compute first

$$\rho_{ik} = \min_{c_i^T x_i = y_{ik}} V_i(x_i), \qquad k = 1, 2, \tag{7.33}$$

and note that the minimum is at the point $x_{im} = y_{ik} H_i^{-1} c_i / c_i^T H_i^{-1} c_i$ on the hyperplane $c_i^T x_i = y_{ik}$ at which grad $V_i(x_i)$ is orthogonal to the hyperplane. Therefore,

$$\rho_{ik} = y_{ik}^2 (c_i^T H_i^{-1} c_i)^{-1} + \zeta_i \int_0^{y_{ik}} \varphi_i(y_i)\, dy_i. \tag{7.34}$$

On the basis of the new limits y'_{ik}, $k = 1, 2$, we use (7.34) to find $V_i^0 = \min_{k=1,2} \rho_{ik}$ as

$$V_i^0 = \min_{k=1,2}\{(\lambda - \zeta_i^{-1})y_{ik}'^2 + \zeta_i\mu_i[\cos\delta_{in}^0 - \cos(y'_{ik} + \delta_{in}) - y'_{ik}\sin\delta_{in}^0]\}. \tag{7.35}$$

An estimate $\tilde{\mathfrak{X}}_i$ of the stability region $\mathfrak{X}_i$ containing the origin ($x_i^* \in \tilde{\mathfrak{X}}_i$) is the connected open region contained in $\mathfrak{X}_i$ ($\tilde{\mathfrak{X}}_i \subseteq \mathfrak{X}_i$), which is determined for each individual subsystem by

$$\tilde{\mathfrak{X}}_i = \{x_i\colon V_i(x_i) < V_i^0\}, \qquad i = 1, 2, \ldots, n-1, \tag{7.36}$$

where V_i^0 is given by (7.35).

Before we use the estimates $\tilde{\mathfrak{X}}_i$ of (7.36) to compute estimates $\tilde{\mathfrak{X}}$ for the stability region $\mathfrak{X}$ ($\tilde{\mathfrak{X}} \subseteq \mathfrak{X}$) of the overall system using the material of Section 2.8, let us show that the following inequalities involving the function $v_i(x_i) = V_i^{1/2}(x_i)$,

$$\begin{aligned} \eta_{i1}\|x_i\| \leqslant v_i(x_i) \leqslant \eta_{i2}\|x_i\|, \\ \dot{v}_i(x_i)_{(7.15)} \leqslant -\eta_{i3}\|x_i\|, \end{aligned} \qquad i = 1, 2, \ldots, n-1, \tag{7.37}$$

are true throughout the region $\tilde{\mathfrak{X}}_i$, where $\eta_{i1}, \eta_{i2}, \eta_{i3}$ are all positive numbers. This fact implies that $\tilde{\mathfrak{X}}_i$ is a region of exponential absolute stability, which was established and explored in a number of papers by Šiljak and Sun (1972) and by Šiljak and Weissenberger (1969, 1970a, b), as well as by Seltzer and Šiljak (1970) and by Karmarkar and Šiljak (1975).

Following the last part of Section 2.5, we note that

$$x_i^T \underline{H}_i x_i \leqslant V_i(x_i) \leqslant x_i^T \overline{H}_i x_i, \tag{7.38}$$

where

$$\underline{H}_i = \frac{1}{2}\begin{bmatrix} \zeta_i & 1 \\ 1 & \lambda \end{bmatrix}, \qquad \overline{H}_i = \frac{1}{2}\begin{bmatrix} \zeta_i & 1 \\ 1 & \lambda + \zeta_i\mu_i\cos\delta_{in}^0 \end{bmatrix}. \tag{7.39}$$

The matrix $\underline{H}_i$ is obtained from H_i of (7.26) when $\alpha = 0$, and $\overline{H}_i = \underline{H}_i + \frac{1}{2}\zeta_i\kappa_i' c_i c_i^T$, where κ_i' is the slope of the function $\varphi_i(y_i)$ at $y_i = 0$. Since the integral in (7.27) is positive on $[y_{i1}', y_{i2}']$, the first inequality in (7.38) is obvious. The choice of $\overline{H}_i$ comes from the fact that $y_i\varphi_i(y_i) \leqslant \kappa_i' y_i^2$ and the fact that the integral in $V_i(x_i)$ of (7.27) can be majorized by $\frac{1}{2}\kappa_i' y_i^2$. This confirms the second inequality in (7.38).

Finally, from (7.32) and (7.38) we get the inequalities (7.37) for the function $v_i(x_i) = V_i^{1/2}(x_i)$ with

$$\eta_{i1} = \lambda_m^{1/2}(\underline{H}_i), \qquad \eta_{i2} = \lambda_M^{1/2}(\overline{H}_i), \qquad \eta_{i3} = \frac{\lambda_m(G_i)}{2\lambda_M^{\frac{1}{2}}(\overline{H}_i)}, \tag{7.40}$$

where λ_m and λ_M are the minimum and maximum eigenvalues of the indicated matrices. This completes the subsystem analysis which is necessary for the aggregation procedure outlined in the next section.

7.4. STABILITY REGION

The purpose of this section is to use the Liapunov functions $v_i(x_i)$ with estimates (7.37) computed for the individual (free) subsystems (7.15), and calculate certain constraints on the interconnections $h_i(x)$ in the overall system

$$\dot{x}_i = \begin{bmatrix} -\lambda & 0 \\ 1 & 0 \end{bmatrix} x_i + \begin{bmatrix} -1 \\ 0 \end{bmatrix} \varphi_i(y_i) + h_i(x), \qquad i = 1, 2, \ldots, n-1, \quad (7.10)$$
$$y_i = [0 \quad 1] x_i,$$

to construct the aggregate model.

From (7.12) and the fact that grad $v_i(x_i) = \frac{1}{2} V_i^{-1/2}(x_i)$grad $V_i(x_i)$, we compute

$$[\text{grad } v_i(x_i)]^T h_i(x) = \tfrac{1}{2} v_i^{-1}(x_i)(\zeta_i x_{i1} + x_{i2}) \left[f_i + \sum_{\substack{j=1 \\ j \neq i}}^{n-1} (M_n^{-1} f_{nj} - M_i^{-1} f_{ij}) \right],$$
$$i = 1, 2, \ldots, n-1, \qquad (7.41)$$

where

$$\begin{aligned} f_i &= (M_n^{-1} - M_i^{-1}) A_{in} \cos\theta_{in} [\cos(x_{i2} + \delta_{in}^0) - \cos\delta_{in}^0], \\ f_{ij} &= A_{ij} [\cos(x_{i2} - x_{j2} + \delta_{ij}^0 - \theta_{ij}) - \cos(\delta_{ij}^0 - \theta_{ij})], \qquad (7.42) \\ f_{nj} &= A_{jn} [\cos(x_{j2} + \delta_{jn}^0 + \theta_{jn}) - \cos(\delta_{jn}^0 + \theta_{jn})]. \end{aligned}$$

Applying the obvious inequalities

$$\begin{aligned} \cos(x_{i2} + \delta_{in}^0 + \theta_{in}) - \cos(\delta_{in}^0 + \theta_{in}) &\leqslant |\sin(\delta_{in}^0 + \theta_{in})| \, |x_{i2}|, \\ x_{i1} + x_{i2} &\leqslant 2^{1/2} \|x_i\|, \qquad (7.43) \\ |x_{i2} - x_{j2}| &\leqslant |x_{i2}| + |x_{j2}| \end{aligned}$$

to (7.41), we obtain the interconnection constraints

$$\|[\text{grad } v_i(x_i)]^T h_i(x)\| \leqslant \sum_{j=1}^{n-1} \xi_{ij} \|x_j\|, \qquad i = 1, 2, \ldots, n-1, \quad (7.44)$$

with nonnegative numbers ξ_{ij} defined as

$$\xi_{ij} = \begin{cases} \hat{\zeta}_i \eta_{i1}^{-1} \Big[|M_n^{-1} - M_i^{-1}| A_{in} \cos \theta_{in} |\sin \delta_{in}^0| \\ \qquad + \sum_{\substack{j=1 \\ j \neq i}}^{n-1} M_i^{-1} A_{ij} |\sin(\delta_{ij}^0 - \theta_{ij})| \Big], & i = j, \\ \hat{\zeta}_i \eta_{i1}^{-1} [M_n^{-1} A_{jn} |\sin(\delta_{jn}^0 + \theta_{jn})| + M_i^{-1} A_{ij} |\sin(\delta_{ij}^0 - \theta_{ij})|], & i \neq j, \end{cases} \tag{7.45}$$

where $\hat{\zeta}_i = \max\{1, \zeta_i\}$.

Taking the time derivative of the function $v_i(x_i)$ with respect to (7.10), and using the inequalities (7.37) and (7.44) with η_{ik} $(k = 1, 2, 3)$ and ξ_{ij} computed in (7.40) and (7.44), we get

$$\begin{aligned} \dot{v}_i(x_i)_{(7.10)} &= \dot{v}_i(x_i)_{(7.15)} + [\text{grad } v_i(x_i)]^T h_i(x) \\ &\leqslant -\eta_{i3} \|x_i\| + \sum_{j=1}^{n-1} \xi_{ij} \|x_j\| \\ &\leqslant -\eta_{i2}^{-1} \eta_{i3} v_i(x_i) + \sum_{j=1}^{n-1} \xi_{ij} \eta_{j1}^{-1} v_j(x_j), \qquad i = 1, 2, \ldots, n-1. \end{aligned} \tag{7.46}$$

By introducing the vector Liapunov function $v(x) = [v_1(x_1), v_2(x_2), \ldots, v_{n-1}(x_{n-1})]^T$, the scalar inequalities (7.46) can be rewritten as a vector differential inequality

$$\dot{v} \leqslant Wv, \tag{7.47}$$

which represents an aggregate model for the system (7.10). In (7.47), the $(n-1) \times (n-1)$ constant aggregate matrix $W = (w_{ij})$ is defined by

$$w_{ij} = \begin{cases} -\eta_{i2}^{-1} \eta_{i3} + \xi_{ii} \eta_{i1}^{-1}, & i = j, \\ \xi_{ij} \eta_{j1}^{-1}, & i \neq j. \end{cases} \tag{7.48}$$

To calculate an estimate for the overall system stability region using the aggregate model, we have to make sure that the matrix W is a stable matrix, i.e., it satisfies the conditions

$$(-1)^k \begin{vmatrix} w_{11} & w_{12} & \cdots & w_{1k} \\ w_{21} & w_{22} & \cdots & w_{2k} \\ \cdots & \cdots & \cdots & \cdots \\ w_{k1} & w_{k2} & \cdots & w_{kk} \end{vmatrix} > 0, \qquad k = 1, 2, \ldots, n-1. \tag{7.49}$$

It is easy to see that the conditions (7.49) imply that the diagonal elements

w_{ii} of W are all negative. By (7.48), this further implies that

$$\eta_{i2}^{-1}\eta_{i3} > \xi_{ii}\eta_{i1}^{-1}, \qquad i = 1, 2, \ldots, n-1. \tag{7.50}$$

Using (7.40) and (7.45), we can rewrite (7.50) as

$$\lambda_m(G_i) > 2\hat{\xi}_i \frac{\lambda_M(\overline{H}_i)}{\lambda_m(\underline{H}_i)} \Big[|M_n^{-1} - M_i^{-1}| A_{in} \cos\theta_{in} |\sin\delta_{in}^0| + M_i^{-1} \sum_{\substack{j=1 \\ j\neq i}}^{n-1} A_{ij} |\sin(\delta_{ij}^0 - \theta_{ij})| \Big], \qquad i = 1, 2, \ldots, n-1. \tag{7.51}$$

For a given ξ_i, a suitable choice for $\varepsilon_i (0 < \varepsilon_i < \cos\delta_{ij}^0)$ in (7.30) is such that

$$\lambda\xi_i - 1 > \varepsilon_i \mu_i, \tag{7.52}$$

and $\lambda_m(G_i) = \varepsilon_i \mu_i$ for G_i defined in (7.32). Now, we can express (7.51) in terms of ε_i as

$$\varepsilon_i > 2 \frac{\hat{\xi}_i}{\mu_i} \frac{\lambda_M(\overline{H}_i)}{\lambda_m(\underline{H}_i)} \Big[|M_n^{-1} - M_i^{-1}| A_{in} \cos\theta_{in} |\sin\delta_{in}^0| + M_i^{-1} \sum_{\substack{j=1 \\ j\neq i}}^{n-1} A_{ij} |\sin(\delta_{ij}^0 - \theta_{ij})| \Big], \qquad i = 1, 2, \ldots, n-1. \tag{7.53}$$

From (7.53) we can conclude that the smaller the values of A_{ij}, the easier it is to come up with suitable numbers ε_i. This, in turn, means that the decomposition of the power system model should be performed in such a way that the resulting subsystems are weakly coupled. This conclusion confirms the general property of decomposition principle as applied to stability analysis of dynamic systems.

To determine an estimate $\tilde{\mathscr{X}}$ of the overall stability region $\mathscr{X}$ using the aggregate model, we follow the general analysis of Section 2.8. We first choose the overall Liapunov function

$$\nu_2(x) = \max_{1 \leq i \leq n-1} \frac{|v_i(x_i)|}{d_i}, \tag{7.54}$$

where $d = (d_1, d_2, \ldots, d_{n-1})^T$ is a vector with positive components (that is, $d > 0$). We recall that for any given positive constant vector $c \in \mathscr{R}_+^{n-1}$, we can compute a vector d as

$$d^T = -c^T W^{-1}. \tag{7.55}$$

We choose $c = (\varepsilon_1, \varepsilon_2, \ldots, \varepsilon_{n-1})^T$ and compute the vector d from (7.55). Then we calculate $v_i^0 = (V_i^0)^{1/2}$ from (7.35) and determine

$$\gamma_2 = \min_{1 \leqslant i \leqslant n-1} \frac{v_i^0}{d_i}. \tag{7.56}$$

The estimate $\tilde{\mathfrak{X}}$ defined in (2.281),

$$\tilde{\mathfrak{X}} = \{x \in \mathfrak{R}^{2(n-1)}: \nu_2(x) < \gamma_2\}, \tag{7.57}$$

is found as

$$\begin{aligned} \tilde{\mathfrak{X}} = \Big\{ x \in \mathfrak{R}^{2(n-1)}: & \tfrac{1}{2}\zeta_i x_{i1}^2 + x_{i1} x_{i2} + \tfrac{1}{2}\lambda x_{i2}^2 \\ & + \zeta_i \mu_i \int_0^{x_{i2}} [\sin(y_i + \delta_{in}^0) - \sin \delta_{in}^0]\, dy_i < (\gamma_2 d_i)^2, \\ & i = 1, 2, \ldots, n-1 \Big\}. \end{aligned} \tag{7.58}$$

In terms of the original physical variables of the power system (7.1), the inequalities in (7.58) are

$$\begin{aligned} & \tfrac{1}{2}\zeta_i \omega_{in}^2 + \omega_{in}(\delta_{in} - \delta_{in}) + \tfrac{1}{2}\lambda(\delta_{in} - \delta_{in})^2 \\ & \quad + \zeta_i \mu_i [\cos \delta_{in}^0 - (\delta_{in} - \delta_{in}^0) \sin \delta_{in}^0 - \cos \delta_{in}] < (v^0 d_i)^2, \\ & \qquad i = 1, 2, \ldots, n-1. \end{aligned} \tag{7.59}$$

For estimating the overall stability region $\mathfrak{X}$, the Lyapunov functions $\nu_1(x)$ and $\nu_3(x)$ of Section 2.8 can also be used. Our choice of $\nu_2(x)$ here was guided by simplicity in interpreting the estimate $\tilde{\mathfrak{X}}$ in terms of system variables and parameters.

Furthermore, the estimate $\tilde{\mathfrak{X}}$ of the region $\mathfrak{X}$ is obtained by a systematic procedure which is suitable for computer implementation. This fact can be used to provide information about the subtle interrelationship among various parameters and their influence on the size of the estimate. Namely, computer algorithms can be used to determine the parameters ζ_i, ε_i which are involved in the trade-off between the size of the subsystem stability-region estimates and the degree of stability of the aggregate model. This could justify our intuitive choice of the vector c in the above calculations, and provide an optimal choice of the parameters ζ_i and ε_i with respect to the size of the estimate $\tilde{\mathfrak{X}}$ for the overall stability region.

7.5. THREE-MACHINE SYSTEM

In this section, we present a simple illustration of the decomposition-aggregation approach in case of three interconnected machines. The machines are specified as follows:

$$
\begin{aligned}
M_1 &= 0.01,\ M_2 = 0.01,\ M_3 = 2.00 \\
E_1 &= 1.017,\ E_2 = 1.005,\ E_3 = 1.033 \\
Y_{12} &= 0.98 \times 10^{-3},\ Y_{13} = 0.114,\ Y_{23} = 0.106 \\
\theta_{12} &= 86\,°,\ \theta_{13} = 88\,°,\ \theta_{23} = 89\,° \\
\delta_{12}^0 &= 5\,°,\ \delta_{13}^0 = -2\,°,\ \delta_{23}^0 = -3\,° \\
\lambda &= 100.
\end{aligned}
\tag{7.60}
$$

With the third machine as a reference (see Figure 7.1), two subsystems are formed as described in Section 7.2. We choose two positive numbers $\varepsilon_1 = \varepsilon_2 = 0.99$ to be slightly less than cos δ_{13}^0 and cos δ_{23}^0, and $\zeta_1 = \zeta_2 = 1$. By using the formulas of Section 7.3, we get the following subsystems estimates

$$
\begin{aligned}
\eta_{11} &= 0.704,\ \eta_{12} = 7.488,\ \eta_{13} = 0.863 \\
\eta_{21} &= 0.704,\ \eta_{22} = 7.454,\ \eta_{23} = 0.800.
\end{aligned}
\tag{7.61}
$$

From Equations (7.45) - (7.48), we determine the interaction bounds as

$$\xi_{11} = 0.0160,\ \xi_{12} = 0.0615,\ \xi_{21} = 0.0565,\ \xi_{22} = 0.0114 \tag{7.62}$$

With the numbers given in (7.60) - (7.62), we compute aggregate matrix

$$W = \begin{bmatrix} -0.0930 & 0.0874 \\ 0.0827 & -0.0910 \end{bmatrix}. \tag{7.63}$$

Since the matrix W satisfies conditions (7.49), it is a stable matrix and we can procede to determine a stability region estimate for the overall system.

First, we solve Equation (7.30) for the values of y_{ij}''s, which when introduced into (7.35) produce the values of the Liapunov functions at the limits of the stability region estimates for the decoupled subsystems

$$V_1^0 = 0.13325,\ V_2^0 = 0.12326. \tag{7.64}$$

Then, we choose the vector $r = (1, 1)^T$, and compute $\nu^0 = 0.00247$. Finally, an estimate $\tilde{\mathfrak{X}}$ of the stability region $\mathfrak{X}$ for the overall system is obtained from (7.59) as

$$
\begin{aligned}
&0.5\omega_{13}^2 + \omega_{13}(\delta_{12} - 0.035) + 50(\delta_{12} - 0.035)^2 + 12.053 \cos \delta_{12} \\
&\qquad - 0.42(\delta_{12} - 0.035) < 0.1270 \\
&0.5\omega_{23}^2 + \omega_{23}(\delta_{23} - 0.052) + 50(\delta_{23} - 0.052)^2 + 11.053 \cos \delta_{13} \\
&\qquad\qquad\qquad - 0.578(\delta_{23} - 0.052) < 0.1232.
\end{aligned} \tag{7.65}
$$

This completes our numerical example.

Before we conclude the exposition of transient stability analysis of multimachine power systems, several comments are in order. First, we should note from the above example that the machines are assumed to be weakly coupled. This may be an unrealistic assumption, because weak coupling is likely to occur between group of strongly coupled machines rather than between individual machines. For this reason, a preliminary equivalencing of the strongly connected machines may be necessary. Equivalencing procedures can be carried-out as proposed by Darwish, Fantin, and Richetin (1976). Second, the aggregation procedure used in the above example can be readily replaced by the less conservative one described in Section 2.5, Theorem 2.11, Equations (2.133) - (2.143). Alternatively, the aggregation process of Araki (1975) described in Section 2.6, can be advantageously applied to our decomposed model to reduce further the conservativeness in the results because the norms can be replaced by less restrictive positive definite functions. Finally, we should mention the fact that the more general case of nonuniformly damped machines can be treated in pretty much the same way (Jocić and Šiljak, 1977) as the uniform damping one considered here.

7.6. AUTOMATIC GENERATION CONTROL

Under normal operating conditions, the automatic generation control (AGC) of an interconnected power system has two principal objectives:

(1) To maintain *frequency* and *power exchange* between areas at scheduled values.

(2) To monitor *power allocation* among generating units.

The first objective is often termed the load and frequency control (LFC), while the other is called economic dispatch (ED). The LFC accomplishes the area generation-consumption power balance by proportional-plus-integral control adjustments of the regulating units, which depend on deviations of frequency and tie-line exchange. The purpose of ED is to achieve a response to area load demand with least cost. This is accomplished by a static optimization process which provides base values for load and participation of units in AGC as functions of area load. Although LFC and ED have different and often conflicting goals, they are both achieved by pulsing a governor motor at each regulating unit. A reasonable trade-off between the two control objectives is maintained by appropriate coordinating actions.

The AGC have evolved over the past forty years from simple slide-wire-based devices to sophisticated applications of modern digital computers. This evolution process is recorded in the papers by Wild (1941) and Concordia, Crary, and Parker (1941) and the books by Kirchmayer (1959) and Cohn (1966), as well as a recent survey by Ewart (1975). Despite the fact that classical AGC systems have performed satisfactorily, a considerable effort has been made recently by a large number of people, notably by Elgerd and Fosha (1970), Fosha and Elgerd (1970), Ćalović (1972), Glover and Schwepe (1972) and Kwatni, Kalnitsky, and Bhatt (1975), as surveyed by Reddoch (1975), to develop new algorithms for AGC using modern optimal-control theory. This effort is justified by the fact that the classical design faces considerable difficulty in coping with the changes that AGC has been undergoing in recent times, especially in control hardware and a variety of generating types including highly engineered nuclear units.

Although the new multivariable control approach definitely offered more freedom in choosing the best AGC, it lost some of the important advantages of the conventional design based upon the area autonomy of models and controls. In modern control schemes with a centralized design and/or operation, the whole interconnected model is manipulated "in one piece" in order to obtain a global control for the overall system. This creates both conceptual and computational difficulties whenever control of large power systems is attempted, mainly because centralized information has to be obtained from power areas and generating plants which are spread over large geographic territories and, at the same time, processed in a "one shot" design procedure.

The purpose of the rest of this chapter is to outline a new decentralized AGC scheme for interconnected power systems, which was presented in a paper by Ćalović, Djorović, and Šiljak (1976). In the scheme, the AGC regulators are designed "piece by piece" by the decomposition-aggregation method of Chapters 1 and 2, utilizing the individual area models and applying the concept of autonomous local control to each area and the associated tie-lines. This requires a decomposition of the interconnected system into *overlapping subsystems* each representing an area and all the tie-lines originating from that area. By overlapping of the subsystems, the decomposition leads to a decentralized area control of proportional-plus-integral type which can be designed as a linear-quadratic regulator with complete subsystem feedback. It retains the area-control concept and load-distribution property of conventional regulators, while at the same time offering a possibility of improved system transients and stability margins to be achieved for the overall interconnected system.

The main features of the proposed decentralized AGC scheme are as follows:

(1) Each decentralized regulator applies only local feedback based upon the measurements accessible from its own area.

(2) In the local-regulator design, only models of the corresponding area model and the associated tie-lines are used.

(3) The local feedback control law is linear, constant, and independent of the area load disturbance.

(4) The steady-state errors of the frequency and tie-line exchange are reduced to zero.

(5) In comparison with actually used conventional AGC, the decentralized AGC regulators may produce improved transients and increased stability margins for the overall system.

(6) A desired steady-state distribution of the generated power is incorporated in the decentralized AGC scheme as an independent process, which implies the existence of only one independent local control per area.

7.7. A DECENTRALIZED MODEL

A linearized model of an interconnected power system can be described by the equations

$$\begin{aligned} \dot{x} &= Ax + Bu + Fz, \\ y &= C^T x, \end{aligned} \tag{7.66}$$

where $x(t) \in \mathfrak{R}^n$, $u(t) \in \mathfrak{R}^m$, $z(t) \in \mathfrak{R}^s$, and $y(t) \in \mathfrak{R}^r$ are state, input, unmeasurable disturbance, and output vectors; A, B, F, and C are constant matrices of appropriate dimensions; and $x(t_0) = x_0$ is the initial state at initial time t_0. The model (7.66) is composed of a set of s area models and the associated tie-line models, which can be presented in a decomposed form as

$$\begin{aligned} \dot{x}_i &= A_i x_i + a_{ti} \Delta p_{ei} + B_i u_i + f_i z_i, \\ \Delta \dot{p}_{ei} &= \alpha_{1i} \sum_{\substack{j=1 \\ j \neq i}}^{s} (m_{ij}^T x_i - m_{ji} x_j), \qquad i = 1, 2, \ldots, s. \\ y_i &= C_i^T x_i, \end{aligned} \tag{7.67}$$

Here $x_i(t) \in \mathcal{R}^{n_i}$, $u_i(t) \in \mathcal{R}^{m_i}$, $z_i(t) \in \mathcal{R}$, $y_i(t) \in \mathcal{R}^{r_i}$, and $\Delta p_{ei}(t) \in \mathcal{R}$ are the state, the input, the disturbance, the output, and the variation of the total power exchange corresponding to the ith area. We also have

$$\Delta p_{ei} = \alpha_{1i} \sum_{\substack{j=1 \\ j \neq i}}^{s} \Delta p_{ij}, \qquad i = 1, 2, \ldots, s, \tag{7.68}$$

where Δp_{ij} is the fractional power-exchange deviation between the ith and the jth area. The matrices A_i, B_i, F_i, and C_i are constant and have appropriate dimensions; a_{ti}, $m_{ij} \in \mathcal{R}^{n_i}$, $m_{ji} \in \mathcal{R}^{n_j}$ are constant coupling vectors; and $\alpha_{1i} = P_{10}/P_{i0}$ is a normalization factor with the steady-state load of area 1 taken as reference (thus $\alpha_{ii} = 1$). Finally, in addition to $x_i(t_0) = 0$, we assume $\Delta p_{ei}(t_0) = 0$.

The particular properties of the model represented by Equations (7.66)–(7.62) imply the relations

$$n = \sum_{i=1}^{s} n_i + s - 1, \qquad m = \sum_{i=1}^{s} m_i, \qquad r = \sum_{i=1}^{s} r_i, \tag{7.69}$$

and

$$\sum_{j=1}^{s} \frac{\Delta p_{ej}}{\alpha_{ij}} = 0, \tag{7.70}$$

where $\alpha_{ij} = P_{i0}/P_{j0}$ is the size-ratio coefficient for the jth area relative to the steady-state load of the ith area taken as reference.

To design the decentralized area AGC regulators, the decoupled model (7.67) should be suitably rearranged. First, in order to achieve the design objectives (3) and (4) listed in Section 7.6, it is necessary to augment the model (7.67) with the new state defined as an integral of the dynamic area control error (ACE) $y_i^0(t)$, as proposed by Ćalović (1972). We have

$$v_i(t) = v_{i0} + \int_{t_0}^{t} y_i^0(t)\, dt, \qquad i = 1, 2, \ldots, s, \tag{7.71}$$

where

$$y_i^0(t) = \Delta p_{ei}(t) + b_{si} \Delta f_i, \tag{7.72}$$

b_{si} is the bias factor, and Δf_i is the frequency deviation in the ith area (expressed in per unit). The expression for $y_i^0(t)$ can be rewritten as

$$y_i^0(t) = \Delta p_{ei}(t) + d_i^T x_i(t), \tag{7.73}$$

where $d_i \in \mathcal{R}^{n_i}$.

The second rearrangement is a consequence of the design objective (5), which is known as the area control concept (ACC). A desired (unique) area control $w_i(t)$ is obtained by imposing the equality constraints on the unit control vectors $u_i(t)$ in (7.67) of the type

$$u_i(t) = c_i w_i(t), \tag{7.74}$$

where $c_i \in \mathcal{R}^{m_i}$ is the constant participation vector (Ćalović, 1972), the elements c_k of which satisfy the relation

$$\sum_{k=1}^{m_i} c_k = 1, \tag{7.75}$$

so that it determines the relative steady-state load distribution in the ith area.

With regard to Equations (7.71) and (7.75), the interconnected power system model (7.67) can be rewritten as

$$\begin{aligned} \dot{x}_i &= A_i x_i + a_{ti} \Delta p_{ei} + B_i c_i w_i + f_i z_i, \\ \dot{v}_i &= d_i^T x_i + \Delta p_{ei}, \\ \Delta \dot{p}_{ei} &= \alpha_{1i} \sum_{\substack{j=1 \\ j \neq i}}^{s} (m_{ij}^T x_i - m_{ji}^T x_j), \end{aligned} \qquad i = 1, 2, \ldots, s, \tag{7.76}$$

where $x_i(t_0) = 0$, $v_i(t_0) = v_{i0}$, and $\Delta p_{ei}(t_0) = 0$.

The equations (7.76) define an $(n_i + 2)$-dimensional model of one subsystem in the sense of the decentralized system representation. Assuming that $z(t)$ is constant, the model (7.76) can be reduced to the standard form (see Chapter 3) of a decomposed decentralized system:

$$\dot{\hat{x}}_i = \hat{A}_i \hat{x}_i + \hat{b}_i \hat{w}_i + \sum_{\substack{j=1 \\ j \neq i}}^{s} \hat{A}_{ij} \hat{x}_j, \qquad i = 1, 2, \ldots, s, \tag{7.77}$$

where

$$\hat{x}_i = [(x_i - x_{iss})^T, v_i - v_{iss}, \Delta p_{ei}]^T, \qquad \hat{w}_i = w_i - w_{iss} \tag{7.78}$$

and

$$\hat{A}_i = \begin{bmatrix} A_{ii} & 0 & a_{ti} \\ d_i^T & 0 & 1 \\ \alpha_{1i} \sum\limits_{\substack{j=1 \\ j \neq i}}^{s} m_{ij}^T & 0 & 0 \end{bmatrix}, \qquad \hat{b}_i = \begin{bmatrix} B_i v_i \\ 0 \\ 0 \end{bmatrix}, \qquad \hat{A}_{ij} = \begin{bmatrix} 0 & 0 & 0 \\ 0 & 0 & 0 \\ -\alpha_{1i} m_{ji}^T & 0 & 0 \end{bmatrix} \tag{7.79}$$

with $\hat{x}_{i0} = (-x_{iss}^T, v_{i0} - v_{iss}, 0)^T$, where the subscript "ss" designates a steady-state value of the indicated variable.

From (7.79), we see there is an overlapping of the subsystem state vectors $\hat{x}_i(t)$ due to the fact that each subsystem in (7.79) incorporates the corresponding tie-line models. Therefore, the subsystem state vector contains all accessible states from the individual area it represents. Such definition of a system is quite natural. Subsystems without overlapping would lead to a complete separation of an interconnected power system, which would aggravate the problem of decentralized AGC regulator design.

7.8. REGULATOR DESIGN

Having in mind the properties (1)–(3) outlined in Section 7.6, we require that the AGC regulator should be of a decentralized type with linear constant feedback. Using the complete state feedback applied to the decomposed system model (7.77), we choose the local control for the ith area as

$$\hat{w}_i(t) = -\hat{k}_i^T \hat{x}_i(t), \tag{7.80}$$

where

$$\hat{k}_i = (k_{Pi}^T, k_{Ii}, k_{Ti})^T \tag{7.81}$$

is an $(n_i + 2)$-vector of feedback gains associated with the ith area. In (7.81),

k_{Pi} is the n_i vector of proportional control gains,
k_{Ii} is the integral control gain,
k_{Ti} is the proportional control gain for the net interchange load deviation.

By substituting the local control (7.80) into (7.77), we get the closed-loop interconnected system as

$$\dot{\hat{x}}_i = (\hat{A}_i - \hat{b}_i \hat{k}_i^T)\hat{x}_i + \sum_{\substack{j=1 \\ j \neq i}}^{s} \hat{A}_{ij}\hat{x}_j, \qquad i = 1, 2, \ldots, s. \tag{7.82}$$

A choice of the gain vector $\hat{k}_i$ which produces a closed-loop system (7.82) with a satisfactory degree of exponential stability can be made by using the stabilization procedures of Chapter 3. Since results are already available for the power-system load and frequency control of an individual area by utilizing the optimum linear regulator with proportional-plus-integral feedback (Ćalović, 1972), we choose $\hat{k}_i$'s in (7.82) to optimize each subsystem locally as decoupled (see Section 3.5). With this choice of gains $\hat{k}_i$, we establish stability of the overall closed-loop system (7.82), applying either the vector Liapunov function or the suboptimal design of Section 3.5.

With each decoupled subsystem

$$\dot{\hat{x}}_i = \hat{A}_i \hat{x}_i + \hat{b}_i \hat{w}_i, \qquad i = 1, 2, \ldots, s, \tag{7.83}$$

we associate a quadratic performance index

$$\hat{J}_i(t_0, \hat{x}_{i0}, \hat{w}_i) = \int_{t_0}^{\infty} e^{2\pi t} (\hat{x}_i^T \hat{Q}_i \hat{x}_i + \hat{r}_i \hat{w}_i^T \hat{w}_i)\, dt, \tag{7.84}$$

where $\hat{Q}_i$ is a symmetric nonnegative definite matrix ($\hat{Q}_i = \hat{Q}_i^T \geqslant 0$) and $\hat{r}_i$ is a positive number ($\hat{r}_i > 0$). We choose the matrix $\hat{Q}_i$ as a quasidiagonal matrix

$$\hat{Q}_i = \begin{bmatrix} Q_{Pi} & 0 & 0 \\ 0 & q_{Ii} & 0 \\ 0 & 0 & q_{Ti} \end{bmatrix}, \tag{7.85}$$

where $Q_{Pi} = Q_{Pi}^T \geqslant 0$ is a constant $n_i \times n_i$ matrix and $q_{Pi} > 0$, $q_{Ti} \geqslant 0$ are scalars. In (7.84), π is a positive number which is a measure of the degree of exponential stability.

It is a well-known fact (e.g. Anderson and Moore, 1971) that under the assumption that the pair $(\hat{A}_i, \hat{b}_i)$ is completely controllable, there is a unique optimal gain vector $\hat{k}_i^0$ for the feedback control law in (7.80), which is given as

$$\hat{k}_i^{0T} = \hat{r}_i^{-1} \hat{b}_i^T \hat{P}_i, \tag{7.86}$$

where $\hat{P}_i$ is an $(n_i + 2) \times (n_i + 2)$ symmetric positive definite matrix which is the solution of the algebraic Riccati equation

$$(\hat{A}_i + \pi I_i)^T \hat{P}_i + \hat{P}_i(\hat{A}_i + \pi I_i) - \hat{r}_i^{-1} \hat{P}_i \hat{b}_i \hat{b}_i^T \hat{P}_i + \hat{Q}_i = 0. \tag{7.87}$$

Here I_i is the $(n_i + 2) \times (n_i + 2)$ identity matrix. The optimal control for the ith subsystem is

$$\hat{w}_i^0(t) = \hat{k}_i^{0T} \hat{x}_i(t), \tag{7.88}$$

which results in the optimal cost

$$\hat{J}_i^0(t_0, \hat{x}_{i0}) = e^{2\pi t_0} \hat{x}_{i0}^T \hat{P}_i \hat{x}_{i0}. \tag{7.89}$$

It is equally well known (e.g. Anderson and Moore, 1971) that if $\hat{Q}_i$ can be factored as $\hat{Q}_i = \hat{M}_i \hat{M}_i^T$, where $\hat{M}_i$ is an $(n_i + 2) \times (n_i + 2)$ matrix, so that the pair $(\hat{A}_i, \hat{M}_i)$ is completely observable, then each closed-loop subsystem without interactions,

$$\dot{\hat{x}}_i = (\hat{A}_i - \hat{r}_i^{-1}\hat{b}_i\hat{b}_i^T\hat{P}_i)\hat{x}_i, \qquad i = 1, 2, \ldots, s, \tag{7.90}$$

is globally exponentially stable. That is, there exists a positive number Π_i such that the solution $\hat{x}_i(t; t_0, \hat{x}_{i0})$ satisfies the inequality

$$\|\hat{x}_i(t; t_0, \hat{x}_{i0})\| \leqslant \Pi_i \|\hat{x}_{i0}\| \exp[-\pi(t - t_0)] \tag{7.91}$$

for all t_0, $\hat{x}_0$, and $t \geqslant t_0$.

Now we examine the influence of the interconnections $\hat{A}_{ij}\hat{x}_j(t)$ in (7.82), and from exponential stability of each subsystem (7.90), infer the same stability property of the overall system (7.82). Before we set out to do this, let us identify the individual components in (7.81) for the optimal gain vector $\hat{k}_i^0$ given by (7.86).

First, we partition the matrix $\hat{P}_i$ as

$$\hat{P}_i = \begin{bmatrix} P_{i11} & p_{i12} & p_{i13} \\ p_{i12}^T & p_{i22} & p_{i23} \\ p_{i13}^T & p_{i23} & p_{i33} \end{bmatrix} \begin{matrix} n_i \\ 1. \\ 1 \end{matrix} \tag{7.92}$$
$$\qquad n_i \quad 1 \quad 1$$

Then the components of $\hat{k}_i^0$ are obtained from (7.86) and (7.92) as

$$k_{Pi}^{0T} = \hat{r}_i^{-1} c_i^T B_i^T P_{i11}, \qquad k_{Ii}^0 = \hat{r}_i^{-1} c_i^T B_i^T p_{i12}, \qquad k_{Ti}^0 = \hat{r}_i^{-1} c_i^T B_i^T p_{i13}. \tag{7.93}$$

The optimal control law (7.88) reduces to the standard proportional-plus-integral autonomous area control

$$\hat{w}_i^0 = k_{Pi}^{0T} x_i - k_{Ii}^0 \int_{t_0}^t [d_i^T x_i + \Delta p_{ei}]\, dt - k_{Ti}^0 \Delta p_{ei} + w_{i0}^0, \tag{7.94}$$

where the initial control is $w_{i0}^0 = \hat{r}_i^{-1} c_i^T B_i^T P_i v_{i0}$.

A block-diagram representation of a decentralized AGC regulator for the ith area is shown in Figure 7.3, where $C_i = I_i$.

After completing the local optimization procedure it is necessary to show that the local controls are not conflicting and that the overall system is stable. This can be accomplished either by applying the suboptimal design of Section 3.5 or by using the aggregate model involving a vector Liapunov function as shown in Section 3.3.

With the partition (7.81) and the components (7.93) of the optimal gain vector of (7.86), the overall closed-loop system

$$\dot{\hat{x}}_i = (\hat{A}_i - \hat{r}_i^{-1}\hat{b}_i\hat{b}_i^T\hat{P}_i)\hat{x}_i + \sum_{\substack{j=1 \\ j \neq i}}^{s} \hat{A}_{ij}\hat{x}_j, \qquad i = 1, 2, \ldots, s, \tag{7.95}$$

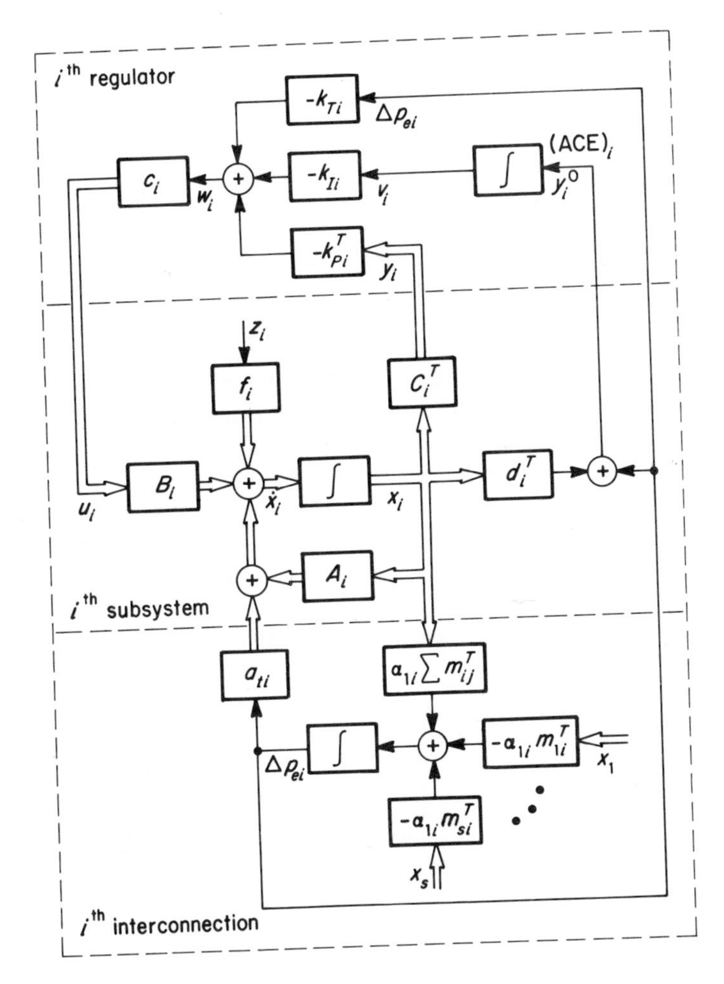

FIGURE 7.3. The ith AGC regulator.

can be rewritten as

$$\dot{\hat{x}}_i = \begin{bmatrix} A_i - B_i c_i k_{Pi}^{0T} & -B_i c_i k_{Ii}^0 & a_{ti} - B_i c_i k_{Ti}^0 \\ d_i^T & 0 & 1 \\ \alpha_{1i} \sum\limits_{\substack{j=1 \\ j\neq i}}^{s} m_{ij}^T & 0 & 0 \end{bmatrix} \hat{x}_i + \sum_{\substack{j=1 \\ j\neq i}}^{s} \begin{bmatrix} 0 & 0 & 0 \\ 0 & 0 & 0 \\ -\alpha_{1i} m_{ji}^T & 0 & 0 \end{bmatrix} \hat{x}_j, \tag{7.96}$$

$i = 1, 2, \ldots, s.$

From Section 3.5, we recall that if there exists a number $\varepsilon > 0$ such that

$$\sum_{i=1}^{s}\sum_{j=1}^{s}\lambda_M^{1/2}(\hat{A}_{ij}^T\hat{A}_{ij}) \leqslant \frac{1}{2}\frac{\varepsilon}{1+\varepsilon}\frac{\min_i \lambda_{mi}(\hat{S}_i)}{\max_i \lambda_{Mi}(\hat{P}_i)}, \tag{7.97}$$

then the closed-loop system (7.96) is globally asymptotically stable and ε is the index of suboptimality. In (7.97), the matrix $\hat{S}_i$ is defined by

$$\hat{S}_i = \hat{r}_i^{-1}\hat{P}_i\hat{b}_i\hat{b}_i^T\hat{P}_i + \hat{Q}_i, \tag{7.98}$$

and λ_{mi}, λ_{Mi} are the minimum and maximum eigenvalues of the indicated matrices.

As shown by Šiljak and Sundareshan (1976), inequality (7.97) guarantees that the overall system (7.96) is in fact globally exponentially stable with degree π. Having in mind that the choice of the performance index (7.84) implies that each decoupled subsystem (7.90) is globally exponentially stable with degree π, we conclude that the inequality implies the overall system has the same degree of exponential stability as the decoupled subsystems. This is a fairly strong implication, and (7.97) may be regarded as overrestrictive. An alternative approach is to apply the aggregation process of Section 2.5, Theorem 2.11, as it was proposed by Šiljak (1977b). This is shown next.

With each optimal subsystem

$$\dot{\hat{x}} = (\hat{A}_i - \hat{r}_i^{-1}\hat{b}_i\hat{b}_i^T\hat{P}_i)\hat{x}_i, \qquad i = 1, 2, \ldots, s \tag{7.90}$$

we associate a scalar Liapunov function

$$v_i(\hat{x}_i) = (\hat{x}_i^T\hat{P}_i\hat{x}_i)^{1/2}, \tag{7.99}$$

where the symmetric positive definite matrix $\hat{P}_i$ is computed as a solution of Equation (7.87). We note that $v_i(\hat{x}_i)$ is a Lipschitz function and

$$|v_i(\hat{x}_i) - v_i(\hat{y}_i)| \leqslant \kappa_i\|\hat{x}_i - \hat{y}_i\|, \tag{7.100}$$

where $\kappa_i = \lambda_M(\hat{P}_i)\lambda_m^{1/2}(\hat{P}_i)$. By computing $\dot{v}_i(\hat{x}_i)_{(7.95)}$, we get

$$\begin{aligned}\dot{v}_i(\hat{x}_i)_{(7.95)} &\leqslant \frac{1}{2}V_i^{-1/2}(\hat{x}_i)\dot{V}_i(\hat{x}_i)_{(7.90)} + \kappa_i\sum_{\substack{j=1\\ j\neq 1}}^{s}\hat{\xi}_{ij}\|\hat{x}_j\| \\ &\leqslant -\phi_i(\|\hat{x}_i\|) + 2\kappa_i\sum_{\substack{j=1\\ j\neq i}}^{s}\lambda_M^{1/2}(\hat{P}_j)\lambda_m^{-1}(\hat{S}_j + 2\pi\hat{P}_j)\hat{\xi}_{ij}\phi_j(\|\hat{x}_j\|)\end{aligned} \tag{7.101}$$

where $\dot{V}_i(\hat{x}_i)_{(7.90)}$ is the total time derivative of the function $V_i(\hat{x}_i) = \hat{x}_i^T\hat{P}_i\hat{x}_i$

computed with respect to the corresponding decoupled system (7.90),

$$\phi_i(\|\hat{x}_i\|) = \tfrac{1}{2}\lambda_M^{-1/2}(\hat{P}_i)\lambda_m(\hat{S}_i + 2\pi\hat{P}_i)\|\hat{x}_i\|, \tag{7.102}$$

and $\hat{\xi}_{ij} = \lambda_M^{1/2}(\hat{A}_{ij}^T\hat{A}_{ij})$.

From (7.101), we can get the aggregate model $\hat{\mathscr{A}}$ for the system $\hat{\mathbb{S}}$ of (7.95), which is described by

$$\dot{v} \leqslant \hat{W}z, \tag{7.103}$$

where the constant aggregate $s \times s$ matrix $\hat{W} = (\hat{w}_{ij})$ is defined as

$$\hat{w}_{ij} = \begin{cases} -1, & i = j \\ 2\kappa_i\lambda_M^{1/2}(\hat{P}_j)\lambda_m^{-1}(\hat{S}_j + 2\pi\hat{P}_j)\hat{\xi}_{ij}, & i \neq j \end{cases} \tag{7.104}$$

and $z = (\phi_1, \phi_2, \ldots, \phi_s)^T$. By applying Theorem 2.11 to the aggregate model $\hat{\mathscr{A}}$ of (7.103), we conclude that stability of the aggregate matrix $\hat{W}$ implies stability of the overall system $\hat{\mathbb{S}}$ of (7.95). From Section 2.5 we have a simple test (7.49) for stability of $\hat{W}$. Furthermore, it is easy to show that stability if established, is exponential. The degree of exponential stability, however, is generally smaller than π, which is the degree of decoupled subsystems. Therefore, when the aggregation procedure is compared with the suboptimal design, one concludes that the two alternatives offer a trade-off between the degree of exponential stability and the size of interactions among power areas.

In the proposed decentralized scheme, a complete state feedback is used to control each interconnected area. This may be an overrestrictive requirement. Further research is needed to alleviate this problem and either build decentralized estimators, or replace the state feedback by a output decentralized scheme (Figure 7.3) using the results of Chapter 3.

There is a subtle interplay between the choice of the performance indicies for the subsystems and the nature of their interconnections. A suitable balance between the two aspects of the optimization should be achieved in order to come up with a successful design. Furthermore, we know that the systems stabilized by the decomposition-aggregation method are connectively stable. At this time we do not know how to interpret connective stability in the context of the interconnected power systems due to overlapping of the subsystems. Some attempts and results are already reported along these lines by Zaborsky and Subrammanian (1977), but additional efforts are needed to fully exploit the connectivity aspect of the decomposition-aggregation method and come up with a satisfactory decentralized design of dynamically reliable automatic generation control of large-scale interconnected power systems.

REFERENCES

Anderson, B. D. O., and Moore, J. B. (1971), *Linear Optimal Control*, Prentice-Hall, Englewood Cliffs, New Jersey.

Anderson, P. M., and Fouad, A. A. (1977), *Power System Control and Stability*, Iowa State University Press, Ames, Iowa.

Araki, M. (1975), "Application of $\mathfrak{M}$-Matrices to the Stability Problems of Composite Nonlinear Systems", *Journal of Mathematical Analysis and Applications*, 52, 309–321.

Bitsoris, G., and Burgat, C. (1976), "Stability Conditions and Estimates of the Stability Regions of Complex Systems", *International Journal of Systems Science*, 7, 911–928.

Ćalović, M. S. (1972), "Linear Regulator Design for a Load and Frequency Control", *IEEE Transactions*, PAS-91, 2271–2285.

Ćalović, M. S., Ćuk, N. M., and Djorović, M. V. (1977), "An Autonomous Area Generation Control of Interconnected Power Systems", *Proceedings IEE*, 124, 393–402.

Ćalović, M. S., Djorović, M. V., and Šiljak, D. D. (1976), "Automatic Generation Control: A Decentralized Approach", *Proceedings of the Fourteenth Annual Allerton Conference on Circuit and System Theory*, University of Illinois, Monticello, Illinois.

Cohn, N. (1966), *Control of Generation and Power Flow of Interconnected Systems*, Wiley, New York.

Concordia, C., Crary, S. B., and Parker, E. E. (1941), "Effects of Prime Mover Speed Governor Characteristics in Power-System Frequency Variations and Tie-Line Power Swings", *AIEE Transactions*, PAS-60, 559–567.

Darwish, M. Fantin, J., and Richetin, M. (1976), "Decomposability and Reduction of Dynamical Models of Large Power Systems with Application to Stability Studies", *Proceedings of the IFAC Symposium on Large Scale Systems Theory and Applications*, G. Guardabassi and A. Locatelli (eds.), Udine Italy, 19–26.

Dharma Rao, N. (1962), "A New Approach to the Transient Stability Problem", *AIEE Transactions*, PAS-81, 186–190.

DiCaprio, U., and Saccomanno, F. (1970), "Non-Linear Stability Analysis of Multimachine Electric Power Systems", *Ricerche di Automatica*, 1, 2–29.

El-Abiad, A. H., and Nagappan, K. (1966), "Transient Stability Regions of Multimachine Power Systems", *IEEE Transactions*, PAS-85, 169–179.

Elgerd, O. I., and Fosha, Jr., C. E. (1970), "Optimum Megawatt-Frequency Control of Multiarea Electric Energy Systems", *IEEE Transactions*, PAS-89, 556–563.

Ewart, D. N. (1975), "Automatic Generation Control", *Proceedings of the Engineering Foundation Conference on Systems Engineering for Power: Status and Prospects*, Henniker, New Hampshire, 1–14.

Fallside, F., and Patel, M. R. (1966), "On the Application of the Lyapunov Method to Synchronous Machine Stability", *International Journal of Control*, 4, 501–513.

Fink, L. H. (1971), "Concerning Power System Control Structures", *ISA Advances in Instrumentation*, 26, 1–11.

Fink, L. H., and Carlsen, K. (eds.) (1975), *Systems Engineering for Power: Status and Prospects*, U.S. Energy Research and Development Administration, Electric Power Research Institute, Washington, D.C.

Fosha, Jr., C. E., and Elgerd, O. I. (1970), "The Megawatt-Frequency Control Problem: A New Approach Via Optimal Control Theory", *IEEE Transactions*, PAS-89, 563–577.

Fouad, A. A. (1975), "Stability Theory—Criteria for Transient Stability", *Proceedings of the Engineering Foundation Conference on Systems Engineering for Power: Status and Prospects*, Henniker, New Hampshire, 421–450.

Gless, G. E. (1966), "Direct Method of Liapunov Applied to Transient Power System Stability", *IEEE Transactions*, PAS-85, 159–168.

Glover, J. D., and Schweppe, F. C. (1972), "Advanced Load Frequency Control", *IEEE Transactions*, PAS-91, 2095–2103.

Grujić, Lj. T., and Šiljak, D. D. (1973), "Asymptotic Stability and Instability of Large-Scale Systems", *IEEE Transactions*, AC-18, 636–645.

Henner, V. E. (1974), "A Multi-Machine Power System Liapunov Function Using the Generalized Popov Criterion", *International Journal of Control*, 19, 969–976.

Jocić, Lj. B., Ribbens-Pavella, M., and Šiljak, D. D. (1977), "On Transient Stability of Multimachine Power Systems", *Proceedings of the 1977 JACC Conference*, San Francisco, California, 627–632 (to appear in: *IEEE Transactions*, AC-23, 1978).

Jocić, Lj. B., and Šiljak, D. D. (1977), "On Decomposition and Transient Stability of Multimachine Power Systems", *Ricerche di Automatica*, 8, 42–59.

Kalman, R. E. (1963), "Lyapunov Functions for the Problem of Lur'e in Automatic Control", *Proceedings of the National Academy of Sciences*, USA, 49, 201–205.

Karmarkar, J. S., and Šiljak, D. D. (1975), "Maximization of Absolute Stability Regions by Mathematical Programming Methods", *Regelungstechnik*, 23, 59–61.

Kirchmayer, L. K. (1959), *Economic Control of Interconnected Systems*, Wiley, New York.

Kwatny, H. G., Kalnitsky, K. C., and Bhatt, A. (1975), "An Optimal Tracking Approach to Load Frequency Control", *IEEE Transactions*, PAS-94, 1635–1643.

Pai, M. A., and Murthy, P. G. (1973), "On Lyapunov Functions for Power Systems with Transfer Conductances", *IEEE Transactions*, AC-18, 181–183.

Pai, M. A., and Murthy, P. G. (1974), "New Lyapunov Functions for Power Systems Based on Minimal Realizations", *International Journal of Control*, 19, 401–415.

Pai, M. A., and Narayana, C. L. (1975), "Stability of Large Scale Power Systems", *Proceedings of the Sixth IFAC Congress*, Boston, Massachusetts, 31.6: 1–10.

Popov, V. M. (1973), *Hyperstability of Control Systems*, Springer, New York.

Reddoch, T. W. (1975), "Automatic Generation Control—Normal: Load-Fraequency Control Performance Criteria with Reference to the Use of Advanced Control Theory", *Proceedings of the Engineering Foundation Conference on Systems Engineering for Power: Status and Prospects*, Henniker, New Hampshire, 15–26.

Ribbens-Pavella, M. (1971), "Critical Survey of Transient Stability Studies of Multimachine Power Systems by Liapunov's Direct Method", *Proceedings of the Ninth Allerton Conference on Circuit and System Theory*, University of Illinois, Monticello, Illinois, 751–767.

Seltzer, S. M., and Šiljak, D. D. (1972), "Absolute Stability Analysis of Attitude Control Systems for Large Boosters", *Journal of Spacecraft and Rocketts*, 9, 506–510.

Šiljak, D. D. (1969), *Nonlinear Systems*, Wiley, New York.

Šiljak, D. D. (1972), "Stability of Large-Scale Systems Under Structural Perturbations", *IEEE Transactions*, SMC-2, 657–663.

Šiljak, D. D. (1977a), "On Pure Structure of Dynamic Systems", *Nonlinear Analysis, Theory, Methods, and Applications*, 1, 397–413.

Šiljak, D. D. (1977b), "On Decentralized Control of Large-Scale Systems", *Proceedings of the IUTAM Simposium on Multibody Systems*, Munchen, Germany (to be published by Springer, New York, 1978).

Šiljak, D. D. (1977c), "Vulnerability of Dynamic Systems", *Proceedings of the IFAC Workshop on Control and Management of Integrated Industrial Complexes*, Toulouse, France, 133–144.

Šiljak, D. D., and Sun, C. K. (1972), "On Exponential Absolute Stability", *International Journal of Control*, 16, 1003–1008.

Šiljak, D. D., and Sundareshan, M. K. (1976), "A Multilevel Optimization of Large-Scale Dynamic Systems", *IEEE Transactions*, AC-21, 79–84.

Šiljak, D. D., and Weissenberger, S. (1969), "Regions of Exponential Stability for the Problem of Lur'e", *Regelungstechnik*, 17, 27–29.

Šiljak, D. D., and Weissenberger, S. (1970a), "Regions of Exponential Ultimate Boundedness for the Problem of Lur'e", *Regelungstechnik und Process-Dataverarbeitung*, 18, 69–71.

Šiljak, D. D., and Weissenberger, S. (1970b), "A Construction of the Lur'e-Liapunov Function", *Regelungstechnik und Process-Dataverarbeitung*, 18, 455–456.

Walker, J. A., and McClamroch, N. H. (1967), "Finite Regions of Attraction for the Problem of Lur'e", *International Journal of Control*, 6, 331–336.

Weissenberger, S. (1966), "Stability-Boundary Approximations for Relay-Control Systems Via a Steepest-Ascent Construction of Lyapunov Functions", *ASME Transactions*, 88, 419–428.

Weissenberger, S. (1968), "Application of Results from the Absolute Stability Problem to the Construction of Finite Stability Domains", *IEEE Transactions*, AC-13, 124–125.

Weissenberger, S. (1973), "Stability Regions of Large-Scale Systems", *Automatica*, 9, 653–663.

Wild, E. (1941), "Methods of System Control in a Large Interconnection", *AIEE Transactions*, PAS-60, 232–236.

Willems, J. L. (1968), "Improved Lyapunov Function for Transient Power-System Stability", *IEE Proceedings*, 115, 1315–1317.

Willems, J. L. (1969), "The Computation of Finite Stability Regions by Means of Open Liapunov Surfaces", *International Journal of Control*, 10, 537–544.

Willems, J. L. (1970), "Optimum Lyapunov Functions and Stability Regions for Multimachine Power Systems", *IEE Proceedings*, 117, 573–577.
Willems, J. L. (1971), "Direct Methods for Transient Stability Studies in Power System Analysis", *IEEE Transactions*, AC-16, 332–341.
Willems J. L. (1974), "A Partial Stability Approach to the Problem of Transient Power System Stability", *International Journal of Control*, 19, 1–14.
Zaborsky, J., and Subramanian, A. K. (1977), "On the Control of a Class of Large, Nonlinear, Variable Structure Systems Typified by the Power System in Emergency", *Proceedings of the 1977 JACC Conference*, San Francisco, California, 737–746.

APPENDIX

Matrices

The purpose of this Appendix is to collect in one place all the basic facts about various kinds of matrices and their applications which are scattered throughout the book. The attention is centered on the $\mathfrak{M}$-matrices, due to their obvious importance in developing criteria for connective stability of dynamic systems. Their connection with Metzler and Hicks matrices is straightforward. Probably the most useful property of $\mathfrak{M}$-matrices is the dominant diagonal, since it can also serve satisfactorily as a measure of complexity in stable large-scale systems. For more results on various classes of matrices mentioned here, the reader should review the classical books by Gantmacher (1960), Bellman (1960), and Varga (1962), as well as more recent ones by Nikaido (1968), Shisha (1970), and Seneta (1973). The two fundamental papers in this context are those by Newman (1959) and Fiedler and Pták (1962).

By $A = (a_{ij})$ we denote a real $n \times n$ matrix with a_{ij} as its i, jth element and $i, j \in N$, where N is the set of indices $\{1, 2, \ldots, n\}$. A^T is the transpose of A, and det A is its determinant. The principal submatrix of A is denoted by $A(M)$, where $M \subseteq N$. By $\lambda_i(A)$ we denote the ith eigenvalue of A, and $\lambda_M(A)$, $\lambda_m(A)$ are the maximum and minimum eigenvalues of A, respectively.

Now we state several definitions concerning the matrix A. We start with the following:

Definition A.1. *An $n \times n$ matrix $A = (a_{ij})$ is said to be dominant diagonal if*

$$|a_{ii}| > \sum_{\substack{j=1 \\ j \neq i}}^{n} |a_{ij}| \qquad \forall i \in N. \tag{A.1}$$

If, in addition, $a_{ii} > 0$ ($a_{ii} < 0$) for all $i \in N$ then A is *dominant positive (negative) diagonal.*

A generalization of Definition A.1 is the following (McKenzie, 1966; Newman, 1959):

Definition A.2. *An $n \times n$ matrix $A = (a_{ij})$ is said to be quasidominant diagonal if there exist positive numbers d_j, $j \in N$, such that either*

$$d_i |a_{ii}| > \sum_{\substack{j=1 \\ j \neq i}}^{n} d_j |a_{ij}| \qquad \forall i \in N \tag{A.2}$$

or

$$d_j |a_{jj}| > \sum_{\substack{i=1 \\ i \neq j}}^{n} d_i |a_{ij}| \qquad \forall j \in N. \tag{A.3}$$

Again, if all a_{ii}'s are positive (negative) and either (A.2) or (A.3) is true, then A is *quasidominant positive (negative) diagonal.* If in (A.2), $d_i = 1$ for all $i \in N$, then it reduces to (A.1). It will be shown later that (A.2) and (A.3) are equivalent, whereas an analogous statement for (A.1) is not true in general.

We also need

Definition A.3. *An $n \times n$ matrix $A = (a_{ij})$ is said to be reducible (decomposable) if there exists a nonvoid proper subset M of N ($M \subset N$) such that $a_{ij} = 0$ for $i \in M$, $j \in N - M$. Otherwise the matrix A is said to be irreducible (indecomposable).*

In other words, a matrix A is called reducible if by identical row and column transpositions it can be brought into the form

$$\begin{bmatrix} A_{11} & A_{12} \\ 0 & A_{22} \end{bmatrix},$$

where A_{11} and A_{22} are square matrices. Note that an irreducible matrix A cannot have a zero row or column.

Finally, we state the following:

Definition A.4. *An $n \times n$ matrix $A = (a_{ij})$ is said to be nonnegative (positive) if $a_{ij} \geqslant 0$ ($a_{ij} > 0$) for all $i, j \in N$.*

By $A \geqslant 0$ ($A > 0$) we denote that the matrix A is nonnegative (positive). Similarly, an n-vector $x = (x_1, x_2, \ldots, x_n)^T$ is nonnegative (positive) if $x_i \geqslant 0$ ($x_i > 0$) for all $i \in N$. Furthermore, $B \geqslant A$ means $A - B \geqslant 0$.

The following is the important Perron-Frobenius theorem:

Theorem A.1. *A nonnegative $n \times n$ matrix A always has a nonnegative eigenvalue $\lambda_p(A)$, the "Perron root of A", such that $|\lambda_i(A)| \leqslant \lambda_p(A)$ for all $i \in N$. If A is irreducible, then the Perron root $\lambda_p(A)$ is positive and simple, and the corresponding eigenvector can be chosen as a positive vector.*

The proof of this theorem can be found in the books by Bellman (1960), Gantmacher (1960), Varga (1962), and Seneta (1973). According to this theorem, the *spectral radius* $\sigma(A) = |\lambda_M(A)|$ of A is equal to $\lambda_p(A)$.

Now we are in a position to consider the important class of matrices with nonpositive off-diagonal elements:

Definition A.5. *By $\mathfrak{N}$ we denote the class of all $n \times n$ matrices $A = (a_{ij})$ such that $a_{ij} \leqslant 0$ for all $i, j \in N$, $i \neq j$.*

If we add the restriction that the a_{ii}'s are nonnegative, then we have a class of Minkowski matrices (Ostrowski, 1937).

For this kind of matrices we can prove the following theorem following Fiedler and Pták (1962):

Theorem A.2. *If $A \in \mathfrak{N}$, then the following conditions are equivalent:*

(1) *There exists a vector $x \geqslant 0$ such that $Ax > 0$.*
(2) *There exists a vector $x > 0$ such that $Ax > 0$.*
(3) *There exists a positive $n \times n$ diagonal matrix $D = \operatorname{diag}\{d_1, d_2, \ldots, d_n\}$ such that $ADe > 0$, where e is the n-vector $(1, 1, \ldots, 1)^T$.*
(4) *There exists a positive diagonal $n \times n$ matrix D such that AD is positive dominant diagonal.*
(5) *For each diagonal matrix R such that $R \geqslant A$, the inverse R^{-1} exists and $\sigma[R^{-1}(P - A)] < 1$, where P is the diagonal of A ($P = \operatorname{diag}\{a_{11}, a_{22}, \ldots, a_{nn}\}$).*
(6) *If $B \in \mathfrak{N}$ and $B \geqslant A$, then B^{-1} exists.*
(7) *Each real eigenvalue of A is positive.*
(8) *All principal minors of A are positive.*
(9) *There exists a strictly increasing sequence $0 \neq M_1 \subset M_2 \subset \cdots \subset M_n = N$ such that the principal minors $\det A(M_i)$ are positive.*
(10) *There exists an $n \times n$ permutation matrix P such that PAP^{-1} may be written in the form RS, where R is an $n \times n$ lower triangular matrix with positive diagonal elements such that $R \in \mathfrak{N}$, and S is an $n \times n$ upper triangular matrix with positive diagonal elements such that $S \in \mathfrak{N}$.*
(11) *The inverse A^{-1} exists and $A^{-1} \geqslant 0$.*
(12) *The real part of each eigenvalue of A is positive.*

Proof. The symbol $\Rightarrow$ means "implies".

(1) $\Rightarrow$ (2). From (1) we conclude that $x + \varepsilon e > 0$ and $A(x + \varepsilon e) = Ax + \varepsilon Ae$ is positive if e is the n-vector $(1, 1, \ldots, 1)^T$ and $\varepsilon > 0$ is sufficiently small number.

(1) $\Rightarrow$ (3) by setting $x = De$.

(3) $\Rightarrow$ (4). Let us denote $B = AD$. Then

$$b_{ii} > -\sum_{\substack{j=1\\ j\neq i}}^{n} b_{ij} \qquad \forall i \in N. \tag{A.4}$$

Because $A \in \mathfrak{N}$ and $d_i > 0$ for all $i \in N$, we have $B \in \mathfrak{N}$, and (A.4) can be written as

$$b_{ii} > \sum_{\substack{j=1\\ j\neq i}}^{n} |b_{ij}| \qquad \forall i \in N. \tag{A.5}$$

By Definition A.1, the matrix $B = AD$ is diagonal dominant.

(4) $\Rightarrow$ (5). Let us first establish that if B is a dominant diagonal matrix, then $\sigma(I - Q^{-1}B) < 1$, where I is the $n \times n$ identity matrix and Q is the diagonal of B. To see this, let $\lambda_k = \lambda_k(I - Q^{-1}B)$. Then there exists a vector $x \neq 0$ such that $\lambda_k x = x - Q^{-1}Bx$. Let also $|x_i| = \max_{j\in N}|x_j| > 0$. Then

$$\lambda_i x_i = \sum_{\substack{j=1\\ j\neq i}}^{n} w_{ii}^{-1} w_{ij} x_j \tag{A.6}$$

and

$$|\lambda_k||x_i| \leqslant \left(\sum_{\substack{j=1\\ j\neq i}}^{n} |w_{ii}|^{-1}|w_{ij}|\right)|x_i| < |x_i|. \tag{A.7}$$

Therefore, $|\lambda_k| < 1$. Since $k \in N$ was arbitrary, we have $\sigma(I - Q^{-1}B) < 1$.

Now, suppose that D is a positive diagonal matrix, that is, (A.5) is true. Thus, the b_{ii}'s and a_{ii}'s are all positive. From the above, it follows that $\sigma(I - Q^{-1}B) < 1$, where $Q = PD$, and $\sigma(I - P^{-1}A) = \sigma[D^{-1}(I - P^{-1}A)D] = \sigma(I - D^{-1}P^{-1}AD) = \sigma(I - Q^{-1}B) < 1$. From (5), we conclude that $r_i \geqslant a_{ii} > 0$ and R^{-1} exists. Furthermore, $R^{-1} \leqslant P^{-1}$. Using Theorem A.1, we have $\sigma(M) = \lambda_p(M)$ whenever $M \geqslant 0$. Since $P - A \geqslant 0$, we have $\sigma[R^{-1}(P - A)] = \lambda_p[R^{-1}(P - A)] \leqslant \lambda_p[P^{-1}(P - A)] = \lambda_p(I - P^{-1}A) = \sigma(I - P^{-1}A) < 1$.

(5) ⇒ (6). Let P and Q be the diagonals of A and B, respectively. Then, Q^{-1} exists and has positive diagonal elements, and $\sigma[Q^{-1}(P - A)] < 1$. From $B \in \mathfrak{N}$ and $B \geqslant A$, we conclude that $0 \leqslant Q - B \leqslant P - A$, and $0 \leqslant Q^{-1}(Q - B) \leqslant Q^{-1}(P - A)$. Thus, $\sigma[Q^{-1}(Q - B)] = \lambda_p[Q^{-1}(Q - B)] \leqslant \lambda_p[Q^{-1}(P - A)] = \sigma[Q^{-1}(P - A)] < 1$, so that the series $I + (I - Q^{-1}B) + (I - Q^{-1}B)^2 + \cdots$ converges to $(Q^{-1}B)^{-1}$. This implies that B^{-1} exists.

(6) ⇒ (7). Let $B = A - \alpha I$, where $\alpha \leqslant 0$. Since $B \in \mathfrak{N}$ and $B \geqslant A$, we have that B^{-1} exists and α cannot be an eigenvalue of A.

(7) ⇒ (8). Let us first establish that if $A, B \in \mathfrak{N}$, $A \leqslant B$, and each real eigenvalue of A is positive, then both A^{-1} and B^{-1} exist, $A^{-1} \geqslant B^{-1} \geqslant 0$, each real eigenvalue of B is positive, and $\det B \geqslant \det A > 0$. It is obvious that there exists a number $\beta > 0$ such that $G = I - \beta B \geqslant 0$ and $H = I - \beta A \geqslant I - \beta B = G \geqslant 0$, so that $\det\{[1 - \lambda_p(H)]I - \beta A\} = \det[H - \lambda_p(H)I] = 0$. Since each real eigenvalue of A is positive, we have $1 - \lambda_p(H) > 0$ and $0 \leqslant \lambda_p(H) < 1$. The series $I + H + H^2 + \cdots$ converges to $(I - H)^{-1} = (\beta A)^{-1} \geqslant 0$. Similarly, from $0 \leqslant G^k \leqslant H^k$ for $k = 1, 2, \ldots$, we conclude that $I + G + G^2 + \cdots$ converges to $(I - G)^{-1} = (\beta B)^{-1} \geqslant 0$. Thus $(\beta A)^{-1} \geqslant (\beta B)^{-1} \geqslant 0$, and $A^{-1} \geqslant B^{-1} \geqslant 0$. For $\gamma \leqslant 0$, we have $C = B - \gamma I \geqslant A$. As we just proved, this last inequality implies that C^{-1} exists and all real eigenvalues of B are positive. Finally, we prove $\det B \geqslant \det A > 0$ by induction. For $n = 1$, the statement follows trivially. We assume that $n > 1$ and that all pairs of $k \times k$ matrices satisfy the statement for $1 \leqslant k \leqslant n$. The principal submatrices $\hat{A} = A(M)$, $\hat{B} = B(M)$, $M = \{1, 2, \ldots, n - 1\}$ belong to $\mathfrak{N}$, and $\hat{A} \leqslant \hat{B}$. We consider a matrix $\tilde{A}$ defined as

$$\tilde{A} = \begin{bmatrix} \hat{A} & 0 \\ 0 & a_{nn} \end{bmatrix}, \tag{A.8}$$

and conclude that $A \leqslant A \in \mathfrak{N}$ and that each real eigenvalue of $\tilde{A}$, and thus of $\hat{A}$, is positive. It follows that $\det \hat{B} \geqslant \det \hat{A} > 0$. From $A^{-1} \geqslant B^{-1} \geqslant 0$, it follows that

$$\frac{\det \hat{A}}{\det A} \geqslant \frac{\det \hat{B}}{\det B} \geqslant 0 \tag{A.9}$$

and $\det A > 0$, $\det B > 0$,

$$\det B \geqslant \frac{\det \hat{B}}{\det \hat{A}} \det A \geqslant \det A > 0. \tag{A.10}$$

Now we are ready to show that statement (7) implies (8). Assume that each real eigenvalue of A is positive, and prove that for $M \subset N$ we have $\det A(M) > 0$. For this purpose, we define an $n \times n$ matrix $B = (b_{ij})$ as

$$b_{ij} = \begin{cases} a_{ij}, & i, j \in M, \\ a_{ii}, & i, j \notin M, \quad i = j, \\ 0, & i, j \notin M, \quad i \neq j. \end{cases} \tag{A.11}$$

Obviously, $B \in \mathfrak{N}$ and $B \geqslant A$. From what was just established above, this implies that $\det B > 0$ and all real eigenvalues of B are positive. Positivity of the a_{ii}'s for $i \notin M$, and the fact that $\det B$ is the product of $\det A(M)$ and a_{ii}'s for $i \notin M$, imply that $\det A(M) > 0$. Since M is arbitrary, (8) is established.

(8) $\Rightarrow$ (9) is trivial.

(9) $\Rightarrow$ (10). We start by showing that for an $n \times n$ matrix $A = (a_{ij})$ the sequence of principal minors $\det A(M_i)$, where $M = 1, 2, \ldots, i$, is positive if and only if there exist a lower triangular matrix U and an upper triangular matrix V, both with positive diagonal elements, such that $A = UV$. The "if" part is established by induction. For $n = 1$, the hypothesis follows trivially. We assume that the hypothesis is true for $n - 1$ and show that it is also true for n. We write

$$A = \begin{bmatrix} A_{n-1} & a \\ b & a_{nn} \end{bmatrix} \tag{A.12}$$

and assume $A_{n-1} = \tilde{U}\tilde{V}$, where $\tilde{U}$ ($\tilde{V}$) is a lower (upper) triangular matrix with positive diagonal elements. From (A.12), we have

$$a_{nn} - bA_{n-1}^{-1}a = \frac{\det A}{\det A_{n-1}} > 0. \tag{A.13}$$

The choice

$$U = \begin{bmatrix} \tilde{U} & 0 \\ b\tilde{V}^{-1} & 1 \end{bmatrix}, \qquad V = \begin{bmatrix} \tilde{V} & \tilde{U}^{-1}a \\ 0 & a_{nn} - bA_{n-1}^{-1}a \end{bmatrix} \tag{A.14}$$

establishes the "if" part of the statement. The "only if" part follows immediately by observing that the principal minors $\det A(M_i)$ are equal to the product of the first i diagonal elements of U and V.

We also need to show that if $A \in \mathfrak{N}$, and $A = UV$ as above, then $U, V \in \mathfrak{N}$. To see this, let $U = (u_{ij})$ and $V = (v_{ij})$, so that $u_{ij} = 0$ for $i < j$ and $v_{ij} = 0$ for $i > j$ with $u_{ii}, v_{ii} > 0$ for $i, j \in N$. We again use induction and show that $u_{ij} \leqslant 0$, $v_{ij} \leqslant 0$, $i \neq j$. For $i + j = 3$, we have $a_{12} = u_{11}v_{12}$ and $a_{21} = u_{21}v_{11}$, and $u_{21} \leqslant 0$, $v_{12} \leqslant 0$. We assume that $u_{kl} \leqslant 0$, $v_{kl} \leqslant 0$ for $k + l < i + j$, where $i + j > 3$, $i \neq j$. If $i < j$, then from

$$a_{ij} = u_{ii} v_{ij} + \sum_{k<i} u_{ik} v_{kj} \tag{A.15}$$

and the assumption $u_{ik} \leqslant 0$, $v_{kj} \leqslant 0$, $i + k < i + j$, $k + j < i + j$, we have $a_{ij} \leqslant 0$, $\sum_{k<i} u_{ik} v_{kj} \geqslant 0$, and thus $v_{ij} \leqslant 0$. Similarly, when $i > j$ we get $u_{ij} \leqslant 0$, and (10) follows automatically.

(10) ⇒ (11). From (10) we conclude readily that R^{-1}, $S^{-1} \geqslant 0$. Thus, $A^{-1} = S^{-1}R^{-1} \geqslant 0$.

(11) ⇒ (1). Let $x = A^{-1}e$, so that $x \geqslant 0$ and $Ax = e > 0$.

This establishes the equivalence of statements (1)–(11). To complete the proof of the theorem, we show that

(12) ⇔ (7). It is obvious that (12) implies (7). To show the converse, let us assume that all real eigenvalues of A are positive. We choose a sufficiently large number $\xi > 0$ so that $\xi I - A \geqslant 0$. Then $|\xi - \eta| \leqslant \lambda_p(\xi I - A)$ for each eigenvalue η of A. Moreover, there is a real eigenvalue $\eta_0 > 0$ of A such that $\xi - \eta_0 = \lambda_p(\xi I - A)$. Therefore, $|\xi - \eta_0| \leqslant \xi - \eta_0 < \xi$ for any eigenvalue η of A.

The proof of Theorem A.2 is completed.

Theorem A.2 can be used to define the class of $\mathfrak{M}$-matrices introduced by Ostrowski (1937, 1956):

Definition A.6. *By $\mathfrak{M}$ we denote the class of all matrices $A \in \mathfrak{N}$ which satisfy one of the properties of Theorem A.2.*

This is a somewhat stronger definition of $\mathfrak{M}$-matrices than the one introduced by Ostrowski, but it is in common current usage. A number of important properties of $\mathfrak{M}$-matrices have been obtained, notably by Fan (1957, 1958, 1959, 1960) as well as by many others; they are summarized by Fiedler and Pták (1962). Some of them we will outline here, following reference Fiedler and Pták (1962).

Theorem A.3. *If $A \in \mathfrak{M}$, then there is a positive eigenvalue $\lambda_q(A)$ such that* $\operatorname{Re} \lambda_i(A) \geqslant \lambda_q(A)$, $i \in N$.

Proof. We can choose $\beta > 0$ so that $\beta I - A \geqslant 0$. Set $\lambda_q(A) = \beta - \lambda_p(\beta I - A)$. From $A \in \mathfrak{M}$, we get $\lambda_q(A) > 0$. Since $\beta - \lambda_i(A)$ is an eigenvalue of $\beta I - A$, we have $|\beta - \lambda_i(A)| \leqslant \lambda_p(\beta I - A) = \beta - \lambda_q(A)$. This proves Theorem A.3.

We also prove the following result (Fiedler and Pták, 1962) concerning a pair of matrices A and B:

Theorem A.4. *If $A \in \mathfrak{M}$, $B \in \mathfrak{N}$ and $B \geqslant A$, then*

(1) $B \in \mathfrak{M}$;

(2) $0 \leqslant B^{-1} \leqslant A^{-1}$;

(3) $\det B \geqslant \det A > 0$;

(4) $A^{-1}B \geqslant I$, $BA^{-1} \geqslant I$;
(5) $B^{-1}A, AB^{-1} \in \mathfrak{M}$ *and* $B^{-1}A \leqslant I, AB^{-1} \leqslant I$;
(6) $\sigma(I - B^{-1}A) < 1, \sigma(I - AB^{-1}) < 1$;
(7) $\lambda_q(B) \geqslant \lambda_q(A)$.

Proof.

(1) From the condition of the theorem and the proof of (7) $\Rightarrow$ (8) in Theorem A.2, we have that B^{-1} exists and $B^{-1} \geqslant 0$. Thus, $B \in \mathfrak{M}$.

(2) From the proof of (7) $\Rightarrow$ (8) in Theorem A.2.

(3) From the proof of (7) $\Rightarrow$ (8) in Theorem A.2.

(4) Since $A^{-1} \geqslant 0$ and $B - A \geqslant 0$, we get $A^{-1}(B - A) \geqslant 0$ and $A^{-1}B \geqslant I$. Analogously, $BA^{-1} \geqslant I$.

(5) Since $B^{-1} \geqslant 0$ and $B - A \geqslant 0$, we obtain $B^{-1}(B - A) \geqslant 0$, which implies that $B^{-1}A \leqslant I$. Thus, $B^{-1}A \in \mathfrak{N}$. The inverse of $B^{-1}A$ is $A^{-1}B$, which is shown in (4) to be $A^{-1}B \geqslant I$, and thus $A^{-1}B \geqslant 0$. From (11) of Theorem A.2, we have that $B^{-1}A \in \mathfrak{M}$. Similarly, $AB^{-1} \in \mathfrak{M}$.

(6) Since $I - B^{-1}A \geqslant 0$, we have $\sigma(I - B^{-1}A) = \lambda_p(I - B^{-1}A) = 1 - \lambda(B^{-1}A)$, where $\lambda(B^{-1}A)$ is an eigenvector of $B^{-1}A$. Furthermore, $\lambda(B^{-1}A)$ is real, and $B^{-1}A \in \mathfrak{M}$. From (7) of Theorem A.2, we have that $\lambda(B^{-1}A) > 0$ and $\sigma(I - B^{-1}A) < 1$. Similarly, $\sigma(I - AB^{-1}) < 1$.

(7) We prove this statement by contradiction. We show that $\gamma < \lambda_q(A)$ implies $\gamma I - B$ is not a singular matrix and γ cannot be an eigenvalue of B. Since $A - \gamma I \in \mathfrak{N}$, we have $\lambda(A) - \gamma \geqslant \lambda_q(A) - \gamma$ for each real eigenvalue $\lambda(A)$ of A. Thus, from (7) of Theorem A.2, it follows that $A - \lambda(A)I \in \mathfrak{M}$. Since $B - \gamma I \geqslant A - \gamma I$ and $B - \gamma I \in \mathfrak{N}$, we have from (1) that $B - \gamma I \in \mathfrak{M}$, and thus $B - \gamma I$ is nonsingular. This proves Theorem A.4.

We also show the following (Fiedler and Pták, 1962):

Theorem A.5. *If $A \in \mathfrak{M}$, then $\lambda_q(A) \leqslant a_{ii}$ for all $i \in N$.*

Proof. Let $B = (b_{ij})$ be defined as

$$b_{ij} = \begin{cases} a_{ii}, & i = j \\ 0, & i \neq j. \end{cases} \tag{A.16}$$

Thus, $B \in \mathfrak{N}$ and $B \geqslant A$. By Theorem A.4, $B \in \mathfrak{M}$ and $\lambda_q(B) \geqslant \lambda_q(A)$. This implies $\lambda_q(B) = \min_i a_{ii}$. This proves Theorem A.5.

Another class of matrices that are closely related to $\mathfrak{M}$-matrices are the so-called $\mathcal{P}$-matrices, defined as follows (Fiedler and Pták, 1962):

Definition A.7. *By $\mathcal{P}$ we denote the class of all $n \times n$ matrices A which satisfy one of the following equivalent conditions*:

(1) *All principal minors of A are positive.*

(2) *Every real eigenvalue of A as well as of each principal minor of A is positive.*

(3) *For every vector $x \neq 0$ there exists a diagonal matrix D_x with positive diagonal elements such that the scalar product $(Ax, D_x x) > 0$.*

The class of $\mathcal{P}$-matrices was studied by Fiedler and Pták (1962), Sandberg and Willson (1969), and others (see Nikaido, 1968). An extended class of $\mathcal{P}\mathcal{N}$-matrices was considered by Uekawa, Kemp and Wegge (1972) and Maybee (1976). In this connection, we only prove the following simple results:

Theorem A.6. *If $A \in \mathcal{P}$ and D is a diagonal matrix with positive elements, then $DA, AD \in \mathcal{P}$.*

Proof. Follows trivially from (1) of Definition A.7.

Theorem A.7. *If $A \in \mathcal{M}$ and D is a diagonal matrix with positive elements, then $DA, AD \in \mathcal{M}$.*

Proof. We merely notice that $\mathcal{M} = \mathcal{N} \cap \mathcal{P}$ and apply Theorem A.6.

In the context of $\mathcal{M}$-matrices, we also mention the concept of D-stability introduced by Arrow and McManus (1958):

Definition A.8. *An $n \times n$ matrix A is D-stable if for $D = \operatorname{diag}\{d_1, d_2, \ldots, d_n\}$, DA is stable (that is, all eigenvalues of DA have negative real parts) if and only if $d_i > 0$, $i \in N$.*

D-stability was studied by Enthoven and Arrow (1956) and by Johnson (1974), where the "only if" part of Definition A.8 is dropped to prove that if $-A \in \mathcal{M}$, then A is D-stable. This follows directly from (8) of Theorem A.2.

Let us now turn our attention to Metzler matrices, which are of so much importance throughout this book. The Metzler matrix was first introduced in economics by Mosak (1944), but was named after Metzler (1945), since he gave it its essential development. The economists do not agree on the definition of the Metzler matrix, and we have two distinct definitions:

Definition A.9 (Newman, 1959). *An $n \times n$ matrix $A = (a_{ij})$ is a Metzler matrix if*

$$a_{ij} = \begin{cases} < 0, & i = j, \\ \geqslant 0, & i \neq j, \end{cases} \tag{A.17}$$

for all $i, j \in N$.

Definition A.10 (Arrow, 1966). *An $n \times n$ matrix $A = (a_{ij})$ is a Metzler matrix if $a_{ij} \geqslant 0$, $i \neq j$, $i, j \in N$.*

Obviously, Definition A.1 implies Definition A.2, but not vice versa. As we will see shortly, a necessary condition for stability of Metzler matrices is negativity of the diagonal elements a_{ii} of A. Thus, with respect to stability, the two definitions are equivalent. We also note that according to Definition A.10, if A is a Metzler matrix, then $-A \in \mathfrak{M}$, and vice versa. This simple fact opens up a possibility of establishing numerous properties of Metzler matrices using the properties of $\mathfrak{M}$-matrices listed in Theorem A.2.

Since our interest is predominantly in stability, we need

Definition A.11. *An $n \times n$ matrix A is called stable (or Hurwitz) if* Re $\lambda_i(A) < 0$ *for all $i \in N$.*

We also recall (Newman, 1959)

Definition A.12. *A matrix A is called a Hicks matrix if all odd-order principal minors of A are negative and all even-order principal minors of A are positive.*

Then we have the fundamental result of Metzler (1945):

Theorem A.8. *A Metzler matrix A is stable if and only if it is Hicks.*

Proof. By noting that $-A \in \mathfrak{M}$, the theorem follows from (8) and (13) of Theorem A.2.

It is of interest to note that to test for stability of a Metzler matrix one does not need to test all principal minors as required by Theorem A.8, but only the leading principal minors of A. This result was proved by Kotelyanskii (1952) using the results of Sevastyanov (1951). Thus, we have

Theorem A.9. *An $n \times n$ Metzler matrix $A = (a_{ij})$ is stable if and only if*

$$(-1)^k \begin{vmatrix} a_{11} & a_{12} & \cdots & a_{1k} \\ a_{21} & a_{22} & \cdots & a_{2k} \\ \cdots & \cdots & \cdots & \cdots \\ a_{k1} & a_{k2} & \cdots & a_{kk} \end{vmatrix} > 0 \qquad \forall k \in N. \tag{A.18}$$

Proof. For the proof of this theorem see Gantmacher (1960).

It is obvious that Kotelyanskii's result (Theorem A.9) can now be used to conclude that (8) of Theorem A.2 can be replaced by: All leading principal minors of A are positive. In mathematical economics this result is known as the Hawkins-Simon conditions (see Nikaido, 1968).

Another fundamental result used throughout this book, which is due to McKenzie (1966), is the following:

Theorem A.10. *A Metzler matrix A is stable if and only if it is quasidominant negative diagonal.*

Proof. The theorem follows immediately from (3) and (12) of Theorem A.2.

We can dispense with the Metzler structure of A and still use the diagonal dominance to conclude stability of A. For this we need McKenzie's "diagonal form" of an $n \times n$ matrix $A = (a_{ij})$ with a negative diagonal ($a_{ii} < 0$, $i \in N$), which is defined as the $n \times n$ matrix $B = (b_{ij})$ with

$$b_{ij} = \begin{cases} a_{ii}, & i = j, \\ |a_{ij}|, & i \neq j, \end{cases} \tag{A.19}$$

for all $i, j \in N$. Then we have the following (McKenzie, 1966):

Theorem A.11. *A matrix A with a negative diagonal is stable if its diagonal form B is Hicks.*

Proof. The matrix B is Metzler by construction. Thus, from Theorems A.8 and A.10, we have that B is negative quasidominant diagonal. It is obvious that A is quasidominant diagonal if and only if B is. From (4) and (12) of Theorem A.2 we can conclude that the quasidominant negative diagonal of A implies stability of A. This proves Theorem A.11.

We should also mention the fact that for a negative diagonal $n \times n$ matrix $A = (a_{ij})$, either

$$|a_{ii}| > \sum_{\substack{j=1 \\ j \neq i}}^{n} |a_{ij}| \qquad \forall i \in N \tag{A.20}$$

or

$$|a_{jj}| > \sum_{\substack{i=1 \\ i \neq j}}^{n} |a_{ij}| \qquad \forall j \in N \tag{A.21}$$

implies stability of A. This follows from above choosing $d_i = 1$, $i \in N$ in (A.2) and (A.3).

When a matrix is not Metzler, we still can establish stability by quasidominance conditions. However, the conditions are only sufficient. We can recover the necessity part even if some of the off-diagonal elements of a matrix are negative, so long as the matrix is of Morishima's (1952) type.

Definition A.13. *A matrix A is called a Morishima matrix if A is irreducible matrix and can be permuted into the form*

$$\begin{bmatrix} A_{11} & A_{12} \\ A_{21} & A_{22} \end{bmatrix},$$

where $A_{11} \geqslant 0$, $A_{22} \geqslant 0$ *are square matrices and* $A_{12} \leqslant 0$, $A_{21} \leqslant 0$.

Bassett, Habibagahi, and Quirk (1967) proved the following:

Theorem A.12. *Let* $B = A - \alpha I$, *where* A *is an* $n \times n$ *Morishima matrix and* $\alpha > a_{ii}$ *for all* $i \in N$. *Then* A *is a stable matrix if and only if* A *is a quasidominant negative diagonal matrix.*

For more results on diagonal-dominance conditions, one should review the books by Varga (1962) and by Quirk and Saposnik (1968), as well as the papers by McKenzie (1966), Pearce (1974), Johnson (1974), and Beauwens (1976).

There is another interesting problem in this context, which combines the properties of nonnegative, Metzler, and $\mathfrak{M}$-matrices. To present this problem we need some preliminary considerations.

We first recall the well-known (Bellman, 1960)

Definition A.14. *A real* $n \times n$ *matrix* A *is called positive definite if* $x^T A x > 0$ *for all n-vectors* $x \neq 0$.

By writing $A = B + C$, where $B = (A + A^T)/2$ is the symmetric and $C = (A - A^T)/2$ the skew-symmetric part of A, we can show that $x^T C x = (x^T C x)^T = -x^T C x$ and $x^T C x = 0$, so that $x^T A x = x^T B x$. Therefore, A is positive definite if and only if its symmetric part $B = (A + A^T)/2$ is positive definite.

We also recall the classical Liapunov result (see Gantmacher, 1960):

Theorem A.13. *A matrix* A *is stable if and only if for any positive definite symmetric matrix* G *there is a positive definite symmetric matrix* H *such that*

$$A^T H + HA = -G. \tag{A.22}$$

In the stability analysis of large-scale systems (Grujić and Šiljak, 1973) it is of interest to find conditions under which the solution H of the Liapunov matrix equation (A.22) is a positive matrix, that is, its elements are all positive numbers. This problem was partially resolved by Grujić and Šiljak (1973) as follows:

Theorem A.14. *If a matrix* A *is Metzler and stable, then for any positive and positive definite symmetric matrix* G *there is a positive and positive definite symmetric matrix* H *as a solution of the corresponding Liapunov matrix equation.*

Proof. As is well known (e.g. Gantmacher, 1960), Liapunov's matrix equation (A.22) can be rewritten as a linear matrix equation

$$Bh = -g, \tag{A.23}$$

where

$$B = A^T \otimes I + I \otimes A \tag{A.24}$$

is an $n^2 \times n^2$ matrix with $\otimes$ denoting the Kronecker product; $H = (h_{11}, h_{12}, \ldots, h_{1n}, h_{21}, \ldots, h_{nn})^T$, $g = (g_{11}, g_{12}, \ldots, g_{1n}, g_{21}, \ldots, g_{nn})^T$; and $A = (a_{ij})$, $H = (h_{ij})$, $G = (g_{ij})$ are all $n \times n$ matrices. The matrix B is stable, and by construction it is also Metzler. By (2) of Theorem A.2, we conclude that for any $g > 0$ we have $h > 0$. That H is positive definite follows from Theorem A.13. This completes the proof of Theorem A.14.

It is fairly obvious that an "only if" part can be proved by Theorem A.14. It is more important, however, to try to dispense with the Metzlerian structure of A. To this effect, the following conjecture was stated by Šiljak (1972).

Conjecture. *If a matrix A is stable, then there is a positive definite symmetric matrix G such that the matrix H as a solution of the corresponding Liapunov matrix equation is a positive and positive definite symmetric matrix.*

In a private communication to the author, Professor V. M. Popov from the University of Florida pointed out that some additional assumptions are needed regarding the matrix A, since otherwise

$$A = \begin{bmatrix} 1 & -2 \\ 3 & -4 \end{bmatrix} \tag{A.25}$$

is a counterexample to the conjecture. It is of interest to point out that A^T of (A.25) satisfies the conjecture. Conditions on A for the conjecture to be true, as well as some important generalizations, are presented by Womack and Montemayor (1975), Montemayor and Womack (1976), and Datta (1977).

REFERENCES

Arrow, K. J. (1966), "Price Quantity Adjustments in Multiple Markets with Rising Demands", *Proceedings of the Symposium on Mathematical Methods in the Social Sciences*, K. J. Arrow, S. Karlin, and P. Suppes (eds.), Stanford University Press, Stanford, California, 3–15.

Arrow, K. J., and McManus, M. (1958), "A Note on Dynamic Stability", *Econometrica*, 26, 448–454.

Bassett, L., Habibagahi, H., and Quirk, J. (1967), "Qualitative Economics and Morishima Matrices", *Econometrica*, 35, 221–233.

Beauwens, R. (1976), "Semistrict Diagonal Dominance", *SIAM Journal of Numerical Analysis*, 13, 109–112.

Bellman, R. (1960), *Introduction to Matrix Analysis*, McGraw-Hill, New York.

Datta, B. N. (1977), "Matrices Satisfying Šiljak's Conjecture", *IEEE Transactions*, AC-22, 132–133.

Debreu, G., and Herstein, I. N. (1953), "Nonnegative Square Matrices", *Econometrica*, 21, 597–607.

Enthoven, A. C., and Arrow, K. J. (1956), "A Theorem on Expectations and the Stability of Equilibrium", *Econometrica*, 24, 288–293.

Fan, K. (1957), "Inequalities for the Sum of Two $\mathcal{M}$-Matrices", *Inequalities*, O. Shisha (ed.), Academic, New York, 105–117.

Fan, K. (1958), "Topological Proofs for Certain Theorems on Matrices With Non-Negative Elements", *Monatshefte für Mathematik*, 62, 219–237.

Fan, K., and Householder, A. S. (1959), "A Note Concerning Positive Matrices and $\mathcal{M}$-Matrices", *Monatshefte fur Mathematik*, 63, 265–270.

Fan, K. (1960), "Note on $\mathcal{M}$-Matrices", *Quarterly Journal of Mathematics*, 11, 43–49.

Fiedler, M., and Pták, V. (1962), "On Matrices with Non-Positive Off-Diagonal Elements and Positive Principal Minors", *Czechoslovakian Mathematical Journal*, 12, 382–400.

Fiedler, M., and Pták, V. (1966), "Some Generalizations of Positive Definitness and Monotonicity", *Numerische Mathematik*, 9, 163–172.

Gantmacher, F. R. (1960), *The Theory of Matrices*, Vols. I and II, Chelsea, New York.

Grujić, Lj. T., and Šiljak, D. D. (1973), "Asymptotic Stability and Instability of Large-Scale Systems", *IEEE Transactions*, AC-18, 636–645.

Hawkins, D., and Simon, H. (1949), "Note: Some Conditions of Macroeconomic Stability", *Econometrica*, 17, 53–56.

Johnson, C. R. (1974), "Sufficient Conditions for $\mathcal{D}$-Stability", *Journal of Economic Theory*, 9, 53–62.

Kotelyanskii, I. N. (1952), "On Some Properties of Matrices with Positive Elements" (in Russian), *Mathematicheski Sbornik*, 31, 497–506.

Maybee, J. S. (1976), "Some Aspects and Solutions of $\mathcal{PM}$-Matrices", *SIAM Journal of Applied Mathematics*, 31, 397–410.

McKenzie, L. (1966), "Matrices with Dominant Diagonals and Economic Theory", *Proceedings of the Symposium on Mathematical Methods in the Social Sciences*, K. J. Arrow, S. Karlin, and P. Suppes (eds.), Stanford University Press, Stanford, California, 47–62.

Metzler, L. A. (1945), "Stability of Multiple Markets: The Hicks Conditions", *Econometrica*, 13, 277–292.

Montemayor, J. J., and Womack, F. B. (1976), "More on a Conjecture by Šiljak", *IEEE Transactions*, AC-21, 805–806.

Morishima, M. (1952), "On the Laws of Change of the Price-System in an Economy which Contains Complementary Commodities", *Osaka Economic Papers*, 1, 101–113.

Morishima, M. (1964), *Equilibrium, Stability, and Growth*, Clarendon, Oxford, England.

Mosak, J. L. (1944), *General Equilibrium Theory in International Trade*, Cowles Commission *Monograph* 7, Principia, Bloomington, Indiana.

Newman, P. K. (1959), "Some Notes on Stability Conditions", *Review of Economic Studies*, 72, 1–9.

Nikaido, H. (1968), *Convex Structures and Economic Theory*, Academic, New York.

Ostrowski, A. (1937), "Über die Determinanten mit überwiegender Hauptdiagonale", *Commentarii Mathematici Helvetici*, 10, 69–96.

Ostrowski, A. (1956), "Determinanten mit überwiegender Hauptdiagonale und die absolute Konvergenz von linearen Iterationsprozessen", *Commentarii Mathematici Helvetici*, 30, 175–210.

Pearce, I. F. (1974), "Matrices with Dominant Diagonal Blocks", *Journal of Economic Theory*, 9, 159–170.

Quirk, J., and Saposnik, R. (1968), *Introduction to General Equilibrium Theory and Welfare Economics*, McGraw-Hill, New York.

Sandberg, I. W., and Wilson, A. N. Jr. (1969), "Some Theorems on Properties of DC Equations of Nonlinear Networks", *The Bell System Technical Journal*, 48, 1–34.

Seneta, E. (1973), *Non-Negative Matrices*, Wiley, New York.

Sevastyanov, B. A. (1951), "Theory of Branching Stochastic Processes" (in Russian), *Uspekhi Matematicheskih Nauk*, 6, 47–99.

Shisha, O. (ed.) (1970), "*Inequalities*", Vols. I and II, Academic, New York.

Šiljak, D. D. (1972), "Stability of Large-Scale Systems", *Proceedings of the Fifth IFAC Congress*, Part IV, Paris, C-32: 1–11.

Uekawa, Y., Kemp, M. C., and Wegge, L. L. (1972), "$\mathcal{P}$- and $\mathcal{PN}$-Matrices, Minkowski- and Metzler-Matrices, and Generalizations of the Stolper-Samuelson and Samuelson-Rybczynski Theorems", *Journal of International Economics*, 3, 53–76.

Varga, R. S. (1962), "*Matrix Iterative Analysis*", Prentice-Hall, Englewood Cliffs, New Jersey.

Willems, J. C. (1976), "Lyapunov Functions for Diagonally Dominant Systems", *Automatica*, 12, 519–523.

Womack, B. F., and Montemayor, J. J. (1975), "On a Conjecture by Šiljak", *IEEE Transactions*, AC-20, 512–513.

INDEX